ADSORPTION AND THE GIBBS SURFACE EXCESS

ADSORPTION AND THE GIBBS SURFACE EXCESS

D. K. Chattoraj
Jadavpur University
Calcutta, India

and

K. S. Birdi
The Technical University of Denmark
Lyngby, Denmark

SPRINGER SCIENCE+BUSINESS MEDIA, LLC

Library of Congress Cataloging in Publication Data

Chattoraj, D. K., 1930–
  Adsorption and the Gibbs surface excess.

  Includes bibliographical references and indexes.
  1. Adsorption. 2. Gibbs free energy. 3. Surface energy. I. Birdi K. S., 1934–   . II. Title.
QD547.C43   1984   541.3′453   84-1906
ISBN 978-1-4615-8335-6   ISBN 978-1-4615-8333-2 (eBook)
DOI 10.1007/978-1-4615-8333-2

© 1984 Springer Science+Business Media New York
Originally published by Plenum Press, New York in 1984
Softcover reprint of the hardcover 1st edition 1984
A Division of Plenum Publishing Corporation
233 Spring Street, New York, N.Y. 10013

Dedicated to Professor B. N. Ghosh for his manifold contribution and inspiring leadership in the field of colloid and surface sciences

PREFACE

The subject matter of this book is not entirely new in that much has been written about the state of the interfacial chemistry in other textbooks. However, the authors have found that there is really a need for a book which can lead an investigator, a student, or a beginner through the in-depth understanding and descriptions involved in the derivations and usage of the adsorption equations based on the unified concept of the Gibbs surface excess. The derivation of the original Gibbs equation for the liquid interface has been amplified with reformulations and criticisms. Attempts have been frequently made to relate the surface excess quantities with the thermodynamic properties of the physically defined interfacial phase. In the last two decades, useful applications of the Gibbs equations have been made for the study of adsorption at solid–liquid as well as liquid–gas interfaces formed in the presence of the multicomponent solutions in the bulk. These recent treatments become useful for the experimental studies of the adsorption of electrolytes and interfacial properties related to neutral and charged monolayers. The consistency of the experimental results with those to be expected from the theories of the electrical double layer needs careful examination. An attempt has been made to present in a single volume all these developments based on the unified concept of the Gibbs surface excess.

The study of the adsorption from solution is of considerable importance in the field of biology and in many other branches of applied science. The discussion of the concept of the Gibbs surface excess in this book for the interpretation of the hydration and binding phenomena in the biopolymer system is relatively new. It is hoped that this will initiate further research work in the various important fields of physical biochemistry, biophysics, and biology in general. Relevant discussions on the adsorption at biosurfaces, model biomembranes, etc., have been included in this book to demonstrate that the concept of the Gibbs excess is very useful to the life sciences in general.

The treatment in different parts of the book is based on the fundamental concept of classical thermodynamics. A large amount of experimental results have been included throughout the book for the critical examination of the theoretical treatments. Results based on statistical thermodynamics have been discussed in a few sections to clarify certain basic aspects of the adsorption phenomena. The notations used in the mathematical treatment have been kept uniform as far as possible.

The references are as up-to-date as is technically possible. The book does not cover all other aspects of surface science where the Gibbs surface excess concept is not essential since these are well covered by the existing textbooks. The aim of this book is to point out those specific areas of surface phenomena where the application of the Gibbs surface excess concept is essential or not yet fully appreciated. Topics of great controversy are also mentioned in brief with comments at the end of each chapter in order to provide a basis for future research for a newcomer in the field.

More than two hundred illustrations have been included in the eleven chapters of this book. We express our thanks and gratitude to the following publishers and societies for permitting us to use the figures first published in their journals and books:

The Chemical Society, London; The American Chemical Society, Washington, D.C.; The Chemical Society of Japan, Tokyo; The Council of Scientific and Industrial Research, New Delhi; F. A. Davis Company, Philadelphia; Dr. Dietrich Steinkoff Verlag, Darmstadt; Academic Press Inc., New York; Wiley-Interscience Publications and John Wiley and Sons, New York; Elsevier Biomedical Press and Elsevier Scientific Press, Amsterdam; Marcel Dekker Inc., New York; Pergamon Press, London; Butterworth Scientific Ltd., Borough Green, England.

In addition, many authors have given their consent to the use of their illustrations in our book for which we express our sincere thanks. The appropriate references for these illustrations in specific journals and books have been mentioned in the figure captions. Many of these figures have been redrawn with major or minor changes to suit the system of notation used in this book.

The authors would like to thank their many colleagues, friends, and students for helpful suggestions and discussions during the course of writing this book. They are particularly grateful to Professor S. K. Mukherjee and Dr. K. P. Das, who read the complete manuscript of the book and made valuable comments. The authors had many discussions with the late Professor J. Koefoed which were very useful and important. The technical assistance received from Dr. B. K. Sadhukhan, Mr. Subrata Basu, and Mr. D. Thakurta is also acknowledged with thanks. One of the authors (DKC) expresses his indebtedness to the Danish Natural Science Research Council for the award of a travel grant and would also like to acknowledge the hospitality offered by the staff of the Fysisk Kemisk Institute during his stay in Denmark from 1979 to 1980 as the Guest Professor of the Danish Technical University.

D. K. CHATTORAJ
K. S. BIRDI

CONTENTS

CHAPTER 9: ADSORPTION OF WATER VAPOR BY BIOPOLYMERS

CHAPTER 10: BINDING INTERACTIONS IN BIOLOGICAL SYSTEMS

CHAPTER 11: MISCELLANEOUS SYSTEMS

CHAPTER 1

INTRODUCTION

1.1. ADSORPTION FROM SOLUTION

Adsorption at various types of interfaces will take place when a liquid phase is in contact with another immiscible phase which may be solid, liquid, or gas. For adsorption to occur in excess, one or both of the phases should contain more than one component. Two immiscible bulk phases must be in contact with each other so that some of these components are accumulated in excess at the interfacial region and the process of transfer of the component from the bulk to the surface will continue until a state of adsorption equilibrium is reached. The component which is preferentially accumulated at the interface is frequently termed an adsorbate. The term adsorbent is also used for the powdered solid on the surface of which adsorbates from the liquid phase are accumulated in excess. The types of the interfaces are usually designated by names of the bulk phases in contact with each other such as air–water, benzene–water, mercury–water, or alumina–water interfaces.

The adsorption of a gas by powdered solid materials has been extensively studied from the experimental and theoretical standpoints. The kinetic and thermodynamic interpretations of the process are relatively simple because the concentration of the gas can be altered by change of pressure. The adsorption of gases has been extensively reviewed by Adamson[1] and others,[2] and this will not be discussed here in further detail. The water vapor adsorption by biopolymers will be discussed in Chapter 9 in brief to show some aspects of adsorption which appear to be similar when it takes place either in the gas or the solution phases.

The adsorption from the solution depends on its composition. Unlike the gaseous system, the concentration of the solute component in the solution can be reduced only by increasing the amount of the solvent component in the bulk. Further both solute and solvent components in the bulk medium compete with each other for excess accumulation at the interfacial region. The surface region between two bulk phases is thin and its property is very sensitive to the change in properties of the bulk. In fact, the properties of the surface region can be altered to a significant extent by the change in the properties of the bulk system. The thermodynamics of adsorption at the fluid interface were elaborately considered by Gibbs[3] and an attempt will be made in this book to treat different types of adsorption phenomena from the unified concept of "the Gibbs surface excess."

1.2. NATURE OF THE INTERFACE

The gas–liquid and liquid–liquid systems are mobile so that the interfaces attain an equilibrium shape quickly. Because of this, thermodynamic treatment of the adsorption for such systems becomes simple.

The boundary region between a liquid and a solid is distinctly different in this respect since the surface structure and shape for this system are mostly controlled by the arrangement of the molecules in the crystal lattice at the interfacial region of the solid. The shape of the surface therefore is not smooth like the liquid–gas interface. In most cases, the solid–liquid interface is highly heterogeneous in structure in multiple directions.[1] Further, the solid surface possesses a tendency to assume minimum surface area after rearrangement of the crystal forces. For the attainment of equilibrium shape therefore at the solid interface, a large amount of time may be needed. A true state of thermodynamic equilibrium is usually not attained at the solid–liquid interface. The extent of adsorption of a component at the solid–liquid interface usually changes with time but it reaches a constant value within a period varying between a few hours and few days depending on the nature of the system. A state of adsorption equilibrium is thus reached so that the application of the thermodynamic principles appears to be justified. However, it is not clear how far this state of equilibrium is dependent on the dynamic structure of the underlying solid surface.

Adsorption and desorption phenomena frequently occurring at a biosurface are of considerable interest in the field of biology. Cell-membrane, liposome, ribosome, proteins, carbohydrates in insoluble forms, bones, etc. are examples of biosurfaces. In living systems, the biosurfaces are frequently formed due to the complex mixing of biopolymers and lipids. These surfaces are in contact with water containing organic solutes and inorganic salts so that the phenomenon of adsorption occurs in the biological systems. Because of the extensive interaction of the biosurface with the adjacent bulk-water phase, the biosystem in many cases becomes swollen gel so that the phase boundary region is difficult to detect even qualitatively. Many workers in their thermodynamic treatment assume the whole biogel system as a single phase without an interface. Further, the structure of the biosurface in many cases becomes flexible due to the conformational alterations of the partly rigid and partly flexible biopolymers present in the system. The adsorption sites on the biosurfaces are also highly heterogeneous since the chemical structures of different monomers forming the biopolymers are widely different from each other. The adsorption of solutes from the solution by synthetic polymers and textile fibers also possesses many of the complexities which may occur in the case of the biosurfaces just discussed.

Many biopolymers like proteins, nucleic acids, and carbohydrates actually dissolve in the aqueous medium, apparently forming a single phase system.

However, small solute as well as water molecules forming the aqueous phase may accumulate close to the boundary region of the biocolloid particle due to the binding interaction. In fact the binding phenomena may be identified with the phenomena of adsorption if one assumes that there exists a physical boundary between the aqueous and biocolloid phases.

1.3. NATURE OF THE ADSORBATE

There exists a large number of organic compounds called the surfactants which lower the surface tension of water significantly. These solutes accumulate at the liquid interface in large excess. The surfactants are frequently adsorbed at various types of solid–liquid, gas–liquid, liquid–liquid interfaces and also at the biosurfaces. More recently a surfactant is termed an amphiphile[4] since its molecule contains an elongated hydrocarbon chain (termed a hydrophobic tail) and an ionic or polar group (termed a polar head group). When the concentration of the surfactant in the aqueous medium exceeds a critical value (called the critical micelle concentration or cmc), the amphiphile molecules aggregate forming colloidal micelles.[4,5] The amphiphiles may be cationic, anionic, or nonionic in nature. A list of few surfactants and their cmc values are given in Table 1.1.

There exist many organic compounds such as alcohols, acids, urea, dyes, etc., a molecule of which contains a polar head group attached to a short-chain hydrophobic tail. Because of this, the affinities of these substances for various surfaces are relatively low. In many systems, these short-chain compounds compete with water solvent for excess accumulation at the interface.

Many organic macromolecules soluble in water may accumulate at various types of interfaces provided the monomeric units contain hydrophobic and hydrophilic groups in a balanced form. Globular proteins soluble in water

TABLE 1.1. Critical Micelle Concentrations (cmc) of Some Amphiphiles in Aqueous Medium at 25°C

Amphiphiles	Name	cmc (M)
$C_8H_{17}OCH_2CH_2OH$	Octyl glycol ether	4.9×10^{-3}
$C_8H_{17}-C_6H_4-O(CH_2CH_2O)_{10}OH$	Triton X100	2×10^{-4}
$C_8H_{17}N^+(CH_3)_2CH_2COO^-$	Octyl N-beatine	2.5×10^{-1}
$C_8H_{17}OCHC_5H_{10}O_5$	β-D-octylglucoside	2.5×10^{-2}
$C_{16}H_{33}N^+(CH_3)_3Br^-$	Cetyltrimethyl ammonium bromide	7.6×10^{-4}
$C_{12}H_{25}SO_4^-Na^+$	Sodium dodecyl sulphate	8.0×10^{-3}
$C_{12}H_{25}NH_3^+Cl^-$	Dodecyl ammonium chloride	1.38×10^{-2}
$C_8H_{17}COO^-K^+$	Potassium nonanoate	2.0×10^{-1}

possess surface-active properties. The formulas of an amino acid and a dipeptide, respectively, may be written as

$$\underset{\underset{H}{|}}{\overset{\overset{R}{|}}{H_2N-C-COOH}} \qquad \underset{\underset{H}{|}}{\overset{\overset{R}{|}}{H_2N-C-CONH-}}\underset{\underset{H}{|}}{\overset{\overset{R_1}{|}}{C-COOH}}$$

The side-chains represented by R, R_1 etc. of many amino acids contain hydrophobic groups. Sometimes many hydrophilic groups are also incorporated in the side-chain of an amino acid. Two amino acids are linked with each other by a —CONH— or "peptide bond" whereby a dipeptide is formed.[6,7] During protein synthesis within the living system, the polypeptide chains are formed in perfect sequence.[7] The number of the amino acids in a polypeptide chain may vary between 50 to 200 or more depending on the nature of the protein. The sequence of amino acids in the polypeptide chain of a particular protein is thus fixed and this represents the primary structure of the biopolymer. The globular proteins egg albumin, bovine serum albumin, and ribonuclease each contain only one polypeptide chain. The proportion of hydrophobic and hydrophilic groups in this chain will control the surface-active property of a protein with respect to a particular interface.

A polypeptide chain in a protein can undergo helical or planar folding due to the intrachain or interchain hydrogen bonding between —NHCO— groups.[6,8,9] This is sometimes regarded as the secondary structure. A polypeptide can also fold further due to the interaction between side-chain groups within the chain whereby a relatively rigid globular structure of a protein molecule is formed.[6] This is termed the tertiary structure of the protein. Various forces responsible for the globular structure have been discussed elsewhere.[10] Many proteins such as hemoglobin contain more than one polypeptide chain and these remain associated due to the specific forces present thus forming quaternary structure in a protein molecule.

Conformation of globular protein molecules of rigid structure has been discussed in several reviews.[10,11] When an aqueous solution of a globular protein is heated, the forces responsible for the compact structure are affected by thermal agitation until a stage is reached when the biopolymer molecule unfolds extensively and the phenomenon is known broadly as denaturation.[6,12] Denaturation of protein also occurs at ordinary temperature due to the addition of acid, alkali, urea, guanidine hydrochloride, etc. in sufficient amount. Many soluble proteins such as gelatin, histone, etc. are flexible in structure at ordinary temperature and these are regarded as denatured proteins.

A globular protein adsorbed at an interface may undergo denaturation in various degrees due to the impact of the boundary forces on its structure. This alteration of the structure at the interface is time dependent[13] and the phenomenon itself is of considerable biological interest.

The nucleic acids are also biopolymers which control the genetic properties of the living systems. The monomer unit of deoxyribonucleic acid (DNA) is called deoxyribonucleotide or simply nucleotide which contains a deoxyribose unit attached to a purine or pyrimidine base on one side and a phosphate group on another side. The polymerization of nucleotides takes place through the formation of phosphodiester bonds.[7] The nucleotide unit occurring in DNA polymer is anionic because of the presence of a negative charge on the phosphate group. The sodium salt of DNA polymer thus behaves as a polyelectrolyte. The purine and pyrimidine bases attached to two DNA chains are hydrogen bonded with each other, whereby a double helix structure is formed. This rigid structure is extremely important in controlling the genetic properties within the living system.[7] The nucleotide unit of ribonucleic acid (RNA) contains a ribose sugar attached to an organic base and a phosphate group. Soluble RNA behaves as a flexible polyelectrolyte and is single stranded in structure.

1.4. NATURE OF THE BULK PHASE

For preferential or the excess adsorption to occur from the solution, at least one of the bulk phases must remain in the liquid state. The adsorbate should be soluble to some extent in this liquid solvent. The common solvent widely used in the study of adsorption is water. For the study of adsorption of an organic electrolyte, the aqueous solvent sometimes contains inorganic neutral salt whose affinity for the surface is negligible. Many important properties of the organic and inorganic solutes dissolved in water are already known. These known properties (such as activity coefficients, solubilities, cmc) are very useful in dealing with thermodynamic treatment of the adsorption phenomena. The study of adsorption from an aqueous solution to the biosurface is of considerable biological importance.[4,13] In this book also, we shall discuss mostly the results of the adsorption from the aqueous phase.

The second bulk phase in the adsorption system may be gas (air, vapor, etc.), organic liquid practically insoluble in water (e.g., benzene, heptane, etc.), powdered solid (e.g., alumina, glass, etc.), or insoluble biopolymers. The role of this second bulk phase has been neglected apparently in dealing with the theories of adsorption from the aqueous solution. The effect of the second phase has, however, been implicitly included in the experimental data.

1.5. THERMODYNAMICS OF THE BULK PHASE

We have previously pointed out that the thermodynamic properties of the interfacial region are controlled to a large extent by the thermodynamic functions related to the adjacent bulk phases. The chemical thermodynamics of the system containing the bulk phase or phases at equilibrium have been developed quantitatively. We shall discuss some relevant points of this subject in concise form. For details, the readers are referred to books dealing with chemical thermodynamics.[14-16]

Let us assume that the bulk phase α is composed of n_1, n_2, ..., n_i moles of components so that the differential change in internal energy dE due to the change in heat, dQ, and work, dW, in a reversible process can be expressed from the first and second laws of thermodynamics by the equation

$$dE = dQ - dW + \sum_i \mu_i dn_i$$

$$= TdS - pdV + \sum_i \mu_i \, dn_i \tag{1.1}$$

Here dS and dV represent the differential change in entropy and volume, respectively, and T and p are the temperature and pressure in the system under consideration undergoing mechanical work only. The last term in equation (1.1) takes into account the variation in the moles n_1, n_2, ..., n_i of components during the differential change occurring in the open system. From equation (1.1), the chemical potential μ_i of the ith component can be defined thus:

$$\mu_i = \left(\frac{\partial E}{\partial n_i}\right)_{S, V, n_j} \tag{1.2}$$

Except n_i, moles of all other components represented by n_j remain constant on the right-hand side of equation (1.2). If there is no variation in the mass in phase α, $\sum_i \mu_i \, dn_i$ becomes zero in equation (1.1), and the relation for the closed system will be obtained.

If H, F, and G stand for the enthalpy, the Helmholtz free energy, and the Gibbs free energy of the system, respectively, these are defined by the equations

$$H = E + pV \tag{1.3}$$

$$F = E - TS \tag{1.4}$$

$$G = H - TS$$

$$= E + pV - TS \tag{1.5}$$

Differentiating these equations and then combining with the fundamental equation (1.1), one finds[15]

$$dH = TdS + Vdp + \sum_i \mu_i dn_i \tag{1.6}$$

$$dF = -SdT - pdV + \sum_i \mu_i dn_i \tag{1.7}$$

$$dG = -SdT + Vdp + \sum_i \mu_i dn_i \tag{1.8}$$

From these equations and also from equation (1.1), it can be shown that

$$\mu_i = \left(\frac{\partial E}{\partial n_i}\right)_{S,V,n_j} = \left(\frac{\partial H}{\partial n_i}\right)_{S,p,n_j} = \left(\frac{\partial F}{\partial n_i}\right)_{T,V,n_j} = \left(\frac{\partial G}{\partial n_i}\right)_{T,p,n_j} \tag{1.9}$$

It may also be noted that V, n_i, E, H, F, G represent the functions of the state whose magnitudes will depend on the amount of the material of the system. These are termed extensive state properties of the system. On the other hand, the state functions T, p, μ_i etc. do not depend on the amount of the substance and these are regarded as intensive properties. The ratio of two extensive properties is regarded as an intensive state function for a system.

Keeping the intensive properties T, p, μ_i constant, the system containing phase α can undergo a reversible process such that the extensive properties S, V, and n_i are altered to KS, KV, Kn_i, etc. Here K is a proportionality constant. Integration of equation (1.1) under this situation[15] leads to the following relation:

$$E = TS - pV + \sum_i \mu_i n_i \tag{1.10}$$

so that one can write also in the light of equations (1.3), (1.4), and (1.5)

$$H = TS + \sum_i \mu_i n_i \tag{1.11}$$

$$F = -pV + \sum_i \mu_i n_i \tag{1.12}$$

$$G = \sum_i \mu_i n_i \tag{1.13}$$

Equation (1.13), on general differentiation and then in combination with equation (1.8), will give the following relation:

$$SdT - Vdp + \sum_i n_i d\mu_i = 0 \tag{1.14}$$

This is known as the Gibbs–Duhem equation.

1.6. PARTIAL MOLAL QUANTITIES

For investigating the thermodynamic properties of the mixture of components forming the bulk phase α, it is extremely useful to introduce the concept of the partial molal quantity Y_i in the form of the mathematical equation

$$Y_i = \left(\frac{\partial Y}{\partial n_i}\right)_{T,p,n_j} \tag{1.15}$$

Here Y is the total property of the system such as V, E, H, S, F, or G, respectively. It can also be shown[15] that

$$Y = \sum_i n_i Y_i \tag{1.16}$$

Replacing Y_i by the partial molal free energy G_i and then comparing equation (1.15) with relation (1.9), one finds that μ_i is equal to G_i. From thermodynamic grounds, one can also derive[15] the following relations:

$$V_i = \left(\frac{\partial \mu_i}{\partial p}\right)_{T,n_i,n_j} \tag{1.17}$$

$$S_i = -\left[\frac{\partial \mu_i}{\partial T}\right]_{p,n_i,n_j} \tag{1.18}$$

$$H_i = -T^2 \left[\frac{\partial(\mu_i/T)}{\partial T}\right]_{p,n_i,n_j} \tag{1.19}$$

V_i, S_i, and H_i represent partial molal volume, entropy, and enthalpy, respectively of the ith component in the mixture. Each of these quantities can be suitably defined by equation (1.15).

1.7. GAS OR VAPOR PHASE

Suppose the bulk phase is composed of a single gaseous (or vapor) component and its chemical potential is μ at a given pressure p and temperature T. The system is regarded as ideal if $RT \ln p$ becomes equal to $\mu - \mu^0$ so that

$$\mu = \mu^0 + RT \ln p \qquad (1.20)$$

Here μ^0 is defined as the chemical potential of the gas at a chosen standard state of temperature T and pressure one atmosphere. Thus μ^0 is a function of temperature only. Assuming the validity of this equation, one can deduce the ideal equation of state $pv = RT$, where v is the volume per mole of a gas.

If the gaseous phase contains more than one component forming the ideal gas mixture, then one can write[15]

$$\mu_i = \mu_i^0 + RT \ln p_i$$

$$= \mu_i^0 + RT \ln X_i + RT \ln p \qquad (1.21)$$

here p is the total pressure of the gas mixture and p_i the partial pressure of the ith component in the gas mixture. When mole fraction X_i and p become unity, the system contains only one gas i and its chemical potential is μ_i^0 from either equation (1.20) or (1.21). This means that the standard chemical potentials of a gas in the ideal mixture and in the pure state are identical.

1.8. BINARY SOLUTION OF A LIQUID

If the bulk phase α is made up of a binary solution containing two volatile liquid components 1 and 2 (e.g., mixture of water and alcohol), then one can conventionally write

$$p_1 = p_1^0 X_1 f_1$$

$$= p_1^0 a_1 \qquad (1.22)$$

$$p_2 = p_2^0 X_2 f_2$$

$$= p_2^0 a_2 \qquad (1.23)$$

Here p_1 and p_2 are the partial vapor pressures of components 1 and 2, respectively, in the vapor adjacent to the liquid phase at a given temperature

T. Also p_1^0 and p_2^0 are the vapor pressures of the pure components 1 and 2, respectively, at the temperature and total pressure of the system. The liquid mixture is made up of X_1 and X_2 mole fractions of components 1 and 2, respectively. Also f_1 and f_2 are the respective activity coefficients of the components and a_1 and a_2 are the corresponding activities in the liquid mixture. Since f_1 and f_2 are, respectively, equal to $p_1/p_1^0 X_1$ and $p_2/p_2^0 X_2$, the values of the activity coefficients in principle can be computed from vapor pressure measurements. When f_1 and f_2 become unity, equations (1.22) and (1.23) become expressions for Raoult's law for the ideal solution. When f_1 and f_2 deviate from unity, there is deviation from Raoult's law and by convention the solution becomes nonideal.[15]

The chemical potential μ_i of the ith component in a solution of a liquid in bulk can be expressed in general[15] by the equation

$$\mu_i = \mu_i^0 + RT \ln X_i + RT \ln f_i \qquad (1.24)$$

Here μ_i^0 is called the standard chemical potential whose value depends on the temperature and the total pressure* associated with the system. When X_i is unity, the liquid contains pure component i. If in this pure state one assumes f_i is also unity, then $\mu_i = \mu_i^0$. Putting f_i equal to unity in equation (1.24) for a given value of X_i, one can also derive ideal equations for vapor pressure of a liquid obeying Raoult's law [vide equations (1.22) and (1.23)]. $RT \ln f_i$ in equation (1.24) is a measure of the deviation from the ideal behavior of the solution as prescribed by Raoult's law. The activity coefficients f_1 and f_2 as well as the standard chemical potentials μ_1^0 and μ_2^0 therefore refer to Raoult's law scale of ideality fixed by a particular convention.[15]

Frequently, for a binary solution, the liquid component (say component 1) remaining in large excess is termed the solvent, whereas component 2 (which may be solid, liquid, or gas), dissolved in the solvent, is termed the solute. The amount of the solute in the mixture is relatively small. The solvent is usually volatile so that its activity coefficient f_1 measuring deviation from Raoult's law can be calculated from the vapor pressure measurement.

The solutes, on the other hand, may be nonvolatile (e.g., sugar, inorganic salts, etc.) and their solubilities in many cases are fixed at a given temperature. To fix the ideal behavior of these solute components on the basis of Raoult's law is most inconvenient.

Following an alternative convention for the activity coefficient f_i of solute present in the solution, one can write

$$p_i = K_i f_i X_i \qquad (1.25)$$

---

* In a slightly different approach, μ_i^0 and f_i are always referred at a given temperature to one atmospheric pressure rather than to the total pressure of the system. The difference in the two cases is, however, trivial provided the pressure is not high.[14]

Here K_i (not equal to p_i^0) is a constant at constant temperature and the total pressure of the system. According to this convention, f_i tends to unity when X_i of the solute in solution tends to zero, so that equation (1.25) will express Henry's law in the mathematical form. Since K_i for a dilute solution is equal to p_i/X_i, its value for a volatile solute can be computed from the vapor pressure measurements. From the knowledge of K_i, X_i, and p_i for a relatively concentrated solution therefore, the value of f_i can be evaluated. Deviation of f_i from unity now measures deviation from the ideal behavior of a solution which in turn has been prescribed by Henry's law.[15] According to equation (1.24), f_i refers to the standard chemical potential μ_i^0 or the chemical potential of a hypothetical solute of X_i and f_i both tending to unity and still obeying Henry's law. The rational scales of activity coefficients according to Henry's law and Raoult's law, respectively, are different although the concentrations of the solutes in both scales are expressed in mole fraction unit.

Following two other conventions, one can also write

$$p_i = K_i' f_i m_i \tag{1.26}$$

or,

$$p_i = K_i'' f_i c_i \tag{1.27}$$

where m_i and c_i are, respectively, molal and molar concentrations of the solute in the solution and K_i' and K_i'' are the proportionality constants. One also assumes that as m_i or c_i tend to zero, f_i in both cases will approach unity and the solution becomes ideal. From the determination of K_i' and K_i'', respectively, in dilute solutions from the vapor pressure measurements, one can compute f_i in the two scales of concentrations. These practical scales are frequently referred as "hypothetical one molal" or "hypothetical one molar" scales[15] if one writes μ_i as

$$\mu_i = \mu_i^0 + RT \ln m_i + RT \ln f_i \tag{1.28}$$

or

$$\mu_i = \mu_i^0 + RT \ln c_i + RT \ln f_i \tag{1.29}$$

When m_i or c_i tend to unity, f_i in both cases hypothetically tends to unity and μ_i^0 will approach μ_i [μ_i^0 in equations (1.28) and (1.29) are not the same].

The activity coefficient of a solute in the solution may be different in different scales but its value in one scale can be converted to that in the other scale by multiplication with a suitable conversion factor.

1.9. ACTIVITY COEFFICIENT OF AN ELECTROLYTE

An electrolyte $P_{\nu_+}Q_{\nu_-}$ dissociates in an aqueous solution thus

$$P_{\nu_+}Q_{\nu_-} = \nu_+ P^{z+} + \nu_- Q^{z-} \tag{1.30}$$

Here ν_+ and ν_- are the number of cations and anions, respectively, of valence z_+ and z_- which result from the complete dissociation of an electrolyte in the aqueous phase. The chemical potential μ of the electrolyte as a whole can be related to the chemical potentials μ_+ and μ_- of the cations and anions, respectively, by the equation

$$\mu = \nu_+ \mu_+ + \nu_- \mu_- \tag{1.31}$$

Inserting in it, values of μ_+ and μ_- from equation (1.28), one finds

$$\mu = (\nu_+ \mu_+^0 + \nu_- \mu_-^0) + \nu RT \ln f_\pm m_\pm \tag{1.32}$$

Here ν is equal to $\nu_+ + \nu_-$. Also $f_\pm$ and $m_\pm$ are, respectively, mean activity coefficient and mean molality of the electrolyte in the solution defined by

$$f_\pm = (f_+^{\nu_+} f_-^{\nu_-})^{1/\nu} \tag{1.33}$$

$$m_\pm = (m_+^{\nu_+} m_-^{\nu_-})^{1/\nu} \tag{1.34}$$

The mean activity coefficient in the hypothetical "one molal" practical scale can be determined by various types of physicochemical experiments. By multiplication by a suitable factor, the value of $f_\pm$ in the rational mole fraction scale can easily be calculated.[15]

1.10. STANDARD FREE ENERGY CHANGE OF A CHEMICAL REACTION

Suppose a chemical reaction occurs in the bulk phase according to the equation

$$\nu_a A + \nu_b B + \cdots = \nu_c C + \nu_d D + \cdots \tag{1.35}$$

Here ν_a, ν_b, ν_c, ν_d are the stoichiometric coefficients for the reactants and the products, respectively. The standard free energy change ΔG^0 for this chemical reaction can be expressed by[15]

$$\Delta G^0 = (\nu_c \mu_c^0 + \nu_d \mu_d^0 + \cdots) - (\nu_a \mu_a^0 + \nu_b \mu_b^0 + \cdots) \qquad (1.36)$$

From a detailed thermodynamic analysis, it can also be shown that[15]

$$\Delta G^0 = -RT \ln K \qquad (1.37)$$

Here K is the equilibrium constant for the reaction at temperature T. If the reaction takes place in the gas phase, then

$$K = \frac{(p_c)^{\nu_c}(p_d)^{\nu_d} \cdots}{(p_a)^{\nu_a}(p_b)^{\nu_b} \cdots} \qquad (1.38)$$

where $p_a, p_b, p_c, p_d, \ldots$, etc., are partial pressures at chemical equilibrium. From the knowledge of the experimental quantity K therefore, ΔG^0 can be computed using equation (1.37). The standard chemical potentials of the individual components in equation (1.36) all refer to temperature T and one atmospheric pressure [vide equation (1.21)].

If, on the other hand, the reaction occurs in the liquid phase, then

$$K = \frac{(f_c)^{\nu_c}(f_d)^{\nu_d}}{(f_a)^{\nu_a}(f_b)^{\nu_b}} \cdot \frac{[C]^{\nu_c}[D]^{\nu_d}}{[A]^{\nu_a}[B]^{\nu_b}} \qquad (1.39)$$

Here $[A]$, $[B]$, $[C]$, and $[D]$ may be molar or molal concentrations of A, B, C, and D at equilibrium and f_a, f_b, f_c, f_d are the, respective, activity coefficients in the "hypothetical" one molar or one molal scales. Inserting the experimental value of K, ΔG^0 for the reaction (1.35) can be calculated using equation (1.37) so that μ_a^0, μ_b^0, μ_c^0, μ_d^0 etc. in relation (1.36) will refer to the appropriate practical scales of activities.

It may be pointed out here that in calculating ΔG^0 in this manner, the role of the aqueous medium in the reaction has been neglected. When the solvent is involved in the chemical reaction, the concentration and activity coefficient terms in equation (1.39) should be expressed in the mole fraction scale with additional inclusion of the mole fraction of water.

Suppose an electrolyte component in solution is dissociated incompletely according to equation (1.30). The dissociation constant K for the reaction can be written as

$$K = \frac{(f_{\pm})^{\nu} \cdot m_+^{\nu_+} m_-^{\nu_-}}{\gamma_u m_u} \qquad (1.40)$$

Here γ_u and m_u represent the activity coefficient and molal concentration of the undissociated electrolyte, respectively. ΔG^0 for the dissociation is expressed as

$$\Delta G^0 = \nu_+\mu_+^0 + \nu_-\mu_-^0 - \mu_u^0 \tag{1.41}$$

The chemical potential and the activity coefficient all refer to the hypothetical one molal scale.

1.11. SOLUTE DISTRIBUTION BETWEEN TWO INSOLUBLE LIQUID PHASES

When a solute i is distributed between two insoluble liquid phases α and β in contact with each other, μ_i^α becomes equal to μ_i^β according to the Gibbs phase rule, so that

$$\mu_i^{\alpha,0} + RT \ln f_i^\alpha X_i^\alpha = \mu_i^{\beta,0} + RT \ln f_i^\beta X_i^\beta \tag{1.42}$$

Here $\mu_i^{\alpha,0}$ and $\mu_i^{\beta,0}$ are standard chemical potentials, f_i^α and f_i^β the activity coefficients, and X_i^α and X_i^β the mole fractions of the ith component in α and β phases, respectively. From equation (1.42), one can easily find[15]

$$\Delta G^0 = \mu_i^{\beta,0} - \mu_i^{\alpha,0}$$

$$= -RT \ln K_d \tag{1.43}$$

where ΔG^0 is the standard free energy change for the transfer of one mole of the solute from phase α to phase β and K_d is the distribution constant given by

$$K_d = \frac{f_i^\beta X_i^\beta}{f_i^\alpha X_i^\alpha} \tag{1.44}$$

K_d determined from the experiment is usually higher than unity (α being taken as the aqueous phase) so that the solute remains in higher concentration in phase β.

In many chemical reactions, the reactants and products in the mixture at equilibrium including the solvent form a heterogeneous system containing more than one phase. ΔG^0 and K here can still be correctly expressed by equations (1.36) and (1.37), respectively. One must be alert, however, to the standard scales involved for μ_a^0, μ_b^0, μ_c^0, μ_d^0, etc., which may be different depending upon the solid, liquid, and gas nature of the compounds as well as on solute–solvent characteristics. For any meaningful comparison of ΔG^0 for closely allied processes, one should carefully check how far these values refer to the identical states of standard scales.

1.12. THE SURFACE ENERGY

The molecules of matter in the gaseous state move individually at random due to the possession of the translational kinetic energy. Besides, the molecules in the gas phase experience forces due to the mutual attraction with each other when the pressure is relatively high. In the condensed liquid and solid phases, the intermolecular attraction forces are enormously high in magnitude due to the dipolar and dispersion interaction. From detailed mathematical calculations based on quantum mechanics, it has been shown that the potential energy of a molecule is decreased when another molecule is brought (by attraction forces) from infinity to a distance extremely close to the first molecule. A molecule in the bulk of the condensed phase of the solid is surrounded by maximum number of neighboring molecules so that its potential energy becomes minimum. When the same molecule is present in the surface layer of the liquid or the solid phase, it is in contact with a lesser number of molecules since there are fewer neighboring molecules in the gaseous region just at the top of the interface. The energy of the molecule in the interfacial phase thus becomes greater than that in the bulk phase. For the formation of one square centimeter of the surface, surface energy γ has to be spent for bringing molecules from the bulk to the surface.

Let us now consider this phenomenon from the macroscopic point of view. If the area of a liquid film stretched in a wire frame (Fig. 1.1) is increased by $d\mathscr{A}$, the surface energy of the film is increased by $\gamma d\mathscr{A}$. This increase of energy further indicates that the motion of the wire attached to the film is opposed by a force f. If dx represents the magnitude of the displacement of the wire, then the energy spent for this is $f dx$. At equilibrium,

$$f dx = \gamma d\mathscr{A} \tag{1.45}$$

or

$$\gamma = f \frac{dx}{d\mathscr{A}}$$

$$= \frac{f}{2l} \tag{1.46}$$

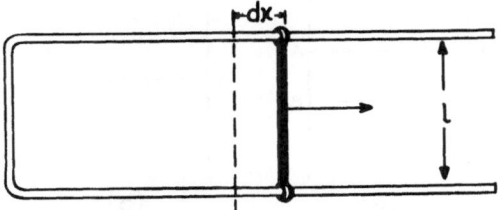

FIGURE 1.1. Surface film of liquid.

TABLE 1.2. Surface Tension of Liquids in mN/m against Air

Substance	Temperature (°C)	Surface tension (mN/m)
Water	25	71.97
Ethanol	20	22.3
n-Butyric acid	20	26.8
Benzene	20	28.88
Acetone	20	23.7
Acetic acid	20	27.6
Methanol	20	22.6
n-Octyl alcohol	20	27.5
Cyclohexane	20	25.3
n-Hexane	20	18.4

where $2l$ is the total length of both sides of the film of the air–liquid system. Here γ represents also the force per unit length of the surface and this force is frequently referred as the surface tension. The values of γ for different liquid systems are given in Table 1.2. It may be mentioned here that the equality between the surface energy and the surface tension becomes valid only when a state of equilibrium is reached both in the bulk and at the interface.

1.13. SURFACE TENSION AND MECHANICAL EQUILIBRIUM

Consider two fluid phases α and β (vide Fig. 1.2a) which are in contact with each other by the interfacial region $AA'BB'$. The pressures p and p' in the α and β phases, respectively, remain uniform up to the planes BB' and AA'. In the inhomogeneous region $AA'BB'$, the contribution of p and p' cannot be known precisely. Young,[17] however, proposed that the real system can be replaced by a model system (Fig. 1.2b) in which the fluids may be imagined to be separated by a plane YY' called the surface of tension. Bulk fluids α and β extend up to this model surface across which the surface

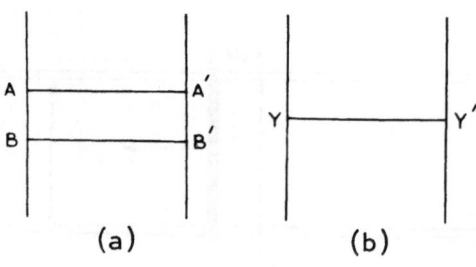

(a)   (b)

FIGURE 1.2. (a) Real and (b) model surface in a liquid column.

tensional force is supposed to act. The equivalence between the real and model systems in terms of the balance of the mechanical and surface tensional forces has been worked out.[17] There exists also a surface of tension for the real surface, but its actual location within $AA'BB'$ is unknown. As will be seen in the next section, the concept of the surface of tension is specially useful for interpreting the mechanical equilibrium at the curved interface.

For the real system undergoing an infinitesimal process, one can write[17]

$$dW = pdV + p'dV' - \gamma d\bar{\mathscr{A}} \qquad (1.47)$$

Here dW is the work done by the system for the change in volume dV and dV' in α and β phases, respectively, and the change of interfacial area by $d\bar{\mathscr{A}}$. The sign of the surface tensional work is written as negative by convention.

A liquid surface at equilibrium may be plane or curved. Let us imagnine that the surface separating two phases is not a plane but part of a sphere of radius r (vide Fig. 1.3). The spherical surface is under the action of the mechanical forces arising from p, p', and γ. From the appropriate resolution of these forces acting on the system near the interface, it can be shown that at mechanical equilibrium[17]

$$p - p' = 2\gamma/r \qquad (1.48)$$

This equation predicts that the radius of curvature for the surface of tension depends on γ, p, and p'. Further the pressure p on the concave side of the curvature is greater than p' remaining on the convex side. When the liquid in the air remains in the form of small droplet of radius 10^{-4} to 10^{-5} cm, values of $p - p'$ will become quite significant. When the surface becomes plane, the radius of curvature r will tend to infinity and p becomes equal to p'.

If the curved surface is nonspherical, an equation in general form derived by Laplace reads[17]

$$p - p' = \gamma\left(\frac{1}{r_1} + \frac{1}{r_2}\right) \qquad (1.49)$$

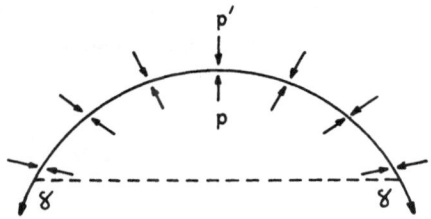

FIGURE 1.3. Spherical fluid surface at equilibrium.

Here r_1 and r_2 are two principal radii of curvature of the surface at a given point. When $r_1 = r_2 = r$, the equation reduces to the form (1.48). Also for the plane surface r_1 and r_2 will both tend to infinity so that $p = p'$.

For a curved surface, γ actually acts at the real surface of tension with definite values of r_1 and r_2 so that equation (1.49) will remain valid. One cannot arbitrarily select any other plane (with different r_1 and r_2) to represent the surface of tension. For the plane surface, this precision is not necessary to fix the position of the dividing plane within $AA'BB'$ exactly on the real surface of tension. Since $p = p'$ for this case of the plane surface, equation (1.47) can be written[17] as

$$dW = pd(V + V') - \gamma d\bar{\mathscr{A}}$$

$$= pdV' - \gamma d\bar{\mathscr{A}} \tag{1.50}$$

Here V' the total volume of the system is taken to be equal to $V + V'$ and p is the pressure of the system in α and β phases.

1.14. SURFACE FREE ENERGY

The thermodynamic treatment of the surface region $AA'BB'$ is complex because this region is thin, inhomogeneous, and not well defined. Guggenheim[18] defined a space of arbitrary thickness τ in the surface region as an imaginary surface phase σ (vide Section 3.16). Both surface and bulk phases are composed of i number of components. Applying thermodynamic consideration, it has been shown that

$$dF^\sigma = -S^\sigma dT - pdV^\sigma + \gamma d\bar{\mathscr{A}} + \sum_i \mu_i dn_i^\sigma \tag{1.51}$$

and

$$dG^\sigma = -S^\sigma dT + V^\sigma dp + \bar{\mathscr{A}} d\gamma + \sum_i \mu_i dn_i^\sigma \tag{1.52}$$

Here $\bar{\mathscr{A}}$ stands for the area of the interface and F^σ, G^σ, S^σ, V^σ, and n_i^σ are the extensive parameters related to the surface phase. From equation (1.51), one finds

$$\gamma = \left(\frac{\partial F^\sigma}{\partial \bar{\mathscr{A}}}\right)_{T, V^\sigma, n_i^\sigma} \tag{1.53}$$

γ thus represents the rate of change of the Helmholtz free energy in the σ phase with change of the surface area $\bar{\mathscr{A}}$.

In the treatment of Gibbs dealt in Chapter 3, the expression for dE' for the air–liquid system as a whole has been given in equation (3.19). Replacing the total internal energy E' by $F' + TS'$ or $G' - pV' + TS'$, respectively, it can be shown that

$$dF' = -S'dT - pdV' + \gamma d\bar{\mathscr{A}} + \sum_i \mu_i dn_i' \qquad (1.54)$$

and

$$dG' = -S'dT + V'dp + \gamma d\bar{\mathscr{A}} + \sum_i \mu_i dn_i' \qquad (1.55)$$

Here S', V', n_i' represent the extensive parameters of the whole system. From these two equations, one can show[17]

$$\gamma = \left(\frac{\partial F'}{\partial \bar{\mathscr{A}}}\right)_{T,V',n_i'} \qquad (1.56)$$

and

$$\gamma = \left(\frac{\partial G'}{\partial \bar{\mathscr{A}}}\right)_{T,p,n_i'} \qquad (1.57)$$

γ here represents the rate of change of the total Helmholtz and the Gibbs free energies of the system with change of $\bar{\mathscr{A}}$ at constant T, V', n_i' and constant T, p, n_i' respectively.

Following an algebraic treatment discussed in Section 3.16, it has been shown by Schay[19] that

$$dF^x = -S^x dT - pdV^x + \gamma d\bar{\mathscr{A}} + \sum_i \mu_i dn_i^x \qquad (1.58)$$

and

$$dG^x = -S^x dT + V^x dp + \gamma d\bar{\mathscr{A}} + \sum_i \mu_i dn_i^x \qquad (1.59)$$

Here S^x, n_i^x are the excess entropies and moles of the ith component at the interface at position x of the dividing plane (vide Section 3.6) and V^x is the excess volume of the interface whose value according to the Gibbs concept is negligible:

$$\gamma = \left(\frac{\partial F^x}{\partial \bar{\mathscr{A}}}\right)_{T,V^x,n_i^x} \tag{1.60}$$

$$= \left(\frac{\partial G^x}{\partial \bar{\mathscr{A}}}\right)_{T,p,n_i^x} \tag{1.61}$$

Various definitions of γ must be equivalent and should not depend on the imaginary picture of the interface.

It may further be pointed out that the chemical potential μ_i for the bulk and the surface phases are equal at equilibrium [vide eq. (1.42)], so that

$$\mu_1^0 + RT \ln f_1 X_1 = \mu_1^{s,0} + RT \ln f_1^{s,0} X_1^s \tag{1.62}$$

Here μ_1^0 and $\mu_1^{s,0}$ are the standard chemical potentials, f_1 and f_1^s the activity coefficients, and X_1 and X_1^s are mole fractions of component 1 in the bulk and surface phases, respectively. Here the value of X_1^s depends upon the exact position of the plane dividing surface phase from the bulk so that its value is uncertain if one follows the Gibbs treatment. Further since $f_1^s X_1^s / f_1 X_1$ is not unity for the unequal distribution of the component 1 between the two phases, $\mu_1^{s,0}$ must be different from μ_1^0. All these point out that the standard scale for the surface activity coefficient is uncertain.

It should be pointed out here that γ for a liquid system can be measured precisely from suitable experiments discussed in Chapter 2. In most of the discussions presented in this book γ is uniformly expressed in the SI unit, milli-Newton per meter (mN/m). However, in some figures, γ has also been expressed in the equivalent cgs unit, dynes per centimeter or ergs per square centimeter.

CHAPTER 2

EXPERIMENTAL METHODS AND PROCEDURES

2.1. INTRODUCTION

Surface science has become very important in the last few decades both in terms of theories as well as in terms of application of its concepts in biological and technological fields. The theoretical aspects are based on the performance of precise and reliable experimental data. However, techniques of measurement of various properties of matter near the boundary regions of liquid, solid, or biocolloid systems are relatively few compared to those in the bulk, but experiments in many cases are highly precise and reproducible. Precision of measurement may be affected due to the trace impurities present in the bulk or inner bodies of the instruments in contact with the systems. Thin boundary regions of a solid, liquid, or biocolloid system are surrounded with the infinitely large portion of the bulk. The presence of a large amount of total solute in the bulk causes great difficulties in the measurements of the properties of this solute accumulated only in small amount in the interfacial phase. In spite of all these difficulties, many relatively precise techniques have been developed only in recent years to fill the need of the surface science. These are reviewed in many books and articles referred to in different sections of this chapter. In the present chapter, we will give a short and elementary description of the methods of measurements of surface tension, surface pressure, spreading pressure, contact angle, direct adsorption by solids, water vapor adsorption, and solute binding to biocolloid systems etc. This description will be helpful in understanding the theoretical and descriptive aspects of adsorption and binding phenomena occurring in solid, liquid, and biocolloid systems as discussed subsequently in various other chapters of this book.

2.2. MEASUREMENT OF BOUNDARY TENSION

In Chapter 1 it was pointed out that the boundary between liquid and air or liquid and another immiscible liquid is in a state of tension. Various types of experimental techniques are available for the measurement of the surface tension γ_0 for a pure liquid, and accurate mathematical analyses are available in most cases for the calculation of their exact values. However, for the studies of adsorption indirectly with the help of the Gibbs equation, one has to measure always the surface tension of a series of binary (or ternary) solutions of different compositions. The accuracy of a particular method of measurement depends on the nature of the solute and solvent as well as composition range, temperature, time set for experiments, etc. Thus most suitable methods of measurement of γ for aqueous solutions of inorganic electrolyte, organic surfactants, and proteins are usually not the same. Several of the methods of measurement of surface tension are regarded as static and useful for thermodynamic analysis without ambiguity. Other methods are dynamic ones. Values of γ measured by the static and dynamic methods are usually close to each other. Although γ is measured in many of these methods within 0.01 dyne/cm accuracy, the extent of adsorption (Γ_R) of the solute per unit area calculated indirectly with the help of the Gibbs adsorption equation (vide Chapter 3)

$$\Gamma_R = -\frac{1}{RT}\frac{d\gamma}{d\ln c_R}$$

$$= -\frac{c_R}{RT}\frac{d\gamma}{dc_R} \tag{2.1}$$

will include error in the range of 3% to 10% or even more. The extent of this error depends on the exact procedure followed for the calculation of the slope $d\gamma/d\ln c_R$ or $d\gamma/dc_R$ for the nonliner plot of γ against the bulk concentration c_R. Previously the slopes were determined graphically in many cases using the differential method.[1] Nowadays, virial equations relating γ and c_R are initially set with three or four coefficients.[2] The values of these coefficients are obtained from computers or precision calculators. From the differentiation of this equation, numerical values of the slopes have been obtained directly.

Several methods of measuring γ frequently used will now be discussed only in elementary level. Brief description of the methods has been presented by Adamson[3] to which the readers are referred.

2.2.1. CAPILLARY RISE METHOD

It is well known that when a capillary is dipped in water or an aqueous solution, the liquid will rise upward in height due to the action of the surface tensional forces equal to $2\pi r\gamma\cos\theta$ (vide Fig. 2.1). Here r is the radius of the capillary and θ the contact angle between the liquid and the glass wall. When equilibrium is attained, this upward force will be balanced by a down-

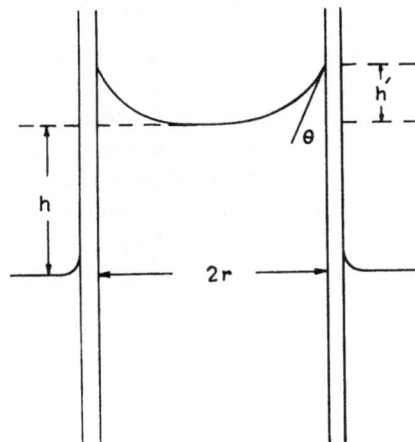

FIGURE 2.1. Capillary rise experiment.

ward gravity force $\pi r^2 h \rho g$ so that

$$\pi r^2 h \rho g = 2\pi r \gamma \cos \theta \qquad (2.2)$$

or

$$\gamma = \frac{r h \rho g}{2 \cos \theta} \qquad (2.3)$$

Here ρ is the density of the liquid and h the height of the capillary at equilibrium. The values of h, ρ, and r can be easily measured from the experiments. Since the aqueous solvent wets the glass surfaces, contact angle θ becomes zero and so $\cos \theta$ becomes unity. γ for a solution can easily be calculated with the help of equation (2.3).

The capillary rise method may become highly accurate for measurement of γ provided certain corrections are taken into account. These corrections mentioned by various workers[4-7] have been thoroughly discussed by Adamson.[3] From Fig. 2.1, it is noted that some amount of liquid remains above the height h in the region of h' of the meniscus region. To take this into account, h in equation (2.2) should be replaced by $h + r/3$. Since the liquid rises and reaches a state of equilibrium within the capillary, this method is regarded as the static method of measurement of γ.

For the precise measurement of the surface tension of the aqueous solutions of inorganic salts, Jones and Ray[8] devised a differential method based on the capillary rise technique where the height of the liquid in the capillary can be measured by direct weighing. Values of many of these data have been analyzed from the thermodynamic standpoint (vide Chapter 3). This differential method has also been followed recently for the measurement of γ of the amino-acid solutions.[9,10]

2.2.2. DROP-WEIGHT METHOD

The drop-weight method is very convenient for the measurement of the surface and interfacial tension of aqueous solution of organic solutes in the presence and absence of inorganic salts. As shown in Fig. 2.2, the aqueous solution is allowed to pass through a narrow tube (sometimes attached to an Agla syringe[1]) so that a drop is allowed to grow slowly until it falls in a closed container (sometimes a stoppered test tube). Several such drops are thus collected and the mass (W) per drop is determined. γ of the solution is then calculated from the following relation, first obtained by Tate[3] from a semiempirical consideration:

$$\gamma = \frac{gW}{2\pi r} f \tag{2.4}$$

Here r is the outer radius of the tip of the tube where the drop is initially formed (vide Fig. 2.2) and the liquid wets the glass surface. Prior to the detachment of the drop, its shape becomes nonspherical, and after detachment of the drop, several tiny droplets follow the main drop. For this reason, Harkins and Brown[11] introduced a dimensionless correction factor f which is a function of $r/V^{1/3}$ where V is the volume of the drop. V is equal to W/ρ, where ρ is the density of the drop. For a series of solutions of surface tensions determined from the capillary rise method, the values of f were estimated by Harkins and Brown as functions of $r/V^{1/3}$ using equation (2.4).

For the measurement of the interfacial tension,[1] the drop of the aqueous solution is allowed to grow in oil phase present in the container until it is detached from the tip. The weight (W) of the drop after correction for the buoyancy has been determined and then the interfacial tension is calculated using equation (2.4). With proper control of temperature, one can estimate γ with deviation of ± 0.05 mN/m or even less. The method is highly suitable for the calculation of the surface excess of amphiphiles at the liquid

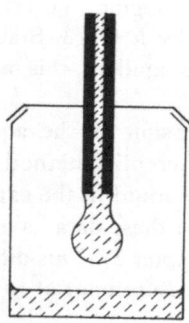

FIGURE 2.2. Fall of a drop in drop-volume method.

interfaces with the help of the Gibbs adsorption equations. Chattoraj and co-workers[1] (vide Chapters 4 and 5) used this method extensively.

This method is a dynamic one in principle and is only suitable when equilibrium at the liquid interface is established within a few minutes.

2.2.3. WILHELMY SLIDE METHOD

The apparatus consists of a chainomatic balance from which the balance pans are removed. From one side of the balance, a thin platinum or glass slide[12,13] (sometimes a battery of slides) is suspended and its weight is counter-balanced in air by adding weights on the other side of the balance in a special type of arrangement (vide Fig. 2.3). The slide or the foil is then just dipped in the solution containing the surface-active solute. By adding further W grams of weight, the system is balanced against surface tensional forces. The surface tension of the solution can be calculated from the equation

$$\gamma = \frac{gWf}{p \cos \theta} \tag{2.5}$$

where p stands for the perimeter of the slide (almost equal to $2L$, where L is the length of the slide). Here f stands for the small correction due to the buoyancy effect. The slide must be wet by the liquid so that the contact angle is zero.

The method gives very good results if a sand-blasted roughened platinum plate is used. A refined method has been proposed,[14] in which many corrections have been included. The Wilhelmy plate method is convenient for the study of γ for the protein solution which varies slowly and continuously with time.[15] The accuracy of this method for measuring surface and interfacial tensions is considerably increased (± 0.001 mN/m) if a modern electrobalance is

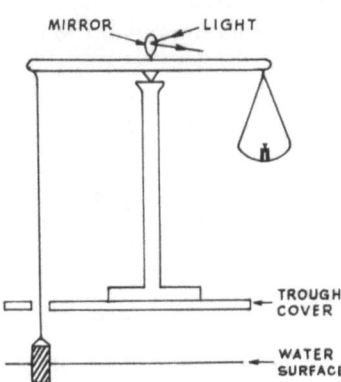

FIGURE 2.3. Wilhelmy balance.

used.[16,17] Various precautions and the theoretical implications connected with this method have recently been discussed.[18,19]

2.2.4. DU NOUY RING METHOD

This method is based on the procedure to measure surface tension from the maximum force required to pull a platinum ring from the surface of a liquid. It is suitable for the quick determination of γ for air–water as well as oil–water interfaces. The surface tension is calculated from the relation

$$\gamma = \frac{Wg}{4\pi R} \tag{2.6}$$

Here Wg is the upward pull of the ring (measured from experiment) and $4\pi R\gamma$ the downward pull. R stands for the radius of the ring.

Certain corrections have to be introduced on the right-hand side of equation (2.6) for the estimation of γ accurately. The extent of this correction will depend both on the ring diameter and the diameter of the wire. The nature of the corrections involved are discussed by several workers.[20–22]

2.2.5. PENDANT DROP METHOD

It is well known that the shape of the drop (in the form of a pendant) hanging from a tip is a function of γ and ρ. The enhanced photographic image of the drop may be taken (vide Fig. 2.4). The surface tension of the solution can be calculated using the equation[3,23,24]

$$\gamma = \rho g d_e^2 f(d_s/d_e) \tag{2.7}$$

Here d_e and d_s are, respectively, the equitorial diameter and the diameter of the drop measured at a distance. An accurate relation between d_s/d_e and $f(d_s/d_e)$ has been worked out so that the value of the latter may be calculated.

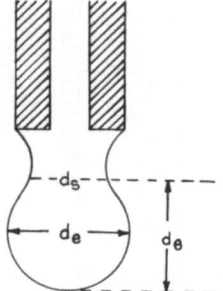

FIGURE 2.4. Hanging drop.

The variation of the interfacial tension with time of a protein solution has been determined most conveniently using this method.[25]

The sessile drop method, flow method, capillary wave method, and many other procedures are frequently used for the measurement of the boundary tension of solutions. These methods have been reviewed by Adamson.[3]

2.3. SURFACE PRESSURE OF INSOLUBLE FILM

It has been known for a long time that a water-insoluble amphiphilic substance may spread in two dimensions when placed in a suitable manner at the air–water interface. The spread monolayer of the amphiphiles thus formed exerts a force per unit length of the liquid surface. This surface tensional force, usually termed the surface pressure, can be measured directly by using horizontal or vertical types of the film balance. The surface pressures of spread monolayers of fatty acids, cholesterol, proteins, and various other substances have been exhaustively studied. This has been reviewed by Gaines.[26]

An important part of all kinds of the surface balance is the film trough T (vide Fig. 2.5), which is a shallow rectangular tray about 2 to 3 feet long and 1 foot wide. The trough, made of brass, glass, Teflon, or steel, must be carefully coated with paraffin so that water poured in the trough may rise to a level a few millimeters above the rim of the trough. The surface of water in the trough must be cleaned by sweeping procedure with the help of a barrier made of paraffined glass or other materials.

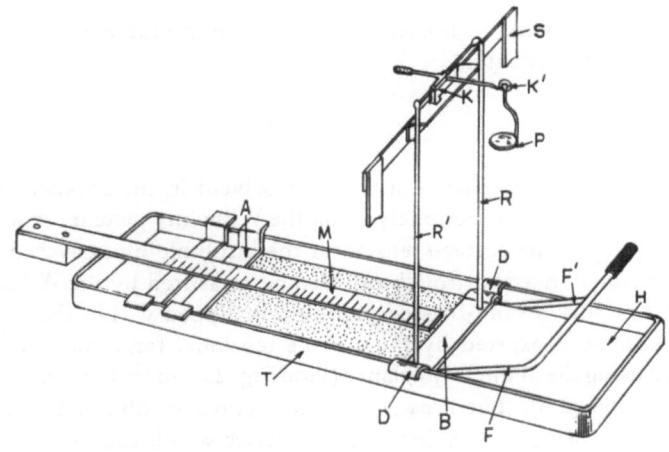

FIGURE 2.5. Langmuir film balance (redrawn from Ref. 27).

2.3.1. HORIZONTAL METHOD

In the original horizontal type of film balance initially used by Langmuir[27] (vide Fig. 2.5), a strip of waxed paper B is placed on the surface of the liquid. The paper is connected to the knife-edge balance device (KK'P) by the rods R and R'. The balance rests on the beam support S. The trough is at first filled with pure liquid. A known quantity of the film material is spread between A and B by a suitable procedure.[26] The region H represents the clean surface. The force (π) exerted by the monolayer per unit length of the strip B can be calculated from the weight used in the balance to keep B in its original position. The area covered by the monolayer containing definite weight of the spread substance can also be calculated with the help of the scale (M) attached. The area A per spread molecule can be thus calculated. Using the movable barrier A, the film area can be gradually decreased and corresponding surface pressure π can be calculated from the increase in the balance weight required to keep the float in its original position.

There exists a gap between the strip D and the wall of the trough. The leakage of the film through this channel is prevented originally by the use of jets of air directed from the tubes F and F' under the metal pieces D. This has been modified later by other workers[26] who used flexible threads or ribbons to prevent leakage. The automatic surface balance in modern form is highly sensitive due to the use of Cahn-type electrobalance, automatic movement system of the barriers, precisional thermostatic control, and many other improved procedures.[3,26]

2.3.2. VERTICAL METHOD

The surface pressure π exerted either by an insoluble or soluble film can be defined by the equation

$$\pi = \gamma_0 - \gamma \tag{2.8}$$

Here γ_0 is the surface tension of the pure solvent in the absence of the film, which can be measured accurately using the Wilhelmy plate method discussed in Section 2.2.3. The surface tension of the insoluble film-covered surface of liquid contained in a film trough can also be measured by the Wilhelmy plate method. For this, a sand-blasted platinum is dipped within the film and the vertical pull on it exerted by the surface tensional force may be accurately measured using the Wilhelmy balance (vide Fig. 2.3) so that π can be calculated using equation (2.8). The area A_m of the monolayer film at the film trough can be altered using a suitable movable barrier which can separate insoluble film from clean liquid to an appropriate extent as desired. Thus π can be measured as function of A_m at a given temperature. The precision of measurement may become very high by the recent use of electrobalance, thermostatic, and mechanical control.[26]

It may be pointed out that in the horizontal type of Langmuir balance, the difference γ_0 for the clear liquid (region H in Fig. 2.5) and γ of the film-covered surface (region M) is registered in the balance directly so that value of π becomes consistent with equation (2.8).

2.4. EQUILIBRIUM SPREADING PRESSURE

For the measurement of π, the insoluble substance at the liquid surface is taken only in very small amount so that all its molecules remain on the extended liquid surface in the monomolecular film. On the other hand, if one introduces a droplet of oleic acid, or a crystal of cetyl alcohol on the liquid surface of relatively small area, molecules from the solid (or liquid) nonaqueous phase will disperse in two dimensions forming insoluble monolayer until a state of saturation is reached. The condensed nonaqueous phase will be in equilibrium with the molecules present in the monolayer. The surface pressure measured at this state with the help of equation (2.8) is called the equilibrium spreading pressure (π_{eq}), which depends upon the molecular forces existing in the condensed and the monolayer phases, respectively.[26] This pressure is the characteristic of the organic substance at a fixed temperature.

For the accurate measurement[28] of the equilibrium surface pressure (π_{eq}), few small crystals of the organic substance are added to the surface of an aqueous solution taken in a circular thermostated dish of about 50 cm^2 surface area. The surface pressure is usually measured by the Wilhelmy plate method using a sand-blasted platinum plate attached to a Cahn electrobalance. Sufficient time should be allowed for the attainment of equilibrium between the crystal and the monolayer phases.

2.5. MEASUREMENT OF CONTACT ANGLE

When a liquid is placed on the surface of a solid, it remains as a drop with formation of an angle of contact between the liquid and solid phases provided the liquid does not wet the solid (vide Figs. 7.2a and 7.2b). In the current literature, various methods have been described which allow one to measure the contact angle θ at the gas–liquid–solid or liquid$_1$–solid–liquid$_2$ interfaces. This has been reviewed by Adamson.[3]

In the most convenient method of measurement, a small drop of liquid (10–20 μl) is carefully placed on the surface of a solid and the value of θ is estimated by looking through a comparator microscope fitted with a goniometer scale (vide Fig. 2.6). The sample is enclosed in a thermostat for temperature control. For the measurement of θ in the case of a liquid$_1$–solid–liquid$_2$ system, one can use a glass cuvette. A sample of solid is placed at the bottom of the cuvette. The liquid L_1 with lower density is placed over the solid. The drop of liquid L_2 is then carefully placed in contact with the solid and the contact angle is measured with the same setup as shown in Fig. 2.6. This method has been extensively used by Birdi for the measurement of the contact angle (Birdi, unpublished data).

The Wilhelmy plate technique is also used conveniently for the measurement of the contact angle with high degree of precision.[3,29] From the measurement of the adhesion tension between a paraffin-coated glass slide dipped in aqueous solution of a protein, Ghosh and co-workers[30] have estimated the contact angle for the solid–liquid system.

2.6. ADSORPTION BY POWDERED SOLID FROM BINARY SOLUTION

2.6.1. INTRODUCTION

Several types of experiments are used for the determination of the extent of adsorption at the solid–liquid interfaces. A balanced review of all these methods has been presented by Kipling.[31] In this section, principles of only few conventional methods will be presented in relation to the results and discussions included in Chapter 8.

In the most widely used analytical method, a definite amount of powdered solid is brought in contact with a definite amount of the binary solution of solid–liquid or liquid–liquid mixtures in a sealed bottle. The dry weight (w_s) of the solid and initial weights (w_1^i and w_2^i) of the two components forming the binary solution must be known. The mixture in the bottle is shaken appropriately at a constant temperature and the period of shaking required for the attainment of the adsorption equilibrium may vary from 24 hr to four days depending upon the system. The solid–liquid system is then centrifuged and the clear supernatant is chemically analyzed for weight composition (molality or mole fraction) of one of the components using a very sensitive analytical technique suitable for the particular system. The techniques used may be refractometry, interferometry, gravimetry, densitometry, volumetric

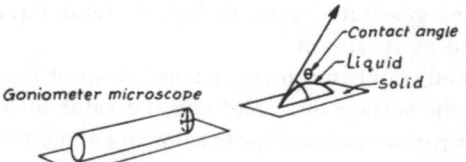

Goniometer microscope

FIGURE 2.6. Setup for contact angle measurement.

analysis, radio-tracer method, cryoscopic measurement, surface tension measurement, etc. The adsorbent, and the components of the binary mixture as well as the containers, must be free of impurities as far as possible so that the methods of analysis will be very accurate. The accuracy in the method of analysis actually controls the level of precision for the determination of the extent of adsorption Γ_2^1 (or Γ_1^2) of a component by the solid powder.

The extent of adsorption Γ_2^1 of component 2 present in the liquid mixture by a gram (or kilogram) of solid powder can be calculated from the analytical equation

$$\Gamma_2^1 = \frac{w_1^t}{1000}(m_2^t - m_2) \tag{2.9}$$

Here m_2^t and m_2 are the molalities of component 2 in the mixture, respectively, before and after adsorption, and w_1^t is the total weight of the solvent component associated per gram (or kilogram) of the solid powder. Values of m_2^t, m_2, and w_1^t may be calculated from the experimental data so that the extent of adsorption can be calculated using equation (2.9).

When the solution is dilute and the relative affinity of the solute component 2 for the solid surface is high, then w_1^t, m_2^t, and m_2 in equation (2.9) can be replaced by v^t, c_2^t, and c_2, respectively, where v^t is the total volume of the solution and c_2^t and c_2 are the respective solute concentrations in the molar scale before and after adsorption. These quantities can be determined rapidly so that Γ_2^1 can be determined in quite convenient manner.

In the case of the adsorption of dyes, ionic surfactants and proteins, respectively, on the surface of powered solid, the solvents used are aqueous solutions of various inorganic salts of fixed concentrations. Experiments on the adsorption of liquid mixture by precipitated gelatin (vide Section 8.2) is carried out after keeping the solid gel inside a cellophane bag.[32] The dispersion of the biocolloid in the presence of water in this situation does not make any problem for the estimation of the composition of the liquid components in the mixture.

For the study of the adsorption from solutions formed by two completely miscible liquid components, the concept of molality of component 2 will not be meaningful when it behaves as a solvent. Equation (2.9) in this situation cannot be used to calculate the extent of adsorption Γ_2^1. For such a case of liquid mixture, a new quantity termed the apparent adsorption Γ_2^n can be calculated from the experimental determination of the mole fractions X_2^t and X_2 before and after adsorption, respectively [vide equation (8.11)].

2.6.2. ADSORPTION FROM VAPOR MIXTURE

For the study of adsorption from binary mixture of two volatile liquid components, the experiments may be set up in a manner such that the solid

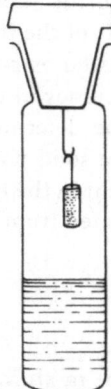

FIGURE 2.7. Adsorption from mixed vapors.

powder is in direct contact with the vapor mixture.[31] For this purpose, a stoppered tube is used (vide Fig. 2.7) in which definite amounts n_1^t and n_2^t in moles of the two liquid components are taken so that the mole fraction (X_2^t) is known. A known amount of solid powder is taken in a bucket placed in the vapor phase. When the uptake of the liquid vapor by the solid powder reaches a state of equilibrium, the total weight w of the vapor associated with the solid can be determined from the increase in weight in the bucket. One can relate w to the weights of individual component bound to the solid from the vapor phase through equation (8.30), discussed in Chapter 8. At the state of equilibrium the mole fraction X_2 of the liquid at the bottom of the tube in Fig. 2.7 is analyzed so that Γ_2^n can be estimated using equation (8.11). It has also been shown in Chapter 8 that Γ_2^n can be related to the absolute amounts of the two components bound per gram of solid powder [vide equation (8.19)]. Solving simultaneously equations (8.19) and (8.30), one can evaluate the amounts of the two components bound to the powdered solid.

From all these discussions, in Section 2.6, one finds that the extent of adsorption of a surface-active substance per gram of the powdered solid can be estimated following suitable analytical procedures. If the adsorption is to be expressed per unit surface area of the solid, then Γ_2^1 is to be divided by A_m, the specific surface area per gram of the powdered material. A_m can be measured independently by the gas-adsorption method[3] using the BET equation. There are many other methods for independent measurement of A_m discussed by Adamson[3] and others.[33]

2.7. DIRECT ANALYSIS OF ADSORPTION AT THE LIQUID INTERFACE

The extent of adsorption of a surfactant at the liquid interface can be evaluated indirectly but quite conveniently from the surface tension concentration data with the help of the Gibbs adsorption equation (2.1). Attempts have been made by several workers to estimate the extent of adsorption at the liquid interface directly from analysis. The major aim of these works is to verify different forms of the Gibbs adsorption equations. A summary of the methods has also been given by Adamson.[3]

In the microtome blade method developed by McBain and co-workers,[34,35] about one square meter of the air–liquid interface of a binary solution of thickness 0.1 mm is taken away with the help of a movable microtome blade and its weight w_t and molal concentration m_2 are determined. The molality m_2^t of the solution as a whole is known. From the measured values of w_t, m_2^t, and m_2, the extent of adsorption Γ_2^1 per unit surface area can be calculated. Γ_2^1 thus determined analytically has been compared with that evaluated with the help of the Gibbs adsorption equation (vide Table 3.1).

For the adsorption at the air–water interface, emulsion technique[36,37] can be suitably used for the estimation of Γ_2^1. Oil and aqueous solution of a surfactant or protein (of concentration c_2^t) may be emulsified to a suitable size in a blender. The stable emulsion is then allowed to cream for several hours and aqueous fluid further clarified by centrifugation, is analyzed for the concentration c_2 of the adsorbate remaining free in aqueous solution. From the difference $c_2^t - c_2$, the extent of adsorption at the oil–water interface of surface area A_m can be calculated. The total surface area of the emulsion droplets can be measured using photomicrographic and spectrophotometric methods.

The radioactive tracer method has frequently been used for the determination of the Gibbs surface excess of a surface-active component for an aqueous solution. The surface-active electrolyte component is previously labeled with ^{14}C or ^{35}S for this purpose.[3,38] Tajima et al.[39] more recently continued this study of adsorption using tritium-labeled SDS. An aqueous solution of tritiated SDS was introduced in a Teflon trough and its surface was cleaned by sweeping technique. A scintillator probe was then placed on the airside very close to the liquid interface, and counts of radioactivity for the adsorbed monolayer were made. Correction for radioactivity count due to the distribution of TSDS in the bulk phase near the interface was also made. The Gibbs surface excess was then calculated from the net count, the extent of liquid surface area, and some other parameters.[40] The whole experiment was carried out in a nitrogen atmosphere.

2.8. WATER VAPOR ADSORPTION BY BIOPOLYMERS

Adsorption of water vapor by proteins, nucleic acids, and many other biopolymers has been studied by various workers both in the absence and presence of electrolytes, urea, and sugar. The importance of such studies for

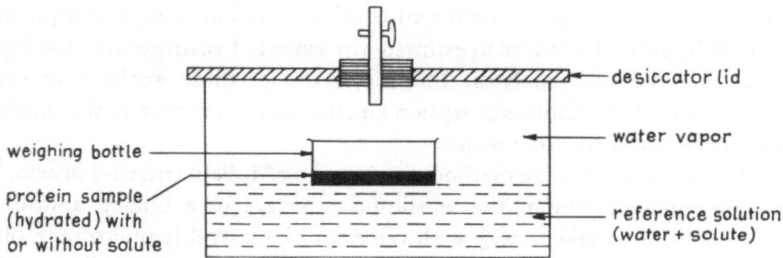

FIGURE 2.8. Experiments on water vapor adsorption by a biopolymer.

the estimation of the thermodynamic parameters related to hydration of biopolymers has been discussed extensively in Chapter 9. Bull and co-workers[41–43] have recently studied the hydration of biopolymers using the isopiestic vapor pressure method, which will be described here in brief.

A definite volume of an aqueous solution of H_2SO_4 is usually taken in a specially designed desiccator (vide Fig. 2.8) of approximately 300 ml volume. This liquid taken in the desiccator is known as a reference solution. A definite amount of the dry protein is taken in a specially designed low form weighing bottle, which is then allowed to float over the reference solution after removal of its lid. This container with the biopolymer sample is termed a sample bottle. The desiccator is closed and carefully evacuated so that exchange of water vapor between the protein sample and magnetically stirred reference solution would take place for four to seven days till isopiestic vapor pressure equilibrium is attained at a fixed temperature. The weighing bottle is removed, fitted with a lid, and quickly weighed. Let this weight be W_1. The sample bottle without its lid is then placed in a vacuum oven at 105°C for 24 hr and weighed with the lid to obtain dry weight (W_2). $W_1 - W_2$ then represents the weight of water vapor adsorbed by a given weight of the dry protein. From this, moles (n_1) of water adsorbed per mole (or per kilogram) of dry biopolymer can easily be calculated. The concentration of H_2SO_4 in the reference vessel is analytically determined so that the relative vapor pressure can be obtained at a given concentration of the acid from the literature. The relative vapor pressure may be varied between unity and zero by suitably varying the acid concentration from one to thirty normal strengths. For the study of desorption of water vapor from the biopolymer, initially some water is added to the protein sample and then the system is allowed to reach isopiestic equilibrium within the desiccator whereby water vapor is desorbed from the sample to the reference solution.

Sometimes the hydration of biopolymers is studied in the presence of a solute with the help of the isopiestic method.[42–44] These solutes used may be highly soluble inorganic salts such as LiCl, NaCl, KCl, Na_2SO_4, $CaCl_2$, etc. The organic solutes used in these studies are urea, sugar, guanidinium salts, which are also highly soluble. Here a known weight of a concentrated solution of the solute is taken in the weighing bottle (vide Fig. 2.8) so that the dry weight (w_s) of the solute in the sample bottle is initially known. A weighed amount of a biopolymer (egg albumin, serum albumin, deoxyribonucleic acid, etc.) is then taken in the sample bottle whereby it is mixed with water and electrolyte components thus forming the sample mixture. The reference solution present in the desiccator also contains usually the same solution of the solute over which the sample bottle is allowed to float. The desiccator is then closed and evacuated so that exchange of water vapor between the sample and reference systems takes place until isopiestic equilibrium is reached at a constant temperature. The weight (w_2) of the sample bottle with the lid is then taken accurately. This bottle without the lid is then placed in a vacuum oven at 105°C for 24 hr and its weight (w_1) with lid is determined again. Moles (n_1^t) of water and moles of salt (n_2^t) present per mole of biopolymer in the sample vessel can then be calculated using the equations

$$n_1^t = \frac{w_2 - w_1}{w_1 - (w_s + w_b)} \frac{M_p}{M_w} \tag{2.10}$$

$$n_2^t = \frac{w_s}{w_1 - (w_s + w_b)} \frac{M_p}{M_s} \tag{2.11}$$

Here w_b stands for the weight of the dry and empty weighing bottle. M_p is the molecular weight of the biopolymer, and M_w and M_s the molecular weights of water and salt, respectively.

After the attainment of the isopiestic equilibrium, a known weight of the reference solution is taken in a weighing bottle which is then dried at 105°C till moisture is freed. The molality m_2 of the reference solution can then be obtained from these data so that its mole ratio compositions can be calculated with the help of the relations

$$\frac{X_1}{X_2} = \frac{55.5}{m_2} \tag{2.12}$$

$$\frac{X_2}{X_1} = \frac{m_2}{55.5} \tag{2.13}$$

Here X_1 and X_2 stand for mole fractions of the solvent and solute in the reference solution, respectively. At a given mole ratio composition, the extent of adsorption, Γ_2^1, may be calculated[44] from the values of n_1^t, n_2^t, and X_1/X_2 [vide equations (9.22) and (9.23)].

Many solutes such as urea, KSCN, Gu-HCl, may decompose at 105–110°C so that these salts cannot be dried by heating. For such a system special procedures for drying required[44] for isopiestic experiments are to be followed. In an indirect but accurate isopiestic method of measurement,[43,44] the solute in the reference solution is always NaCl whereas those in the sample may be different. The method of calculation of Γ_1^2 and Γ_2^1 for such a system has also been described.[44]

The same types of isopiestic experiments have been used for the study of water vapor adsorption of powdered detergents and powdered solids (e.g., $BaSO_4$, Al_2O_3). Some results thus obtained have been discussed in Chapter 11.

2.9. DETERGENT–BIOPOLYMER BINDING FROM EQUILIBRIUM DIALYSIS

There exist various methods for the study of binding of ligands and small inorganic ions to several types of biopolymers in buffer media. These methods have been extensively reviewed by Steinherdt *et al.*[45] The dialysis equilibrium method serves as a direct procedure for the calculation of the extent of binding of ligands to proteins and nucleic acids in the aqueous solution. Results of such measurements for several systems are analyzed in detail in Chapter 10. The principle of this method will be discussed here in brief.

In an equilibrium dialysis experiment, a definite volume (V_i liter) of a buffer solution containing a definite weight (w_p) of protein is taken in a clean dialysis tubing. The bag is tightly knotted and dropped into a Pyrex glass bottle containing a definite volume (V_0 liter) of the detergent solution of known weight (w_R) of the detergent (vide Fig. 2.9). The bottle is shaken in a thermostatic shaker at a constant temperature until dialysis equilibrium for the detergent inside and outside the membrane is attained. The concentration c_R of the detergent in the outside dialysate solution is determined by a suitable analytical method. These methods may be dye partition technique, direct

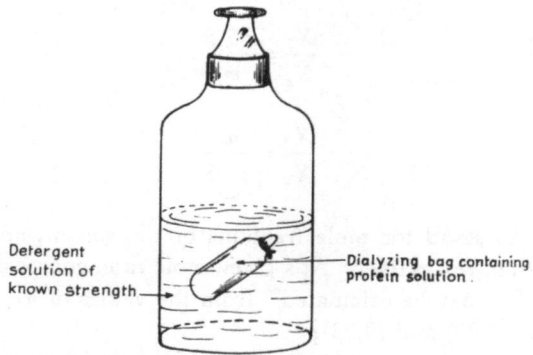

Detergent solution of known strength —

— Dialyzing bag containing protein solution .

FIGURE 2.9. Equilibrium dialysis experiment.

titration, measurement of surface tension, radio-tracer technique, and many other procedures.[45] Accuracy of the method of this analysis controls mostly the precision in the binding measurement.

The number of moles Γ_R of the detergent ions bound per mole of (or kilogram of) biopolymer unit can be calculated from the expression[46,47]

$$\Gamma_R = \frac{c'_R - c_R}{c_p} \frac{V_i + V_0}{V_i} \tag{2.14}$$

Here c_p is the molar (or weight) concentration of the biopolymer originally present in V_i liters of the solution. Also c^t_R stands for $w_R/M(V_i - V_0)$ whose value can easily be calculated. M stands for the molecular weight of the detergent.

The equilibrium dialysis techniques have close similarities with adsorption experiments discussed in Section 2.6. In the adsorption experiments, however, the adsorbents remain uniformly distributed with appropriate shaking arrangements. In binding experiments, the protein confined inside the bag is completely unable to distribute itself in the dialysate solution since the large macromolecule is impermeable to the membrane wall, whose pore size is relatively small. Small molecules of the ligand, water, and buffer components which are not bound to the macromolecule, however, distribute uniformly throughout the system provided the effect due to the osmotic pressure is neglected.

Akinrimisi *et al.*[48] have used a modified equilibrium dialysis technique for the study of binding of basic proteins histones to deoxyribonucleic acid. Here the pore size of the dialysis membrane has been increased significantly by treatment with salt so that histones become permeable to it. The molecular weight of DNA is of the order of 10^6 or more so that this biopolymer remains impermeable to the membrane wall. From the study of histone concentration in the dialysate solution at equilibrium, the extent or binding of the basic protein to DNA remaining inside the bag can easily be calculated using equation (2.14).

Equation (2.14) is quite suitable for the calculation of Γ_R when the affinity between the ligand and biopolymer is considerably high. Many solutes such as urea, guanidine hydrochloride, sugar, etc. have relatively low affinities for a biopolymer. Further, the affinity of many inorganic electrolytes to a protein is relatively lower than that of water so that the measured value of Γ_R becomes negative (vide Chapter 10). For such cases, equation (2.14) based on the determination of the molar concentrations c^t_R and c_R will not be suitable for the evaluation of Γ_R. The recent method of Bull and Breese[49] appears to be quite suitable for this purpose.

In this method, w^i_1 and w^i_2 grams of (dry) water and solute components, respectively, are mixed with w_p grams of a (dry) biopolymer within a cellophane bag of definite weight. The bag knotted appropriately is then placed in a weighing bottle containing w_1^0 and w_2^0 grams of the water and solute, respectively. The whole arrangement is shaken for 72 hr at a constant temperature whereby dialysis equilibrium is reached. The molal concentration (m_2) of the outside solution is then determined by evaporating a definite weight of the dialysate solution at 105°C so that the dry weight of the solute may be obtained. The excess amount of solute bound per mole of protein can be calculated from the equation (vide Chapter 10)

$$\Gamma_2^1 = (n_2^t - n_1^t)(X_2/X_1) \qquad (2.15)$$

where X_2/X_1 is equal to $m_2/55.51$. Also n_1^t and n_2^t are, respectively, equal to $(w_1^i + w_1^0)M_p/w_pM_1$ and $(w_2^i + w_2^0)M_p/w_pM_2$, respectively. M_1, M_2, and M_p are molecular weights of water, solute component, and protein, respectively, so that Γ_2^1 can be calculated from experimental values of n_1^t, n_2^t, and X_1/X_2.

Unlike protein, DNA is observed not to affect the surface tension of an aqueous solvent at a given temperature. Long-chain amine completely bound to DNA is also found to be surface inactive. The measurement of surface tension of the amine in the presence and absence of DNA, respectively, may lead to the calculation of the amount of the detergent bound to DNA. The method does not require the use of a dialyzing membrane and it is the case of direct measurement of adsorption of the surfactant to the surface of the DNA molecule.[46] The assumption involved is that the ligand bound to DNA is surface inactive whereas that remaining free in solution lowers the surface tension of the aqueous system. The extent of binding of amines to DNA directly determined is found to become the same as that obtained from the equilibrium dialysis experiment.[46] This direct method was originally used for the calculation of the extent of binding of various types of histone to deoxyribonucleic acid.[51]

CHAPTER 3

ADSORPTION AT LIQUID INTERFACES AND THE GIBBS EQUATION

3.1. INTRODUCTION

In Section 1.12, it was pointed out that the molecules at the boundary region of a solid in contact with a liquid possess higher energy than those present in the interior bulk regions of the condensed phases. It is a common experience that a powdered solid material is able to adsorb many gases with which the former is kept in contact at high pressure. At the interface of the solid powder, the gas molecules accumulate in excess whereby the interfacial energy is decreased. The extent of such accumulation (or adsorption) is experimentally measurable and it is found to depend on temperature, pressure, and other experimental conditions.[1] The amount of such adsorption of a gas by a solid material may be positive or zero but never negative.

When powdered solid materials such as alumina or silica are brought in contact with a liquid solvent such as water or alcohol, the particles may be solvated due to the accumulation of the solvent molecules at the solid interface. The phenomenon has similarity with the adsorption of a gas close to the region of saturated vapor pressure. If the liquid phase contains two or more components, the adsorption becomes competitive so that one of these components may preferentially accumulate at the interfacial region. Thus when a large amount of powdered charcoal is added to an aqueous solution of acetic acid and the mixture is shaken for hours, the acid is preferentially accumulated or adsorbed at the interfacial region of the solid. Concentration (c_2) of the acid solution after adsorption, on estimation, is usually found to be lower than its initial concentration (c_2^i) prior to adsorption so that the extent of positive adsorption per gram of the solid powder can be calculated. Positive adsorption of the fatty acid on the surface of charcoal powder increases with the increase of the concentration of the acid in the aqueous solution. If the same type of adsorption experiment is carried out with butanol and water in the presence of dry gelatin powder, the bulk concentration (c_2) of butanol after adsorption becomes greater than its initial concentration (c_2^i) before adsorption. The extent of adsorption will thus become negative in this case. In Fig. 8.3, the extent of this negative adsorption is observed to increase with increase of the bulk concentration of butanol in the mixture. The analytical method of measurement thus indicates that unlike gas adsorption phenomena, adsorption of one of the components from a solution may become positive or negative depending upon the nature of both the components of the solution and the adsorbent. The concepts of positive and negative adsorption will be clear when the relative adsorption at the liquid–gas, liquid–liquid, and solid-liquid interfaces is examined in terms of the Gibbs adsorption equation.[2]

3.2. ADSORPTION AT LIQUID INTERFACES

The surface tension (γ) of water at a constant temperature changes significantly with the addition of organic or inorganic solutes into the aqueous medium. In Fig. 3.1, such changes for aqueous solutions of butyric acid and sodium chloride are shown graphically. Addition of butyric acid decreases the surface tension (γ) of the solution. On the other hand, γ is observed to increase with increase of the concentration of sodium chloride in the medium.

More than one hundred year ago, Willard Gibbs[2] derived, on clear thermodynamic grounds, a differential equation relating surface tension of a solution with its solute concentration. This, in its simplest form, reads

$$\Gamma_2^1 = -\frac{c_2 d\gamma}{RT dc_2} = -\frac{1}{RT}\left(\frac{d\gamma}{d\ln c_2}\right) \tag{3.1}$$

Here Γ_2^1 is the excess amount of solute accumulated (or adsorbed) per unit surface area of the liquid and c_2 the molar (or molal) concentration of the solute component 2 in the medium. From Fig. 3.1 it appears that the slope $d\gamma/dc_2$ for the butyric acid solution is negative so that Γ_2^1 increases with increase of the concentration c_2 of the solute in the solution. Thus it is possible to calculate the extent of solute adsorption at the liquid interface from the surface tension data with the help of the Gibbs adsorption equation. Direct analytical measurement of adsorption with the help of equation (2.9) becomes unnecessary in such a case.

On the other hand, the slope of the curve for the sodium chloride solution in Fig. 3.1 becomes positive. Γ_2^1 according to equation (3.1) thus becomes negative. The magnitudes of both positive and negative adsorption of the solute have been plotted against the solute concentration in Fig. 3.2. In Section 3.13, certain problems associated with the calculation of the magnitude of Γ_2^1 for an electrolyte solution from the Gibbs equation are discussed.

The Gibbs adsorption equation is one of the most fundamental relations for the thermodynamic interpretation of positive and negative adsorption

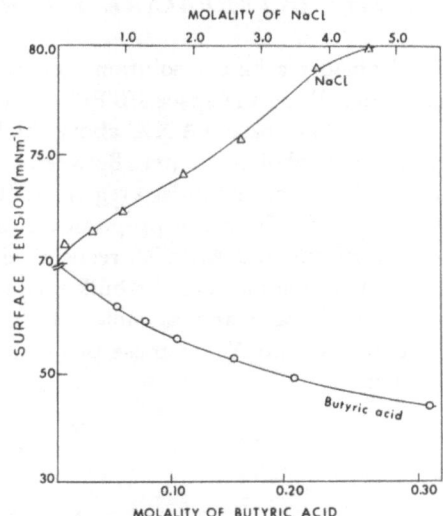

FIGURE 3.1. Plot of surface tension vs. concentration of butyric acid and NaCl.

occurring at gas–liquid, liquid–liquid, and solid–liquid interfaces. Its application to the phenomena of adsorption of gases by solid powder and of binding of organic molecules to biological polymers can lead to the evaluation of many useful parameters of thermodynamic importance. The concept of the Gibbs treatment will be discussed with clarity in Section 3.6.

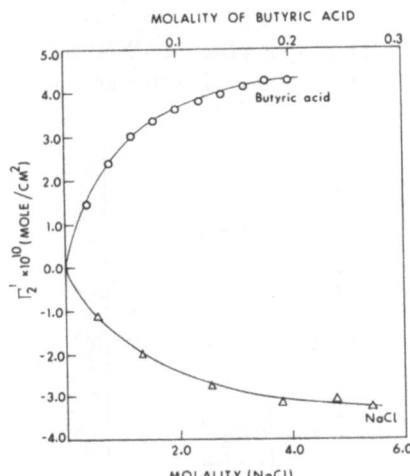

FIGURE 3.2. Plot of Γ_2^1 against molal concentration.

3.3. THE INTERFACIAL PHASE

Consider a binary solution containing two components such as alcohol and water filling the space BB'PP' of a column of 1 cm² of surface area (vide Fig. 3.3). The space AA'XX' above the liquid column remains filled with the vapors of alcohol and water. Between the bulk vapor and bulk liquid phases, there exists a thin boundary region AA'BB' whose properties may be thought to be different from the properties of either liquid or vapor present in the regions BB'PP' and AA'XX', respectively.

Let us assume that the bulk phase of the liquid BB'PP' is composed of n_1 moles of water and n_2 moles of alcohol so that the corresponding mole fractions, X_1 and X_2, of these two components are defined by the following relations:

$$X_1 = \frac{n_1}{n_1 + n_2} \tag{3.2}$$

$$X_2 = \frac{n_2}{n_1 + n_2} \tag{3.3}$$

Mole ratio compositions of the solution can be defined as

$$\frac{X_1}{X_2} = \frac{n_1}{n_2} \tag{3.4}$$

$$\frac{X_2}{X_1} = \frac{n_2}{n_1} \tag{3.5}$$

The vapor phase AA'XX' may similarly be imagined to be composed of n_1' and n_2' moles of water and alcohol, respectively, so that corresponding

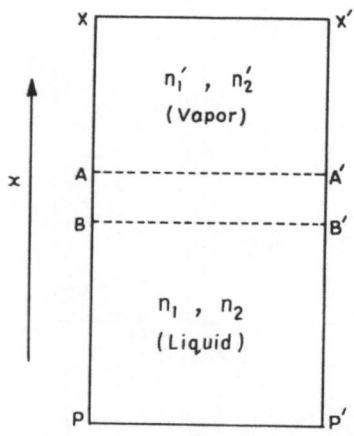

FIGURE 3.3. Composition of the liquid and vapor phases in a liquid column.

mole fractions and mole ratio compositions may be defined by the following relations:

$$X_1' = \frac{n_1'}{n_1' + n_2'} \tag{3.6}$$

$$X_2' = \frac{n_2'}{n_1' + n_2'} \tag{3.7}$$

$$\frac{X_1'}{X_2'} = \frac{n_1'}{n_2'} \tag{3.8}$$

$$\frac{X_2'}{X_1'} = \frac{n_2'}{n_1'} \tag{3.9}$$

Since alcohol is more volatile than water, we expect that

$$n_2'/n_1' > n_2/n_1 \qquad \text{or} \qquad n_1/n_2 > n_1'/n_2'$$

At phase equilibrium, any small (but macroscopic) region of AA'XX' possesses uniform composition n_2'/n_1'. Similarly the composition n_2/n_1 of the liquid phase BB'PP' is uniform throughout in the macroscopic scale. As the distance x from PP' in the vertical direction is increased (vide Fig. 3.3), the composition n_2/n_1 of the liquid remains uniform until the boundary plane BB' is reached. On increasing x further gradually, let us assume that the ratio of the moles of alcohol per mole of water begins to increase until plane AA' is reached. On increasing the distance further, from AA' inside the vapor phase, the mole ratio composition n_2'/n_1' becomes again uniform. We may represent these observations qualitatively in the form of an extended curve shown in Fig. 3.4.

Let us imagine that the surface phase AA'BB' (Fig. 3.3) is made up of $1, 2, 3, \ldots, J$ number of homogeneous layers. Each of these J layers contains Δn_1^J and Δn_2^J moles of components 1 and 2, respectively. If Δn_1 and Δn_2 represent the total number of moles of these two components in the interfacial phase AA'BB', respectively, then

$$\Delta n_1 = \Delta n_1^1 + \Delta n_1^2 + \cdots + \Delta n_1^J \tag{3.10}$$

$$\Delta n_2 = \Delta n_2^1 + \Delta n_2^2 + \cdots + \Delta n_2^J \tag{3.11}$$

Since the surface phase is inhomogeneous,

$$\frac{\Delta n_2^1}{\Delta n_1^1} \neq \frac{\Delta n_2^2}{\Delta n_1^2} \neq \cdots \neq \frac{\Delta n_2^J}{\Delta n_1^J} \tag{3.12}$$

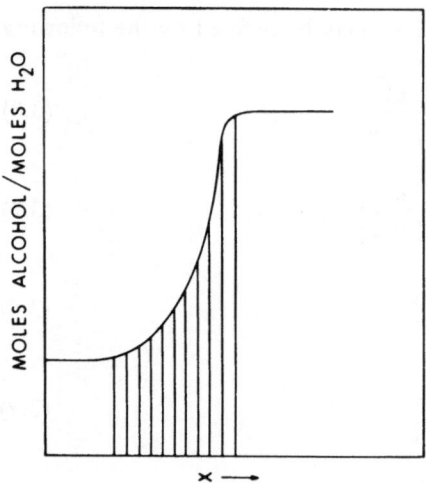

FIGURE 3.4. The interfacial region between the bulk phases.

Further since the composition of the surface phase is different from that of the bulk phase, $\Delta n_2^J/\Delta n_1^J$ (or even $\Delta n_2/\Delta n_1$ in most cases) will not be equal to either n_2/n_1 or n_2'/n_1'.

We can divide the bulk phase BB'PP' into K number of layers so that

$$n_1 = n_1^1 + n_1^2 + \cdots + n_1^K \tag{3.13}$$

$$n_2 = n_2^1 + n_2^2 + \cdots + n_2^K \tag{3.14}$$

since the composition of the bulk phase is uniform, it is legitimate to write

$$\frac{n_2}{n_1} = \frac{n_2^1}{n_1^1} = \frac{n_2^2}{n_1^2} = \cdots = \frac{n_2^K}{n_1^K} \tag{3.15}$$

The difference between equations (3.12) and (3.15) for surface and bulk phases, respectively, is noted with interest. If S and ΔS stand for the entropies of the bulk and interfacial regions of the liquid column in Fig. 3.3, then $\Delta S^J/\Delta n_1^J$ for different J layers in the interfacial phase AA'BB' are different from each other. Further, $\Delta S/\Delta n_1$ is also different from S/n_1. In all the K-layers of the bulk phase BB'PP', however, S^K/n_1^K terms become equal to S/n_1.

We can then physically define the inhomogeneous region between the two homogeneous phases in Fig. 3.4 to be the boundary region. Regions which are homogeneous are regarded as bulk phases. The inhomogeneous boundary region is frequently termed the "bound phase," "surface phase," or "interfacial phase."*

---

* In the case of the study of adsorption at solid–liquid interfaces from binary solution, the presence of surface azeotrope[3] has been recognized in a very special situation. Here, $\Delta n_2/\Delta n_1$ will be equal to n_2/n_1, but equation (3.12) may still remain valid for the inhomogeneous surface phase. This situation will be further discussed in Chapter 8.

3.4. PHYSICAL MODEL FOR THE SURFACE PHASE

It is extremely difficult to establish from the experimental investigation the exact structure of the inhomogeneous surface phase remaining in contact with homogeneous bulk solution of a liquid containing more than one component. However, many imaginary models have been proposed from time to time for the inhomogeneous surface phase during the thermodynamic treatment of such liquid system. Gibbs[2] regarded the surface phase in an idealized model as possessing zero thickness. This picture will be discussed in detail in the next section. The spatial variation of the thermodynamic properties within the surface region had been earlier discussed by van der Waals[4] and Bakker.[5] Verschaffelt[6] presented a detailed thermodynamic theory for a multicomponent surface phase system having some physical thickness. Guggenheim[7] presented rigorous derivation for the different thermodynamic parameters of the interfacial phase. He also regarded the surface phase as possessing a thickness of about 1 to 10 nm. The physical, mathematical or quasithermodynamic pictures for the surface phase have been also discussed by Tolman,[8] Fowkes,[9,10] Hansen,[11] Rusanov,[12] Eriksson,[13] Goodrich,[14] Motomura,[15] and others.[16,17]

In the previous section, the inhomogeneous features of the surface region have been discussed in terms of the variation of the mole ratio composition of component 1 (water) and component 2 (alcohol) in the interfacial region. Adam[18] and particularly Adamson[1] expressed these features in terms of the variation of the molar concentrations c_1 and c_2 for the binary liquid mixture in contract with its vapor phase. In any region of BB'PP' (Fig. 3.3) of the liquid phase, c_1 and c_2 will have constant values for a given composition of the liquid. Similarly in the bulk vapor phase AA'XX', the molar concentrations c_1' and c_2' of the components will remain constant throughout. However, concentration of the ith component in the interfacial region can neither be expressed by c_i nor c_i'. The molar concentrations in this region will vary with the distance x (vide Fig. 3.5) as shown graphically on an enlarged scale. The figure further indicates that although dc_i/dx varies between AA'BB' continuously, the nature of the gradient for each component is asymmetric and

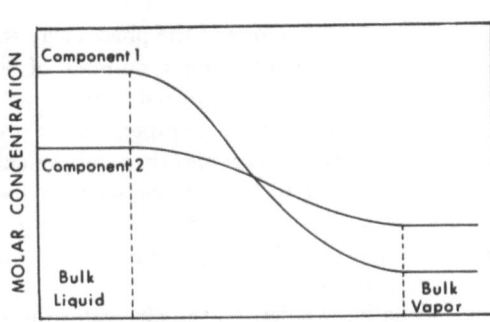

FIGURE 3.5. The gradient picture at the interface.

distinctly different from the other. Defay and Prigogine[16] showed the picture of the gradients in the surface region of a multicomponent solution by means of mechanical drawing to indicate the inhomogeneous feature of the surface phase. The picture of the concentration gradient in the surface region may be far from simple when a surface-active electrolyte RNa instead of alcohol is dissolved in water. In this situation, an electrical double layer is believed to be formed at the interfacial region, as will be discussed in Chapter 4. The molar concentration c_R of the anion in the adsorbed layer becomes maximum near charged plane AA' (Fig. 4.6). Just below this plane, the concentration of Na^+ is enormously high and that of R^- practically zero due to electrostatic interaction. Further in the x direction, towards the bulk aqueous side, the concentration of Na^+ cation decreases and that of the anion increases asymptotically until their concentrations become the same as those in bulk. Keeping all these facts in mind, it is difficult to precisely draw the exact physical picture of the interfacial region of a multicomponent solution.

In spite of these uncertainties regarding the exact nature and properties of the inhomogeneous surface phase, it is often necessary to assume a model for it containing nonelectrolytic or electrolytic solute components in the presence of a solvent component. Most of these pictures are approximate, and as such their limitations and drawbacks should be kept in mind while applying them to the experimental data. We may broadly assume that two phases α and β are in contact with an inhomogeneous surface region which is under the influence of boundary tension γ. The composition of different parts of this interfacial region in terms of mole ratio or molar concentrations of the ith component is nonuniform. Further, the compositions of i components in the surface phase are different from those of the bulk phases. The boundary tension at the liquid interface can be measured accurately as a function of the composition n_i/n_1 or c_i, so that a relation between these measurable quantities should be established on theoretical grounds.

3.5. THE BULK PHASES

In Section 3.3, one of the phases (say the α phase) is always referred to as a liquid system containing i number of components. Component 1 in the α-phase may be liquid (usually water) in which other components remain dissolved. In many cases, component 1 is present in large excess and may thus behave as a solvent. Various other components may be mixed with (or dissolved in) component 1 to form a homogeneous solution of composition expressed either by n_i/n_1 or c_i. The components 2, 3, ..., i may be solid or liquid, electrolyte or nonelectrolyte and should completely dissolve in component 1, i.e., the solvent.

Usually, γ refers to the air–water interface, the other bulk phase (say β phase) being considered to be vapor which may contain all or some of the components present in the α-phase. The vapor will contain component 1 because of the volatility of water. If a solute component (such as alcohol) is volatile, it also should be present in the vapor phase. Nonvolatile solutes such as sodium chloride, sucrose, etc. will not be present in the β phase in any significant amount. Further, the partial vapor pressure (p_i) of any of the i components present in the β phase is very low so that its molar concentration c_i' (equal to p_i/RT) will be negligibly small compared to its concentration c_i in the α-phase (vide Fig. 3.5). For many practical purposes, c_i' may be neglected compared to c_i.

The β-phase may alternatively be inert solid fine particles separating themselves from the phase α by a thin phase boundary region. Any component i cannot penetrate the interior bulk of the solid particles so that c_i' becomes equal to zero. The thin boundary region in this case contains i number of components whose concentrations are different from those in the bulk phase. The boundary tension for such a solid–liquid system is difficult to measure directly but it can be computed on the basis of certain types of analytical measurements; these will be discussed in Chapter 8 of this book.

The boundary tension γ of the oil–water system can be measured with precision as a function of the composition of the α-phase. If the component 1 is water and the other solute components are electrolytes, the latter becomes usually insoluble in the oil phase and $c_i \gg c_i'$. However, if the solute component is a nonelectrolyte (e.g., alcohol), then it distributes itself in the oil phase in considerable amount, so that c_i' may be much greater than c_i.

3.6. THE GIBBS ADSORPTION EQUATION

In his classical paper entitled "Equilibrium of Heterogeneous Substances," Gibbs[2] used for the derivation of the adsorption equation many thermodynamic notations which are altered at present. The concept involved by this original derivation of Gibbs has been discussed by Adam[18] and more recently by Defay et al.[16] using current familiar symbols of thermodynamics. A liquid column containing i number of components shown in Fig. 3.6 has been considered in the treatment of Gibbs in which two bulk phases α and β are separated from each other by surface region AA′BB′. Gibbs realized that this surface region is inhomogeneous and difficult to define so that he considered side by side an idealized liquid column (Fig. 3.7) in which the two phases α and β are separated not by an actual interfacial phase AA′BB′ but by a mathematical plane GG′ placed at an arbitrary position parallel to AA′ or BB′. In the actual system in Fig. 3.6, the bulk composition of the ith component in α and β phases are c_i and c_i', respectively. In the idealized system in Fig. 3.7, the chemical compositions of the α and β phases are imagined to remain unchanged right up to the dividing surface so that their concentrations in the two imaginary phases are also c_i and c_i', respectively. If $\bar{n}_i$ and $\bar{n}_i'$ stand for

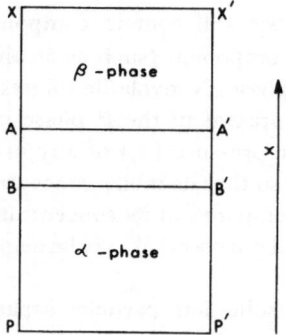

FIGURE 3.6. The liquid column in a real system.

the total moles of the ith component in the two phases of the idealized system, then the Gibbs surface excess n_i^x (defined as superficial density by Gibbs himself) of the ith component can be defined as

$$n_i^x = n_i^t - \bar{n}_i - \bar{n}_i'$$ (3.16)

For a given composition of the liquid system, the total moles of the ith component n_i^t remain fixed for the real system. However, values of $\bar{n}_i + \bar{n}_i'$ of the ideal liquid system may be varied by arbitrarily placing the dividing plane GG′ at various positions perpendicular to the x-direction. Guggenheim and Adam[19] had actually shown by numerical calculation that $\bar{n}_i + \bar{n}_i'$ for a binary liquid mixture may be greater, equal to, or less than n_i^t depending upon the position of the plane GG′. Thus the value of the surface excess n_i^x for a given composition of the liquid may become positive, zero, or negative depending upon the arbitrary placement of the plane GG′ inside the idealized system.

   In an exactly similar manner, one can define the respective (surface) excess internal energy E^x and entropy S^x by the mathematical relations

$$E^x = E' - \bar{E} - \bar{E}'$$ (3.17)

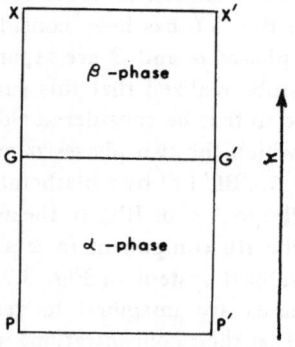

FIGURE 3.7. The liquid column in ideal system.

and

$$S^x = S' - \bar{S} - \bar{S}' \tag{3.18}$$

Here E' and S' are the total internal energy and entropy, respectively, of the system as a whole for the actual liquid system in Fig. 3.6. $\bar{E}$ and $\bar{S}$ are the internal energy and entropy respectively for the α phase of the idealized system in Fig. 3.7 whereas $\bar{E}'$ and $\bar{S}'$ are the values of these quantities in the β-phase. E^x and S^x may become positive, zero, or negative depending upon the position of the plane GG'.

The real and idealized systems are both open so that following the arguments given in Chapter 1 [vide equations (1.1) and (1.47)], it can be shown that

$$dE' = TdS' - (pdV + p'dV' - \gamma d\bar{\mathscr{A}}) + \mu_1 dn_1' + \mu_2 dn_2' + \cdots + \mu_i dn_i' \tag{3.19}$$

Here V and V' are the actual volumes of the α and β phases of the real system and p and p' are the respective pressures. Here $\bar{\mathscr{A}}$ stands for the area of the liquid surface and γ the surface energy per unit area. Since the volume V_s of the interfacial region AA'BB' in the actual system is negligibly small, total volume V' of the system is taken to be equal to $V + V'$. If the surface is almost planar, then p should be equal to p' from the standpoint of the mechanical equilibrium (vide Section 1.13). $pdV + p'dV'$ will thus become equal to pdV'.

The changes in the internal energy for idealized phases α and β (Fig. 3.7) may similarly be expressed as follows:

$$d\bar{E} = Td\bar{S} - pd\bar{V} + \mu_1 d\bar{n}_1 + \cdots + \mu_i d\bar{n}_i \tag{3.20}$$

and

$$d\bar{E}' = Td\bar{S}' - p'd\bar{V}' + \mu_1 d\bar{n}_1' + \cdots + \mu_i d\bar{n}_i' \tag{3.21}$$

Here $\bar{V}$ and $\bar{V}'$ are the volumes of the α and β phases in the idealized system so that $\bar{V} + \bar{V}'$ is equal to V' since the total volume of the liquid in Figs. 3.6 and 3.7 is assumed to remain the same. Since p is equal to p' in the ideal system also, $pd\bar{V} + p'd\bar{V}'$ is equal to pdV'. In the real system, the contribution due to the change of the surface energy, $\gamma d\bar{\mathscr{A}}$, is included as an additional work. Such contribution is absent in the idealized system containing only two bulk phases without existence of any physical interface. One may then subtract the sum of equations (3.20) and (3.21) from equation (3.19) whereby the following relations are obtained:

$$d(E^t - \bar{E} - \bar{E}') = Td(S^t - \bar{S} - \bar{S}') + \gamma d\bar{\mathcal{A}} + \mu_1 d(n_1^t - \bar{n}_1 - \bar{n}_1')$$
$$+ \cdots + \mu_i d(n_i^t - \bar{n}_i - \bar{n}_i') \qquad (3.22)$$

or

$$dE^x = TdS^x + \gamma d\bar{\mathcal{A}} + \mu_1 dn_1^x + \cdots + \mu_i dn_i^x \qquad (3.23)$$

Following the principle of the Gibbs integration discussed in Chapter 1 [vide equation (1.10)], one can integrate this differential equation by varying only the extensive properties E, S, $\bar{\mathcal{A}}$, n_i^x, etc. K times, keeping the intensive parameters T, γ, μ_i, etc. constant so that

$$E^x = TS^x + \gamma\bar{\mathcal{A}} + \mu_1 n_1^x + \cdots + \mu_i n_i^x \qquad (3.24)$$

Equation (3.24) may be differentiated in general so that

$$dE^x = TdS^x + \gamma d\bar{\mathcal{A}} + \sum_i \mu_i dn_i^x + \sum_i n_i^x d\mu_i$$
$$+ \bar{\mathcal{A}}d\gamma + S^x dT \qquad (3.25)$$

Combining equations (3.23) and (3.25), we obtain

$$-\bar{\mathcal{A}}d\gamma = S^x dT + \sum_i n_i^x d\mu_i \qquad (3.26)$$

Let S_σ^x and Γ_i^x stand for surface excess entropy and moles of the ith component per unit surface area, respectively. Then

$$S_\sigma^x = \frac{S^x}{\bar{\mathcal{A}}} \qquad (3.27)$$

$$\Gamma_i^x = \frac{n_i^x}{\bar{\mathcal{A}}} \qquad (3.28)$$

and

$$-d\gamma = S_\sigma^x dT + \Gamma_1^x d\mu_1 + \Gamma_2^x d\mu_2$$
$$+ \cdots + \Gamma_i^x d\mu_i \qquad (3.29)$$

This equation bears similarity with the Gibbs–Duhem equation for the bulk liquid system. Values of S_σ^x, Γ_1^x, $\Gamma_2^x, \ldots, \Gamma_i^x$ are not very meaningful since the magnitudes and signs of these quantities are dependent on the arbitrary position x of the plane GG'.

In order to make this equation more meaningful, Gibbs further pointed out that the position of the plane may be shifted parallel to GG' along the x-direction and fixed in a particular location when n_1^i becomes equal to $\bar{n}_1 + \bar{n}_1'$. Under this extrathermodynamic condition, n_1^x (or Γ_1^1 by convention) become zero. Equation (3.29) will then be converted to the form

$$-d\gamma = S_{.\sigma}^1 dT + \Gamma_2^1 d\mu_2 + \cdots + \Gamma_i^1 d\mu_i$$

$$= S_\sigma^1 dT + \sum_{i=1}^{i} \Gamma_i^1 d\mu_i \qquad (3.30)$$

Γ_2^1, $\Gamma_3^1, \ldots, \Gamma_i^1$ and S_σ^1 are now all defined as relative surface excess quantities per unit surface area when the surface excess Γ_1^1 of component 1 becomes zero by appropriate fixation of the plane. For a two-component system ($i = 2$) at constant temperature, Γ_2^1 becomes equal to $-d\gamma/d\mu_2$ which will lead to equation (3.1) for dilute solution. The expression Γ_2^1 thus becomes equal to the amount of component 2 accumulated in excess (adsorbed) with the assumption that similar accumulation of component 1 is zero. In an alternative procedure, one may fix the mathematical plane GG' arbitrarily at another position such that n_2^i becomes equal to $\bar{n}_2 + \bar{n}_2'$ so that Γ_2^x (now termed as Γ_2^2) becomes zero. The Gibbs equation will now read as

$$-d\gamma = S_\sigma^2 dT + \Gamma_1^2 d\mu_1 + \cdots + \Gamma_i^2 d\mu_i$$

$$= S_\sigma^2 dT + \sum_{i=1}^{i} \Gamma_i^2 d\mu_i \qquad (3.31)$$

Here S_σ^2, Γ_1^2, $\Gamma_3^2, \ldots$, etc., are relative excesses when Γ_2^2 is zero. It may be pointed out here that Γ_i^1 and Γ_i^2 are totally different quantities but they are related to each other by certain equations to be discussed later (Section 3.11).

3.7. RELATIVE SURFACE EXCESSES

Following Defay *et al.*,[16] it will now be shown that the relative surface excesses are more meaningful than the arbitrary surface excess quantities Γ_i^x. The Gibbs–Duhem equations for the idealized liquid system containing α and β phases may be written as

$$-\bar{V}dp + \bar{S}dT + \bar{n}_1 d\mu_1 + \cdots + \bar{n}_i d\mu_i = 0 \qquad (3.32)$$

and

$$-\bar{V}'dp + \bar{S}'dT + \bar{n}_1'd\mu_1 + \cdots + \bar{n}_i'd\mu_i = 0 \tag{3.33}$$

T, p, μ_i must be the same for α and β phases at equilibrium. Eliminating dp from both sides of equations (3.32) and (3.33), it may be shown

$$-d\mu_1 = \frac{s - s'}{c_1 - c_1'}\,dT + \frac{c_2 - c_2'}{c_1 - c_1'}\,d\mu_2 + \cdots + \frac{c_i - c_i'}{c_1 - c_1'}\,d\mu_i \tag{3.34}$$

where s and s' are entropy densities $\bar{S}/\bar{V}$ and $\bar{S}'/\bar{V}'$, respectively. Inserting these values of $d\mu_1$ in the generalized Gibbs equation (3.29), it may be shown that

$$-d\gamma = \left(S_\sigma^x - \Gamma_1^x \frac{s - s'}{c_1 - c_1'}\right)dT + \left(\Gamma_2^x - \Gamma_1^x \frac{c_2 - c_2'}{c_1 - c_1'}\right)d\mu_2$$

$$+ \cdots + \left(\Gamma_i^x - \Gamma_1^x \frac{c_i - c_i'}{c_1 - c_1'}\right)d\mu_i \tag{3.35}$$

The forms of equations (3.35) and (3.30) are exactly similar. Unlike equation (3.30), there is no necessity to fix the GG′ plane at Γ_1^x equal to zero arbitrarily in deriving equation (3.35). We can then set the following relations, after comparing these two equations:

$$S_\sigma^1 = S_\sigma^x - \Gamma_1^x \frac{s - s'}{c_1 - c_1'} \tag{3.36}$$

$$\Gamma_2^1 = \Gamma_2^x - \Gamma_1^x \frac{c_2 - c_2'}{c_1 - c_1'} \tag{3.37}$$

$$\Gamma_i^1 = \Gamma_i^x - \Gamma_1^x \frac{c_i - c_i'}{c_1 - c_1'} \tag{3.38}$$

If the relative excesses are written in this manner, then it is clear that their values should not depend on the arbitrary fixation of the plane GG′. Their values are definite for any chosen position of x for the Gibbs dividing plane.

Thus although values of S_σ^x, Γ_i^x, etc. are dependent on the position of x, the relative excesses S_σ^1, Γ_i^1, etc. are independent of it. The relative quantities are thus definite also for the real system presented in Fig. 3.6. This point has been further established in Section 3.8.

3.8. TOTAL CONCENTRATION AND SURFACE EXCESS

Since the volume of the surface phase in the idealized system is zero we can put V' equal to $\bar{V} + \bar{V}'$. Further the bulk concentrations of c_i and c_i' and entropy densities s and s' for the bulk phases of the idealized and real liquid systems presented in Figs 3.6 and 3.7 are same. Following Defay et al.,[16] it may be shown from the mass balance equation (3.16) that

$$n_i^x = n_i^t - \bar{V}c_i - \bar{V}'c_i'$$

$$= n_i^t - V^t c_i + \bar{V}'(c_i - c_i') \qquad (3.39)$$

and

$$n_1^x = n_1^t - V^t c_1 + \bar{V}'(c_1 - c_1') \qquad (3.40)$$

Eliminating $\bar{V}'$ from equations (3.39) and (3.40) and dividing by $\bar{\mathscr{A}}$, one gets the relation

$$\frac{1}{\bar{\mathscr{A}}}\left[n_i^x - n_1^x \frac{c_i - c_i'}{c_1 - c_1'} \right] = \frac{1}{\bar{\mathscr{A}}}\left[\left(n_i^t - n_1^t \frac{c_i - c_i'}{c_1 - c_1'} \right) - V^t\left(c_i - c_1 \frac{c_i - c_i'}{c_1 - c_1'} \right) \right] \qquad (3.41)$$

Combining equation (3.28) with relation (3.38), and then comparing the resulting equation with relation (3.41), it can be shown that

$$\Gamma_i^1 = \frac{1}{\bar{\mathscr{A}}}\left[\left(n_i^t - n_1^t \frac{c_i - c_i'}{c_1 - c_1'} \right) - V^t\left(c_i - c_1 \frac{c_i - c_i'}{c_1 - c_1'} \right) \right] \qquad (3.42)$$

Since all the terms in the right side of equation (3.42) may also refer to the real liquid system presented in Fig. 3.6, the concept of Γ_i^1 is real, meaningful, and independent of the idealized picture of the liquid system presented in Fig. 3.7. It can similarly be shown by combining the entropy balance equation

$$S^x = S^t - \bar{V}s - \bar{V}'s' \qquad (3.43)$$

with the mass balance equation (3.40) that

$$S_\sigma^1 = \frac{1}{\bar{\mathscr{A}}}\left[\left(S^t - n_1^t \frac{s - s'}{c_1 - c_1'} \right) - V^t\left(s - c_1 \frac{s - s'}{c_1 - c_1'} \right) \right] \qquad (3.44)$$

All the terms on the right-hand side of equation (3.44) are meaningful in terms of the properties of the real liquid system in contact with its vapor.

3.9. ADSORPTION AND SURFACE EXCESS

For the adsorption of solutes at air–water or solid–water interfaces, or adsorption of an electrolyte at an oil–water interface, $c_i \gg c_i'$ and $s \gg s'$. Further c_i/c_1 and s/c_1 for the α-phase may be written as n_i/n_1 and S/n_1 so that equations (3.42) and (3.44) have been shown[16,20,21] to assume the following simple forms:

$$\Gamma_i^1 = \frac{1}{\mathscr{A}}\left[n_i^t - n_1^t \frac{n_i}{n_1} \right] \tag{3.45}$$

and

$$S_\sigma^1 = \frac{1}{\mathscr{A}}\left(S^t - n_1^t \frac{S}{n_1} \right) \tag{3.46}$$

Here n_1 and n_i are moles of components 1 and i present in the α-phase of the real system and S is the entropy of this phase. All quantities on the right-hand side of these equations are the properties of the real liquid system (Fig. 3.6). As we will see in Chapter 4, these equations are extremely useful in understanding the surface excess properties of the systems containing many electrolyte components.

Equation (3.45) may be written for $i = 2$ in the form

$$\Gamma_2^1 = \frac{n_1^t}{\mathscr{A}}\left(\frac{n_2^t}{n_1^t} - \frac{n_2}{n_1} \right) \tag{3.47}$$

If m_2^t is the total molality of component 2 (or component i) in the whole liquid system and m_2 the molality of the same component in the bulk phase after accumulation of component 2 in excess at the interfacial phase, then n_2^t/n_1^t and n_2/n_1 become equal to $m_2^t M_1/1000$ and $m_2 M_1/1000$, respectively. M_1 is the molecular weight of component 1. n_1^t in equation (3.47) may be replaced by w_1^t/M_1, where w_1^t is the total weight of component 1 in the liquid system. Replacing all these terms in equation (3.47) in this manner, relation (3.48) is obtained

$$\Gamma_2^1 = \frac{1}{\mathscr{A}}\left[\frac{w_1^t}{1000}(m_2^t - m_2) \right] \tag{3.48}$$

If the solution is not too concentrated, m_2^t, m_2, and w_1^t may be replaced by c_2^t, c_2, and V^t. Here c_2^t and c_2 are total and bulk molar concentrations of the component 2, respectively, and V^t is the total volume of the real liquid system so that

$$\Gamma_2^1 = \frac{1}{\mathcal{A}} \left[\frac{V'}{1000} (c_2^t - c_2) \right] \tag{3.49}$$

The term inside the brackets in equation (3.48) is the analytical formula (2.9) for calculating the amount of a solute component adsorbed at a definite amount of surface formed by dispersing a definite amount of solid powder or nonaqueous liquid droplets in the aqueous liquid medium. For the solid powder, $\bar{\mathcal{A}}$ may be measured by separate experiments, so that Γ_2^1 can be evaluated analytically from the estimation of the concentration of the component in the bulk before and after adsorption. Adsorption or excess accumulation of component 2 at the air–water or oil–water interfaces has been directly measured for aqueous solutions of various solutes by the microtome blade method,[22,23] the emulsion method,[24] and radioactive tracer methods[25,26] using essentially equation (3.49). Thus we find that the operational definition for the relative surface excess as expressed by equation (3.49) may actually agree in form with the thermodynamic expression for the adsorption of a solute from a solution with some reasonable approximations.

For a two-component system, Γ_2^1 can be indirectly calculated with the help of equation (3.1), from the determination of the slope $d\gamma/d \ln c_2$. Some of the calculated values of Γ_2^1 from the direct analytical method and the indirect surface tension method have been compared in Table 3.1. The agreement of the data establishes the correct thermodynamic basis for the derivation of the Gibbs equation. Further, it gives firm support for the thermodynamic definition of the adsorption in terms of the surface excess quantities.

TABLE 3.1. Comparison of Γ_2^1 Calculated from the Gibbs Equation with that Obtained from the Direct Experimental Method[22,23]

Component	Concentration (molality)	Γ_2^1 Microtome $\times 10^{10}$ (mol/cm^2)	Γ_2^1 Gibbs $\times 10^{10}$ (mol/cm^2)
p-Toluidine	0.0187	5.7	4.9
p-Toluidine	0.0164	4.3	4.6
Phenol	0.2179	4.4	5.1
Caproic acid	0.0223	5.9	5.4
Caproic acid	0.0259	4.4	5.6
Caproic acid	0.0452	5.3	5.4
Hydrocinnamic acid	0.0100	3.7	3.4
Hydrocinnamic acid	0.0300	3.6	5.3

3.10. ABSOLUTE COMPOSITION OF THE INTERFACIAL PHASE

Although equation (3.45) for the relative excess is quite useful in understanding the analytical aspect of the adsorption phenomenon in a real liquid system, this as well as relation (3.38) for Γ_i^1 cannot give any effective information about the inhomogeneous phase AA'BB' (Fig. 3.6). This inhomogeneous phase is composed of i components each of amount Δn_i, so that values of Δn_i may be zero or positive. The absolute value of Δn_i cannot be negative. In contrast to this, Γ_i^1 may be positive, zero, or negative depending upon the magnitude of c_2^t and c_2 in equation (3.49). Chattoraj[21] has pointed out that it is quite logical to write equation (3.50) for the real system as

$$n_i^t = n_i + n_i' + \Delta n_i$$

$$= c_i V + c_i' V' + \Delta n_i \tag{3.50}$$

and

$$n_1^t = c_1 V + c_1' V' + \Delta n_1 \tag{3.50a}$$

where n_i and n_i' are total moles of the component i present in the α- and β-phases BB'PP' and AA'XX' (Fig. 3.6), respectively. As before, V and V' stand for the total volumes, respectively, of BBPP' and AA'XX'. Unlike equation (3.16), this equation for the real system does not depend on the distance x of the dividing plane in the idealized system (Fig. 3.7). Inserting values of n_i^t and n_1 from (3.50) and (3.50a) in equation (3.42) and putting V' equal to $V + V'$ as before after neglecting volume AA'BB' of the interfacial phase, it has been shown by Chattoraj[21] that

$$\Gamma_i^1 = \frac{1}{\mathscr{A}} \left[\Delta n_i - \Delta n_1 \frac{c_i - c_i'}{c_1 - c_1'} \right] \tag{3.51}$$

If $c_i \gg c_i'$ and $c_1 \gg c_1'$, then one can further simplify and write

$$\Gamma_i^1 = \frac{1}{\mathscr{A}} \left[\Delta n_i - \Delta n_1 \frac{n_i}{n_1} \right] \tag{3.52}$$

We are thus able now to relate the experimentally measurable relative surface excess Γ_i^1 with the absolute compositions Δn_i and Δn_1 of the interfacial phase AA'BB'. If Δn_i moles of component i is present in the bulk, it should have been associated with $\Delta n_1(n_i/n_1)$ moles of component 1. In the surface phase having $\mathscr{A}$ cm^2 facial area, Δn_1 moles of component 1 is associated with Δn_i moles of component i. The quantity $\Delta n_i - \Delta n_1(n_i/n_1)$ in equation (3.52) represents the excess amount of component i bound to Δn_1 moles of component 1 at the interfacial phase. Its value may be positive or negative.

One can similarly write for the real system

$$S^t = S + S' + \Delta S$$

$$= sV + s'V' + \Delta S \qquad (3.53)$$

Here ΔS is the entropy associated with the interfacial phase AA'BB'. S and S' are the entropies associated with α and β phases, respectively. Combining equations (3.53) in a similar manner with equation (3.44), it may be shown that

$$S_\sigma^1 = \frac{1}{\mathscr{A}}\left(\Delta S - \Delta n_1 \frac{s - s'}{c_1 - c_1'} \right) \qquad (3.54)$$

For systems where $s \gg s'$ and $c_1 \gg c_1'$, we get

$$S_\sigma^1 = \frac{1}{\mathscr{A}}\left(\Delta S - \Delta n_1 \frac{S}{n_1} \right) \qquad (3.55)$$

Like Γ_i^1, S_σ^1 also represent excess entropy associated with Δn_1 moles of component 1.

3.11. BINARY MIXTURE OF LIQUIDS

The surface tension γ of water in the presence of various organic and inorganic additives has been measured extensively by one of the precise experimental procedures discussed in Chapter 2. In Fig. 3.8, γ of the binary

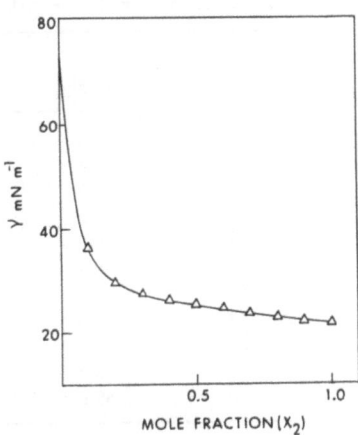

FIGURE 3.8. Plot of surface tension against mole fraction of ethanol in ethanol–water mixture.

mixture of alcohol and water has been plotted against the mole fraction of alcohol in the mixture at temperature 25°C using the results presented by Guggenheim and Adam.[19] The relative excess Γ_2^1 for alcohol at constant temperature can be calculated with the help of the Gibbs equation (3.30) which now reads

$$\Gamma_2^1 = -\frac{d\gamma}{d\mu_2} \tag{3.56}$$

The chemical potential μ_2 is related to the activity a_2 of alcohol by the equation

$$\mu_2 = \mu_2^0 + RT \ln a_2 \tag{3.57}$$

According to Raoult's law, a_2 is equal to p_2/p_2^0, where p_2 is the partial vapor pressure of this component in the mixture and p_2^0 the vapor pressure of the pure component which is constant at a constant temperature. Differentiating equation (3.57) therefore, $d\mu_2$ becomes equal to $RTd \ln p_2$ so that equation (3.56) now assumes the form[19]

$$\Gamma_2^1 = -\frac{p_2}{RT}\frac{d\gamma}{dp_2} \tag{3.58}$$

The partial vapor pressures p_2 for various mixtures of ethyl alcohol and water had been computed from the gas analysis of vapor mixtures in contact with the binary liquid system. Values of γ, X_2, and p_2 for the binary mixture are also given in Table 3.2. In Fig. 3.9, a plot of γ versus p_2 for ethyl alcohol is shown. From the slope $d\gamma/dp_2$ of the plot, Γ_2^1 for various values of X_2 (or p_2)

TABLE 3.2. Relative Adsorption of Ethyl Alcohol–Water Mixture[19]

$X_2 = 1 - X_1$	p_2 (mm Hg)	γ (mN/m)	$\Gamma_2^1 \times 10^{10}$ (mol/cm^2)	$\Gamma_1^2 \times 10^{10}$ (mol/cm^2)
0	0.0	72.2	0	0
0.1	17.8	36.4	6.33	−56.9
0.2	26.8	29.7	6.42	−25.6
0.3	31.2	27.6	5.75	−13.4
0.4	34.2	26.3	5.07	−7.60
0.5	36.9	25.4	4.25	−4.25
0.6	40.1	24.6	3.40	−2.26
0.7	43.9	23.8	2.56	−1.10
0.8	48.3	23.2	2.40	−0.60
0.9	53.3	22.6	2.25	−0.25
1.0	59.0	22.0	0	0

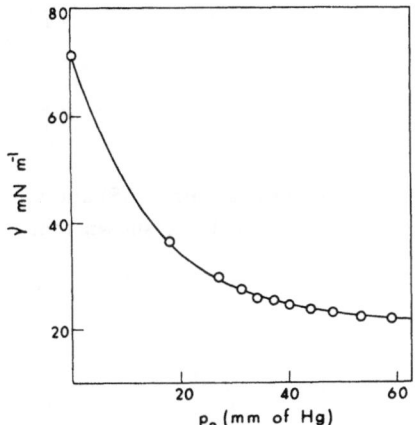

FIGURE 3.9. Plot of surface tension against the vapor pressure of ethanol over ethanol–water mixture.

for the binary solution had been calculated[19] using the Gibbs equation (3.58). These values of Γ_2^1 are also included in Table 3.2.

From equations (3.38), (3.45), and (3.52), Γ_2^1 may be written in various equivalent forms:

$$\Gamma_2^1 = \Gamma_2^x - \Gamma_1^x \frac{n_2}{n_1} \tag{3.59}$$

$$\Gamma_2^1 = \frac{1}{\mathscr{A}}\left(n_2^s - n_1^s \frac{n_2}{n_1} \right) \tag{3.60}$$

$$\Gamma_2^1 = \frac{1}{\mathscr{A}}\left(\Delta n_2 - \Delta n_1 \frac{n_2}{n_1} \right) \tag{3.61}$$

In the original derivation of the Gibbs adsorption equation (3.30), Γ_2^1 is the relative surface excess when the dividing plane GG′ is placed in Fig. 3.7 in such a manner that Γ_1^1 becomes zero. If the dividing plane is placed in such a manner that Γ_2^2 becomes zero for the binary mixture, then from equation (3.31) we find, at constant temperature,

$$-d\gamma = \Gamma_1^2 d\mu_1$$

$$= RT\Gamma_1^2 d \ln p_1 \tag{3.62}$$

Thus from the plot of γ versus p_1, it is also possible to calculate relative surface excess Γ_1^2. However, following the same procedure used for the derivation of equations (3.59), (3.60), and (3.61) it may easily be shown that

$$\Gamma_1^2 = \Gamma_1^x - \Gamma_2^x \frac{n_1}{n_2} \tag{3.63}$$

$$\Gamma_1^2 = \frac{1}{\mathscr{A}} \left(n_1^t - n_2^t \frac{n_1}{n_2} \right) \tag{3.64}$$

$$\Gamma_1^2 = \frac{1}{\mathscr{A}} \left(\Delta n_1 - \Delta n_2 \frac{n_1}{n_2} \right) \tag{3.65}$$

Combining equations (3.65) and (3.61) or (3.63) and (3.59) or even (3.64) and (3.60), it can easily be shown that

$$n_1 \Gamma_2^1 + n_2 \Gamma_1^2 = 0 \tag{3.66}$$

or

$$\Gamma_1^2 = -\frac{n_1}{n_2} \Gamma_2^1$$

$$= -\frac{X_1}{X_2} \Gamma_2^1 \tag{3.67}$$

Equation (3.67) first deduced by Guggenheim and Adam[19] indicates that the relative excesses Γ_2^1 and Γ_1^2 are not independent quantities but are related to each other quantitatively. If the value of one is determined at a given composition of the liquid, the value of the other is automatically fixed. The two values of excesses are also opposite in sign. Using equation (3.67), values of Γ_2^1 for various values of Γ_1^2 have been calculated. These are also included in Table 3.2.

It has been shown by Chattoraj and Moulik[20] that Γ_1^2 for alcohol–water mixture varies linearly with n_1/n_2 (or X_1/X_2) for a wide range of composition of the liquid (vide Fig. 3.10). According to equation (3.65), the slope and intercept of this plot are equal to Δn_1 and Δn_2. These two quantities represent the total amount of the components 1 and 2 forming the interfacial phase AA'BB' (vide Fig. 3.6) of unit surface area so that $\mathscr{A}$ becomes unity. These values are also given in Table 3.3.

Values of Γ_2^1 and Γ_1^2 for water–methanol, water–propanol, water–acetone, water–pyridine, water–formic acid, and water–glycerol mixtures have been similarly evaluated from the respective literature values of γ, X_2, and p_2 at a fixed temperature.[20] Γ_1^2 for these mixtures plotted against n_1/n_2 (or X_1/X_2) in Figs. 3.10, 3.11, and 3.12 are found to be linear in a wide range of the liquid composition. Values of Δn_1 and Δn_2 thus evaluated from the slope and intercept of the linear plot for various systems using equation (3.65) have been presented in Table 3.3. It is apparent that values of Δn_2 for all these liquid mixtures are positive and their magnitudes lie in the range of 10^{-9} to 10^{-10} moles cm^{-2}. Δn_1 is also positive and the values of Δn_1 given in Table 3.3 range between 10^{-9} and 10^{-11} moles cm^{-2}.

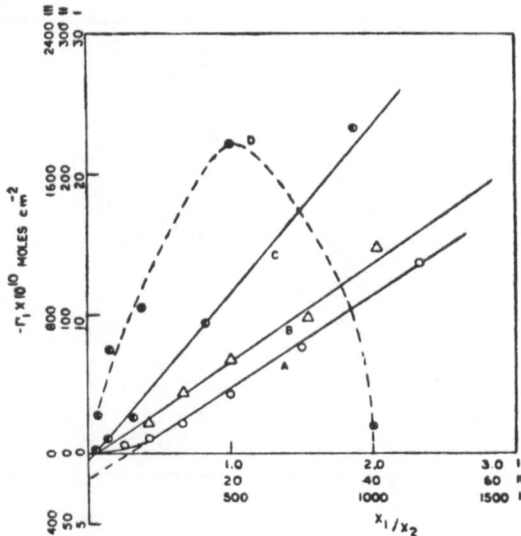

FIGURE 3.10. Surface excess Γ_1^2 (written as Γ_1) of water[20] as a function of mole ratio composition. (A) Ethanol–H_2O (scale I–I); (B) methanol–H_2O (scale I–I); (C) propanol–H_2O, low X_1/X_2 (scale II–II), and (D) propanol—H_2O, high X_1/X_2 (scale III–III). (Reproduced with permission from the *Indian Journal of Chemistry*.)

It may be mentioned here that Γ_1^x and Γ_2^x in equations (3.63) or $n_1^t/\mathscr{A}$ and $n_2^t/\mathscr{A}$ in equations (3.64) cannot represent the slope and intercept of the linear plots shown in Figs. 3.10, 3.11, and 3.12. This is because of the fact that n_1^t and n_2^t are equal to $\Delta n_1 + n_1 + n_1'$ and $\Delta n_2 + n_2 + n_2'$, respectively, and can never remain constant when n_1/n_2 is varied. From equation (3.16) it is obvious that Γ_1^x and Γ_2^x also include the terms n_1' and n_2', so that these two excess terms must always vary with n_1/n_2. Δn_1 and Δn_2 terms in equation (3.65) represent the composition of the interfacial phase AA'BB' which may

TABLE 3.3. Values of Δn_1 and Δn_2 from Linear Plots[20]

Liquid system	$\Delta n_2 \times 10^{10}$ (mol/cm^2)	$\Delta n_1 \times 10^{10}$ (mol/cm^2)	σ_2 (nm^2/molecule)
1. Pyridine + H_2O	3.72 ± 0.01	0.20 ± 0.15	0.444
2. Formic acid + H_2O	3.87 ± 0.07	9.38 ± 0.58	0.188
3. Methyl alcohol + H_2O	6.89 ± 0.13	0.44 ± 0.23	0.242
4. Ethyl alcohol + H_2O	6.60 ± 0.20	1.91 ± 0.34	0.224
5. Propyl alcohol + H_2O	5.98 ± 0.47	4.35 ± 0.68	0.200
6. Acetone + H_2O	3.40	5.00	0.345
7. Glycerol + H_2O	0.89	0.60	1.81

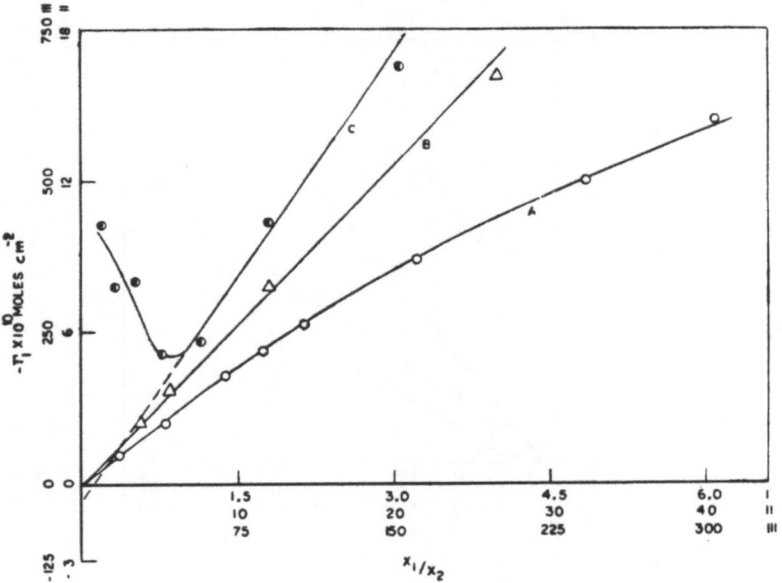

FIGURE 3.11. Surface excess Γ_1^2 of water[20] (written as Γ_1) as function of the mole ratio composition X_1/X_2. (A) Acetone–H_2O, low X_1/X_2 (scale I–I); (B) acetone–H_2O, high X_1/X_2 (scale III–III); and (C) glycerol–H_2O (scale II–II). (Reproduced with permission from the *Indian Journal of Chemistry*.)

physically remain unchanged under certain circumstances when n_1/n_2 is varied. The linear plot thus indicates that either (3.65) or the identical (3.61) are the most fundamental equations relating surface excess with the actual composition of the surface phase. Either of these equations contains two unknown parameters both of which can be evaluated from the linear plot shown in Figs. 3.10, 3.11, and 3.12. However, for a given value of Γ_1^2 and X_1/X_2, the two unknown parameters may be given numerous sets of values. These arbitrary values will be inconsistent in terms of the linear plot.

We can also place an imaginary plane LL′ within AA′BB′ so that the interfacial layer will now be divided into two sublayers 1 and 2. Δn_1 and Δn_2 become, respectively, equal to $\Delta n_1^1 + \Delta n_1^2$ and $\Delta n_2^1 + \Delta n_2^2$. Equation (3.65) for unit surface area will then read[20]

$$\Gamma_1^2 = \left(\Delta n_1^1 - \Delta n_2^1 \frac{n_1}{n_2} \right) + \left(\Delta n_1^2 - \Delta n_2^2 \frac{n_1}{n_2} \right) \tag{3.68}$$

The constants obtained from the slope and intercept of the linear plot shown in Figs. 3.10, 3.11, and 3.12 cannot be identified with Δn_1^1 and Δn_2^1, respectively, since $\Delta n_1^2/\Delta n_2^2 \neq n_1/n_2$ according to equations (3.12), so that the

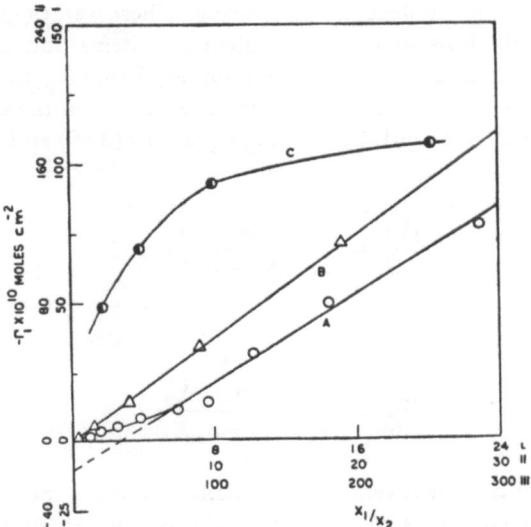

FIGURE 3.12. Surface excess Γ_1^2 of water[20] (written as Γ_1) as a function of the mole ratio composition X_1/X_2. (A) Formic acid–H_2O, low X_1/X_2 (scale I–I); (B) pyridine–H_2O (scale II–I); and (C) formic acid–H_2O, high X_1/X_2 (scale III–II). (Reproduced with permission from *Indian Journal of Chemistry.*)

second term in the parentheses is not equal to zero. The two constants of the linear plot must therefore represent the complete compositions Δn_1 and Δn_2 of the physically defined inhomogeneous interfacial phase.[20,21] For the state of the system containing surface azeotrope,[3] some clarification of equations (3.61), (3.65), and (3.68) will be further necessary. This will be discussed in Chapter 8.

3.12. MONOLAYER MODEL

From the geometric standpoint, let us now make an extrathermodynamic assumption that the bound phase AA′BB′ of $\bar{\mathscr{A}}$ cm^2 facial area is a monomolecular layer[20] so that

$$N(\sigma_1 \Delta n_1 + \sigma_2 \Delta n_2) = \bar{\mathscr{A}} \qquad (3.69)$$

Here σ_1 and σ_2 are the effective cross sectional areas per single surface bound water molecule and organic molecule, respectively. N is the Avogadro number. σ_1 may be put equal to 0.1 nm^2 per molecule[27] so that by inserting appropriate values of Δn_1 and Δn_2 from Table 3.3 in equation (3.69), the values of σ_2 for methanol, ethanol, and propanol are calculated to be 0.24,

0.22, and $0.20\,\text{nm}^2$ per molecule, respectively. These values given in Table 3.3 agree well with those of σ_2 for the alcohol systems obtained by others after using various surface chemical techniques. From all these results, it appears that the assumption about the monolayer nature of the surface bound phase may be essentially valid. Combining equations (3.69) and (3.65), it may be shown [20] that

$$\frac{\Delta n_1}{\bar{\mathscr{A}}} = \frac{(1 - \Gamma_2^1 \sigma_2 N)X_1}{N(\sigma_1 X_1 + \sigma_2 X_2)} \tag{3.70}$$

and

$$\frac{\Delta n_2}{\bar{\mathscr{A}}} = \frac{(1 - \Gamma_1^2 N\sigma_1)X_2}{N(\sigma_2 X_2 + \sigma_1 X_1)} \tag{3.71}$$

Values of Δn_1 and Δn_2 for several compositions of a liquid mixture calculated with the help of equations (3.70) and (3.71) are presented in Table 3.4, assuming plausible values of σ_2. From a scrutiny of the results in the table, it appears that Δn_1 increases with increase in the water content of the bulk medium and simultaneously Δn_2 is reduced in appropriate proportions.

TABLE 3.4. Values of Δn_1 and Δn_2 in the Nonlinear Region[20]

X_1/X_2	$-\Gamma_1^2 \times 10^{10}$ (mol/cm^2)	$\Delta n_2 \times 10^{10}$ (mol/cm^2)	$\Delta n_1 \times 10^{10}$ (mol/cm^2)	σ_2 (nm^2/molecule)	$\Gamma_2^1 \times 10^{10}$ (mol/cm^2)
\multicolumn{6}{c}{Propyl alcohol + H$_2$O}					
999.0	163.0	0.18	18.0	0.20	0.16
499.0	1777.5	3.58	10.0	0.20	3.55
199.0	835.6	4.24	8.9	0.20	4.20
82.3	594.9	7.24	2.0	0.20	7.23
37.5	234.4	6.29	4.1	0.20	6.20
\multicolumn{6}{c}{Methyl alcohol + H$_2$O}					
39.0	122.0	3.34	8.60	0.24	3.13
19.0	73.5	4.21	6.58	0.24	3.87
9.0	50.0	5.85	2.65	0.24	5.56
4.0	27.5	6.87	0.44	0.24	6.87
\multicolumn{6}{c}{Formic acid + H$_2$O}					
253.0	170.7	0.74	15.8	0.18	0.67
99.4	178.5	1.64	14.0	0.18	1.49
48.4	110.8	2.54	12.2	0.18	2.29
22.9	77.5	3.80	9.9	0.18	3.37
\multicolumn{6}{c}{Glycerol + H$_2$O}					
46.0	22.9	0.83	15.3	1.81	0.50

From equations (3.61) and (3.65), it is of interest to note[20] that

$$\Gamma_1^2 = \frac{\Delta n_1}{\bar{\mathcal{A}}}\left[1 - \frac{X_1/X_2}{\Delta n_1/\Delta n_2}\right] \tag{3.72}$$

$$\Gamma_2^1 = \frac{\Delta n_2}{\bar{\mathcal{A}}}\left[1 - \frac{X_2/X_1}{\Delta n_2/\Delta n_1}\right] \tag{3.73}$$

For a relatively strong surface-active component such as propanol, $\Delta n_2/\Delta n_1 \gg X_2/X_1$ so that Γ_2^1 according to equation (3.73) is very close to $\Delta n_2/\bar{\mathcal{A}}$ (vide Table 3.4). For strong adsorbents therefore Γ_2^1 may be identified as $\Delta n_2/\bar{\mathcal{A}}$ without making a significant error. Inserting this value in equation (3.69), the corresponding value of Δn_1 may be approximately calculated. However, for weak surface-active components, viz., methanol, formic acid, and glycerol, Γ_2^1 is less than Δn_2 because of some contribution of the second term in equation (3.73). For all the liquid mixtures considered here, the term $(X_1/X_2)/(\Delta n_1/\Delta n_2)$ is significantly greater than unity so that Γ_1^2 in equation (3.72) is always significantly negative. Γ_1^2 and Δn_1 thus differ in magnitude and sign.

3.13. BOUNDARY TENSION OF SOLUTIONS OF INORGANIC ELECTROLYTES

The inorganic electrolytes are known to increase the surface tension of the aqueous solvent particularly when the solute concentration is relatively high. In Fig. 3.13, the values of surface tension of water for different molal concentrations (m_i) of inorganic electrolytes have been presented. It appears from this figure that γ for the aqueous solution is dependent on the concentration as well as nature of the electrolyte. The slope of each curve in Fig. 3.13 is positive at a given value of m_i. According to the Gibbs adsorption equation, therefore, Γ_i^1 will be negative at a given value of m_i so that the electrolyte concentration in the inhomogeneous surface region must be less than that of the adjacent bulk phase of uniform composition.

Onsager and Samaras[28] indicated that the deficiency of these strong electrolytes in the interfacial region is mainly due to the repulsion of ions from the surface by the electrostatic image forces. This force will be appreciable only within distances from the surface that are of the same order as the Debye–Hückel thickness of the ion atmosphere. On the basis of the Debye–Hückel theory of electrolytes, they also deduced an equation to account for the extent of increase of surface tension of water due to the addition of an inorganic electrolyte in the liquid medium. An attempt will be made in the

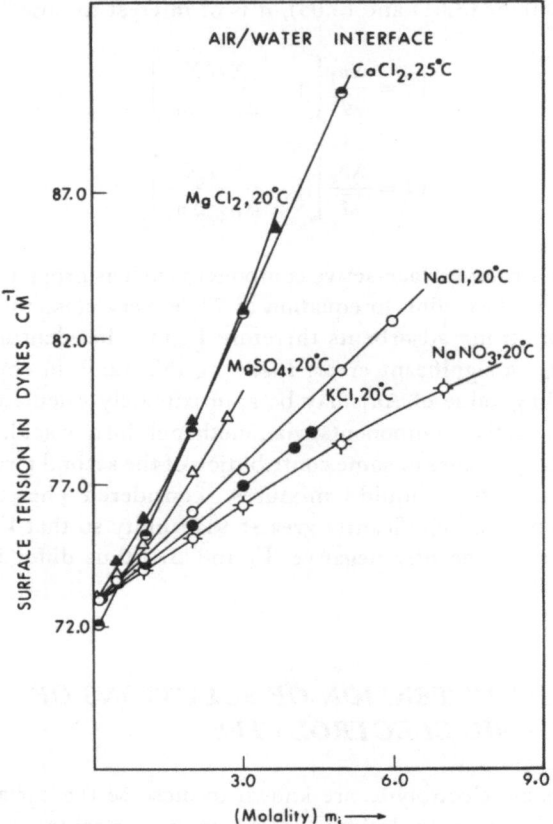

FIGURE 3.13. Plot of surface tension of electrolyte solutions against molality.

next section to obtain the composition of the interfacial phase associated with the aqueous solution by the application of the Gibbs adsorption equation.

3.14. NEGATIVE ADSORPTION OF AN ELECTROLYTE

Let us consider an aqueous solution containing an inorganic electrolyte $P_{\nu_+}Q_{\nu_-}$. This electrolyte in the surface as well as in bulk phases may be ionized completely as

$$P_{\nu_+}Q_{\nu_-} \rightarrow \nu_+ P^{z+} + \nu_- Q^{z-} \tag{3.74}$$

Here $z+$ and $z-$ are the valence of the cations and anions, respectively. A molecule of the electrolyte on complete dissociation thus gives ν_+ and ν_- number of cations and anions, respectively. The Gibbs equation (3.30) for the binary solution of an electrolyte component i (equal to 2) at a constant temperature may be written in the form

$$-d\gamma = \Gamma_i^1 d\mu_i \qquad (3.75)$$

The chemical potential μ_i of the electrolyte in the bulk may be written in the form [vide equation (1.31)]

$$\mu_i = \nu_+\mu_+ + \nu_-\mu_- \qquad (3.76)$$

where μ_+ and μ_- are the chemical potentials of the cation and anion, respectively. Combining equations (3.75) and (3.76), it has been shown by Chattoraj[21] that

$$-d\gamma = RT\Gamma_i^1(\nu_+ d\ln f_+ X_+ + \nu_- d\ln f_- X_-)$$

$$= RT\Gamma_i^1 d\ln (f_+^\nu f_-^\nu)(X_+^\nu X_-^\nu)$$

$$= RT(\nu_+ + \nu_-)\Gamma_i^1 d\ln f_\pm X_\pm \qquad (3.77)$$

The mole fractions X_+ and X_- of cation and anion, respectively, can be calculated from the relations

$$X_+ = \frac{\nu_+ m_i}{(\nu_+ + \nu_-)m_i + 55.51} \qquad (3.78)$$

$$X_- = \frac{\nu_- m_i}{(\nu_+ + \nu_-)m_i + 55.51} \qquad (3.79)$$

where m_i stands for the molality of the inorganic electrolyte which for a solution may be known. The mean mole fraction $X_\pm$ of the electrolyte in the solution can be calculated using the following relation:

$$X_\pm = (X_+^\nu X_-^\nu)^{1/(\nu_+ + \nu_-)} \qquad (3.80)$$

The mean activity coefficient $f_\pm$ is related to the activity coefficients f_+ and f_- of the cation and anion, respectively, by the equation

$$f_\pm = (f_+^\nu f_-^\nu)^{1/(\nu_+ + \nu_-)} \qquad (3.81)$$

From the table of the mean activity coefficients of various inorganic electrolytes

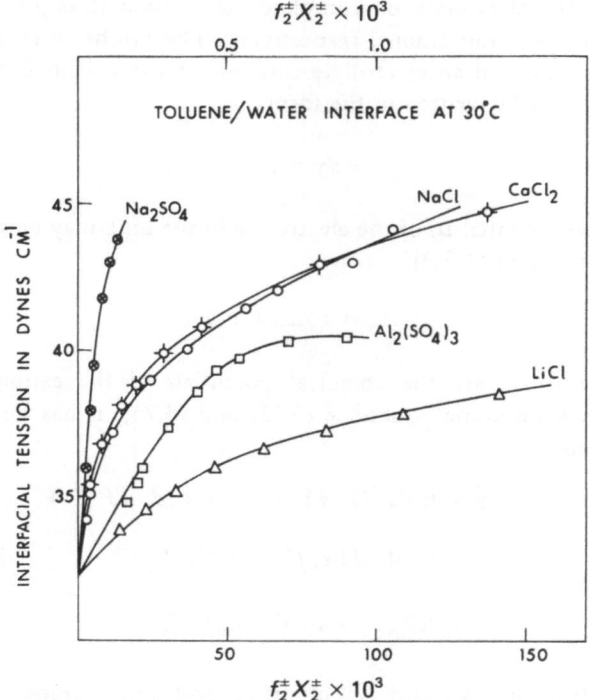

FIGURE 3.14. Plot of interfacial tension against the mean activity of the electrolytes. Upper scale for $Al_2(SO_1)$ only. (L. N. Ghosh and D. K. Chattoraj, unpublished.)

in the practical scale,[29,30] the rational activity coefficients $f_\pm$ can be computed by multiplication with appropriate conversion factors. In Fig. 3.14, γ for various inorganic electrolytes has been plotted against the solute activity $f_\pm X_\pm$ for the toluene–water interface (Ghosh and Chattoraj, unpublished data). In all cases, the slope of the curve is found to be positive. The apparent surface excess Γ_i^a for the electrolyte has been calculated from this plot[31] using the following relation:

$$\Gamma_i^a = -\frac{f_\pm X_\pm}{RT}\frac{d\gamma}{df_\pm X_\pm} \qquad (3.82)$$

Comparing relations (3.82) and (3.77), we can write[21]

$$\Gamma_i^1 = \frac{\Gamma_i^a}{\nu_+ + \nu_-} \qquad (3.83)$$

If the interfacial phase of an electrolyte solution is composed of Δn_1 and Δn_i moles of water and electrolyte components, respectively, then from equations (3.52) and (3.83)

$$\Gamma_i^a = (\nu_+ + \nu_-)\Gamma_i^1$$

$$= (\nu_+ + \nu_-)\frac{\Delta n_i}{\bar{\mathscr{A}}} - (\nu_+ + \nu_-)\frac{\Delta n_1}{\bar{\mathscr{A}}}\frac{n_i}{n_1} \qquad (3.84)$$

In Figs. 3.15 and 3.16, plots of Γ_i^a against n_i/n_1 (i.e., X_i/X_1) for several inorganic electrolytes at air-water[31,21] and oil–water interfaces (Ghosh and Chattoraj, unpublished data) are presented. Over a wide range of electrolyte concentrations, such a plot has been found to be linear. Values of Δn_i ($i = 2$) and Δn_1 calculated from the slope and intercept of the plots are given in Tables 3.5 and 3.6.

Compared to Δn_1, values of Δn_i for different electrolytes are indeed quite · small. The observed experimental error for the evaluation of Δn_i are quite high[21,31] so that the signs and magnitudes of Δn_i for these electrolytes have no significance. They may be taken as zero for all practical purpose. In general, it may be concluded that a negligibly small amount of electrolyte is associated with the surface bound water even when a large amount of it remains dissolved in the bulk water phase. Interfacial water has a strong dislike for inorganic electrolytes; this is quite contrary to its behavior in the bulk liquid phase.[31]

Since Δn_i is negligible, Δn_1 becomes very close to Γ_1^i as expected. Assuming the effective cross-sectional area of a water molecule to be 0.1 nm^2 at the interface,[27] the moles Δn_1^* of water required to form 1 cm^2 of an interfacial monolayer is 1.7×10^{-9}. Putting Δn_i to be zero for all practical purposes, the number (L_S) of water layers in the interfacial phase may be estimated from the ratio $\Delta n_1/\Delta n_1^*$. The values of L_S so calculated for various electrolytes[21] are shown in Tables 3.5 and 3.6. It is of interest to note that for many electrolytes, the water layer bound to the interfacial region of the air–water

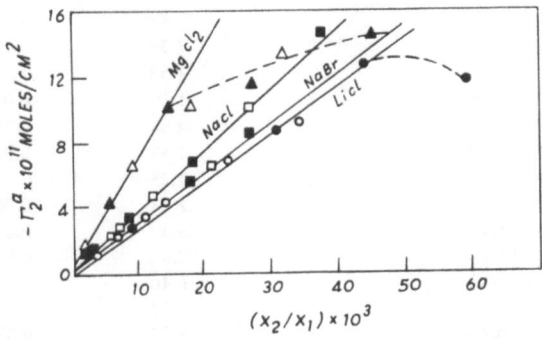

FIGURE 3.15. Plot of Γ_2^a against X_2/X_1 of electrolytes at air–water interface. (Redrawn with changes from Ref. 31.)

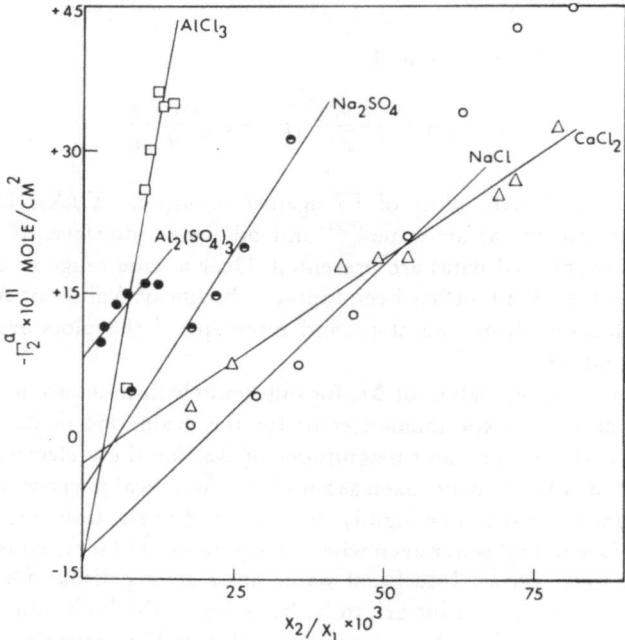

FIGURE 3.16. Plot of Γ_2^a against X_2/X_1 of electrolytes at benzene–water interface (L. N. Ghosh and D. K. Chattoraj, unpublished data).

TABLE 3.5. The Values of Δn_1 and Δn_2 and Thickness of the Aqueous Layer[21]

Serial no.	Solutes	$[\nu_+ + \nu_-]$	$\Delta n_1 \times 10^9$ (mol/cm^2)	$\Delta n_2 \times 10^9$ (mol/cm^2)	L_s
1	LiCl	2	1.45	0.0017	0.89
2	NaCl	2	1.85	−0.0007	1.10
3(a)	KCl (dil)	2	2.52	−0.0065	1.51
(b)	KCl (conc)	2	2.78	−0.0075	1.66
4	NaBr	2	1.53	0	0.92
5(a)	NaNO$_3$(dil)	2	1.64	−0.0020	0.99
(b)	NaNO$_3$(conc)	2	4.68	0.288	1.72
6(a)	K$_2$SO$_4$(dil)	3	1.57	−0.0003	0.94
(b)	K$_2$SO$_4$(conc)	3	1.25	0	0.75
7	MgCl$_2$	3	2.22	0.0017	1.33
8(a)	MgSO$_4$(dil)	2	26.3	0.0005	15.8
(b)	MgSO$_4$(conc)	2	77.5	0.0002	46.5
9(a)	Al$_2$(SO$_4$)$_3$(dil)	5	6.40	0	3.84
(b)	Al$_2$(SO$_4$)$_3$(conc)	5	3.52	0.0093	2.1
10(a)	Sucrose (dil)	1	4.34	0.0262	2.60
(b)	Sucrose (conc)	1	15.1	0.188	9.00

TABLE 3.6. Values of Δn_1 and Δn_2 at the Benzene–Water Interface

Salt	$\Delta n_1 \times 10^9$ (mol/cm^2)	$\Delta n_2 \times 10^9$ (mol/cm^2)	L_s
NaCl	5.12 ± 0.87	+0.12 ± 0.02	1.53
Na$_2$SO$_4$	10.0 ± 1.41	+0.05 ± 0.01	2.00
Al$_2$(SO$_4$)$_3$	7.53 ± 1.07	−0.08 ± 0.01	0.90
AlCl$_3$	36.80 ± 5.81	+0.14 ± 0.02	5.43
CaCl$_2$	4.24 ± 0.27	+0.03 ± 0.0	0.85

or the oil–water system is roughly monomolecular[21] (Ghosh and Chattoraj, unpublished data), in agreement with the observation made for the organic liquid–water system discussed earlier. In a few cases of electrolytes such as MgSO$_4$, AlCl$_3$ at an air–water interface and at a benzene–water interface, the water in the interfacial phase is observed to form polymolecular layers also (vide Tables 3.5 and 3.6). More experiments should be further carried out in the light of the monolayer or multilayer nature of the interfacial water before a definite conclusion can be drawn on this matter.

3.15. DISCUSSION ON THE DERIVATION OF THE GIBBS EQUATION

While applying the Gibbs adsorption equation (3.56) to experimental data, $d\mu_2$ is put equal to $RTd \ln a_2$. The activity a_2 in equation (3.57) may acquire different numerical values in the mole fraction, molality, or molar scales used for representing μ_2^0. When the binary solution is dilute, the molar, molal, and mole fraction concentrations are nearly proportional to each other and $RTd \ln a_2$ will remain almost the same in the three scales. This proportionality will not remain valid when the concentration of the solute is high so that values of $d\gamma/d \ln a_2$ in the three scales will be different to some extent for the same composition of the liquid. For concentrated solutions of nonelectrolytes and electrolytes, rational mole fraction scales have been used for the treatment of the experimental data discussed in Sections 3.12 and 3.14. A reference to standard scale of activity may be of some importance when the free energy change for the transfer of a component from the bulk to surface region is calculated using the Gibbs adsorption equation. This point will be discussed in Chapter 10.

During the derivation of the adsorption equation, Gibbs[2] assumed that the liquid interface is almost planar with negligible curvature. He also pointed out that equation (3.19) for dE' may be written in the following form, if one wants to take the curvature effect into consideration:

$$dE' = TdS' - pdV - p'dV' + \gamma d\bar{\mathscr{A}} + C_1 dC_1^r + C_2 dC_2^r + \sum_i \mu_i dn_i'$$

$$(3.85)$$

Here C_1^r and C_2^r are, respectively, equal to $1/r_1$ and $1/r_2$, where r_1 and r_2 are the radii of curvatures of the surface in two directions, and C_1 and C_2 are constants. By algebraic operations, equation (3.85) may be written in the form,

$$dE' = TdS' - pdV - p'dV' + \gamma d\bar{\mathscr{A}} + \tfrac{1}{2}(C_1 + C_2)d(C_1^r + C_2^r)$$

$$+ \tfrac{1}{2}(C_1 - C_2)d(C_1^r - C_2^r) + \sum_i \mu_i dn_i' \qquad (3.86)$$

In such a case of curved liquid interface, Gibbs[2] put the argument that the $C_1 + C_2$ term will be zero, if the dividing plane GG' (now curved) is placed on the surface of tension where actual boundary tension acts. For a spherical surface, $C_1^r = C_2^r$ since $r_1 = r_2$. For interfaces of other types of shape, C_1^r is nearly equal to C_2^r so that equation (3.86) will become the same as equation (3.19) for a plane liquid interface.

On the basis of the physical concept of the interfacial phase, Guggenheim[7] indicated that the fifth and sixth terms of the right side of equation (3.85) are negligibly small if the radii of curvature r_1 and r_2 (which are equal for spherical surface) are greater than 10^{-5} cm so that their values are considerably greater than the physical thickness of the interface. Attempt has been made by Defay[32] and Dufour[33] to consider more critically the effect of curvature on the Gibbs equation. Finally, a general formulation of the effect of simultaneous variations of curvature and compositions to the surface tension of the liquid mixture has been presented by Koenig.[34] This is, however, difficult to apply when values of actual curvatures remain unknown.

3.16. ALTERNATIVE TREATMENT FOR THE ADSORPTION EQUATION

Several attempts have been made to derive an adsorption equation in which the abstract concept of the Gibbs surface excess has been avoided. Thus Guggenheim[7] has placed two planes A_1A_1' and B_1B_1' in the bulk phases AA'XX' and BB'PP', respectively (vide Fig. 3.17) of the real liquid system. The space between $A_1A_1'B_1B_1'$ has been arbitrarily defined as the surface phase σ of thickness τ. The work done in increasing the volume and the area of this surface phase by dV^σ and $d\bar{\mathscr{A}}$, respectively, is equal to $pdV^\sigma - \gamma d\bar{\mathscr{A}}$. The

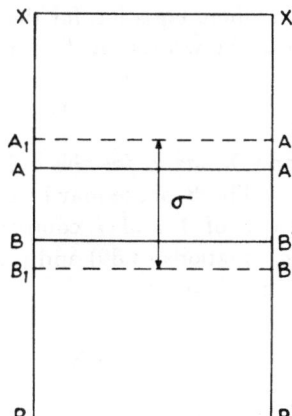

FIGURE 3.17. A model liquid column.

volume change dV in the α-phase in the bulk will lead to the amount of work pdV. If E^σ and S^σ represent the total internal energy and entropy of the σ-phase, then

$$dE^\sigma = TdS^\sigma + \gamma d\bar{\mathscr{A}} - pdV^\sigma + \sum_i \mu_i dn_i^\sigma \qquad (3.87)$$

Integrating this equation without change of intensive parameters, we obtain

$$E^\sigma = TS^\sigma + \gamma\bar{\mathscr{A}} - pV^\sigma + \sum_i \mu_i n_i^\sigma \qquad (3.88)$$

Differentiating this equation and then combining with equation (3.87), we obtain

$$-d\gamma = \frac{S^\sigma}{\bar{\mathscr{A}}} dT - \frac{V^\sigma}{\bar{\mathscr{A}}} dp + \sum_i \frac{n_i^\sigma}{\bar{\mathscr{A}}} d\mu_i$$

$$= s^\sigma dT - \tau dp + \sum_i \Gamma_i^\sigma d\mu_i \qquad (3.89)$$

Here s^σ and Γ_i^σ are, respectively, total entropy and moles of the ith component per unit area of the surface. Compared to the generalized Gibbs equation (3.29), equation (3.89) contains an additional term τdp. Also τ_i^σ and τ_i^x in these two equations should not be taken as identical.

Let us now assume that the binary liquid system in the α-phase contains two components 1 and i. At constant temperature and pressure, respectively, the Gibbs–Duhem equation for the bulk phase liquid in $B_1B_1'PP'$ at constant T and p may be written in the form

$$X_1 d\mu_1 + X_i d\mu_i = 0 \tag{3.90}$$

Here X_1 and X_i stand for the bulk mole fractions of components 1 and i, respectively. The β-phase may be regarded as inert (e.g., air) so that the bulk concentrations of 1 and i components in this phase is negligibly small. Combining equations (3.89) and (3.90) keeping T and p constant[7] equation (3.91) may be obtained:

$$-d\gamma = \left(\Gamma_i^\sigma - \Gamma_1^\sigma \frac{X_i}{X_1} \right) d\mu_i \tag{3.91}$$

Comparing this with equation (3.30), we note that

$$\Gamma_i^1 = \Gamma_i^\sigma - \Gamma_1^\sigma \frac{X_i}{X_1}$$

$$= \frac{1}{\mathscr{A}} \left(n_i^\sigma - n_1^\sigma \frac{X_i}{X_1} \right) \tag{3.92}$$

From examination of Fig. 3.17, it is clear that n_i^σ and n_1^σ will be equal to $\Delta n_i + c_i V_\alpha^\sigma$ and $\Delta n_1 + c_1 V_\alpha^\sigma$, respectively, where V_α^σ is the volume of the space $BB'B_1B_1'$ in the α-phase which are arbitrarily associated with the volume V^S of $AA'BB'$ of the real interfacial phase. Since the β-phase is regarded as inert (e.g., air), contribution of the components in the space $AA'A_1A_1'$ is neglected. With this substitution for n_i^σ and n_1^σ in equation (3.92), it can easily be shown [21,31] that equation (3.92) will be converted into the fundamental equation (3.52).

Following an algebraic approach, Hansen[11] has shown that the thermodynamics of the surface region may be handled satisfactorily without ever introducing the concept of mathematical surface at all. The treatment has been widely used by Goodrich[14] and Motomura.[15] Applying the algebraic method to the real liquid system containing two components 1 and i, Goodrich[14] first wrote the Gibbs–Duhem equation (3.93) for the whole system at constant temperature as follows:

$$-V' dp + \mathscr{A} d\gamma + n_1^1 d\mu_1 + n_i^1 d\mu_i = 0 \tag{3.93}$$

Far from the interfacial region of the real liquid system (vide Fig. 3.6) in the bulk α and β phases, the Gibbs–Duhem equation will read

$$-V dp + n_1 d\mu_1 + n_i d\mu_i = 0 \tag{3.94}$$

$$-V'dp + n_1'd\mu_1 + n_i'd\mu_i = 0 \tag{3.95}$$

Dividing equations (3.94) and (3.95) by V and V', respectively, one will obtain

$$-dp + c_1 d\mu_1 + c_i d\mu_i = 0 \tag{3.96}$$

$$-dp + c_1' d\mu_1 + c_i' d\mu_i = 0 \tag{3.97}$$

Multiplying equations (3.96) and (3.97) by multipliers $\bar{x}$ and $\bar{y}$, respectively, and then subtracting from equation (3.93), we obtain

$$-(V^t - \bar{x} - \bar{y})dp + \bar{\mathscr{A}}d\gamma + (n_1^t - \bar{x}c_1 - \bar{y}c_1')d\mu_1$$

$$+(n_i^t - \bar{x}c_i - \bar{y}c_i')d\mu_i = 0 \tag{3.98}$$

Equation (3.98) is valid for any arbitrary choice of the multipliers x and y. The coefficients of dp or $d\mu_1$ can be eliminated by using the following equations:

$$\bar{x} + \bar{y} = V^t \tag{3.99}$$

$$\bar{x}c_1 + \bar{y}c_1' = n_1^t \tag{3.100}$$

Solving for $\bar{x}$ and $\bar{y}$ on the basis of equations (3.99) and (3.100), and putting their values in equation (3.98), we obtain

$$-d\gamma = \frac{1}{\bar{\mathscr{A}}} (n_i^t - \bar{x}c_i - \bar{y}c_i')d\mu_i$$

$$= \frac{1}{\bar{\mathscr{A}}} \left(n_i^t - \frac{n_1^t - V^t c_1'}{c_1 - c_1'} c_i - \frac{V^t c_1 - n_1^t}{c_1 - c_1'} c_i' \right)d\mu_i \tag{3.101}$$

Comparing the Gibbs equations (3.30) with equation (3.101),

$$\Gamma_i^1 = \frac{1}{\bar{\mathscr{A}}} \left(n_i^t - \frac{n_1^t - V^t c_1'}{c_1 - c_1'} c_i - \frac{V^t c_1 - n_1^t}{c_1 - c_1'} c_i' \right) \tag{3.102}$$

Inserting the values of n_i^t and n_1^t given by equations (3.50) and (3.50a) putting V^t equal to $V + V'$ as before, equation (3.102) will be at once found to be the same as the fundamental relation (3.51).

We thus find that the fundamental relations (3.51) and (3.52) for the surface excess are consistent with the thermodynamic treatments put forward independently by Gibbs,[2] Defay et al.,[16] Guggenheim,[7] Hansen,[11] and Goodrich.[14]

3.17. SURFACE ACTIVITY COEFFICIENTS

From all the analyses given in the previous sections, it is clear that the fundamental equation (3.52) for Γ_i^1 is a composite quantity whose value will depend upon the absolute magnitudes of Δn_i, Δn_1, and n_i/n_1. The surface layer for a binary solution is mostly monomolecular, although in a few cases, the polymolecular nature of the interfacial water layer has been established. On the basis of the monolayer model, Δn_1 and Δn_2 can be calculated in general using equations (3.70) and (3.71). It may also be pointed out here that the monolayer value of L_S is calculated on the basis of the close packing of the molecules existing in the bulk of the condensed phase. The surface region is inhomogeneous where such close packing must be absent in the strict sense, so that the monolayer model is only approximate. The exact location of the physical boundary between interfacial and adjacent bulk phases for the same reason may not be well defined from this standpoint.

Butler[35] assumed that the interfacial phase associated with the binary solution or even with a pure liquid component is, in general, monomolecular in nature. One can imagine that this monolayer phase of interfacial area $\mathscr{A}$ is made up of Δn_1 and Δn_2 moles of components 1 and 2, respectively. The mole fractions X_1^s and X_2^s in the interfacial phase are $\Delta n_1/(\Delta n_1 + \Delta n_2)$ and $\Delta n_2/(\Delta n_1 + \Delta n_2)$ respectively, so that with the help of the general equations (3.70) and (3.71), it can be shown

$$X_1^s = \frac{(1 - \Gamma_2^1 \sigma_2 N) X_1}{1 - N(\Gamma_1^2 \sigma_1 X_2 + \Gamma_2^1 \sigma_2 X_1)} \tag{3.103}$$

and

$$X_2^s = \frac{(1 - \Gamma_1^2 \sigma_1 N) X_2}{1 - N(\Gamma_1^2 \sigma_1 X_2 + \Gamma_2^1 \sigma_2 X_1)} \tag{3.104}$$

Values of X_1^s and X_2^s can thus be calculated using experimental values of Γ_1^2, Γ_2^1, X_1 and X_2 with reasonable assumptions for the magnitudes of σ_1 and σ_2. Schay[36] has shown a rigorous procedure for the calculation of X_1^s and X_2^s from the experimental data.

The interfacial phase like the adjacent bulk phase may be regarded as open system. The Gibbs free energy G^s of this thin phase at constant temperature and pressure can be expressed[35] by

$$dG^s = \mu_1 d(\Delta n_1) + \mu_2 d(\Delta n_2) + \gamma d\mathscr{A} \tag{3.105}$$

At equilibrium μ_1 and μ_2 for the bulk and surface phases are the same.

Following the Gibbs method of integration discussed in Chapter 1, we find

$$G^s = \mu_1 \Delta n_1 + \mu_2 \Delta n_2 + \mathcal{A}\gamma \qquad (3.106)$$

Differentiating this equation, in general and then equating with (3.105), the analogous Gibbs–Duhem equation for the surface phase may be obtained:

$$\Delta n_1 d\mu_1 + \Delta n_2 d\mu_2 + \mathcal{A}d\gamma = 0 \qquad (3.107)$$

At constant T, equation (3.107) is consistent with the Guggenheim equation (3.89) when τdp is negligible because of the negligible values of τ or dp or both.

Butler then defined the partial molar free energies μ_1^s and μ_2^s for the interfacial phase by equations (3.108) and (3.109)

$$\mu_1^s = \left[\frac{\partial G^s}{\partial(\Delta n_1)} \right]_{T,p,\Delta n_2} \qquad (3.108)$$

and

$$\mu_2^s = \left[\frac{\partial G^s}{\partial(\Delta n_2)} \right]_{T,p,\Delta n_1} \qquad (3.109)$$

μ_1^s and μ_2^s referred to frequently as surface chemical potentials of components 1 and 2, respectively, are different from the Gibbs chemical potentials μ_1 and μ_2. From equation (3.105)

$$\left[\frac{\partial G^s}{\partial(\Delta n_1)} \right]_{T,p,\Delta n_2} = \mu_1 + \gamma \left[\frac{\partial \mathcal{A}}{\partial(\Delta n_1)} \right]_{T,p,\Delta n_2} \qquad (3.110)$$

The contribution of the second term on the right-hand side of equation (3.110) will actually occur since at constant value of Δn_2, change of Δn_1 will alter the total interfacial area $\mathcal{A}$ of the monolayer. Butler[35] defined the partial molar area A_1 of the interfacial phase by the equation

$$A_1 = \left[\frac{\partial \mathcal{A}}{\partial(\Delta n_1)} \right]_{T,p,\Delta n_2} \qquad (3.111)$$

The partial molar area A_2 of the monolayer can also be defined in a similar fashion. Combining equations (3.110) and (3.111), the following relation for the Gibbs chemical potential and surface chemical potential has been obtained:

$$\mu_1^s = \mu_1 + \gamma A_1 \qquad (3.112)$$

Similarly,

$$\mu_2^s = \mu_2 + \gamma A_2 \tag{3.113}$$

From equations (3.112) and (3.113), therefore,

$$\gamma = \frac{\mu_1^s - \mu_1}{A_1} = \frac{\mu_2^s - \mu_2}{A_2} \tag{3.114}$$

The interfacial area $\bar{\mathscr{A}}$ is a function of T, p, Δn_1, and Δn_2. At constant temperature and pressure, therefore,

$$d\bar{\mathscr{A}} = A_1 d(\Delta n_1) + A_2 d(\Delta n_2) \tag{3.115}$$

Applying the technique of the Gibbs integration, it can be shown that

$$\bar{\mathscr{A}} = A_1 \Delta n_1 + A_2 \Delta n_2 \tag{3.116}$$

This equation becomes consistent with relation (3.69) if A_1 and A_2 are taken as $N\sigma_1$ and $N\sigma_2$, respectively.

Differentiating equations (3.112) and (3.113) and then combining with equations (3.107), (3.115) and (3.116), we obtain

$$\Delta n_1 d\mu_1^s + \Delta n_2 d\mu_2^s = 0 \tag{3.117}$$

From the analogy of the thermodynamic treatment of the bulk solution, μ_1^s and μ_2^s may be defined by[35,36]

$$\mu_1^s = \mu_1^{s,0} + RT \ln f_1^s X_1^s \tag{3.118}$$

$$\mu_2^s = \mu_2^{s,0} + RT \ln f_2^s X_2^s \tag{3.119}$$

Here, f_1^s and f_2^s are the respective surface activity coefficients of components 1 and 2. When X_1^s is unity, f_1^s becomes unity (by convention) and the standard chemical potential $\mu_1^{s,0}$ becomes equal to μ_1^s. The surface and bulk phases in this situation contain only pure component 1 so that from equation (3.112)

$$\mu_1^{s,0} = \mu_1^0 + \gamma_1^0 A_1 \tag{3.120}$$

Here γ_1^0 is the surface tension of pure component 1 whose molar surface area $N\sigma_1$ is taken equal to the partial molar area A_1. This involves a certain approximation. Equation (3.120) indicates that the magnitudes of $\mu_1^{s,0}$ and μ_1^0 are different so that the reference states for f_1^s and bulk activity coefficient f_1

are not the same. For the bulk solution

$$\mu_1 = \mu_1^0 + RT \ln a_1 \tag{3.121}$$

Replacing the μ_1^s and $\mu_1^{s,0}$ terms in equation (3.118) by the terms present on the right-hand sides of equations (3.112) and (3.120) and then combining with equation (3.121), one will get

$$\ln f_1^s = \ln a_1 + \frac{A_1(\gamma - \gamma_1^0)}{RT} - \ln X_1^s \tag{3.122}$$

Similarly,

$$\ln f_2^s = \ln a_2 + \frac{A_2(\gamma - \gamma_2^0)}{RT} - \ln X_2^s \tag{3.123}$$

These equations for the surface activity coefficients have been derived and used by Schay *et al.*[36-38] For alcohol–water system, a_1 and a_2 are known from the values of p_1/p_0 and p_2/p_0; corresponding values of γ, γ_1^0, and γ_2^0 for various values of a_1 and a_2 have been measured (vide Section 3.11 and Table 3.2). Reasonable values of σ_1 and σ_2 for water and alcohol molecules can also be fixed (vide Section 3.12) so that X_1^s and X_2^s for various liquid compositions can be estimated using equations (3.103) or (3.104) or other appropriate relations. Values of surface-activity coefficients[36] f_1^s and f_2^s for such system have been presented in Table 3.7. The surface activity coefficient of water deviates significantly from unity when X_2 is higher than 0.3. The surface activity coefficients of alcohol remain close to unity for X_2 varying between 0.3 and 1.

TABLE 3.7. Activity Coefficient[36] of Water (Component 1) + Ethanol (Component 2) at the Liquid–Vapor Interface at 25°C

X_2	X_2^s	f_2^s	X_1^s	f_1^s
0.1	0.62	0.93	0.38	1.02
0.2	0.69	0.93	0.31	1.00
0.3	0.69	0.98	0.31	0.92
0.4	0.69	1.02	0.31	0.84
0.5	0.70	1.03	0.30	0.80
0.6	0.73	1.04	0.27	0.80
0.7	0.79	1.02	0.21	0.83
0.8	0.85	1.01	0.15	0.88
0.9	0.92	1.005	0.08	0.94

The definitions and physical features of other partial molar quantities for the interfacial phase have in fact been examined by various workers.[9,10,13,16,38-42] It has been shown by Eriksson[43] that if ψ^σ denotes any extensive property of the surface phase such as surface area, surface entropy, and $n_1^\sigma, \ldots, n_i^\sigma$ are the moles of the components $1, 2, \ldots, i$ in the arbitrarily defined surface phase then the partial molal quantity ψ_i^σ in this phase may be defined by the equation

$$\psi_i^\sigma = \left(\frac{\partial \psi^\sigma}{\partial n_i^\sigma}\right)_{T,n_j^\sigma} \tag{3.124}$$

The quantity ψ_i^σ thus defined becomes dependent on the thickness and the position of the boundary of the imaginary interfacial phase.

Recently surface activity coefficients of binary mixtures of cyclohexane + benzene, cyclohexane + dioxane, etc. based on the models of Hoar and Melford[40] and Sprow and Prausnitz[44] have been reported.[43] Patel *et al.* have calculated the excess surface entropy $\Delta S_\sigma'$ from the use of Eriksson's equation[45]

$$S_\sigma^1 = -\sum_{i=1}^{i=2} X_i^s \frac{\partial}{\partial T}(RT \ln f_i^s)_{X^s} \tag{3.125}$$

The sign of the surface excess entropy has been related to the geometry of these molecules and their mixing characteristics.

The ideal value of the surface tension γ of the binary mixture of liquids can be calculated by the following equation given by Belton and Evans[42]:

$$\gamma = \gamma_1^0 X_1 + \gamma_2^0 X_2 \tag{3.126}$$

For benzene–xylene and a few other mixtures, this equation is found to be obeyed throughout the whole range of solution composition. Deviation from this equation has been observed for which the concept and magnitude of the activity parameters are useful. The usefulness of the concepts of the surface activity coefficients at the solid–liquid interface will be further discussed in Chapter 8.

Ono and Kondo[17] have developed theories for the surface tension of the multicomponent liquid system on the basis of the precise approach of statistical thermodynamics in which the gradient nature of the interfacial region has been recognized. Motomura[15] has treated the partial molar quantities of the interfacial region in the light of this quasithermodynamic approach. Motomura *et al.*[46] have also applied this approach to examine their data on boundary tensions and thus established that evaluation of these partial molar quantities will lead to the elucidation of important physical properties of the interfacial phase (vide Chapter 5).

3.18. SUMMARY AND COMMENTS

The physical interface occurring at the contact region between two homogeneous and immiscible phases is recognized to be inhomogeneous. When one of these phases is a liquid system containing more than one component, one or several of these components are accumulated at the interface as positive or negative excesses. The change of boundary tension γ of the liquid due to the change in its composition has been related to the surface excess quantities $\Gamma_2^1, \Gamma_3^1, \ldots, \Gamma_i^1$, etc. by the Gibbs adsorption equation. For a binary solution Γ_2^1 becomes equal to $d\gamma/d\mu_2$ so that its value can be estimated using experimental data on γ and solute activities.

However, in deriving this useful relation, Gibbs assumed that the surface is an imaginary mathematical plane whose position can be suitably fixed near the region of separation between two phases such that accumulation of one of the components will vanish. This approach is extrathermodynamic but no mathematical error is involved in the postulate. Because of this abstract approach followed by Gibbs, the relation between the surface excess and the composition of the actual interfacial phase is not obvious. For this reason, many alternative and rigorous derivations of the Gibbs equation have been proposed from time to time.

In this chapter, the original derivation of the Gibbs equation has been presented in detail. The difficulties of the concept of the Gibbs surface excess in relation to the interfacial phase have been further pointed out explicitly. Recent mathematical analysis of Defay *et al.*[16] based on the Gibbs approach indicates that Γ_2^1 is related to the composition of the binary liquid mixture as a whole. Extension of this approach further leads to explicit and fundamental relations (3.61) and (3.65) between the surface excess, mole ratio composition (X_1/X_2) of the bulk phase, and moles (Δn_1 and Δn_2) of the two components forming the actual interfacial phase. With alteration of X_1/X_2, Δn_1 and Δn_2 may vary, but in a certain range of composition their values become independent of X_1/X_2 so that from a linear plot, the values of these quantities can be computed. It is highly desirable to find out reasons for this kind of behavior of the interfacial phase on the basis of intermolecular interaction occurring at the surface and bulk phases.

From the evaluated values of Δn_1 and Δn_2, the interfacial phase of liquid appears to be composed of mixture of molecules thus forming a monomolecular layer. There exist few exceptions to this behavior when the binary solution contains inorganic electrolytes. Many older experiments with inorganic electrolytes are not very accurate. It is hoped that more precise measurements of solutions of inorganic electrolytes in aqueous media will be made in future for further thermodynamic analysis.

Butler[35] had earlier given an elegant equation for the chemical potential of a component in the surface phase of monolayer thickness. This has subsequently led to the derivation of the thermodynamic expression for the surface activity coefficients, so these parameters can be calculated from the available experimental data. In the derivation of the Gibbs adsorption equation, the contribution of the β-phase (vide Fig. 3.6) to the interfacial phase is implicitly neglected. The surface activity coefficients thus evaluated in all probability include this contribution. There should be more theoretical and experimental work in this field so that a comparative study of this contribution for various surfaces may be presented.

One must recognize that the standard chemical potential in the Butler equation and that included implicitly in the Gibbs adsorption equation are different from each other. Using the Butler concept for the surface phase, Nassonov[47,48] has made an attempt to show that the form of the adsorption equation derived by Gibbs is inadequate and inconsistent with the principle of thermodynamics. In all probability, this conclusion is misleading because the author has not considered the difference in the standard chemical potentials involved in the two cases.[49]

From the appropriate use of the Gibbs adsorption equation (3.30), one can calculate excess surface entropy S_σ^1 for the interfacial phase from the variation of γ with variation of temperature. The analysis of the entropy data may give valuable information about the structure and mixing characteristics of the components in the interfacial phase. Significant research in this direction has been done only in recent years. Further, γ of a binary solution changes also with change of pressure (vide Chapter 5). This will lead to the evaluation of the excess volume or thickness of the surface phase according to the thermodynamic analysis put forward by Hansen[11] and Guggenheim.[7] In the concept of Gibbs, the volume of the surface phase is treated as an insignificant quantity.

From the analysis of the data on the surface tension of glycerine–water system, the change in orientation of the organic molecules in the interfacial monolayer with change of the composition of the bulk liquid has been detected.[20] More research in this direction is highly desirable. The shapes of the adsorption isotherms for the liquid mixtures are found to be similar to those obtained for the adsorption of the components at the surface of various types of solid powder.[36] From the analysis of the linear portion of these isotherms, Schay[36] has calculated the absolute composition of the liquid–gas interface for the systems using algebraic approach of analysis (vide also Chapter 8). This treatment is consistent[20] with that discussed in Section 3.11.

CHAPTER 4

ADSORPTION AT THE LIQUID INTERFACE FROM THE MULTICOMPONENT SOLUTION

4.1. INTRODUCTION

In many technical and natural phenomena, adsorption at the interface actually takes place when the bulk solution adjacent to the interface contains several solute components. The organic and inorganic solutes present in the bulk may be nonelectrolytes or electrolytes, so that the solution also contains ionic and nonionic species of solutes. The aqueous solvent on slight dissociation may also introduce H^+ and OH^- in the bulk system. The organic solutes in the bulk system are usually surface-active, so that most of the time they accumulate in the interfacial phase as positive excesses. In the absence of organic solute, inorganic salts usually accumulate at the interface as negative excesses. But in the presence of organic amphiphiles, the inorganic cations and anions may accumulate either as negative or positive excesses depending upon the nature of the surface charge in the so-called interfacial double layer.

In Chapter 3, we have shown that in an aqueous solution consisting of a single nonelectrolyte component, the extent of positive or negative accumulation of solute per unit area of the liquid interface may be calculated using the Gibbs equation (3.1) at a constant temperature. If the bulk solution contains more than one solute component, the Gibbs equation in its general form (3.30) remains valid. But this equation in linear form cannot be utilized for the calculation of surface excess or excesses from surface tension measurements unless experiments are specially designed. In this chapter, we propose some suitable forms of the Gibbs adsorption equation for multicomponent aqueous solutions, which are useful for the calculation of the extent of adsorption.

4.2. SURFACE EXCESS FOR MULTICOMPONENT SOLUTIONS

In Chapters 4 and 5, we intend to discuss various features associated with the adsorption of·surface-active amphiphiles at the air–water or oil–water interfaces in the absence or presence of inorganic electrolytes in the aqueous medium. The water in one of the bulk phases is always present in large excess, so that it behaves as a solvent. The water component in the present and next chapters will be represented by w instead of 1 used in Chapter 3. The aqueous phase may contain i number of solutes or ions where i may be 1, 2, 3, etc. The relative excess of the ith solute in the system may then be given by

$$\Gamma_i^w = \Delta n_i - \Delta n_w \frac{X_i}{X_w}$$

$$= \Delta n_i \left(1 - \frac{\Delta n_w / \Delta n_i}{X_w / X_i} \right) \tag{4.1}$$

For a dilute solution of surface-active solute, the term inside the parentheses is practically equal to unity so that Γ_i^w is almost equal to Δn_i. If some inorganic electrolytes are present in the aqueous phase containing the surface-active organic solute, the relative surface excesses of the inorganic ions may be positive or negative depending upon the nature and sign of the surface charge. For convenience the relative surface excesses of the solutes or ions in the liquid system from now on will be represented by $\Gamma_1, \Gamma_2, \ldots, \Gamma_i$ (instead of Γ_1^w, Γ_2^w, etc.). All these relative excesses are expressed in terms of Γ_w (i.e., Γ_w^w) equal to zero by setting the Gibbs dividing plane in appropriate manner.

The measurements of the boundary tension in the case of multicomponent solutions are usually carried out at constant temperature. Further, the chemical potential μ_i of the ith solute in the multicomponent solution is conveniently expressed by

$$\mu_i = \mu_i^0 + RT \ln f_i c_i \tag{4.2}$$

where c_i is the molar concentration of the solute in the bulk solution, f_i and μ_i^0 are the activity coefficient and standard chemical potential, respectively, of the ith component in the molar scale. As pointed out in Chapter 3, $d\mu_i$ will not differ significantly in the molar, molal, and mole fraction scales of activity if the solution is dilute. Equation (4.2) may be differentiated to give

$$d\mu_i = RT\xi_i d \ln c_i \tag{4.3}$$

Here

$$\xi_i = 1 + \frac{d \ln f_i}{d \ln c_i} \tag{4.4}$$

ξ_i is regarded to be an activity coefficient parameter for the ith component in the bulk solution. The Gibbs equation (3.30) for the solution containing i number of solute components may then be written in the form

$$- d\gamma = \Gamma_1 d\mu_1 + \Gamma_2 d\mu_2 + \cdots + \Gamma_i d\mu_i$$

$$= RT(\Gamma_1 \xi_1 d \ln c_1 + \Gamma_2 \xi_2 d \ln c_2$$

$$+ \cdots + \Gamma_i \xi_i d \ln c_i)$$

$$= RT \sum_i \Gamma_i \xi_i d \ln c_i \qquad (4.5)$$

If the system contains a single nonionic solute component 1, the equation reduces to the convenient form

$$\Gamma_1 = -\frac{1}{\xi_1 RT} \frac{d\gamma}{d \ln c_1} \qquad (4.6)$$

From the values of γ obtained from experiments for various values of c_1, $d\gamma/d \ln c_1$ may be evaluated at a given value of c_1. Putting ξ_1 equal to unity for dilute solutions, the relative surface excess may be directly calculated with the help of equation (4.6).

Unlike equation (4.6), which is valid strictly for a single solute component, equation (4.5) for two or more solute components cannot be used as such for the calculation of the surface excess from the experimental data unless experiments are properly designed. Several such experiments will be discussed in the following sections.

4.3. GIBBS EQUATION FOR THE MIXTURE OF NONELECTROLYTES

For a dilute aqueous solution containing a mixture of several nonionic organic solutes, one may reasonably assume the activity coefficient parameter ξ_i to be unity in equation (4.5). The boundary tension experiments may be carried out,[1] so that the bulk concentration c_i of the solute component 1 may be varied during the measurement of γ keeping the concentrations $c_2, c_3, \ldots, c_i$ etc. of the other solutes fixed. Under this condition, $d \ln c_2$, $d \ln c_3$, etc. will be zero, so that equation (4.5) will now be reduced to the form of equation (4.6). From the $\gamma - c_1$ plot for such multicomponent solution, it will be possible to calculate Γ_1, although Γ_2, Γ_3, etc., have always finite values. Following this type of experimental designs, Γ_1 for n-pentanol and n-butanol, respectively, for various values of c_1 have been calculated from the measurement of γ for

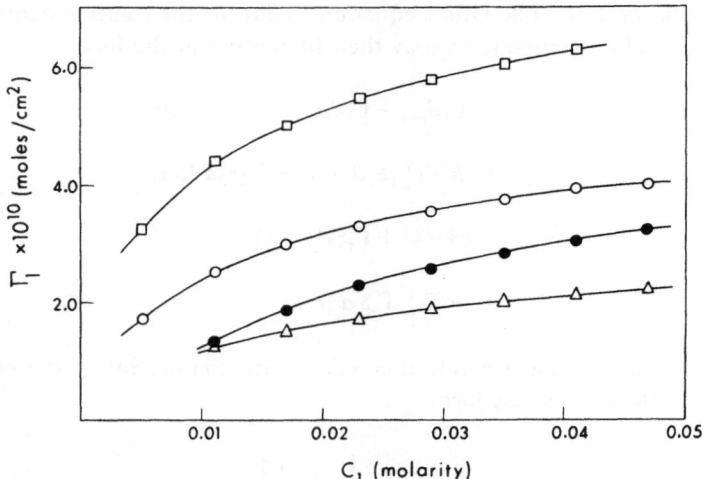

FIGURE 4.1. Plot of Γ_1 against c_1. $\square$, hexanol in 0.066 M propanol in water; $\bigcirc$, pentanol in 0.066 M propanol in water; $\bullet$, butanol in pure water; $\triangle$, butanol in 0.066 M propanol in water. (Das, Birdi, and Chattoraj, unpublished data.)

aqueous solutions of propanol[2] of fixed concentration 0.066 M. These adsorption isotherms have been compared with those obtained for pure water used as a solvent (vide Fig. 4.1). At a given value of c_1, Γ_1 for butanol in propanol solution is significantly lower than that for water as expected. Adsorption of organic component 1 is thus hindered when propanol is present in the bulk as well as at the air–water interface.

In an alternative experimental procedure,[1] the concentrations c_1, c_2, c_3 etc. for different nonionic solutes may be varied in constant proportions during the measurement of γ, so that one can write

$$k_1' c_1 = k_2' c_2 = \cdots = k_i' c_i \tag{4.7}$$

Here k_1', k_2' etc. are respective proportionality constants. From the examination of this relation, one can obtain

$$d \ln c_1 = d \ln c_2 = \cdots = d \ln c_i \tag{4.8}$$

Further, if the total solute concentration c^t is expressed by the relation

$$c^t = c_1 + c_2 + \cdots + c_i$$

$$= k_1 c_1 \left(1 + \frac{1}{k_2'} + \frac{1}{k_3'} + \cdots \right) \tag{4.9}$$

or

$$d \ln c^t = d \ln c_1 = \cdots = d \ln c_i \qquad (4.10)$$

Equation (4.5) in combination with relation (4.10) may then assume the form

$$- d\gamma = (\Gamma_1 + \Gamma_2 + \cdots + \Gamma_i)RTd \ln c^t$$

$$= RT \left(\sum_i \Gamma_i \right) d \ln c^t \qquad (4.11)$$

An amphiphile may remain mixed with another nonionic amphiphile in constant proportions if the sample is not pure. For the solutions of this impure organic solute component, equation (4.11) may become useful for the calculation of the total surface excesses $\sum_i \Gamma_i$, if γ is measured as function of the concentration c^t. From the $\gamma - c^t$ plots of butanol–pentanol–hexanol mixtures in the respective molar proportions 1:1:1, 2:1:1, 1:2:1, and 1:1:2, $\sum_i \Gamma_i$ have been calculated[2] for various solutions using equation (4.11). Adsorption isotherms of these mixtures are shown in Fig. 4.2. The disagreement between

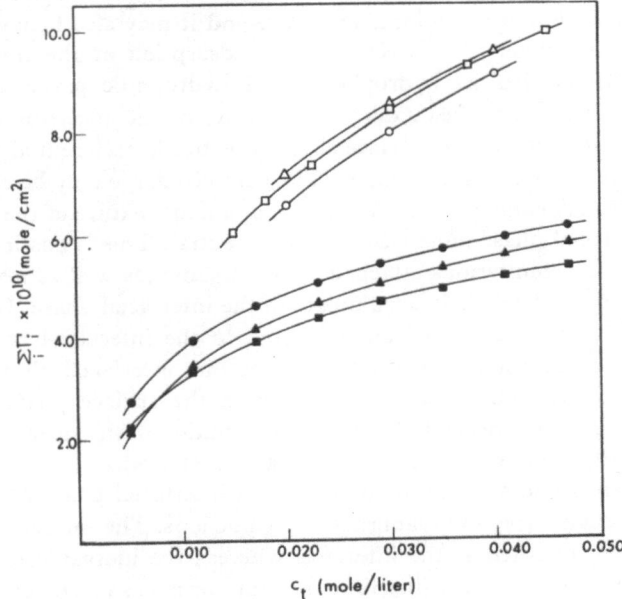

FIGURE 4.2. Adsorption isotherms of butanol, pentanol, hexanol mixtures. ●, ○, $c_4:c_5:c_6 =$ 1:1:2; ▲, △, $c_4:c_5:c_6 = 1:2:1$; ■, □, $C_4:C_5:C_6 = 1:1:1$. Closed symbols are for experiments with mixtures using equation (4.11); open symbols theoretically calculated on the basis of additivity rule from the data of individual alcohols using equation (4.6). (Das, .Birdi, and Chattoraj, unpublished data.)

the theoretical (based on additivity) and experimental curves indicates that the organic components at the interface interact with each other; further study of this interaction may be of interest.

4.4. SOLUTION OF ORGANIC ELECTROLYTES AND THE ELECTRICAL DOUBLE LAYER

In contrast to the behavior of inorganic electrolytes, addition of a trace amount of an organic electrolyte to water lowers its surface tension significantly due to the preferential accumulation of the surface-active solute in the interfacial phase. In the presence of an inorganic electrolyte in the aqueous phase, addition of an organic electrolyte may lower the surface tension further due to its enhanced adsorption at the liquid interface. The interpretation of such behavior in terms of the Gibbs adsorption equation becomes complicated by the dissociation of the electrolyte in the bulk as well as in the interfacial phases. As a result of adsorption, an electrical double layer will be formed at the interface.

It has been pointed out in Section 1.3 that the hydrophilic head group of an amphiphile may be ionic in nature and it may also behave as a strong organic electrolyte. This electrolyte on adsorption at the liquid interface may so orient that its hydrophobic and hydrophilic parts remain in the nonaqueous and aqueous sides, respectively, of the phase-boundary plane. The interface will acquire surface charge due to the preferential accumulation of the organic ions in this manner. The surface charge may be either positive or negative depending upon the cationic or anionic nature of the organic ions. The interfacial phase must also be electroneutral. This is possible only when the inorganic counterions attached to the organic (as well as inorganic) electrolytes also preferentially accumulate in the interfacial phase. Inorganic ions (sometimes called coions) similar in sign to the interfacial charge, will be repelled, so that their concentration in the thin interfacial phase will be less than that in the bulk phase. The water in the surface phase may weakly dissociate into H^+ and OH^- but their contribution to the surface excesses will be so small in many cases that it is usually neglected.

We thus find that the inhomogeneous interfacial phase of a liquid may contain various types of organic and inorganic ions. The surface-active organic ions remain oriented at the interface, whereas the inorganic ions remaining on the aqueous side will be distributed nonuniformly as a result of the combined action of electrostatic, thermal, and other specific and nonspecific forces. The distribution of the inorganic ions in the interfacial phase has been taken into account in the proposed structures of the electrical double layer. The classical pictures of the electrical double layer are based on the simple models proposed by Helmholtz,[3,4] Gouy,[5,6] Chapman[7] and Stern.[8] These models are quite suitable to be dealt with by the Gibbs equation for the adsorption of organic electrolytes, and will be discussed in later sections of this chapter. The electroneutrality conditions of the interfacial phase consistent with these three models are to be carefully examined before applying the Gibbs adsorption equation.

The properties of the electrical double layer existing at the solid–liquid, liquid–liquid, and gas–liquid interfaces are closely associated with many observed phenomena of surface chemistry such as colloid and emulsion stability, electrocapillarity in mercury–water interfaces, electrode phenomena, etc. Highly refined pictures of the electrical double layer have been proposed subsequently to interpret these phenomena.[9–11]

4.5. ELECTRONEUTRALITY IN THE INTERFACIAL PHASE

Consider a liquid system (Fig. 4.3) composed of water to which is added an organic electrolyte RNa_z and inorganic electrolyte $NaCl$. We shall assume further that both these electrolytes are completely dissociated:

$$RNa_z = R^{z-} + zNa^+ \qquad (4.12)$$

$$NaCl = Na^+ + Cl^- \qquad (4.13)$$

Here z stands for the valency (charge) of the organic suface-active anion. For sodium octanoate and sodium dodecyl sulfate (SDS), z is unity, whereas for sodium sebacate, it is 2. The organic ions in all cases are anionic in nature. We shall present mathematical treatments on the basis of the adsorption of

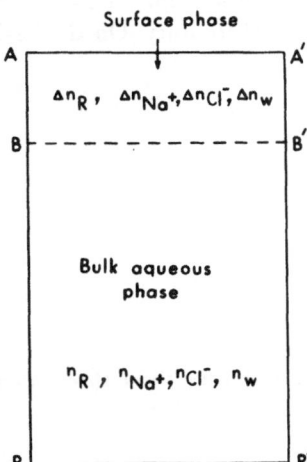

FIGURE 4.3. Liquid column showing the composition of the surface and bulk phases.

the organic anions in the presence of NaCl. There are also many surface-active organic electrolytes RCl_z (e.g., cetyl pyridinium chloride) in which the organic ion is a cation. Although the equations for the electroneutrality in the presence of cationic and anionic organic ions initially appear to be different, their final effect on the Gibbs adsorption equation will be independent of the sign of ionic charge.

Following the arguments given in Section 3.10, we may write equation (4.14) for the *i*th ion in the form

$$n_i^t = n_i + \Delta n_i \qquad (4.14)$$

where n_i^t is the total number of moles of the *i*th component in the system; n_i moles of the ion are present in the homogeneous bulk phase BB'PP', whereas Δn_i moles of it are adsorbed in the interfacial phase. Besides this, the interfacial and bulk phases are composed of Δn_w and n_w moles of water. The presence of ions and water above AA' in the nonaqueous phase (e.g., air or oil) is neglected (Fig. 4.3).

The surface and bulk phases must be individually electroneutral, so that

$$zn_R + n_{Cl^-} = n_{Na^+} \qquad (4.15)$$

and

$$z\Delta n_R + \Delta n_{Cl^-} = \Delta n_{Na^+} \qquad (4.16)$$

Adding equations (4.15) and (4.16), we find

$$zn_R^t + n_{Cl^-}^t = n_{Na^+}^t \qquad (4.17)$$

Suppose 1 cm^2 of surface area is attached to the liquid column so that $\bar{\mathscr{A}}$ is equal to unity. On the basis of equation (3.45), one can write

$$\Gamma_R = n_R^t - n_w^t \frac{n_R}{n_w} \qquad (4.18)$$

$$\Gamma_{Na^+} = n_{Na^+}^t - n_w^t \frac{n_{Na^+}}{n_w} \qquad (4.19)$$

$$\Gamma_{Cl^-} = n_{Cl^-}^t - n_w^t \frac{n_{Cl^-}}{n_w} \qquad (4.20)$$

Combining these three equations with relations (4.15) and (4.17), it may easily be shown that

$$z\Gamma_R + \Gamma_{Cl^-} = \Gamma_{Na^+} \qquad (4.21)$$

This is the electroneutrality equation in terms of the surface excesses. Γ_R and Γ_{Na^+} must be positive excesses, whereas Γ_{Cl^-} will be negative in sign, since its concentration in the negatively charged interface will always be less than that in the bulk. It may be pointed out here that Δn_{Cl^-} in equation (4.16) representing the absolute magnitude of Cl^- in the interfacial phase cannot be negative.

Dividing both sides of equations (4.15) and (4.17) by respective volumes V and V' of the bulk and total liquid systems present in the column in Fig. 4.3, one can write

$$z c_R + c_{Cl^-} = c_{Na^+} \tag{4.22}$$

$$z c_R^t + c_{Cl^-}^t = c_{Na^+}^t \tag{4.23}$$

Here c_i^t and c_i stand for the total and bulk molar concentrations of the ith component in the liquid medium, respectively.

4.6. GIBBS EQUATION FOR ELECTROLYTE ADSORPTION

For the liquid system presented in Fig. 4.3 containing two solute components RNa_z and $NaCl$, the Gibbs equation (4.5) may be written in the form

$$-d\gamma = \Gamma_{RNa_z} d\mu_{RNa_z} + \Gamma_{NaCl} d\mu_{NaCl} \tag{4.24}$$

Further,

$$\mu_{RNa_z} = \mu_R + z\mu_{Na^+} \tag{4.25}$$

$$\mu_{NaCl} = \mu_{Na^+} + \mu_{Cl^-} \tag{4.26}$$

It may also be shown that

$$\Gamma_{NaCl} = n_{NaCl}^t - n_w^t \frac{n_{NaCl}}{n_w}$$

$$= n_{Cl^-}^t - n_w^t \frac{n_{Cl^-}}{n_w}$$

$$= \Gamma_{Cl^-} \tag{4.27}$$

Similarly,

$$\Gamma_{RNa_z} = \Gamma_R \tag{4.28}$$

Inserting equations (4.21), (4.25), (4.26), (4.27), and (4.28) in equation (4.24), it may be shown[12-14] that

$$-d\gamma = \Gamma_R d\mu_R + \Gamma_{Na^+} d\mu_{Na^+} + \Gamma_{Cl^-} d\mu_{Cl^-} \qquad (4.29)$$

This is the form of the Gibbs equation for a liquid system containing three different ionic species. The equation, in fact, does not explicitly contain the term z, the valency of the organic ion. From a similar kind of approach, it can be shown that a general equation may be written for a liquid system containing i number of ionic species of any valency:

$$-d\gamma = \sum_{i=1}^{i} \Gamma_i d\mu_i \qquad (4.30)$$

Here μ_i stands for the chemical potential of the ith ion in the bulk system.

Equation (4.29) in combination with relation (4.3) may assume the form

$$-d\gamma = mRT\Gamma_R d \ln c_R \qquad (4.31)$$

where m, termed the coefficient of the Gibbs adsorption, may be given by the mathematical relation

$$m = \xi_R \left(1 + \frac{\Gamma_{Na^+}}{\Gamma_R} \frac{\xi_{Na^+}}{\xi_R} \frac{d \ln c_{Na^+}}{d \ln c_R} + \frac{\Gamma_{Cl^-}}{\Gamma_R} \frac{\xi_{Cl^-}}{\xi_R} \frac{d \ln c_{Cl^-}}{d \ln c_R} \right) \qquad (4.32)$$

If the value of m can be estimated from the experimental data, the surface excess Γ_R for the electrolyte solutions may be computed from insertion of the value of the slope $d\gamma/d \ln c_R$ in equation (4.31).

An examination of equation (4.32) will further indicate that if the experiments are properly designed, terms related to c_{Na^+}, c_R, and c_{Cl^-} in this equation may be solved using the electroneutrality equation (4.22). Further, using the Debye–Hückel theory for the electrolytes in the bulk system, the parameters ξ_R, ξ_{Na^+}, ξ_{Cl^-} given by equation (4.4) may also be solved under designed conditions of experiments. For the determination of m, therefore, we have to solve for the ratio of the surface excesses occurring in equation (4.32).

Using the electroneutrality equation (4.21), it can be shown that

$$\frac{\Gamma_{Na^+}}{\Gamma_R} = \frac{z}{1 - \Gamma_{Cl^-}/\Gamma_{Na^+}} \qquad (4.33)$$

and

$$\frac{\Gamma_{Cl^-}}{\Gamma_R} = \frac{z}{\Gamma_{Na^+}/\Gamma_{Cl^-} - 1} \qquad (4.34)$$

For the evaluation of m, therefore, in equation (4.32), one needs to know $\Gamma_{Cl^-}/\Gamma_{Na^+}$ on the basis of a certain assumed model for the electrical double layer formed at the interface. Several such convenient models will be examined in the next sections.

4.7. THE HELMHOLTZ DOUBLE LAYER

In Section 4.4, the orientation of the adsorbed long-chain anion, R of amphiphile RNa, at the liquid interface in the presence of Na^+ and Cl^- in the bulk has been discussed qualitatively. The plane AA' will be charged negatively (vide Fig. 4.4) due to the adsorption of R at the air–water or oil–water interface. In the most simple model of the electrical double layer proposed earlier by Helmholtz,[3] Na^+ obtained from the complete dissociation of RNa_z in the interfacial phase may be arranged in a plane BB' towards the aqueous side. The two oppositely charged planes AA' and BB' are parallel according to this model and the distance δ between them is of the order of molecular thickness. The charge densities σ (charge per unit surface area) of the planes BB' and AA' are equal and opposite in magnitude and sign, respectively. The (negative) charge density of the plane AA' is related to the (negative) surface potential ψ_0 at the charged plane by the Helmholtz relation

$$\psi_0 = \frac{4\pi\sigma}{D}\delta \tag{4.35}$$

where δ is the distance of AA' from BB'. With decrease of δ (i.e., increase of the distance x from AA'), the potential ψ sharply decreases from its maximum value ψ_0. At the plane BB' ψ becomes zero as δ becomes zero. The sharp and linear variation of ψ with increase of x for the Helmholtz double layer has been shown in Fig. 4.4.

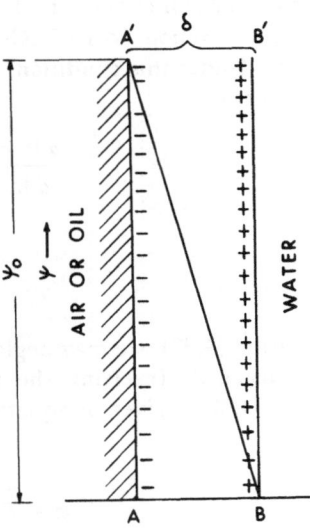

FIGURE 4.4. The Helmholtz double layer.

While applying the Helmholtz model to the adsorption phenomenon at the liquid system, it has been implicitly assumed that RNa_z is highly surface-active, so that Γ_R becomes equal to Δn_R. Further, even in the presence of NaCl in the bulk, the excess accumulation of Cl^- at the interface will be zero according to the Helmholtz model; that is,

$$\Gamma_{Cl^-} = 0 \qquad (4.36)$$

The electroneutrality equation (4.21) for the Helmholtz model will be written in the simple form

$$z\Gamma_R = \Gamma_{Na^+} \qquad (4.37)$$

The electrokinetic and other types of experiments involving charged solid–liquid or liquid–liquid interfaces have subsequently indicated that the double layer is more extended than molecular thickness, and further Γ_{Cl^-} for the charged interface is not in general zero.[15,16] In spite of this serious shortcoming, the Gibbs equation applied to the Helmholtz model of the electrical double layer may give rise to situations which are of some experimental significance.

4.8. HELMHOLTZ MODEL AND THE GIBBS ADSORPTION EQUATION

For the application of the Gibbs equation to the multicomponent solution of RNa_z and NaCl, the experiments are designed in such a manner that the boundary tension of the liquid system may be measured as a function of the bulk concentration (c_R) of RNa_z, keeping the aqueous concentration c_{Cl^-} constant. Under this condition, on differentiation of equation (4.22), one can easily obtain

$$\frac{d \ln c_{Na^+}}{d \ln c_R} = \frac{c_R}{c_{Na^+}} \cdot \frac{dc_{Na^+}}{dc_R}$$

$$= \frac{zc_R}{zc_R + c_{Cl^-}} \qquad (4.38)$$

In equation (4.32), we may neglect the third term in the bracket since c_{Cl^-} is kept constant.[14] Inserting the values of Γ_{Na^+}/Γ_R and $d \ln c_{Na^+}/d \ln c_R$ in equation (4.32) as given by equation (4.37) and (4.38), it can easily be shown[12,13] that

$$m = \xi_R \left[1 + \frac{\xi_{Na^+}}{\xi_R} \cdot \frac{z^2}{z + \bar{x}} \right] \qquad (4.39)$$

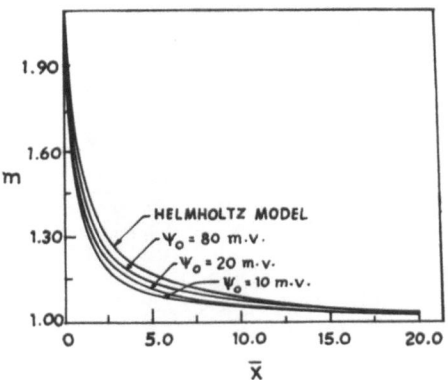

FIGURE 4.5. Variation of coefficient m with the salt ratio $\bar{x}$ at different potentials.[1] (Reproduced with permission from the *Indian Journal of Chemistry*.)

Here $\bar{x}$ stands for c_{Cl^-}/c_R, the ratio of the molar concentrations of the inorganic to organic salts present in the aqueous medium. When the concentration of the organic electrolyte is small, it will be shown in Section 4.11 that ξ_R, ξ_{Na^+} etc. will tend to unity, so that for such simple system

$$m = 1 + \frac{z^2}{z + \bar{x}} \qquad (4.40)$$

If the ratio of the concentration of the neutral salt to that of the organic salt is very high, $\bar{x}$ tends to infinity and m in equation (4.40) becomes unity irrespective of the value of z. On the other hand, if the surface tensions of the solutions are measured for different concentrations of the organic electrolytes in the complete absence of the inorganic salt in the aqueous medium, $\bar{x}$ becomes zero so that m becomes equal to $1 + z$. For uni-univalent organic electrolytes such as SDS, sodium octanoate, etc., m equals 2, whereas for uni-bivalent salt such as sodium sebacate, m will be equal to 3, and so on. Values of m for different values of $\bar{x}$ varying between zero and 20 calculated according to equation (4.40) have been presented graphically in Fig. 4.5. Multiplying $d\gamma/d \ln c_R$ by $(1/m)RT$, one obtains conveniently the values of Γ_R on the basis of the Gibbs equation (4.31). The application of this equation to the experimental data will be further discussed in Section 4.12.

4.9. GOUY–CHAPMAN DOUBLE LAYER

The theory of the diffuse double layer proposed by Gouy[5,6] and Chapman[7] has been frequently used for the interpretation of many experimental data involving surface charge and surface potential. The picture of the Gouy double layer has been discussed elaborately by Kruyt,[17] Voyutsky,[15] and others. The Gouy layer for the adsorption of organic electrolyte RNa_z at the liquid interface in the presence of NaCl is shown in Fig. 4.6. As before, the interfacial plane

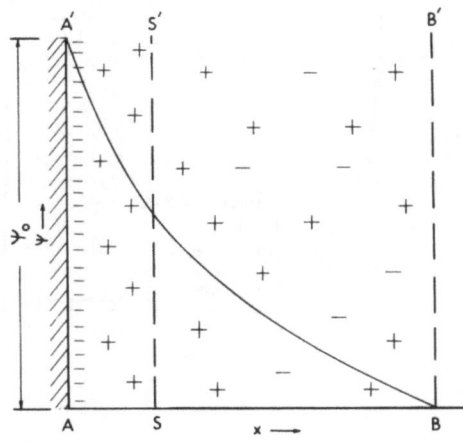

FIGURE 4.6. The Gouy–Chapman double layer.

AA' will be negatively charged due to the adsorption of the organic ions at the interface. As a result of this surface charge, Na^+ and Cl^- in the interfacial region will be distributed nonuniformly owing to the combined action of the electrostatic forces exerted by the negatively charged plane and the thermal forces (of diffusion) acquired by the inorganic ions at the temperature T of the system. If $\dot{c}^s_{Na^+}$ and $\dot{c}^s_{Cl^-}$ represent the number of Na^+ and Cl^-, respectively, per milliliter near the surface region at a perpendicular distance x from the charged plane AA', then using the Boltzmann distribution equations, it may be shown that

$$\dot{c}^s_{Na^+} = \dot{c}_{Na^+}\, e^{-\varepsilon\psi/kT} \tag{4.41}$$

$$\dot{c}^s_{Cl^-} = \dot{c}_{Cl^-}\, e^{+\varepsilon\psi/kT} \tag{4.42}$$

Here, $\dot{c}_{Na^+}$ and $\dot{c}_{Cl^-}$ are the number of sodium and chloride ions per milliliter, respectively, in the bulk phase and k is the Boltzmann constant. The negative electric potential ψ at x tending to zero will acquire the maximum value ψ_0. With increase of x, the value of ψ decreases from ψ_0 slowly with distance and becomes nearly zero at a plane BB' far away from the interface. Inserting a negative sign for ψ in equations (4.41) and (4.42), it may be observed that $\dot{c}^s_{Na^+}$ decreases and $\dot{c}^s_{Cl^-}$ increases as the distance x from the interface increases until their values become equal to $\dot{c}_{Na^+}$ and $\dot{c}_{Cl^-}$ when ψ becomes zero. In Fig. 4.7, the variations of $\dot{c}^s_{Na^+}$ and $\dot{c}^s_{Cl^-}$ with x have been shown for the arbitrary but fixed bulk concentrations of cations and anions, respectively. The extended region of x between AA' and BB' in Fig. 4.6 may be termed the diffuse or the Gouy–Chapman double layer. At $\psi > 0$, $\dot{c}^s_{Na^+}$ and $\dot{c}^s_{Cl^-}$ will,

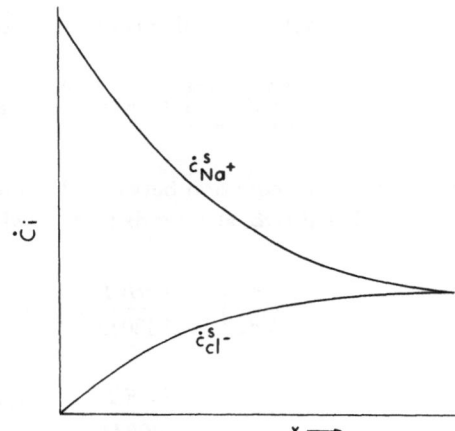

FIGURE 4.7. Variation of the concentration of cation and anion with distance from the interface.

respectively, differ from $\dot{c}_{Na^+}$ and $\dot{c}_{Cl^-}$. The exact physical position of BB' for ψ equal to zero is, however, difficult to locate.

The volume density of charge ρ (per ml) at a position within AA'BB' may be defined as

$$\rho = \varepsilon(\dot{c}^s_+ - \dot{c}^s_-) \tag{4.43}$$

ρ may be alternatively expressed by the Poisson equation

$$\frac{d^2\psi}{dx^2} = -\frac{4\pi\rho}{D} \tag{4.44}$$

with the implicit assumption that the interface is flat, so that it is sufficient to consider change in ψ only in the x direction normal to the surface plane. Here D stands for the dielectric constant of the medium within the space AA'BB'.

Since c_R is usually very small compared to c_{Cl^-}, one may neglect c_R in the equation similar to (4.22) and write $\dot{c}_{Cl^-} \simeq \dot{c}_{Na^+} \simeq cN/1000$ where c (or c_{Cl^-}) stands for the bulk molar concentration of the netural salt in the medium. Combining equations (4.41), (4.42), (4.43), and (4.44), it may be shown

$$\frac{d^2\psi}{dx^2} = \frac{d}{dx}\left(\frac{d\psi}{dx}\right)$$

$$= -\frac{4\pi N\varepsilon}{1000D}c[e^{-\varepsilon\psi/kT} - e^{+\varepsilon\psi/kT}] \tag{4.45}$$

Multiplying both sides of this equation by $(d\psi/dx)dx$, we obtain

$$\frac{d\psi}{dx} d\left(\frac{d\psi}{dx}\right) = -\frac{4\pi N\varepsilon}{1000D} c[e^{-\varepsilon\psi/kT} - e^{+\varepsilon\psi/kT}]d\psi \qquad (4.46)$$

Integrating this equation between the limits 0 and ψ and taking into account the fact that $d\psi/dx$ also tends to zero when ψ approaches zero

$$\frac{1}{2}\left(\frac{d\psi}{dx}\right)^2 = \frac{4\pi RT}{1000D} c\int_0^\psi (e^{\varepsilon\psi/kT} - e^{-\varepsilon\psi/kT})d\frac{\varepsilon\psi}{kT}$$

$$= \frac{4\pi RT}{1000D} c(e^{\varepsilon\psi/2kT} - e^{-\varepsilon\psi/2kT})^2 \qquad (4.47)$$

or

$$\frac{d\psi}{dx} = -\left(\frac{8\pi RTc}{1000D}\right)^{1/2} (e^{\varepsilon\psi/2kT} - e^{-\varepsilon\psi/2kT}) \qquad (4.48)$$

Since ψ decreases with increase of x, $d\psi/dx$ cannot be positive.

If we now draw a circle of unit surface area on the charged plane AA', then the negative charges aquired by adsorbed organic ions within this unit area represents the surface charge density σ. If we draw a cylinder perpendicular to this unit surface area, then the net charges due to the presence of Na$^+$ and Cl$^-$ within this cylindrical space ranging between x equal to zero and infinity (i.e., between the planes AA' and BB') must be equal to $-\sigma$ so that the surface phase as a whole will be electroneutral. Then,

$$\sigma = -\int_0^\infty \rho dx \qquad (4.49)$$

Combining equations (4.44) and (4.49), we obtain

$$\sigma = \frac{D}{4\pi} \int_0^\infty \frac{d^2\psi}{dx^2} dx$$

$$= \frac{D}{4\pi} \int_0^\infty d\left(\frac{d\psi}{dx}\right)$$

$$= -\frac{D}{4\pi} \left(\frac{d\psi}{dx}\right)_{x=0} \qquad (4.50)$$

At x approaching zero, ψ becomes equal to the surface potential ψ_0, so that combining (4.48) and (4.50), we have

$$\sigma = \left(\frac{DRTc}{2000\pi}\right)^{1/2} (e^{\varepsilon\psi_0/2kT} - e^{-\varepsilon\psi_0/2kT})$$

$$= \left(\frac{2DRTc}{1000\pi}\right)^{1/2} \sinh\frac{\varepsilon\psi_0}{2kT} \tag{4.51}$$

The Gouy equation (4.51) may be conveniently derived for symmetrical inorganic electrolytes ($MgSO_4$, $NaCl$, etc.). In the general case, $\varepsilon\psi_0/2kT$ in equation (4.51) is to be replaced by $z_i\varepsilon\psi_0/2kT$, where z_i is the valency of the corresponding inorganic cations or anions. Further, c may be replaced by the ionic strength of the electrolyte in the medium. If c_R is not negligible, its contribution to the ionic strength (and c) is to be taken into account. This may pose slight conceptual difficulties in the non-Boltzmann distribution of the organic ions at the interface, but the effect due to this seems to be trivial.

The average thickness of the electrical double layer $1/\kappa$ is given by the equation

$$\frac{1}{\kappa} = \left(\frac{1000DRT}{8\pi N^2\varepsilon^2 c}\right)^{1/2} \tag{4.52}$$

$1/\kappa$ has the dimension of length and is identical with the Debye–Hückel thickness of the ion atmosphere.[18] In fact, the Gouy–Chapman theory for charged surface is based on the same physical concept as the Debye–Hückel theory of strong electrolytes. At 25°C, for uni-univalent electrolytes

$$\kappa = 3.282 \times 10^7\sqrt{c} \text{ cm}^{-1} \tag{4.53}$$

Values of $1/\kappa$ for various values of c at 25°C are incorporated in Table 4.1.

For small value of ψ, $e^{\varepsilon\psi/2kT}$ and $e^{-\varepsilon\psi/2kT}$ in equation (4.48) may be replaced by $1 + (\varepsilon\psi/2kT)$ and $1 - (\varepsilon\psi/2kT)$, respectively, so that inserting these values in equation (4.48), we have

$$\frac{d\psi}{dx} = -\kappa\psi \tag{4.54}$$

Integrating this equation, we find

$$\ln \psi = -\kappa x + I \tag{4.55}$$

TABLE 4.1. Thickness of the Double Layer at Various Concentrations of 1:1 Electrolyte at 25°C

Concentration (mol/liter)	$1/\kappa$ (nm)
1×10^{-4}	30.5
5×10^{-4}	13.6
1×10^{-3}	9.64
5×10^{-3}	4.31
1×10^{-2}	3.05
5×10^{-2}	1.36
1×10^{-1}	0.96
5×10^{-1}	0.43
1.00	0.31

when $x \to 0$, $\psi \to \psi_0$, I is equal to $\ln \psi_0$. Putting this in equation (4.55),

$$\psi = \psi_0 \, e^{-\kappa x} \tag{4.56}$$

This equation indicates that at a distance x equal to $1/\kappa$, the potential is actually ψ_0/e (or $\psi_0/2.72$). Thus beyond the thickness of the double layer, ψ is small and its value is negligible when x is considerably greater than $1/\kappa$.

Combining equations (4.51) and (4.52), one gets

$$\sigma = \frac{DRT\kappa}{2\pi N\varepsilon} \sinh \frac{\varepsilon \psi_0}{2kT} \tag{4.57}$$

When ψ_0 is small, $\sinh (\varepsilon \psi_0/2kT)$ becomes equal to $\varepsilon \psi_0/2kT$, so that

$$\sigma = \frac{D\kappa}{4\pi} \psi_0 \tag{4.58}$$

Equation (4.58) is also the expression for the relation between the potential and charge for a plane plate condenser where $1/\kappa$ represents the distance between the plates [vide equation (4.35)].

The Gouy equation (4.51) may become quite useful in calculating the surface charge density or the surface potential provided experiments on adsorption or surface potential measurements are carefully carried out. At 25°C, equation (4.51) will assume the following form:

$$\sigma = 0.3514 \times 10^5 \sqrt{c} \sinh 0.0194\psi_0 \tag{4.59}$$

Here ψ_0 is expressed in practical millivolts, whereas in equation (4.51), the electrostatic volt unit is used. If Γ_R is the amount in moles of the organic electrolyte RNa_z adsorbed per unit surface area, then

$$\sigma = z\Gamma_R N\varepsilon \qquad (4.60)$$

Since Γ_R can be experimentally determined directly or indirectly, ψ_0 may be calculated with the help of equations (4.59) and (4.60) from the known value of the salt concentration c in the bulk.

From direct analysis of adsorption, Das and Chattoraj[19] have recently determined values of Γ_R for several types of oil–water interfaces in the presence of SDS or CTAB and NaCl. The variations of charge density σ for these systems calculated with the help of equation (4.60) for various values of c_R are shown in Fig. 4.8. The surface potential of these systems may be calculated with the help of equation (4.59). In Fig. 4.9 are presented values of ψ_0 for different values of c_R.

The electrophoretic mobility of the emulsion droplets formed with these oil–water systems has also been measured by Das and Chattoraj,[19] from which the zeta-potential has been calculated using the Helmholtz–Smoluchowski equation. Zeta-potential has been identified with ψ_0 by many investigators on the basis of adsorption studies. Zeta-potentials of these systems, presented in Fig. 4.9, are found to be considerably lower than ψ_0 when c_R is low but with increasing surface saturation, i.e., at higher values of c_R, the difference is

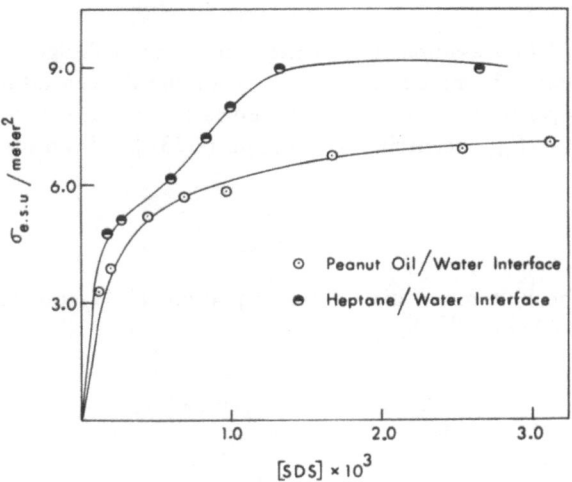

FIGURE 4.8. Plot of charge density of oil–water emulsions against SDS concentration. (Data from the Ph.D. thesis of K. P. Das, Jadavpur University, 1981.)

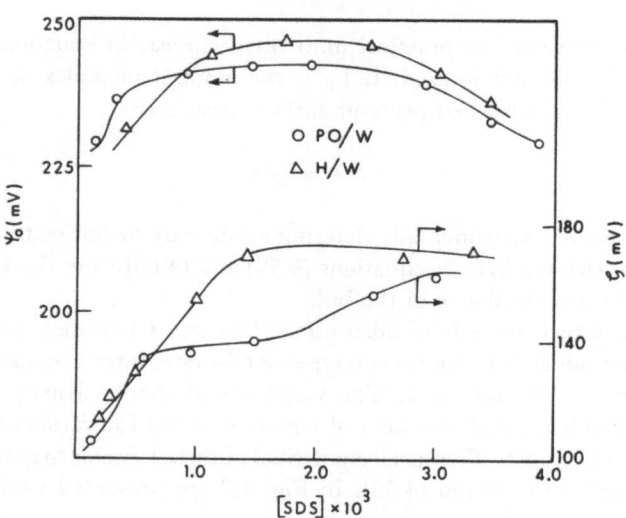

FIGURE 4.9. Plot of ψ_0 and ζ against SDS concentration. (Data from the Ph.D. thesis of K. P. Das, Jadavpur University, 1981.)

gradually reduced. The difference observed between ψ_0 and ζ establishes the hypothesis that the slipping plane SS' for electrokinetic motion is different from the charged plane AA' at the interfacial region (vide Fig. 4.6).

4.10. GOUY MODEL AND THE GIBBS ADSORPTION EQUATION

As before, let us assume that in the experimental design followed, the boundary tension γ is measured as function of the concentration c_R of the organic electrolyte RNa_z at constant concentration c_{Cl^-} (or c) of NaCl. For such system, $\Gamma_{Cl^-}/\Gamma_{Na^+}$ occurring in equation (4.33) has been identified as

$$\frac{\Gamma_{Cl^-}}{\Gamma_{Na^+}} = \frac{\overset{\cdot s}{c}_{Cl^-}}{\overset{\cdot s}{c}_{Na^+}} \tag{4.61}$$

so that using the Boltzmann distribution equations (4.41) and (4.42), it has been shown by Chattoraj[12] that

$$\frac{\Gamma_{Cl^-}}{\Gamma_{Na^+}} = \frac{c_{Cl^-}}{zc_R + c_{Cl^-}} e^{2\varepsilon\psi_0/kT} \tag{4.62}$$

However, Bijsterbosch and van den Hull[20] have pointed out that Γ_{Cl^-} and Γ_{Na^+} are surface excess quantities and not surface concentrations, so that equation (4.62) is not correct. They have shown that the ratio of the excess amounts of Na^+ and Cl^- in the interfacial phase may be calculated from the relation[9]

$$\frac{\Gamma_{Cl^-}}{\Gamma_{Na^+}} = \frac{\int_0^\infty (\dot{c}^s_{Cl^-} - \dot{c}_{Cl^-})dx}{\int_0^\infty (\dot{c}^s_{Na^+} - \dot{c}_{Cl^-})dx} \tag{4.63}$$

$\dot{c}^s_i$ and $\dot{c}_i$ stand for the number of ith ions per milliliter in the interfacial and bulk phases, respectively. $(\dot{c}^s_i - \dot{c}_i)dx$ is the number of ith ions accumulated in excess in the cylindrical volume element dx projected perpendicularly from the unit surface area of the charged surface AA'. By integrating this quantity between the limits x equal to zero (or ψ equal to ψ_0) and x equal to infinity (or ψ equal to zero), the total amounts of excesses for Na^+ and Cl^- in the interfacial phase AA'BB' may be obtained. The equation is consistent with our macroscopic definition of Γ_i given by equation (3.41), provided differences between molality and molarity both in the surface and bulk phases are neglected. Combining equation (4.63) with (4.41) and (4.42),

$$\frac{\Gamma_{Cl^-}}{\Gamma_{Na^+}} = \frac{c_{Cl^-}}{c_{Na^+}} \frac{\int_0^\infty (e^{\varepsilon\psi/kT} - 1)dx}{\int_0^\infty (e^{-\varepsilon\psi/kT} - 1)dx} \tag{4.64}$$

By algebraic operations, it may be shown that

$$\frac{\Gamma_{Cl^-}}{\Gamma_{Na^+}} = \frac{-c_{Cl^-}\int_0^\infty [(e^{-\varepsilon\psi/kT} + e^{\varepsilon\psi/kT} - 2) e^{+\varepsilon\psi/kT}]^{1/2}dx}{+c_{Na^+}\int_0^\infty [(e^{-\varepsilon\psi/kT} + e^{\varepsilon\psi/kT} - 2) e^{-\varepsilon\psi/kT}]^{1/2}dx} \tag{4.65}$$

From equations (4.41), (4.42), and (4.63) and also from Fig. 4.7, it is clear that $\dot{c}^s_{Cl^-} - \dot{c}_{Cl^-}$ is negative and $\dot{c}^s_{Na^+} - \dot{c}_{Na^+}$ is positive for any value of x within the double layer. During square-root operations, negative and positive signs are to be included outside the integrals in the numerator and denominator, respectively, for avoiding confusion in sign after the completion of the integration.[13,21] Since the term $(e^{-\varepsilon\psi/kT} + e^{\varepsilon\psi/kT} - 2)^{1/2}$ is equal to $(e^{\varepsilon\psi/2kT} - e^{-\varepsilon\psi/2kT})$, one can combine (4.48) with (4.65) so that[13,20,22]

$$\frac{\Gamma_{Cl^-}}{\Gamma_{Na^+}} = -\frac{c_{Cl^-}}{c_{Na^+}} \frac{\int_0^\infty e^{\varepsilon\psi/2kT} d\psi}{\int_0^\infty e^{-\varepsilon\psi/2kT} d\psi}$$

$$= -\frac{c_{Cl^-}}{c_{Na^+}} e^{\varepsilon\psi_0/2kT} \tag{4.66}$$

Since c_{Cl^-} is kept constant in the experimental arrangement, the last term in equation (4.32) is zero. Further, the value of Γ_{Na^+}/Γ_R may be obtained by combining equations (4.33) with (4.66). Equation (4.38) will also remain valid, so that, inserting all these in (4.32), m will be solved thus[14]:

$$m = \xi_R\left[1 + \frac{z^2(\xi_{Na^+}/\xi_R)}{z + \bar{x}(1 + e^{\varepsilon\psi_0/2kT})}\right] \tag{4.67}$$

Let us now analyze this equation for dilute solutions of electrolytes such that ξ_{Na^+} and ξ_R become unity. The equation then assumes the form

$$m = 1 + \frac{z^2}{z + \bar{x}(1 + e^{\varepsilon\psi_0/2kT})} \tag{4.68}$$

In the absence of NaCl, $\bar{x}$ becomes zero so that that m becomes equal to $1 + z$. When the concentration of the inorganic salt is very high compared to that of the organic salt, $\bar{x}$ tends to infinity so that m becomes equal to unity.

In Fig. 4.5, values of m corresponding to the Helmholtz and the Gouy models (with ψ_0 equal to -80, -20, and -10 mV) have been compared with each other for a uni-univalent organic electrolyte ($z = 1$). From these curves, it is clear that dependence of m on ψ_0 is not very significant. For example, when $\bar{x}$ equals 2, values of m for ψ_0 equal to -10, -20, and -80 mV are 1.25, 1.26, and 1.29 respectively. Corresponding potential independent value of m for the Helmholtz double layer calculated on the basis of equation (4.40) is 1.32. For ordinary calculation of the surface excess with the help of the Gibbs equation, relation (4.40) may be used which will lead to uncertainties in the value of Γ_R by 10% to 12%, because of the neglect of surface potential. If the magnitude of ψ_0 is high, $1 + e^{\varepsilon\psi_0/2kT}$ tends to unity, so that m for the Helmholtz and Gouy double layers become almost the same. For a more precise estimation of Γ_R, one should first calculate an approximate value of the surface excess from the combination of equations (4.31) and (4.40). Inserting this value in equations (4.60) and (4.59), one obtains an approximate value of ψ_0. Calculation of Γ_R can now be repeated by combining equation (4.31) with (4.67) and (4.66) based on the Gouy theory in which the approximate value of ψ_0 is inserted. The process can be repeated until the correct value of Γ_R independent of ψ_0 is obtained.

4.11. DEBYE–HÜCKEL THEORY AND THE GIBBS ADSORPTION EQUATION

The ionic strength I of an aqueous solution containing electrolytes RNa_z and NaCl can be calculated from the equation

$$I = [\tfrac{1}{2}(z^2 c_R + z c_R) + 2c_{Cl^-}]$$

$$= \left[\frac{z(z + 1)}{2}\right]c_R + c_{Cl^-} \tag{4.69}$$

Here c_R and c_{Cl^-} stand for the molar concentrations of the organic and inorganic electrolytes in bulk. The bulk activity coefficients f_R and f_{Na^+} are related to I by the extended Debye–Hückel equations

$$\ln f_R = -\frac{\bar{A}z^2\sqrt{I}}{1 + a_- \cdot \bar{B}\sqrt{I}} \qquad (4.70)$$

and

$$\ln f_{Na^+} = -\frac{\bar{A}\sqrt{I}}{1 + a_+ \cdot \bar{B}\sqrt{I}} \qquad (4.71)$$

Here $\bar{A}$ is equal to $2.303A_{DH}$. The values of the Debye–Hückel parameters A_{DH} and $\bar{B}$ for water at 25°C are 0.509 and 0.33×10^8, respectively.[23] The respective radii a_- and a_+ of the organic and inorganic ions may be replaced by an average radius $a_\pm$ so that inserting this in equations (4.70) and (4.71), one obtains an additional relation:

$$\ln f_R = z^2 \ln f_{Na^+} \qquad (4.72)$$

Inserting the value of I from equation (4.69) in equation (4.72), and then differentiating $\ln f_R$ with respect to $\ln c_R$ at constant value of c_{Cl^-}, it may be shown that[24,25]

$$\xi_R = 1 - \frac{z^3(z+1)\bar{A}c_R}{[2z(z+1)c_R + 2c_{Cl^-}]^{1/2}\{1 + a_\pm \cdot (\bar{B}/2)[2z(z+1)c_R + 4c_{Cl^-}]^{1/2}\}^2} \qquad (4.73)$$

Combining equations (4.38), (4.71), and (4.72), we obtain

$$\xi_{Na^+} = 1 + \frac{\xi_R - 1}{z^3}\left(z + \frac{c_{Cl^-}}{c_R}\right) \qquad (4.74)$$

Inserting the values of ξ_R and ξ_{Na^+} in equation (4.67) for the Gouy double layer, we find

$$m = \xi_R\left\{1 + \frac{1}{\xi_R}\left[\frac{\xi_R - 1}{z} + \frac{z^2}{z + \bar{x}(1 + e^{\varepsilon\psi_0/2kT})}\right]\right\} \qquad (4.75)$$

For a given value of z, c_R, and c_{Cl^-}, m may be calculated using equation (4.75) for assumed values of the mean radius of the ions. In Fig. 4.10, m for a bolaform electrolyte (e.g., sodium sebacate with z equal to 2) has been calculated using equation (4.75) for various values of c_R at a constant value

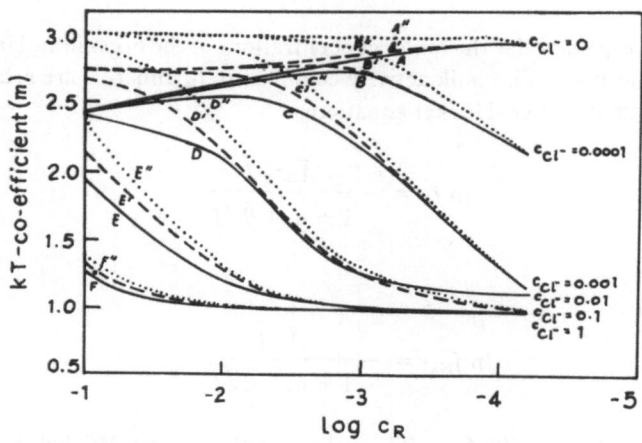

FIGURE 4.10. Plot of m vs. log of the concentration of organic electrolyte.[24] Curves A, B, C, D, E, and F are drawn according to the extended Debye–Hückel equation, $\bar{a}_\pm = 0.5$ nm; curves A′, B′, C′, D′, E′, and F′ are drawn according to the extended Debye–Hückel equation, $\bar{a}_\pm = 1.0$ nm; and curves A″, B″, C″, D″, E″, and F″ are drawn according to the ideal behavior without interionic attractions ($\xi = 1$). (Reproduced with permission from the *Indian Journal of Chemistry*.)

of c_{Cl^-}; $1 + e^{\varepsilon\psi_0/2kT}$ is assumed to be unity in this calculation. The deviation from ideality ($\xi_R = 1$) for the calculated value of m has been noted in this figure over a wide range of concentration. Further, it appears from Fig. 4.10 that the calculated value of m is found to depend on the assumed value of $a_\pm$.

In surface tension measurements, the solutions are usually dilute, so that $\bar{B}$ in equation (4.73) may be put equal to zero in accordance with the Debye–Hückel limiting law. The calculation then becomes relatively simple since there is no need to use an arbitrary value of $a_\pm$ in equation (4.73). Values of m for the kT coefficient of the bolaform electrolyte ($z = 2$) are also given in Fig. 4.11, on the basis of the Debye–Hückel law. Difference in the values of m for certain values of c_R and closeness in some regions may be noted with interest. The significance of this in constructing the pressure-area curves of a charged monolayer will be discussed in the next chapter.

4.12. EXPERIMENTAL VALUES OF THE COEFFICIENT m

If γ of a solution is measured as function of the concentration c_R of the organic electrolyte in the complete absence of the inorganic salt ($c_{Cl^-} = 0$), then from equation (4.75),

$$m = \xi_R \left[1 + \frac{z}{\xi_R} + \frac{\xi_R - 1}{z\xi_R} \right] \qquad (4.76)$$

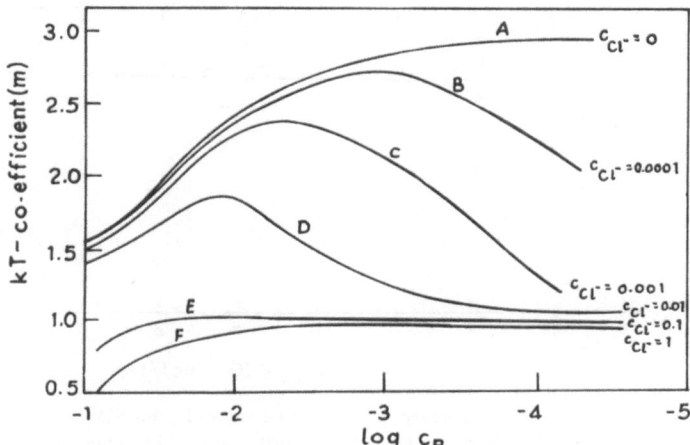

FIGURE 4.11. Plot of m vs. log of the concentration of the organic electrolyte (C_R). Curves A, B, C, D, E, and F are drawn according to the limiting Debye–Hückel equation.[24] (Reproduced with permission from the *Indian Journal of Chemistry*.)

The coefficient in equation (4.76) depends only on the activity parameter ξ_R, which in turn can be calculated with the help of equation (4.73) from a knowledge of c_R and $a_\pm$. ξ_R is close to unity if c_R is low [vide equation (4.73)] so that m tends to $1 + z$. For a uni-univalent electrolyte, m thus tends to 2, which was predicted by Guggenheim[26] on simple thermodynamic grounds. Earlier, several attempts[27–30] were made to determine m for uni-univalent organic electrolyte in the presence and absence of an inorganic salt from direct measurement of Γ_R by radio-tracer or other adsorption experiments together with evaluation of $d\gamma/d \ln c_R$ of the air–water interface using surface tension experiments. For [14]C- and ^{35}S-labeled SDS, such simultaneous measurements[31–34] led to the value of m equal to unity in the absence of NaCl although from the theory the value of the coefficient is expected to be 2.

In more recent works, Tajima *et al.*[35] used tritiated sodium dodecyl sulfate (TSDS) for the direct estimation of Γ_R from radiochemical analysis. These values are shown in Fig. 4.12 for various solution activities of SDS. Γ_R for the same system has also been estimated by these workers from the $\gamma - c_R$ data using m equal to $2\xi_R$ [vide also equation (4.76)]. The two values are found to agree with each other within the limits of experimental error. This proves that tracer analysis in earlier measurements using S- and C-labeled SDS are not reliable, and from the results with tritiated SDS, m is actually close to 2 as predicted from the theory.

Das and Chattoraj[19] have prepared oil–water emulsions in the presence of increasing concentrations of SDS or CTAB. The concentration of the organic electrolyte in the aqueous phase of these systems has been analyzed by them following a special procedure such that the total amount of SDS or

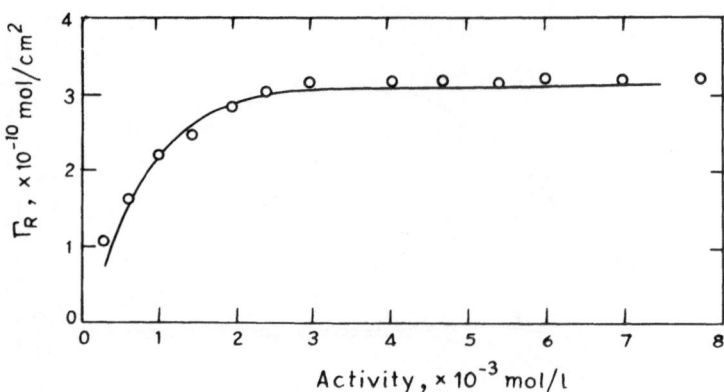

FIGURE 4.12. Comparison of observed and calculated values Γ_R for SDS. $\bigcirc$, observed values by tracer method; ———, calculated values (Wilhelmy plate method) by Gibbs equation. (Redrawn from Ref. 35.)

CTAB adsorbed at the oil–water interface can be calculated. The total surface area of the emulsion droplets has been simultaneously measured by combining microscopic and turbidimetric methods. From all these, it is possible to calculate Γ_R at a given bulk concentration c_R of the surfactant in the aqueous phase. From the measurement of the interfacial tension γ of the oil–water systems as a function of c_R in the aqueous phase, the value of $d\gamma/d \ln c_R$ has been calculated for a given value of c_R and Γ_R. Inserting these values in equation (4.31), m has been calculated directly from experiments. The results are presented in Fig. 4.13 for SDS and CTAB adsorbed at heptane–water and peanut oil–water interfaces. Obviously, the values of m in all cases are very close to 2 in accordance with equation (4.76).

From equation (4.73), it is noted that ξ_R tends to unity when the ratio c_{Cl^-}/c_R becomes considerably high. From equation (4.75), we also note with interest that m tends to unity when $\bar{x}$ is considerably high (i.e., $\bar{x} \to \infty$), as is evident from Fig. 4.5. Tajima[36] has measured Γ_R for tritiated SDS in the presence of a constant concentration of NaCl in the bulk using radiochemical analysis. Values of Γ_R for different values of c_R have been shown in Fig. 4.14. Under similar experimental conditions, Tajima has calculated values Γ_R from the surface tension measurement using the Gibbs equation (4.31) in which m is assumed to be unity. These results, also shown in Fig. 4.14, are found to agree with those obtained from the radiochemical analysis. This proves beyond doubt that the value of m becomes unity at high salt concentrations in agreement with the theory.

Das and Chattoraj[19] have measured Γ_R for oil–water emulsion as a function of c_R in the presence of 1 mM NaCl following the analysis of the surfactant in the aqueous phase and estimation of the interfacial area as described earlier in this section. From the measurement of the interfacial tension γ, values of

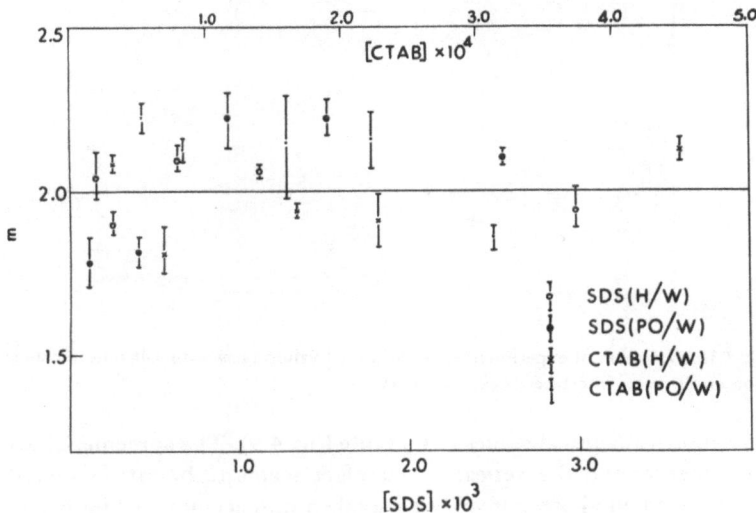

FIGURE 4.13. Variation of m with detergent concentration in absence of added salt.[19]

$d\gamma/d \ln c_R$ for several concentrations of RNa_z have been calculated by them. Inserting these values of Γ_R and $d\gamma/d \ln c_R$ in equation (4.31), values of m for different salt ratios c_{Cl^-}/c_R can be calculated on the basis of experimental data. In Fig. 4.15 values of m obtained from theory for different values of $\bar{x}$ have been compared with those calculated on the basis of equations (4.39) and (4.67), derived, respectively, on the basis of the double layer models of Helmholtz and Gouy. The surface potentials ψ_0 of these systems are known

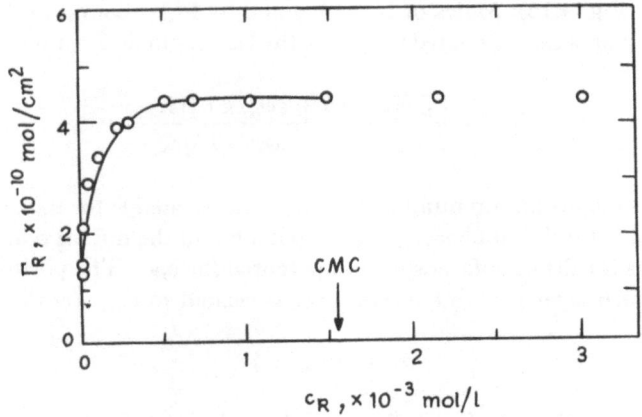

FIGURE 4.14. Adsorption isotherm of SDS solution. $\bigcirc$, observed value by tracer method; ———, calculated using equation (4.31) taking $m = 1$. (Redrawn from Ref. 36.)

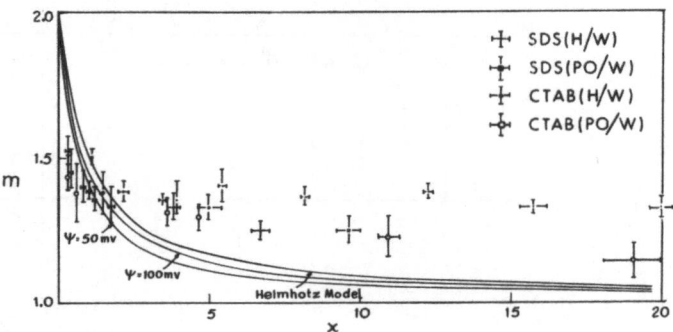

FIGURE 4.15. Variation of experimentally determined values of m with salt ratio for detergent adsorption at oil–water interfaces.[19] ($\bar{x}$ written as x.)

from the known analytical values of Γ_R (vide Fig. 4.9). The agreement between the experimental and theoretical values of m seems to be satisfactory if the limits of experimental error involved are taken into account. At higher values of $\bar{x}$, the experimental values of m tend to reach the limiting value of unity.

4.13. STERN MODEL OF THE DOUBLE LAYER

The formation of the double layer due to the adsorption of the organic electrolyte RNa_z at the liquid interface in the presence of sodium chloride may also be explained on the basis of the more realistic model presented by Stern.[8] According to this model, some fraction of Na^+ counterions in the interfacial phase may remain physically bound to the adsorbed organic anions R under the action of the strong electrostatic as well as other types of specific forces (vide Fig. 4.16). Moles of Na^+ (denoted by $\Gamma^s_{Na^+}$) bound in the double layer per unit area can be related to c_{Na^+} by the Langmuir adsorption equation[15]

$$\Gamma^s_{Na^+} = \frac{\Gamma^m_{Na^+}[\exp{(\theta + \varepsilon\psi_\delta/kT)}]c_{Na^+}}{1 + [\exp{(\theta + \varepsilon\psi_\delta/kT)}]c_{Na^+}} \qquad (4.77)$$

Here $\Gamma^m_{Na^+}$ is the maximum number of charge sites available for the counterion binding at the interfacial phase; ψ_δ is the potential in the diffuse double layer; and θ stands for the specific adsorption potential for Na^+. The positive charge σ_f in the Stern layer per unit surface area is related to $\Gamma^s_{Na^+}$ by the equation

$$\sigma_f = N\varepsilon(\Gamma^s_{Na^+}) \qquad (4.78)$$

The Na^+ ions in the interfacial phase not bound with adsorbed organic ions along with accumulated Cl^- ions may constitute the diffuse double layer due

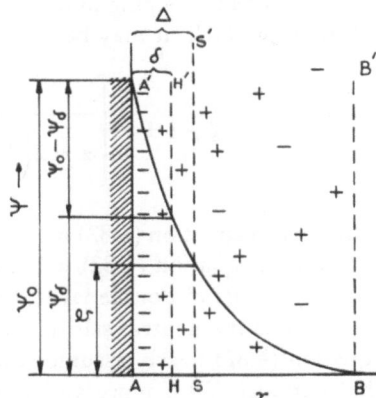

FIGURE 4.16. The Stern double layer.

to the combined action of the thermal and electrostatic forces in the vicinity of the interface. Na^+ and Cl^- in the extended double layer may be distributed following the Boltzmann distribution equations (4.41) and (4.42). From Fig. 4.16, it is clear that the total potential ψ_0 consists of the potentials drop ψ_δ in the diffuse part (in the enclosed space HH'BB') of the double layer and the drop in potential $\psi_0 - \psi_\delta$ within the fixed part (space between AA'HH') of the double layer. The potential ψ_δ is usually higher than the electrokinetic potential ζ since the position of the slipping plane SS' is different from that of the Stern plane (HH') dividing the fixed and diffuse parts of the double layer.

For the application of the Stern model in the Gibbs adsorption equation,[13,14] we write

$$\Gamma_{Na^+} = \Gamma_{Na^+}^s + (\Gamma_{Na^+})_d \tag{4.79}$$

Here $(\Gamma_{Na^+})_d$ is the surface excess of Na^+ in the diffuse double layer. On the basis of this

$$\frac{\Gamma_{Cl^-}}{\Gamma_{Na^+}} = \frac{1}{1+J}\left(\frac{\Gamma_{Cl^-}}{\Gamma_{Na^+}}\right)_d \tag{4.80}$$

where

$$J = \frac{\Gamma_{Na^+}^s}{(\Gamma_{Na^+})_d} \tag{4.81}$$

It is assumed that all Cl^- ions in the interfacial phase exist in the diffuse double layer. The value of $(\Gamma_{Cl^-}/\Gamma_{Na^+})_d$ may be given by equation (4.66) with replacement of ψ_0 by ψ_δ. Combining equations (4.80) with relations (4.32),

(4.33), and (4.38) and keeping in mind that $d \ln c_{Cl^-}/d \ln c_R$ is zero at constant concentration of NaCl, it may be shown that

$$m = \xi_R \left\{ 1 + \frac{1}{\xi_R} \left[\frac{z^2}{z + \bar{x}\left(1 + \dfrac{e^{\varepsilon\psi_\delta/2kT}}{1+J}\right)} + \frac{\xi_R - 1}{z} \right] \right\} \qquad (4.82)$$

The value of J in equation (4.82) will become zero in the absence of counterion binding, so that equation (4.82) will assume the form of relation (4.75) valid for the Gouy model of double layer. On the other hand, with increase of the concentration of the neutral salt in the medium, $1 + J$ may be very high, so that equation (4.82) nearly assumes the form of relation (4.39) based on the Helmholtz model. With more and more counterion binding, therefore, the whole term $e^{\varepsilon\psi_\delta/2kT}/(1 + J)$ is reduced. In the previous section, it has been shown that even in the absence of counterion binding, the effect of the surface potential on m is small and this effect is further reduced with counterion binding within the double layer.

If necessary, the value of J for the double layer may be calculated at 25°C from the approximate relation given by Chattoraj[13] based on the treatment of Grahame[9]:

$$J = 1.47\sqrt{c_{Na^+}} \frac{e^{-\varepsilon\psi_\delta/kT}}{e^{-\varepsilon\psi_\delta/kT} - 1} \qquad (4.83)$$

Although the role of the counterion binding for the computation of m is minor, its effective role for the equation of state of the charged monolayer is quite significant when concentration of NaCl is very high. This will be dealt with in the next chapter.

When RNa_z is a polyelectrolyte, the anionic polyion may bind a considerable amount of Na^+ ions in the bulk as well as in the surface phases in different proportions. The Gibbs equation for such complicated system undergoing polyelectrolyte adsorption has been derived by Chattoraj and Pal,[14] which contains many binding parameters associated with the polymer dissolved in the bulk medium.

4.14. GIBBS EQUATIONS FOR MORE THAN TWO ELECTROLYTE COMPONENTS

It is possible to extend the treatment given in electrolyte mixtures provided the boundary tension experiments are properly designed. For the sake of simplicity, we assume ξ_i to be unity. Let us consider a special case where γ

is measured as a function of increasing concentration of RNa_z at a fixed pH and fixed total concentration of NaCl, KBr, H^+, and OH^-. Under this situation, m in equation (4.31) is given by the relation[14]

$$m = 1 + \frac{z^2}{1 + \dfrac{c_{Cl^-} + c_{Br^-} + c_{OH^-}}{c_R}(1 + e^{\varepsilon\psi_0/2kT})} \tag{4.84}$$

In Fig. 4.17, the adsorption isotherms of NaDS in the presence of $0.01(M)$ NaCl, KBr, or NaCl–KBr mixture in 50:50 molar ratio calculated with the help of the equations (4.68) and (4.84) for the electrolyte or the electrolyte mixtures are compared with each other.[2] The isotherm is found to be independent of the neutral salts used. $e^{\varepsilon\psi_0/2kT}$ in these calculations has been neglected with respect to unity.

Sometimes, one may have to deal with a sample composed of two organic electrolytes RNa_{z_1} and RNa_{z_2} having different anionic valences, z_1 and z_2, respectively, and being present in constant proportions. If γ is measured as function of total concentration c'_R of this sample at a constant excess concentration of NaCl, then it may be shown[14]

$$-d\gamma = mRT(\Gamma_{R_1} + \Gamma_{R_2})d \ln c'_R \tag{4.85}$$

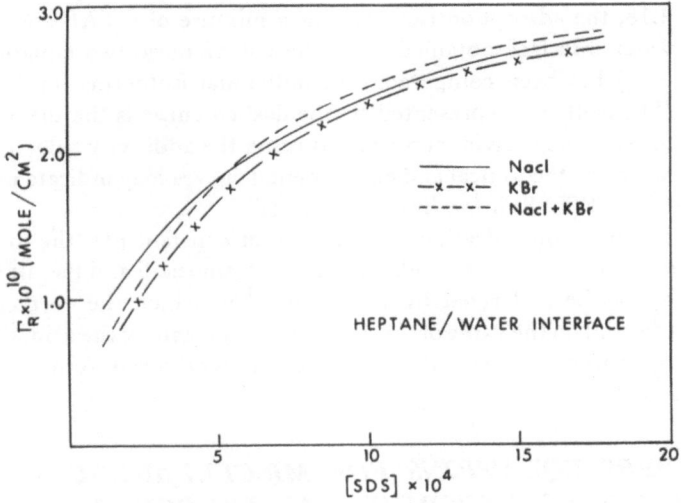

FIGURE 4.17. Plot of Γ_R against detergent (SDS) concentration in presence of salt and salt mixtures. (Das and Chattoraj, unpublished data.)

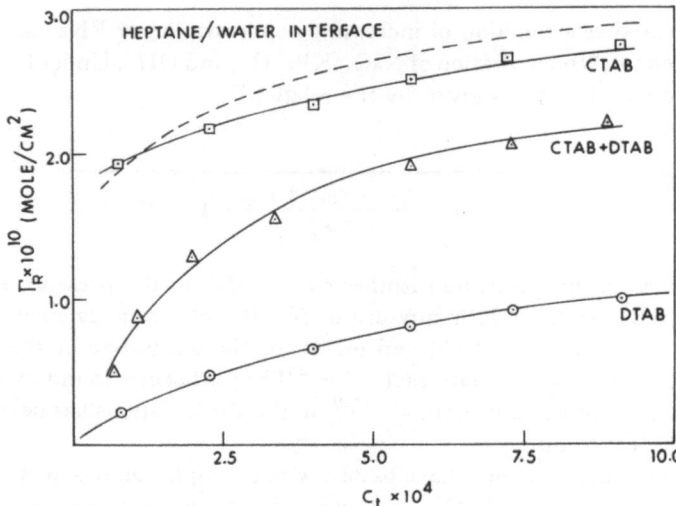

FIGURE 4.18. Isotherms for detergents and detergent mixture. Dashed curve for detergent mixture calculated using the additivity rule. NaCl concentration, 0.1 M. (Das and Chattoraj, unpublished.)

where

$$m = 1 + \frac{z^2}{z + \dfrac{c_{Cl^-}}{c_{R_1} + c_{R_2}}(1 + e^{\epsilon\psi_0/2kT})} \qquad (4.86)$$

In Fig. 4.18, the adsorption isotherm for a mixture of CTAB and DTAB at the oil–water interface obtained from the use of these two equations (with $e^{\epsilon\psi_0/2kT} \ll 1$) has been compared with individual isotherms for CTAB and DTAB. The isotherm represented by the dashed curve is the theoretical one calculated from the individual isotherms using the additivity rule. The difference between the theoretical and experimental curves may indicate interaction of CTAB and DTAB in the interfacial phase.[2]

At constant concentration of NaCl and at constant pH, the form of the Gibbs equation for the mixed adsorption of organic acid and its salt (e.g., RH and RNa) has been derived by Chattoraj[14,37] in which the coefficient m is found to depend on the value of the dissociation constant of the acid, concentrations of c_{RH} and c_{R^-}, ψ_0, etc. The surface excesses obtained from this equation will be $\Gamma_{RH} + \Gamma_{R^-}$.

4.15. GIBBS EQUATIONS FOR MISCELLANEOUS TYPES OF EXPERIMENTAL PROCEDURES

Significant lowering of the boundary tension of oil–water or air–water systems has been noted[38–40] when sodium chloride concentration in the aqueous phase is increased in the presence of constant concentration of sodium dodecyl sulfate in the medium. To treat such a system, equation (4.29) can be written in the form[14]

$$-d\gamma = RT\Gamma_{Cl^-}\xi_{Cl^-}d\ln c_{Cl^-}\left(\frac{\Gamma_{Na^+}}{\Gamma_{Cl^-}}\frac{\xi_{Na^+}}{\xi_{Cl^-}}\frac{d\ln c_{Na^+}}{d\ln c_{Cl^-}} + 1\right) \tag{4.87}$$

Since c_R is kept constant, equation (4.22) on differentiation will lead to the relation

$$\frac{d\ln c_{Na^+}}{d\ln c_{Cl^-}} = \frac{\bar{x}}{1+\bar{x}} \tag{4.88}$$

Organic electrolyte RNa is uni-univalent so that z is taken to be unity. From equation (4.34), we can also write

$$\Gamma_{Cl^-} = \frac{z\Gamma_R}{\Gamma_{Na^+}/\Gamma_{Cl^-} - 1} \tag{4.89}$$

Values of $\Gamma_{Na^+}/\Gamma_{Cl^-}$ can be calculated with the help of equation (4.66) based on the Gouy model provided ψ_0 is determined. Combining relations (4.87), (4.88), and (4.89), the Gibbs equation can be written in the form

$$-d\gamma = \Gamma_R mRTd\ln c_{Cl^-} \tag{4.90}$$

where

$$m = \frac{\xi_{Cl^-}\left(\dfrac{\xi_{Na^+}}{\xi_{Cl^-}}e^{-\varepsilon\psi_0/2kT} - 1\right)}{\dfrac{1+\bar{x}}{\bar{x}}e^{-\varepsilon\psi_0/2kT} + 1} \tag{4.91}$$

The activity parameters calculated with the help of the Debye–Hückel limiting law can be expressed by the relations

$$\xi_{Cl^-} = 1 - \frac{\bar{A}c_{Cl^-}}{2(c_R + c_{Cl^-})^{1/2}} \tag{4.92}$$

$$\xi_{Na^+} = 1 - \frac{\bar{A}}{2}(c_R + c_{Cl^-})^{1/2} \tag{4.93}$$

Assuming $\xi_{Cl^-} = \xi_{Na^+} = 1$, m has been plotted in Fig. 4.19 as function of $\bar{x}$ according to relation (4.91) for several values of ψ_0 in practical millivolts. It

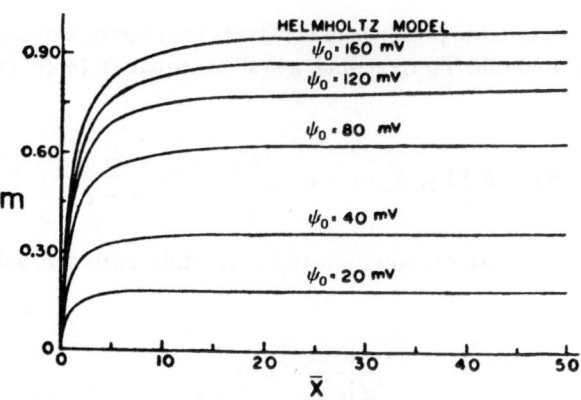

FIGURE 4.19. Variation[1] of the coefficient m with salt ratio $\bar{x}$ at different surface potentials ψ_0. (Reproduced with permission from the *Indian Journal of Chemistry*.)

is evident that m becomes independent of $\bar{x}$ when c_{Cl^-}/c_R is greater than 10. This limiting value of m is observed to be strongly dependent on ψ_0. Only for $\psi_0 > 160 \, \text{mV}$ does m become independent of the surface potential and its limiting value become unity.

In order to explain the effect of sodium chloride on the adsorption of SDS at the air–water interface, Tajima[41] has written the Gibbs equation (4.29) in the form (for z equal to 1)

$$-\frac{d\gamma}{RT} = (\Gamma_R d\ln c_R + \Gamma_{Na^+} d\ln f_{Na^+} + \Gamma_{Cl^-} d\ln c_{Cl^-})$$
$$+ (\Gamma_R d\ln f_R + \Gamma_{Na^+} d\ln f_{Na^+} + \Gamma_{Cl^-} d\ln f_{Cl^-}) \qquad (4.94)$$

Combining this with the electroneutrality equations (4.21) and (4.22), relations (4.95) and (4.96), respectively, can be derived after keeping c_R or c_{Cl^-} constant during the variation of γ:

$$\Gamma_R^a = -\left[\frac{d\gamma}{RTd\ln c_R}\right]_{c_{Cl^-}}$$

$$= \Gamma_R + \frac{\Gamma_{Na^+}}{1+\bar{x}} + 2\Gamma_R \frac{d\ln f_\pm^0}{d\ln c_R} + 2\Gamma_{Cl^-}\frac{d\ln f_\pm^i}{d\ln c_R} \qquad (4.95)$$

$$\Gamma_{Cl^-}^a = -\left[\frac{d\gamma}{RTd\ln c_{Cl^-}}\right]_{c_R}$$

$$= \Gamma_{Cl^-} + \frac{\Gamma_{Na^+}\cdot\bar{x}}{1+\bar{x}} + 2\Gamma_R \frac{d\ln f_\pm^i}{d\ln c_{Cl^-}} + 2\Gamma_{Cl^-}\frac{d\ln f_\pm^0}{d\ln c_{Cl^-}} \qquad (4.96)$$

From the plots of γ against $\ln c_R$ or $\ln c_{Cl^-}$, the apparent values of Γ_R^a and $\Gamma_{Cl^-}^a$ can be calculated at a given liquid composition of the system. The mean activity coefficients of the inorganic and organic salts are $f_\pm^0$ and $f_\pm^i$, respectively. With certain involved assumptions, application of the Debye–Hückel limiting law will lead to the relations

$$\frac{d \ln f_\pm^i}{d \ln c_R} = \frac{d \ln f_\pm^0}{d \ln c_R} = -\frac{\bar{A}c_R}{2(c_R + c_{Cl^-})^{1/2}} \tag{4.97}$$

and

$$\frac{d \ln f_\pm^i}{d \ln c_{Cl^-}} = \frac{d \ln f_\pm^0}{d \ln c_{Cl^-}} = -\frac{\bar{A}c_{Cl^-}}{2(c_R + c_{Cl^-})^{1/2}} \tag{4.98}$$

Combining equations (4.97) and (4.98) with relation (4.95) and (4.96), the following relations can be obtained:

$$\Gamma_{Na^+} = \frac{1}{1+\phi}(\Gamma_{Cl^-}^a + \Gamma_R^a) \tag{4.99}$$

$$\Gamma_R = \frac{1}{1+\phi}\left[\Gamma_{Cl^-}^a\left(1 + \frac{\bar{x}\phi}{1+\bar{x}}\right) - \Gamma_R^a\frac{\phi}{1+\bar{x}}\right] \tag{4.100}$$

$$\Gamma_{Cl^-} = \frac{1}{1+\phi}\left[\Gamma_R^a\left(1 + \frac{\phi}{1+\bar{x}}\right) - \Gamma_{Cl^-}^a\frac{\phi\bar{x}}{1+\bar{x}}\right] \tag{4.101}$$

The value of ϕ can be calculated from the equation

$$\phi = 1 - \bar{A}(c_R + c_{Cl^-})^{1/2} \tag{4.102}$$

Tajima[41] has measured extensively surface tension of water as a function of the increasing concentrations (c_R) of tritiated sodium dodecyl sulfate (TSDS) at constant concentration of NaCl. He has simultaneously measured γ as a function of c_{Cl^-}, keeping c_R constant. The results are presented in Figs. 4.20 and 4.21. From the slopes of these curves at a given value of $\bar{x}$ (equal to c_{Cl^-}/c_R), Γ_R^a and $\Gamma_{Cl^-}^a$ can be estimated so that inserting these with values of ϕ into equations (4.99), (4.100), and (4.101), Γ_{Na^+}, Γ_R, and Γ_{Cl^-} can be estimated. Values of Γ_R and Γ_{Cl^-} evaluated in this manner are presented in Fig. 4.22 for different activities of R in the solution.

Values of Γ_R for TSDS analytically obtained using radiotracer experiments are found to agree quite satisfactorily with those evaluated from the Gibbs adsorption equation. If the activity correction in (4.100) is neglected such that

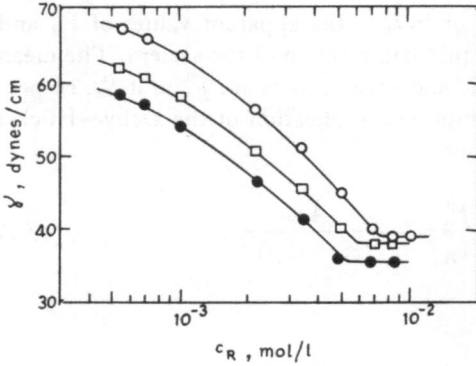

FIGURE 4.20. Surface tensions of SDS solutions as function of c_R of NaCl at 25°C. O, 1.0×10^{-3} M NaCl; □, 5.0×10^{-3} M NaCl; ●, 10.0×10^{-3} M NaCl. (Redrawn from Ref. 41.)

$\phi = 1$, then Γ_R thus evaluated will be slightly lower than the experimental value especially when c_R is high. As expected Γ_{Cl^-} is negligibly small and its value is either slightly negative or zero but never positive.

Sasaki and co-workers[42] have measured the lowering of surface tension of water with increasing concentrations of tritiated sodium dodecyl sulfate NaDS or calcium dodecyl sulfate Ca(DS)$_2$ at constant ionic strengths (I) adjusted by addition of NaCl or CaCl$_2$. The extent of adsorption has also been analytically measured by them under identical conditions using radiotracer methods. At constant ionic strength, the concentration c_M of Na$^+$ or Ca^{++} ions and also the activity coefficients of all the ions do not change during the variation of c_R, so that the Gibbs equation (4.30) will assume the form

$$-\frac{1}{RT}\left(\frac{d\gamma}{d\ln c_R}\right)_I = \Gamma_R - \Gamma_{Cl^-}\frac{c_R}{c_{Cl^-}} \qquad (4.103)$$

In Figs. 4.23 and 4.24, Γ_R, determined directly by the radiotracer method, has been plotted against concentration of the surface-active electrolytes in the

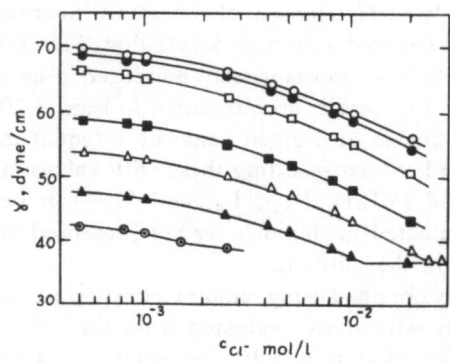

FIGURE 4.21. Surface tensions of NaCl solutions[41] in the presence of SDS at 25°C. O, 0.545×10^{-3} M SDS; ●, 0.698×10^{-3} M SDS; □, 1.01×10^{-3} M SDS; ■, 2.21×10^{-3} M SDS; △, 3.48×10^{-3} M SDS; ▲, 5.01×10^{-3} M SDS; ⊙, 6.65×10^{-3} M SDS. (Redrawn from Ref. 41.)

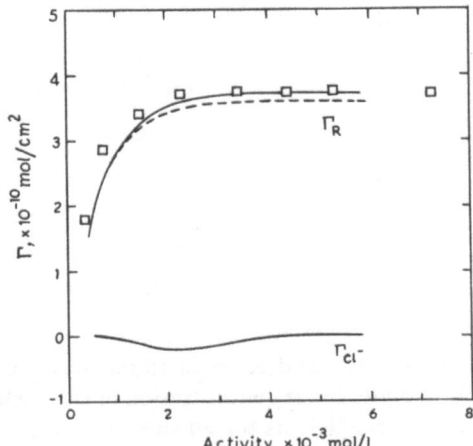

FIGURE 4.22. Amount of adsorption of TSDS anion and chloride ion as functions of TSDS concentration at fixed NaCl concentration of $5.0 \times 10^{-3} M$ NaCl. $\square$, values from the radioactive method; ——, calculated from equations (4.100) and (4.101); - - -, calculated after putting $\phi = 1$. (Redrawn from Ref. 41.)

medium. In the same figure, the numerical values of the left side of equation (4.103) evaluated from the $\gamma - \ln c_R$ plot at constant ionic strength for various values of c_R are also presented in the form of continuous lines. The identity in the two values indicates again that the value of Γ_{Cl^-} is negligibly small in equation (4.103).

Okumura *et al.*[43] have measured the amount of adsorption of a tritiated ampholytic surfactant N-dodecyl-β alanine at the air–water interface using radiotracer as well as surface tension methods. The pH in all cases was kept near 7.0 and measurements were carried out both in the presence and absence of 1(M) NaCl. The Gibbs equation (4.30) for such system in the absence of the neutral salt may be written in the form

$$-\frac{d\gamma}{RT} = \Gamma_{R^\pm} \cdot d \ln c_{R^\pm} + \Gamma_{R^+} \cdot d \ln c_{R^+} + \cdot \Gamma_{R^-} \cdot d \ln c_{R^-}$$

$$+ \Gamma_{H^+} \cdot d \ln c_{H^+} + \Gamma_{OH^-} \cdot d \ln c_{OH^-} \qquad (4.104)$$

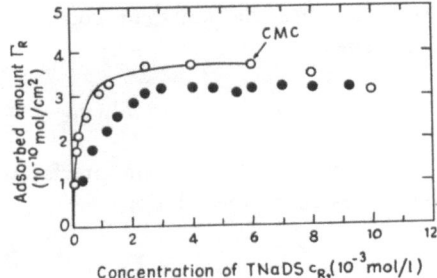

FIGURE 4.23. Adsorption isotherm of TNaDS at constant ionic strength (25°C). $\bigcirc$, TNaDS + NaCl = 1×10^{-2} mol/liter; $\bullet$, without added salt by tracer method; ——, calculated value from Gibbs equation. (Redrawn from Ref. 42.)

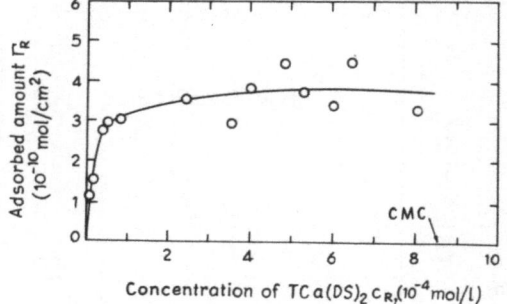

FIGURE 4.24. Adsorption isotherm of TCa(DS)$_2$ at constant ionic strength (55°C). O, observed value (TCa(DS)$_2$ + CaCl$_2$ = 2.5 × 10^{-2} mol/liter); ———, calculated value from Gibbs equation. (Redrawn from Ref. 42.)

Here R$^{\pm}$, R^{+}, and R^{-} refer to zwitterionic, cationic, and anionic forms of the ampholyte. Since concentration of the surfactant used is of the order 10^{-4}(M), activity coefficients for all the ions have been taken to be unity.

If k_1 and k_2 are the first and second dissociation constants of the ampholyte, then[44]

$$k_1 = \frac{c_{R^\pm} \cdot c_{H^+}}{c_{R^+}} \qquad (4.105)$$

or

$$d \ln c_{R^\pm} + d \ln c_{H^+} - d \ln c_{R^+} = 0 \qquad (4.106)$$

and

$$k_2 = \frac{c_{R^-} \cdot c_{H^+}}{c_{R^\pm}} \qquad (4.107)$$

or

$$d \ln c_{R^-} + d \ln c_{H^+} - d \ln c_{R^\pm} = 0 \qquad (4.108)$$

Further, since the ionization constant k_w of water is equal to $c_{H^+} \cdot c_{OH^-}$, we can write

$$d \ln c_{H^+} + d \ln c_{OH^-} = 0 \qquad (4.109)$$

The electroneutrality of the interfacial phase in terms of surface excess may be expressed by the relation

$$\Gamma_{R^+} + \Gamma_{H^+} = \Gamma_{R^-} + \Gamma_{OH^-} \qquad (4.110)$$

Combining equations (4.108), (4.109), and (4.110) with equation (4.104), one can derive relation (4.111) if pH is maintained constant

$$-\frac{d\gamma}{RT} = (\Gamma_{R^\pm} + \Gamma_{R^+} + \Gamma_{R^-})\, d \ln c_{R^\pm}$$

$$= \Gamma_R \, d \ln c_R^\pm \tag{4.111}$$

Here Γ_R stands for the total surface excess of the ampholyte in cationic, anionic, and zwitterionic forms. Okumura *et al.*[43] have further shown that under the prevailing experimental conditions, $d \ln c_R^\pm$ in equation (4.111) can be replaced by $d \ln c_R$, where c_R, equal to $c_{R^\pm} + c_{R^+} + c_{R^-}$, stands for total concentration of the ampholyte in the medium. The Gibbs equation will then assume the simple form

$$\Gamma_R = -\frac{1}{RT}\frac{d\gamma}{d \ln c_R} \tag{4.112}$$

Γ_R calculated from $\gamma - c_R$ plot on the basis of equation (4.112) is observed to agree with that measured directly from the radiotracer experiments.

In the presence of constant and excess concentration of NaCl, $d \ln c_{Na^+} = d \ln c_{Cl^-} = 0$, so that equations (4.104) to (4.109) will remain valid in this case also.

The electroneutrality conditions of the interfacial phase in this case can be expressed by the relation

$$\Gamma_{Na^+} + \Gamma_{R^+} + \Gamma_{H^+} = \Gamma_{Cl^-} + \Gamma_{R^-} + \Gamma_{OH^-} \tag{4.113}$$

so that combining equations (4.108) and (4.109) with relations (4.113) and (4.111) yields

$$-\frac{1}{RT}\frac{d\gamma}{d \ln c_R} = \Gamma_R + \left(\frac{c_{R^-} - c_{R^+}}{c_R}\Gamma_R + \Gamma_{Cl^-} - \Gamma_{Na^+}\right)$$

$$\times \frac{d \ln c_{H^+}}{d \ln c_R} \tag{4.114}$$

If the pH is not altered during the variation of γ with c_R, equation (4.114) will also reduce to the form (4.112) in the presence of excess neutral salt. When calculated from the Gibbs equation (4.112) in the presence of excess neutral salt, Γ_R is found to agree with that estimated from the radiotracer experiments. Γ_R is also found to be independent of pH when excess

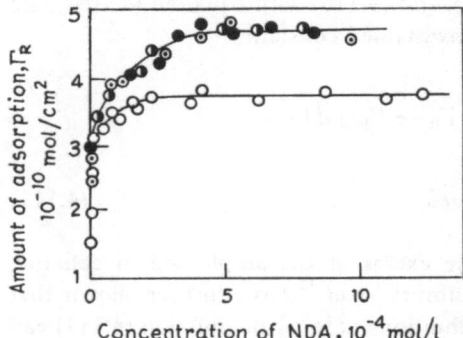

FIGURE 4.25. Adsorption of NDA (by tracer method) at aqueous surface. Addition of $\bigcirc$, 0 mol/liter NaCl near neutral pH; $\bullet$, 1 mol/liter NaCl near neutral pH; $\odot$, 1 mol/liter NaCl at pH 2.0; $\pmb\bullet$, 1 mol/liter NaCl at pH 12.0. (Redrawn from Ref. 43.)

salt is present in the system.[43] Values Γ_R obtained from radiotracer methods are shown in Fig. 4.25.

The work has been further extended by Nakamura *et al.*[45] in which the effect of pH change with HCl on the adsorption of the ampholyte N-dodecyl-alanine has been examined in terms of the Gibbs equation. This equation will now read

$$-\frac{d\gamma}{RT} = \Gamma_{R^\pm} \cdot d\ln c_{R^\pm} + \Gamma_{R^+} \cdot d\ln c_{R^+}$$

$$+ \Gamma_{R^-} \cdot d\ln c_{R^-} + \Gamma_{H^+} \cdot d\ln c_{H^+}$$

$$+ \Gamma_{OH^-} \cdot d\ln c_{OH^-} + \Gamma_{Cl^-} \cdot d\ln c_{Cl^-} \qquad (4.115)$$

The electroneutrality equation of the surface phase will assume the form

$$\Gamma_{R^+} + \Gamma_{H^+} = \Gamma_{OH^-} + \Gamma_{R^-} + \Gamma_{Cl^-} \qquad (4.116)$$

Combining relations (4.115) and (4.116) with equations (4.106), (4.108), and (4.109), it may be shown that

$$-\frac{d\gamma}{RT} = \Gamma_R d\ln c_R + \frac{c_{R^-} - c_{R^+}}{c_R} \Gamma_R d\ln c_{H^+} + \Gamma_{Cl^-} \cdot d\ln c_{H^+}$$

$$+ \Gamma_{Cl^-} d\ln c_{Cl^-} \qquad (4.117)$$

If Γ_R^a and $\Gamma_{Cl^-}^a$ stand for the apparent values of surface excesses at constant

c_R and c_{Cl^-}, respectively, then

$$\Gamma_R^a = -\frac{1}{RT}\left(\frac{\partial \gamma}{\partial \ln c_R}\right)_{c_{Cl^-}}$$

$$= \Gamma_R + \left(\frac{c_{R^-} - c_{R^+}}{c_R}\Gamma_R + \Gamma_{Cl^-}\right)\left(\frac{\partial \ln c_{H^+}}{\partial \ln c_R}\right)_{c_{Cl^-}} \tag{4.118}$$

and

$$\Gamma_{Cl^-}^a = -\frac{1}{RT}\left(\frac{\partial \gamma}{\partial \ln c_{Cl^-}}\right)_{c_R}$$

$$= \Gamma_{Cl^-} + \left(\frac{c_{R^-} - c_{R^+}}{c_R}\Gamma_R + \Gamma_{Cl^-}\right)\left(\frac{\partial \ln c_{H^+}}{\partial \ln c_{Cl^-}}\right)_{c_R} \tag{4.119}$$

Nakamura *et al.*[45] have extensively measured γ for different values of c_R at constant c_{Cl^-} and also for different values of c_{Cl^-} at constant c_R, so that Γ_R^a and $\Gamma_{Cl^-}^a$ can be evaluated from the experimental data. They have also noted variation of $\ln H^+$ (or pH) with variation of c_R at constant c_{Cl^-} or variation of c_{Cl^-} at constant c_R so that the term in the brackets of equations (4.118) and (4.119) can be evaluated. From the known values of the dissociation constants k_1 and k_2 of the amino acids, c_{R^+}, c_{R^-}, and $c_{R^\pm}$ of the amino acid at any composition of the aqueous solution may also be evaluated. Using equations (4.118) and (4.119) judiciously, one can, therefore, determine Γ_R and Γ_{Cl^-} from the experimental data. Values of these quantities thus evaluated at different pH are found to agree satisfactorily with those directly calculated by the radiotracer method (vide Fig. 4.26).

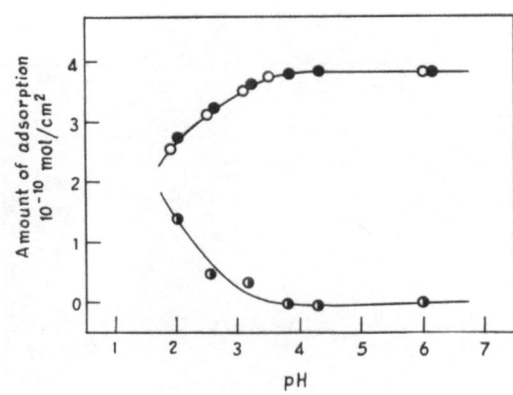

FIGURE 4.26. Amount of adsorption of NDA and chloride ion vs. pH at NDA concentration of 5.0×10^{-4} mol/liter. O, observed values for NDA by radiotracer method; ●, calculated value for NDA; ◑, calculated value for chloride ion, by Gibbs equation. (Redrawn from Ref. 45.)

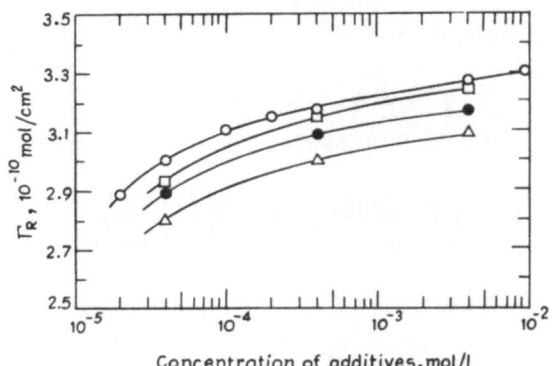

FIGURE 4.27. Effects of various anions on the adsorbed amount Γ_R of TD(EO)$_6$ at 30°C. The bulk concentration of TD(EO)$_6$ is 4.0×10^{-5} mol/liter. O, NaCl; □, NaNO$_3$; ●, NaI; △, NaSCN, by tracer method. (Redrawn from Ref. 47.)

The extent of adsorption of a nonionic tritiated surfactant hexaoxyethylenedodecyl ether TD(EO)$_6$ at air–water interface was directly measured by the radiotracer method by Tajima and coworkers.[46] The amount of adsorption was found to be in excellent agreement with that calculated from the surface tension data applying the Gibbs adsorption equation (3.1) valid for the presence of single nonionic solute in the medium. The effects of additives on the adsorption of tritiated TD(EO)$_6$ was subsequently studied by them[47] from direct radiotracer analysis. In Fig. 4.27, Γ_R is observed to increase with increase of concentration of NaCl, NaNO$_3$, NaI, or NaSCN in agreement with the anionic effect associated with the lyotropic series. Γ_R is also observed to decrease with increasing concentrations of additives such as urea, urea nitrate, and guanidine nitrate, but they remain unaffected in the presence of additives like diethylene glycol, triethylene glycol, etc. (vide Fig. 4.28). These additives are regarded as protein denaturants (vide Chapter 9).

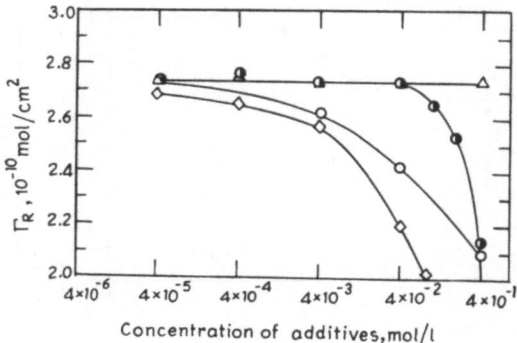

FIGURE 4.28. Effects of various denaturants on the adsorbed amount Γ_R of TD(EO)$_6$ at 30°C. The bulk concentration of TD(EO)$_6$ is 4.0×10^{-5} mol/liter. O, urea; ◑, urea nitrate; ▲, diethylene glycol; △, triethylene glycol; ◇, guanidine nitrate, by tracer method. (Redrawn from Ref. 47.)

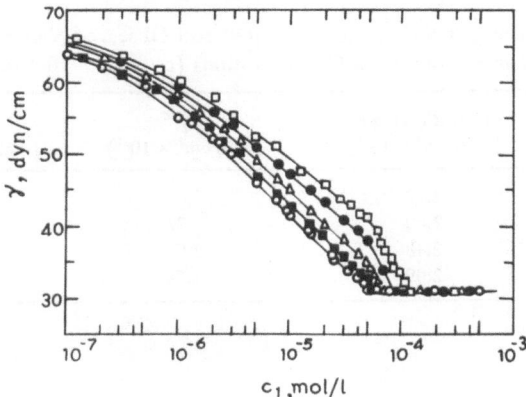

FIGURE 4.29. Surface tension vs. concentration (c_R) curves of $D(EO)_6$ solution at various concentrations of urea. $\bigcirc$, 0 M; $\blacksquare$, 0.04 M; $\triangle$, 0.40 M; $\bullet$, 2.5 M; $\square$, 4.0 M. (Redrawn from Ref. 48.)

Tajima[48] measured γ of water as a function of the concentration c_1 of $TD(EO)_6$ at fixed concentration c_2 of urea. The results of these measurements are shown in Fig. 4.29.

From the Gibbs equation (4.5) for a multicomponent system they deduced equation (4.120) using some reasonable but empirical assumptions:

$$\Gamma_1 = \left(\frac{\partial \gamma}{RT \, \partial \ln c_1}\right)_{c_2} - 4.5 \times 10^{-11} c_2^{0.15} \qquad (4.120)$$

Values of Γ_1 thus calculated from equation (4.120) are given in Table 4.2. They are in satisfactory agreement with those measured directly by means of radiotracer measurements.

From the generalized Gibbs equation (4.5), Tajima also derived an equation

$$\Gamma_2 = \frac{1}{1 - 0.046c_2}\left[-\left(\frac{\partial \gamma}{RT \partial \ln c_2}\right)_{\Gamma_1} \right.$$
$$\left. + 4.5 \times 10^{-11} c_2^{0.15}\left\{ 1 - \frac{\partial \ln c_1}{\partial \ln c_2}\right\} - \Gamma_1\left(\frac{\partial \ln c_1}{\partial \ln c_2}\right)_{\Gamma_1} \right] \qquad (4.121)$$

From the data given in Fig. 4.29, $(\partial \gamma / RT \partial \ln c_2)_{\Gamma_1}$ may be calculated. From separate plots $\partial \ln c_1 / \partial \ln c_2$ and $\partial \ln c_1 / \partial \ln c_2$ may be similarly obtained. Values of Γ_2 for various values of Γ_1 can thus be obtained. The Gibbs equation thus enables us to calculate the amount of urea which may remain adsorbed on $TD(EO)_6$ present in the interfacial monolayer.

TABLE 4.2. Comparison of the Calculated and Observed Values of $D(EO)_6$ Adsorption at Various Urea Concentrations ($c_1 - 4.0 \times 10^{-5}$ mol/liter)

c_2 (mol/liter)	$-(\partial\gamma/RT\partial \log c_1)$ (mol/cm$^2 \times 10^{10}$)	Γ_2^{calc} (mol/cm$^2 \times 10^{10}$)	Γ_1^{obs} (mol/cm$^2 \times 10^{10}$)
0	2.72	2.72	2.73
0.040	2.60	2.32	2.41
0.40	2.48	2.09	2.09
4.0	2.09	1.54	1.57

4.16. SURFACE EXCESSES FOR SMALL CATIONS AND ANIONS

Ikeda and co-workers[49,50] have recently developed a general method of approach based on the Gibbs equation so that the surface excesses of different ionic species forming the multicomponent solution can be evaluated from the surface tension versus concentration data. They have measured γ for aqueous solution of a cationic detergent dodecyldimethyl ammonium chloride as a function of surfactant concentration c_R at different but fixed concentration (c_{Na^+}) of NaCl. Applying the Gibbs equation (4.30) for the adsorption of ions, they have straightforwardly deduced the following relation:

$$-d\gamma = RT(\Gamma_R^a d \ln c_R + \Gamma_{Na}^a d \ln c_{Na^+}) \tag{4.122}$$

such that

$$\Gamma_R^a = 1 + \left(\frac{c_R}{c_R + c_{Na^+}} + \beta c_{Na^+}\right)\Gamma_R$$

$$+ \left(\frac{c_{Na^+}}{c_R + c_{Na^+}} + \beta c_{Na^+}\right)\Gamma_{Na^+} \tag{4.123}$$

and

$$\Gamma_{Na^+}^a = \left(\frac{c_{Na^+}}{c_R + c_{Na^+}} + \beta c_{Na^+}\right)\Gamma_R$$

$$+ \left(1 + \frac{c_{Na^+}}{c_R + c_{Na^+}} + \beta c_{Na^+}\right)\Gamma_{Na^+} \tag{4.124}$$

Here

$$\beta = \frac{d \ln f_{\pm}^2}{d(c_R + c_{Na^+})} \tag{4.125}$$

where $f_{\pm}$ is the mean activity coefficient in the bulk phase whose value can be estimated using the Debye–Hückel limiting law.

Solving equations (4.123) and (4.124), one can obtain the following equations:

$$\Gamma_R = \frac{1}{2 + \beta(c_R + c_{Na^+})} \left[\left(1 + \frac{c_{Na^+}}{c_R + c_{Na^+}} + \beta c_{Na^+}\right) \Gamma_R^a \right.$$

$$\left. - \left(\frac{c_R}{c_R + c_{Na^+}} + \beta c_R\right) \Gamma_{Na^+}^a \right] \tag{4.126}$$

$$\Gamma_{Na^+} = \frac{1}{2 + \beta(c_R + c_{Na^+})} \left[\left(1 + \frac{c_R}{c_R + c_{Na^+}} + \beta c_R\right) \Gamma_{Na^+}^a \right.$$

$$\left. - \left(\frac{c_{Na^+}}{c_R + c_{Na^+}} + \beta c_{Na^+}\right) \Gamma_R^a \right] \tag{4.127}$$

and

$$\Gamma_{Cl^-} = \frac{1}{2 + \beta(c_R + c_{Na^+})} (\Gamma_R^a + \Gamma_{Na^+}^a) \tag{4.128}$$

These equations are consistent with those given by Tajima[41] in the case of anionic surfactant and discussed in the previous section.

Values of Γ_R^a and $\Gamma_{Na^+}^a$ can be obtained from the experimental data using the Gibbs equations,

$$\Gamma_R^a = -\frac{1}{RT} \left(\frac{\partial \gamma}{\partial \ln c_R}\right)_{c_{Na^+}} \tag{4.129}$$

and

$$\Gamma_{Na^+}^a = -\frac{1}{RT} \left(\frac{\partial \gamma}{\partial \ln c_{Na^+}}\right)_{c_R} \tag{4.130}$$

so that the surface excesses of the individual ions can be calculated using equations (4.126), (4.127), and (4.128). Adsorption isotherms for various ions are presented in Fig. 4.30.

In Fig. 4.30 (top), the adsorption of organic cation is observed to increase more sharply with increasing NaCl concentration as a result of the electrostatic effect. The adsorption data at a given value of c_{Na^+} fit the Langmuir adsorption equation. In Fig. 4.30 (bottom), Γ_{Cl^-} is observed to be close to Γ_R when c_R is low so that Γ_{Na^+} is negligibly small. However, at higher concentration of c_R, values of Γ_{Na^+} are found to be significantly positive, which is unexpected from the Gouy model. According to the Boltzmann distribution equations, it should be negative. On further critical analysis, the Gouy model of the double layer is found not to be strictly valid for the charged monolayer formed by the adsorption of this cationic detergent at the interface in the presence of the neutral salt.

The surface tensions for aqueous mixture of HCl, NaCl, and dimethyl dodecylamine oxide (a weak amphoteric electrolyte) have been measured by Ikeda *et al.*[51] extensively at different degrees of neutralization and different concentrations of the neutral salts. The pH-metric titration of the weak

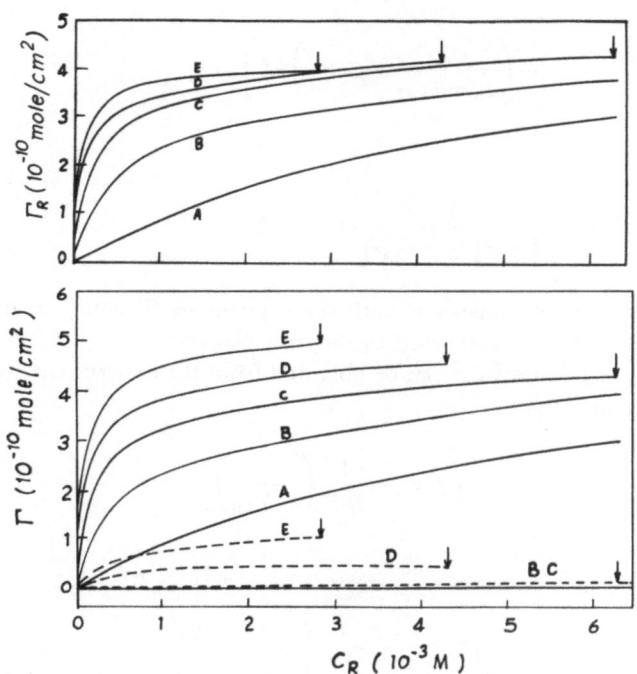

FIGURE 4.30. Adsorption isotherms of ions on the aqueous surface of different NaCl concentrations. (top) Γ_R vs. concentration c_R: A, 0; B, 0.01 M; C, 0.05 M; D, 0.10 M; E, 0.20 M; (bottom) Γ_{Na^+} (- - -) and Γ_{Cl^-} (——) vs. c_R. Arrows indicate the location of c_R. (Redrawn from Ref. 56.)

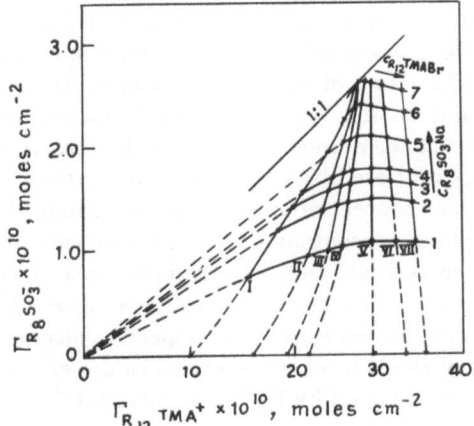

FIGURE 4.31. The correlation between surface excess values of the organic cations and anions. The solutions of a constant concentration of R_{12}TMABr: (I) 2.5×10^{-5}; (II) 5×10^{-5}; (III) 7.5×10^{-5}; (IV) 1×10^{-4}; (V) 2.5×10^{-4}; (VI) 5×10^{-4}; (VII) 7.5×10^{-4} mol dm^{-3}. The solutions of a constant concentration of R_8SO$_3$Na: (1) 2.5×10^{-5}; (2) 5×10^{-5}; (3) 7.5×10^{-5}; (4) 1×10^{-4}; (5) 2.5×10^{-4}; (6) 5×10^{-4}; (7) 7.5×10^{-4} mol dm^3. (Redrawn from Ref. 52.)

electrolyte has been carried out to obtain necessary information of the ionization characteristics of the surfactant in solution. The surface excesses of surfactant, sodium, and chloride ions are obtained as functions of the surfactant concentration by means of the derived forms of the Gibbs equations. On the basis of certain postulates, the extent of surface hydrolysis of the surfactant has been evaluated approximately.

4.17. COADSORPTION OF ORGANIC IONS

The lowering of surface tension of 0.1 M NaBr solution in aqueous medium at a fixed concentration of sodium octyl sulfonate (R_8SO$_3$Na) has been studied by Rodakiwicz-Nowak[52] as a function of increasing concentration of dodecyltrimethylammonium bromide (R_{12}TMABr). From the application of the Gibbs adsorption equation (4.31) for such an ideal system with m taken as unity, $\Gamma_{R_{12}TMABr}$ has been accurately computed for different bulk concentrations of the amine. The adsorption isotherms for the amine at several fixed concentrations of the sulfonate have been obtained from the experimental data. Proceeding in an exactly similar manner, the adsorption isotherm for the octyl sulfonate at several fixed concentrations of the amine and at 0.1 M NaBr can be obtained. From these kinds of complementary studies, the mutual influence between Γ_{amine} and $\Gamma_{sulfonate}$ in the interfacial phase can be described with the help of the curves shown in Fig. 4.31. Here vertical lines describe constant bulk concentrations of the amine while horizontal lines join points of a constant bulk concentration of sulfonate. From this figure, a mutual enhancement of the values of the surface excesses for low concentrations of the components and stabilizing or decreasing tendencies of the surface excess values for relatively concentrated solutions may be observed. The experimental lines have been extrapolated by dotted lines to values of surface excesses for solutions of different concentrations of single surfactant. It is possible that at the region of low total concentrations of the surfactants, the surface excess of the sulfonate at its constant bulk concentration may increase linearly with increase of the surface excess of the amine. The mutual influence of tetradecyl trimethylammonium bromide and the respective alcohol (*m*-hexanol or *n*-octanol) in terms of the Gibbs surface excess values of coadsorption has been further similarly investigated by Rodakiwicz-Nowak.[52]

4.18. SUMMARY AND COMMENTS

From all these discussions, it is clear that there is considerable interest at present in the use of the Gibbs equations in a convenient form for the study of adsorption of solutes from the multicomponent solution to the interface existing at the boundary region of the liquid phase. The application of the Gibbs equation will be easier if the experiments are properly planned. More investigations in this direction for various types of multicomponent systems are highly desirable. The detailed study will be important simply because of the fact that in many natural and in particular living systems, the phenomenon of adsorption is exhibited in the presence of multicomponent aqueous solution.

For a multicomponent nonelectrolyte system, the applicability of the Gibbs equation becomes easier because the activity parameters may be put equal to unity for a dilute solution without making any significant error. For aqueous systems, containing surface-active organic electrolyte in the presence of constant concentration of the neutral salt, the Gibbs equation assumes a convenient form (4.31). This equation contains the coefficient m whose values can be estimated from the bulk concentrations of the, solute and the magnitude of the surface potential to be determined by an independent experiment. Values of m calculated on the basis of the electrical double layer models given by Helmholtz, Gouy, and Stern are found not to differ from each other widely. In the Helmholtz model, the excess concentration of coions is zero, which is proved experimentally to be the case for many surfactant systems. Further, m calculated on the basis of the Helmholtz model does not contain the surface potential term so that it can easily be computed from the values of the bulk concentrations of the components. In the actual derivation of relation (4.40) for m independent of the surface potential, the only requirement is to make the excess concentration of the coions zero so that detailed microscopic structure of the double layer is not needed for its derivation.

When the surface tension of the multicomponent solution is varied by altering the concentration of the neutral salt in the bulk medium at a constant concentration of the organic electrolyte, values of *m* become largely dependent on the surface potential and microscopic picture of the electrical double layer so that the general applicability of the derived Gibbs equation for such a system may be questioned.

Recently Tajima and co-workers have derived suitable forms of the Gibbs equations for the multicomponent solutions containing organic ionic and zwitterionic surfactants, inorganic salts, urea, etc. In these derivations, no assumption about a microscopic model for the charged interface is required. The extents of adsorption of the detergent ions, zwitterions, nonionic solutes, etc. calculated from surface tension measurements using these derived equations are always found to agree satisfactorily with those directly obtained from the radiotracer experiments. The approach seems to be elegant and highly rational. Many of the results may have potential importance in under-standing adsorption phenomena occurring at biosurfaces. More work should be carried out with many other multicomponent systems of importance on the basis of this dual approach of estimation of the extent of adsorption directly and on the basis of the Gibbs adsorption equation so that confidence in such indirect estimation will be firmly established.

CHAPTER 5

ADSORBED MONOLAYERS AND ENERGIES OF ADSORPTION

5.1. INTRODUCTION

In the previous two chapters, it has been shown that an amphiphile dissolved in aqueous medium has a strong tendency to accumulate at the liquid interface, as a result of which the surface free energy of the liquid is lowered. The extent of such lowering of energy usually depends, at a constant temperature and pressure, on the nature and concentration of the solute in the bulk medium. The relation between the lowering of surface tension and the excess adsorption of the solute at the interface is frequently referred to as the equation of state for the adsorbed layer in the interfacial phase. If the experiments on the boundary tension measurements are appropriately made, then the extent of adsorption of a surface-active solute may be calculated using the convenient form of the Gibbs adsorption equation. Experimental evidences given by Langmuir,[1,2] and also those discussed in Chapter 3, have established that the surface-active substances at the interfacial phase remain as a monomolecular layer, so that it is frequently referred to as a soluble or adsorbed monolayer; sometimes it is also mentioned as the Gibbs monolayer.

Because of thermal energy, the molecules accumulated in the interfacial phase exhibit two-dimensional motion parallel to the surface phase. The force exerted by the molecules in motion in the adsorbed film per unit length is termed as surface or interfacial pressure π. If γ_0 is the surface tension of pure liquid in mN/m and γ the lower value of it in the presence of organic solute in the bulk, then[2,3]

$$\pi = \gamma_0 - \gamma \qquad (5.1)$$

This two-dimensional pressure is expressed as force per unit length.

The physical concept of surface pressure for soluble monolayer has been beautifully illustrated[4] with the help of a Langmuir trough divided by a loose and thin rubber membrane into two compartments (Fig. 5.1). One of these compartments is filled with the solvent and the other with a dilute solution

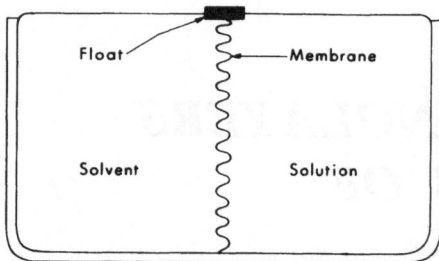

FIGURE 5.1. The compartments of the trough. (Redrawn from Ref. 4.)

of the organic surface-active solute. The membrane will lead to the equalization of any hydrostatic pressure differences between the solutions in the two compartments. Under these conditions, a force will act on the float attached at the top of the membrane and can be measured by applying an opposite force by a lever attached to a torsion wire so that the float will be prevented from moving. This force per unit length is the surface pressure and it can be shown to be equal to $\gamma_0 - \gamma$. Using the surface balance (cf. Section 2.3), π can be directly measured for any insoluble monolayer.

Although π can be calculated from γ_0 and γ, the physicochemical concept of surface pressure in terms of the motion of only the solute molecules within a gaslike monolayer is too simple for its full acceptance. The interfacial phase containing solute as well as solvent molecules is basically inhomogeneous; further there exists a difference in the free energy between this phase and the homogeneous bulk phase. Molecules at the interface may interact with those of neighboring solute and solvent. These interactions also contribute to the surface pressure. As a result of adsorption of organic ions, an electrical double layer will be formed, which also makes a significant contribution to the surface pressure. In subsequent sections, these features of the monolayers will be discussed in detail.

5.2. ADSORBED AND SPREAD MONOLAYERS

Many organic amphiphilic substances, such as stearic acid, are practically insoluble in the bulk water phase, so that the transfer of organic substances from bulk to surface with the formation of adsorbed monolayer will not be effective. However, if this surface-active substance is dissolved in an easily volatile organic solvent and a very small amount of the solution is added to the interface of water and air, the solvent will evaporate allowing the solute to spread out. The monolayer thus formed at the interface will exert surface pressure in two dimensions. The surface pressure, π, and the surface area, A, per molecule can be directly obtained from experiments carried out in a surface balance. Many soluble substances such as proteins, SDS, etc., may

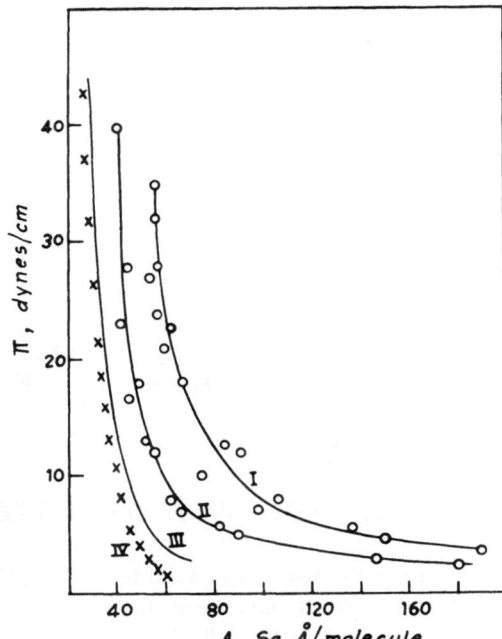

FIGURE 5.2. Force area characteristics of SDS films. I, no salts; II, 0.145 M NaCl; III, 25% $(NH_4)_2SO_4$ spread monolayer. IV, 75% $(NH_4)_2$-SO_4 spread monolayer. (Redrawn from Ref. 5.)

form spread monolayers if the aqueous substrate contains a considerable amount of inorganic neutral salt. The concepts of A and π, however, remain practically unchanged for adsorbed and spread monolayers. Pethica[5] earlier made an attempt to compare π–A curves of the spread and adsorbed monolayers of SDS at the air–water interface (vide Fig. 5.2). Since salt concentrations in the aqueous phases were widely different, π at a given value of A became significantly different. The decrease of π in the case of a spread monolayer is consistent with the reduction of the electrical pressure at the interface in the presence of a considerable amount of neutral salt in the substrate phase. In the case of proteins, e.g., egg albumin and lysozyme, π–A curves for adsorbed and spread monolayers have also been compared.[6,7] The observed difference in the two types of experiments (vide Fig. 5.3) is related to the difference in extent and nature of surface denaturation of the proteins in the adsorbed and spread states. In spite of these differences, it is usually believed that physicochemical behaviors of spread and adsorbed monolayers, whether charged or not, are, in principle, similar to each other at the state of thermodynamic equilibrium. As shown in Chapter 6, spread lipid monolayers can exhibit phase transitions related to gas → liquid → solid. The adsorbed amphiphiles, on the other hand, show only the gas → liquid phase transitions, since these systems are studied at temperature where monolayer can only exist in gas or liquid state.

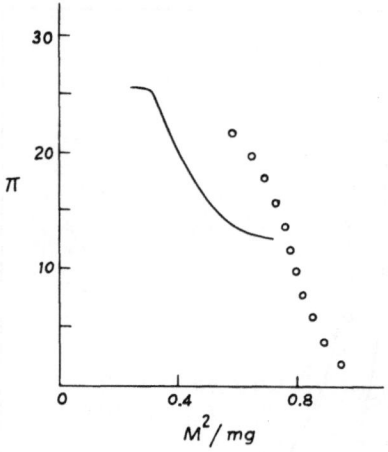

FIGURE 5.3. Force-area curves. Solid line, adsorbed film of egg albumin. Circles, spread film egg albumin, experimental points. $1\,M$ Na$_2$SO$_4$. (Redrawn from Ref. 7.)

5.3. IDEAL EQUATION OF STATE

The surface and interfacial tensions (γ) of many nonionic organic solutes dissolved in water have been measured extensively as function of their concentrations (c_R) in the bulk. We shall first consider γ–c_R data at the oil–water interface. In Fig. 5.4, γ–c_R plots of n-butyric and n-valeric acids (in the

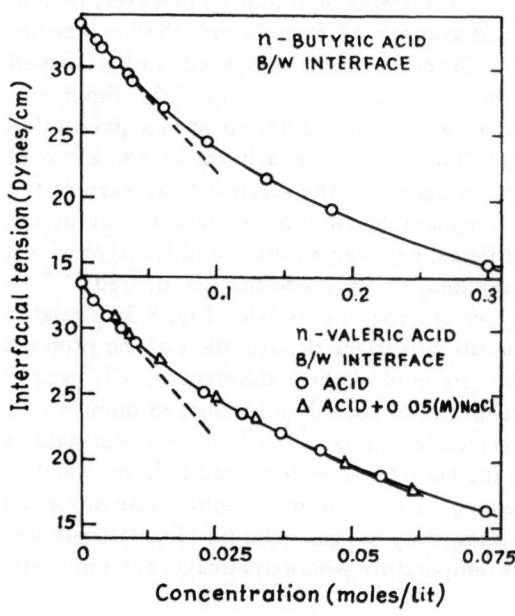

FIGURE 5.4. The interfacial tension as a function of the bulk (aqueous) concentration at 35°C. B/W = benzene–water. (Redrawn from Ref. 8.)

presence and absence of 0.05 M NaCl) are presented for the benzene–water interface.[8] From this figure, we also note with interest that γ for the oil–water interface falls off linearly with increse of c_R when the solute concentration is low, although at higher concentrations, deviations from linearity are observed. At low concentrations, i.e., so long as γ changes linearly, the following empirical equation may be suggested[4,9]:

$$\gamma = \gamma_0 - bc_R \qquad (5.2)$$

Here b is a constant numerically equal to the slope, $d\gamma/dc_R$. From the Gibbs adsorption equation (3.1) for the two-component system, $d\gamma/dc_R$ is equal to $-\Gamma_R RT/c_R$, which in turn will be equal to $-b$. Thus we find

$$\pi = \gamma_0 - \gamma$$

$$= \Gamma_R RT \qquad (5.3)$$

If we define $\mathscr{A}$ and A as respective surface areas occupied by one mole and one molecule of the adsorbed solute, then

$$\mathscr{A} = NA = 1/\Gamma_R \qquad (5.4)$$

Inserting the value of Γ_R in equation (5.3), one gets

$$\pi \mathscr{A} = RT \qquad (5.5)$$

$$\pi A = kT \qquad (5.6)$$

In SI units, kT at 25°C becomes equal to 4.11×10^{-21} joules/molecule. Equations (5.5) and (5.6) have close similarities with the ideal gas equation. Frequently the physicochemical properties of the adsorbed monolayers have been assumed to be quite similar to those of gases. Such comparisons have limitations from the thermodynamic standpoint which will be discussed later.

5.4. NEUTRAL MONOLAYERS AT THE OIL–WATER INTERFACE

The interfacial tension of the oil–water system in the presence of nonionic organic solutes has been extensively measured using drop-volume and other methods.[8,10–14] Oils used in many of these experiments were n-heptane, benzene, and petroleum ether. From the γ–c_R plot shown in Fig. 5.4 both π

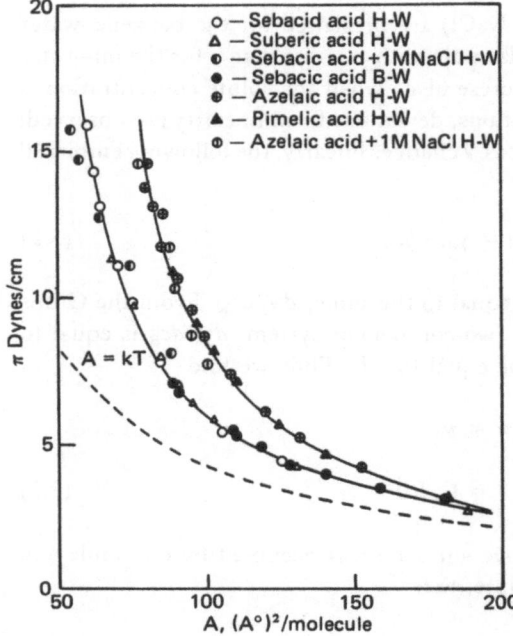

FIGURE 5.5. Pressure-area curves for diabasic acids at oil–water interface.[4] (Reproduced with permission from Academic Press, New York.)

and A at a given concentration c_R may be calculated. In Fig. 5.5 are shown π–A curves for organic dibasic acids (e.g., sebacic, suberic, azeleic, and pimelic) at the benzene–water (B/W) and heptane–water (H/W) interfaces in the absence and presence of NaCl. In the same figure, the π–A curve following the ideal equation (5.6) is also represented by the dashed line. For high value of A (or low value of Γ_R), experimental π is close to the value of that calculated from equation (5.6). With decrease of A, π is found to be significantly greater than its ideal value at a given value of A.

For neutral films at an oil–water interface, Langmuir[1,2] proposed the modified equation of state

$$\pi_{O/w}(A - A_0) = kT \tag{5.7}$$

$\pi_{0/w}$ is the value of surface pressure at the oil–water interface. Here A_0 stands for the area excluded by an adsorbed molecule at the oil–water interface. For an ideal monolayer, A_0 is assumed to be zero, which means that the elongated molecule has zero cross-sectional area. In the derivation of equation (5.7), it is also assumed that the adsorbed molecules in the monolayer film become completely mobile and the two-dimensional forces of cohesional attraction within the monolayer are absent.

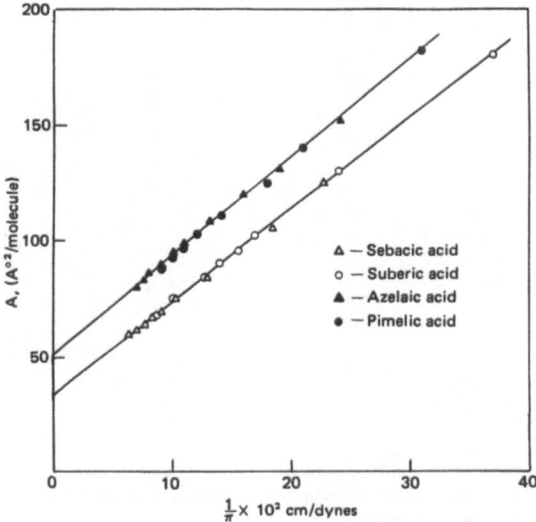

FIGURE 5.6. Plot of area against $1/\pi$ at heptane–water interface.[14] (Reproduced with permission from Academic Press, New York.)

In agreement with the Langmuir equation (5.7), a plot of $1/\pi$ against A for the adsorption of butyric acid at the benzene–water interface has been found to be linear.[12] The linear plots for various dibasic acids at the heptane–water interface[14] are shown in Fig. 5.6. Values of A_0 obtained from the intercept of this plot for homologous series of monobasic fatty acids are found to be 0.24 nm^2 per molecule. For even and odd chain dibasic acids the values thus evaluated are 0.34 and 0.51 nm^2, respectively.

In order to explain the differences in the values of A_0 for different organic acids, let us consider some probable conformations of monobasic and dibasic acids at the oil–water interfaces in Figs. 5.7A, 5.7B, and 5.7C. From boundary tension experiments at the oil–water interface, the effective cross-sectional area of the hydrocarbon chain is shown to be 0.12 nm^2.[13] Since the excluded area per pair of molecules is four times the cross-sectional area, the co-area thus excluded at the interface per molecule is 0.24 nm^2 in agreement with the value of A_0 for monobasic acid.[8] In the case of organic dibasic acids, forming wicket structure at the interface, the excluded area per wicket will be less than 0.24 nm^2 because of the overlapping of the area of exclusion (vide Figs. 5.8B and 5.8C). If the area thus reduced is 0.17 nm^2 then experimental values of A for even- and odd-chain acids may easily be explained.[14] The hard-disk model for A_0 can explain the results only qualitatively, but needs refined and detailed theoretical treatment. It may be pointed out here that A_0 values obtained for long- and short-chain dibasic acids and their esters by different

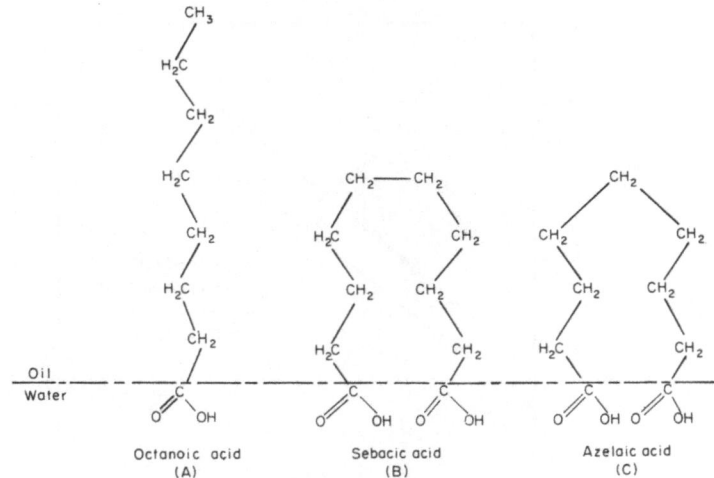

FIGURE 5.7. The conformations of monobasic and dibasic acids at the interface.[14] (Reproduced with permission from Academic Press, New York.)

workers[15–19] show discrepancies. Thus the A_0 value obtained from the study of the insoluble monolayer is the result of multipair interactions whereas in the study of soluble monolayers, pairwise interactions have to be considered. The two sets of A_0 values are expected not to be always same. A_0 in the two-dimensional gaseous monolayer is analogous to van der Waals' co-volume factor b. It is surprising that for an insoluble protein monolayer, the contribution of the actual area of the adsorbed molecule rather than co-area enters into the equation of state.[20] This indicates that the gaslike model of the adsorbed monolayer may have its limitations.

5.5. SURFACE PRESSURE AND OSMOTIC PRESSURE

It may be recalled that the interfacial film is made up of both solute and solvent components thus constituting the surface phase solution. In the treatment of the gaslike monolayer in the previous sections, the role of solvent in

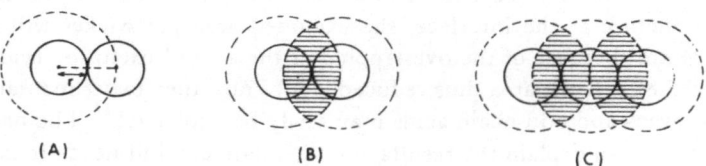

FIGURE 5.8. Overlapping of the area of exclusion.

the interfacial phase has been completely neglected. Let us now assume that the surface phase containing Δn_1 and Δn_2 moles of solvent and solute, respectively, has the total volume $\bar{V}_s$, total area $\mathscr{A}$, and thickness τ. The osmotic pressure π_{os} exerted by the solute in the surface phase may be given by[4,21]

$$\pi_{os} V_1 = RTX_2^s \tag{5.8}$$

Here X_2^s stands for the mole fraction of the solute in the surface phase. If V_1 and V_2 stand for the molar volumes of the solvent and solute, respectively, in the interfacial phase

$$\bar{V}_s = \Delta n_1 V_1 + \Delta n_2 V_2 \tag{5.9}$$

In formulating equations (5.8) and (5.9), ideal behavior of the surface solution has been assumed. If the surface phase of volume V^s is made up to Δn_1^0 moles of solvent when solute is absent in the system,

$$\bar{V}_s = \Delta n_1^0 V_1 \tag{5.10}$$

Further,

$$X_2^s = \frac{\Delta n_2}{\Delta n_1 + \Delta n_2} \tag{5.11}$$

Combining equations (5.9), (5.10), and (5.11), Δn_1 can be eliminated so that

$$X_2^s = \frac{\Delta n_2 V_1}{\Delta n_1^0 V_1 - \Delta n_2 (V_2 - V_1)} \tag{5.12}$$

If we assume that surface pressure π is the measure of the osmotic force per unit length at the interfacial plane, then this force may be assumed to be distributed on τ cm^2 within the interfacial phase so that the interfacial osmotic pressure π_{os} becomes equal to π/τ dyn cm^{-2} in cgs units so that

$$\pi = \tau \pi_{os} \tag{5.13}$$

Inserting this in equation (5.8) and then combining with relations (5.10) and (5.12), it may be shown that

$$\pi = RT \left/ \frac{1}{\Delta n_2} \frac{\Delta n_1^0}{\tau} V_1 - \left(\frac{V_2}{\tau} - \frac{V_1}{\tau} \right) \right.$$

$$= RT \left/ \frac{\mathscr{A}}{\Delta n_2} - (A_2 - A_1) \right. \tag{5.14}$$

Here surface phase is assumed to be monolayer so that $\mathscr{A}_2$ and $\mathscr{A}_1$ are molar surface areas of the solute and the solvent components, respectively. Now, identifying Δn_2 with Γ_2, we may put $\bar{\mathscr{A}}/\Delta n_2$ to be $\mathscr{A}$, the area per mole of adsorbed solute, so that

$$\pi[\mathscr{A} - (\mathscr{A}_2 - \mathscr{A}_1)] = RT \tag{5.15}$$

or

$$\pi\left[A - \frac{1}{N}(A_2 - A_1)\right] = kT \tag{5.16}$$

If one writes A_0 for $(A_2 - A_1)/N$, then equation (5.16) is converted to the form of equation (5.7). Following similar arguments, Bull[20] has derived equation (5.16) for protein monolayer in which A_1 is neglected with respect to A_2. Since equation (5.16) contains A_1 and π (a term derived from π_{os}), the role of solvent and solute both is significant in this relation.[4] The concept of solute–solvent mixture has been developed further by Fowkes.[22] Using the thermodynamic approach of the Gibbs adsorption equation, a relation similar to (5.16) has been derived by Lucassen-Reynders and van den Tempel.[23]

5.6. BINARY SOLUTION AT THE INTERFACIAL PHASE

We thus find from the discussions in the previous chapters that the surface pressure which is a measurable quantity bears analogy with gas pressure as well as osmotic pressure. In the "gas model" the role of solvent present in the interfacial phase has been completely neglected. In the "surface solution" model on the other hand, it is assumed that the solute–solvent mixture in the interfacial phase is homogeneous and ideal in behavior. The inhomogeneous nature of the interfacial phase has been neglected in this model. The physicochemical aspects of the mixture of solute and solvent in the surface phase are not conceptually clear. Fresh ideas and experiments are needed to understand these aspects on a more realistic basis.

An effective attempt has been made in this direction by Motomura et al.,[24] who have measured interfacial tension of the n-hexane–water system for various concentrations of tetradecanol at several temperatures and pressures. These data are given in Figs. 5.9 and 5.11.

Here X_1^0 stands for the mole fraction of tetradecanol in the oil phase. Its concentration in the water phase is small. The slope of the curve in Fig. 5.9 according to the Gibbs equation (3.30) is related to the excess entropy S_σ^0 per

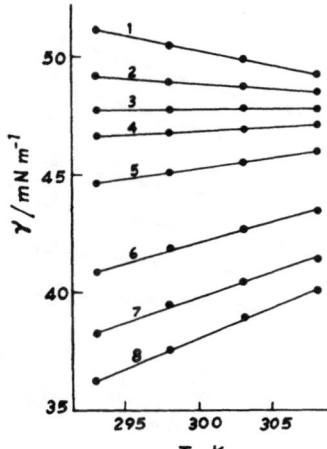

FIGURE 5.9. Interfacial tension vs. temperature curves at constant concentration under atmospheric pressure. (1) $X_1^0 = 0$; (2) 1×10^{-4}; (3) 2×10^{-4}; (4) 3×10^{-4}; (5) 5×10^{-4}; (6) 1×10^{-3}; (7) 1.5×10^{-3}; (8) 2×10^{-3}. (Redrawn from Ref. 24.)

unit surface area by the relation

$$S_\sigma^0 = -\left(\frac{\partial \gamma}{\partial T}\right)_{p, \, X_1^0} \tag{5.17}$$

where the superscript in S_σ^0 indicates that the excess accumulation of the oil (as solvent) at the interface is zero. The concept of excess entropy has been further explained in terms of equation (3.54). Motomura[25] has defined S_σ^0 as entropy of formation of the interface on the basis of the algebraic approach for the Gibbs adsorption equation introduced by Hansen.[26] S_σ^0 thus calculated from the $\gamma - T$ slope decreases from positive to negative values with increase of X_1^0 (vide Fig. 5.9). This fact indicates that with the formation of the binary solution at the interface by transfer of the alcohol from the oil to the interfacial phase, the entropy of the interface formation decreases.

Using the algebraic approach, Motomura[25] has also developed a theoretical equation on the basis of which the partial molal entropy of tetradecanol (denoted by $S_1^H - S_1^0$) in the interfacial phase has been calculated from these experimental data as functions of various values of X_1^0 (vide Fig. 5.10). The

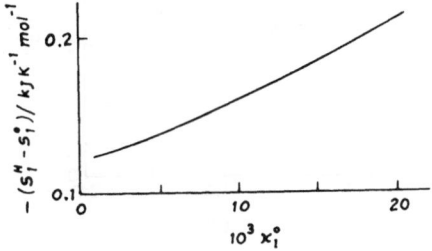

FIGURE 5.10. Partial molar entropy change of tetradecanol vs. mole fraction X_1^0 curve at 303.15 K under atmospheric pressure. (Redrawn from Ref. 25.)

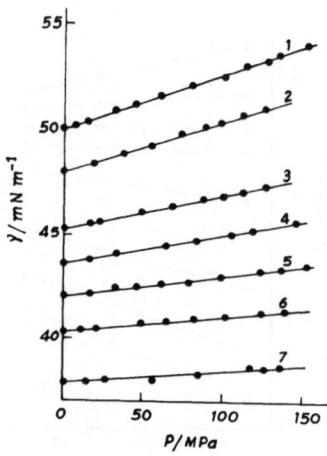

FIGURE 5.11. Interfacial tension vs. pressure curves at 303.15 K and constant concentration: (1) $X_1^0 = 0$; (2) 1.83×10^{-4}; (3) 5.29×10^{-4}; (4) 8.03×10^{-4}; (5) 1.16×10^{-3}; (6) 1.56×10^{-3}; (7) 2.41×10^{-3}. (Redrawn from Ref. 24.)

negative value of this quantity and increase of its magnitude with increase of X_1^0 clearly point out that the tetradecanol molecule is forced to take a more restricted conformation in the interface than in the solution.

The results of the variation of γ with change of external pressure shown in Fig. 5.11 cannot be explained by the Gibbs equation (3.30). From the Guggenheim equation (3.87) and also from the equation developed by the algebraic approach,[25,26] it may be shown that

$$\left(\frac{d\gamma}{dp}\right)_{T,X_1^0} = V^s \tag{5.18}$$

where V^s may be regarded as the volume of the interface formed per unit surface area at constant T and X_1^0. From the slope of the curve in Fig. 5.11,

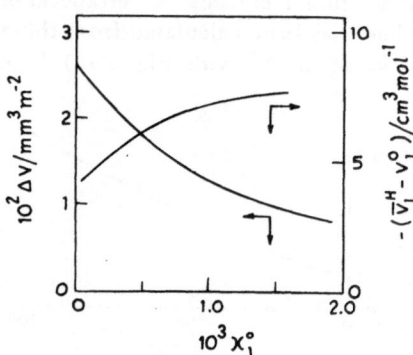

FIGURE 5.12. Volume $\bar{V}_s$ (or ΔV in the figure) of interface formation vs. mole fraction and partial molar volume change $(\bar{V}_1^H - \bar{V}_1^0)$ of tetradecanol vs. mole fraction curves at 303.15 K and 0.1 M Pa. (Redrawn from Ref. 24.)

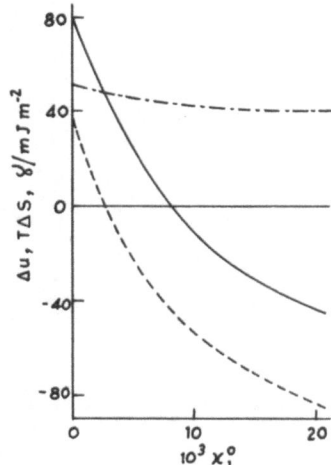

FIGURE 5.13. Thermodynamic quantity of interface formation vs. mole fraction curves at 303.15 K and 0.1 M Pa: (———) Δu; (- - - -) $T\Delta S$; (· · · ·) γ. (Redrawn from Ref. 24.)

V^s may be calculated. Using the same algebraic approach, partial molal volume of tetradecanol in the interfacial phase has also been calculated by Motomura *et al*.[24] These are presented in Fig. 5.12. V^s (denoted as ΔV) decreases with increase of X_1^0 when more solutes are present at the interfacial phase. As a result of this behavior, partial molal volume (denoted by $\bar{V}_1^H - \bar{V}_1^0$) of the solute component becomes negative in the interfacial phase.

Pursuing this treatment more elaborately, the internal energy change (Δu) of surface formation calculated by Motomura *et al*. is found to decrease from positive to negative values with increase of X_1^0 (vide Fig. 5.13). The adsorption of tetradecanol diminishes the energy of hexane–water interface and overcomes the disadvantage brought by the decrease of entropy. From these studies, it has also been suggested that the interface region is composed of a homogeneous mixture of water, hexane, and tetradecanol, except in the direction normal to the plane of the interface. The treatment has been further extended so as to include the case of adsorption of ionic detergents at the oil–water interface.[27,28]

5.7. IONIZED MONOLAYERS AT THE OIL–WATER INTERFACE

The interfacial tension of an oil–water system is considerably lowered in the presence of a surface-active electrolyte (of the type RNa_z) in the aqueous phase. The magnitude of such lowering will be more in the presence of an inorganic salt. Usually, γ is measured at increasing concentrations c_R of the organic electrolyte at constant concentration c_{Cl^-} of the inorganic salt in the

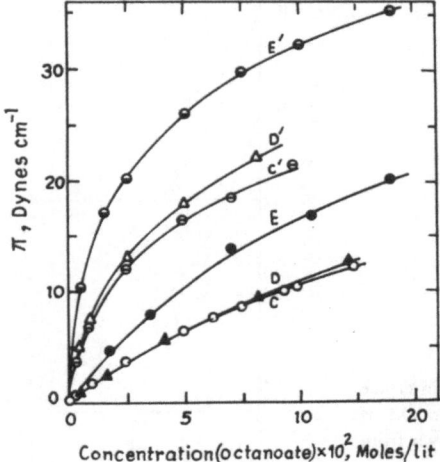

FIGURE 5.14. π–C curves at the oil–water interface at 35°C.[29] ●, $C_7H_{15}COONa$ (P.E/W); ◑, $C_7H_{15}COONa + 1\,N$ NaCl (P.E/W); ○, $C_7H_{15}COONa$ (B/W); ⊖, $C_7H_{15}COONa + 1\,M$ NaCl (B/W); ▲, $C_7H_{15}COOLi$ (B/W); △, $C_7H_{15}COOLi + 1\,M$ LiCl (B/W). (Reproduced with permission from Academic Press, New York.)

medium. The surface pressure π can be calculated using equation (5.1). Here γ_0 may be interfacial tension of the oil–water system either in the presence or in the absence of the inorganic salt in the medium. The surface concentration Γ_R of the organic electrolyte at a given bulk concentration c_R may also be calculated from the γ–c_R data using modified Gibbs equation (4.31). From this value of Γ_R, the area A occupied per adsorbed molecule (or ion) can be evaluated from equation (5.4). In Fig. 5.14, γ–c_R curves for sodium and lithium octanoates with or without excess of neutral salts are shown for benzene–water and petrol-ether–water interfaces.[29] From a critical examination of these graphs, it appears that the surface pressure depends significantly upon the neutral salt added in excess as well as on the nature of the oil phase but to a small extent on the nature of the counterions (e.g., Li^+, Na^+).

In Fig. 5.15, the surface pressures of the octanoic acid and octanoate monolayers at the oil–water interface[29] have been compared for various values of A, the area per organic molecule or ions. When octanoic acid remains adsorbed at the oil–water interface in the un-ionized state, it exerts surface pressure π (equal to π_k) which originates from the kinetic motion of the molecules at the interfacial phase. π–A plot for the neutral acid (previously discussed) is shown by the dashed curve M in Fig. 5.15. At higher pH values, salt of octanoic acid is formed which is completely ionized in the bulk as well as in the surface phases and as a result of this, an electrical double layer is formed at the interfacial region. The general equation of state for the ionized monolayer at the oil–water interface may be written in the form[30,31]

$$\pi_{O/W} = \pi_k + \pi_e \tag{5.19}$$

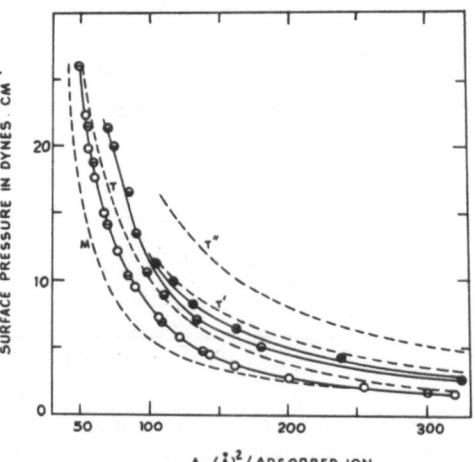

FIGURE 5.15. π–A curves at 35°C $\oplus$, $C_7H_{15}COONa$ (P.E/W); O, $C_7H_{15}COONa + 1\ M$ NaCl (P.E/W); $\bullet$, $C_7H_{15}COONa$ (B/W); $\ominus$, $C_7H_{15}COONa + 1\ M$ NaCl (B/W). Curve T, π^d vs. A ($p = 1$) for C_7H_{15}-COONa $+ 1\ M$ NaCl (B/W); Curve T′, π^d vs. $A(p = 1)$ for $C_7H_{15}COONa$ (B/W); Curve T″, π^d vs. A ($p = 2$) for $C_7H_{15}COONa$ (B/W); Curve M, $kT/(A - A_0)$ vs. A for $C_7H_{15}COOH$ (P.E/W). (Redrawn from Ref. 29.)

Here π_e represents the electrical pressure due to the repulsion between adsorbed ions and work of charging the interfacial double layer.

The kinetic pressure π_k for the adsorbed electrolyte may be given by the relation

$$\pi_k = \frac{pkT}{A - A_0} \qquad (5.20)$$

where p stands for the coefficient of the equation of state. Davies[32–34] assumed p to be unity for ionized monolayer of RNa both in the presence and absence of the excess neutral salt in the medium. Phillips and Rideal,[35] on the other hand, assumed p equal to 2; the experiment on dodecyltrimethylammonium bromide monolayer at the O/W interface in the absence of NaCl supports this assumption.[36] From simple theoretical considerations of the Gibbs equations, Haydon[37] has shown that, like m, p should be equal to unity and 2, respectively, in the presence and absence of neutral salt.

Robb and Alexander[38] made an attempt to solve for the values of p from measurements of π and A of the spread monolayer of decosyltriethylammonium bromide in the presence and absence of 0.1 M NaCl both at the air–water and the oil–water interfaces. The results in Fig. 5.16 indicate that in agreement with Haydon's analysis, πA tends to be equal to kT and $2kT$, respectively, in the presence and absence of salt for low values of π. From a detailed theoretical treatment of the distribution of the organic and inorganic salts in the surface and bulk phases, Chattoraj[39] concluded that p equals 2 irrespective of the presence or absence of the netural salt in the medium.

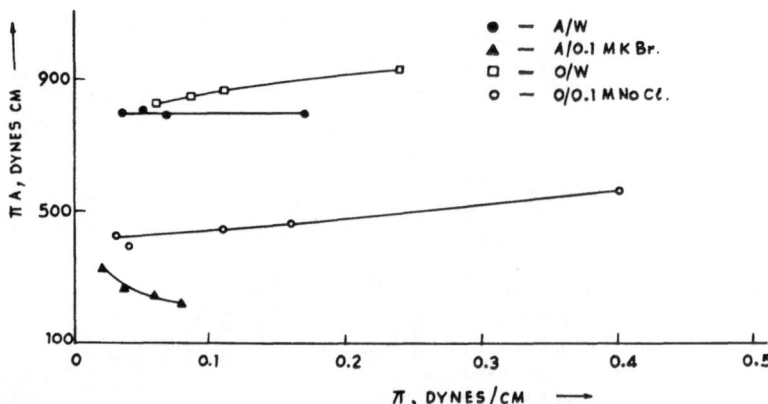

FIGURE 5.16. Plot of πA vs. π for decosyltriethyl ammonium bromide at the air–water and oil–water interfaces in the presence and absence of salts. (Redrawn from Ref. 38.)

Hachisu[40] derived expressions of π_k (and π_e also) in equation (5.20) by three independent methods. The first of these methods was based on the integration of the Gibbs equation, whereas the second was that of an extension of the theory of Deryaguin,[41] Verwey and Overbeek.[42] The third one was based on the extension of the osmotic pressure theory for the interfacial phase. All three analyses indicate that p in equation (5.20) is unity irrespective of the presence or absence of the neutral salt. In the opinion of Hachisu, the experimental value 2 for p in the absence of the neutral salt is due to the contribution in part of π_e to the total pressure π. The matter appears to be settled but it is desirable to verify the conclusion with extensive measurements of π in the low-pressure region.

5.8. THE ELECTRICAL PRESSURE AND THE GOUY MODEL

When an organic salt such as SDS or CTAB is adsorbed at the oil–water or air–water interface, the electrolyte in the interfacial phase will remain ionized so that an electrical double layer will be formed in the interfacial region. The monolayer appears to remain fully ionized in the presence of these salts over a wide range of pH. On the other hand, when an organic fatty acid RH (e.g., octanoic acid) is adsorbed at the liquid interface at low pH, the acid in the bulk as well as at the interface will remain in completely un-ionized form. With increase of pH of the system, the acid begins to ionize in both the phases until the ionization becomes complete at some high pH of the medium. Surface or interfacial pressures at the same time decrease with increase of

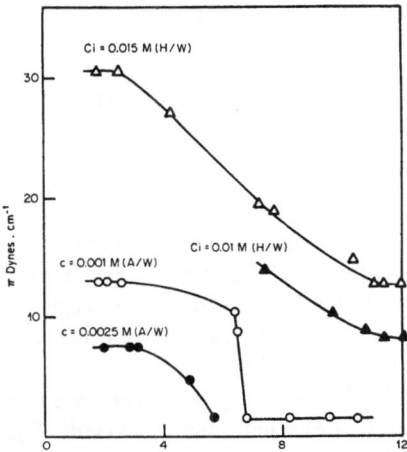

FIGURE 5.17. π–pH plot for sebacic acid at the A–W and H–W interfaces in the presence of 1 M NaCl at 35°C. c, concentration of the acid in water; c_i, concentration of the acid in water before addition of oil.[14] (Reproduced with permission from Academic Press, New York.)

ionization of the acid and reach constant values at high pH corresponding to the formation of fully ionized monolayers (vide Fig. 5.17). It may be pointed out here that the pH-value for the ionization of an organic acid in the surface phase will be different from that in the bulk due to the double layer effect.[43,44] For this reason, complete ionization of the acid at the interface may take place at relatively higher pH than that of the system in bulk.

On complete ionization of the acid at the interface, the surface acquires a charge σ per unit area and the total potential drop due to this will be ψ_0. The electrostatic free energy change due to the formation of the charged surface will be $(-\sigma\psi_0)$. Simultaneously with the charging of the surface, inorganic ions will accumulate in excess as a result of which an electrical double layer will be formed. The free energy change due to the electrostatic work involved in charging the double layer is $\int_0 \psi' d\sigma'$, where σ' is the charge density for any value of ψ' during the progress of the charging process by gradual ionization. The total Helmholtz free energy change ΔF_e per unit surface area due to the ionization of the monolayer will then be given by the equation

$$\Delta F_e = -\sigma\psi_0 + \int_0^\sigma \psi' \, d\sigma'$$

$$= -\int_0^{\psi_0} \sigma' \, d\psi' \tag{5.21}$$

For the charging of the plane surface of a flat plate in this manner with the formation of the Gouy double layer (vide Section 4.9), the right-hand side of equation (5.21) has been integrated by Verwey and Overbeek[42] between the limits 0 to ψ_0 so that

$$-\Delta F_e = \frac{8\dot{c}kT}{\kappa}\left[\cosh\frac{\varepsilon\psi_0}{kT} - 1\right] \qquad (5.22)$$

Here κ is the Debye–Hückel thickness of the ion atmosphere and $\dot{c}$ as before is the number of ions of one kind per milliliter in the bulk away from the surface. The equation is valid for uni-univalent electrolytes. For the sake of simplicity in approach, the distinction between the Gibbs free energy ΔG_e and the Helmholtz free energy ΔF_e may be neglected. ΔF_e includes both electrical work as well as pressure–volume work whereas ΔG_e includes only the first kind of work.

Davies[34] has indicated that the adsorbed ions at the liquid interface may form a charged plane CD (vide Fig. 5.18) in which the ionic head groups are placed at a distance from each other in a discrete manner. The charged plane is located at a distance about 0.3 nm from the phase boundary plane. The adsorbed ions at the interface generate lines of force and a periodic field near the plane CD, whereas at a distance slightly away from CD towards the aqueous phase, the lines of forces are uniformly distributed so that the effect of the electrical field appears to be generated as if from a uniformly charged surface CD. For such apparently uniform charged surface, Davies[34] puts the equation straightforwardly,

$$\pi_e^d = -\Delta F_e \qquad (5.23)$$

From the Gouy equation (4.51), we have also noted that

$$\psi_0 = \frac{2kT}{\varepsilon}\sinh^{-1}\left(\frac{2\dot{c}\sigma DkT}{\pi}\right)^{-1/2}$$

$$= \frac{2kT}{\varepsilon}\sinh^{-1}\frac{\beta_d}{Ac^{1/2}} \qquad (5.24)$$

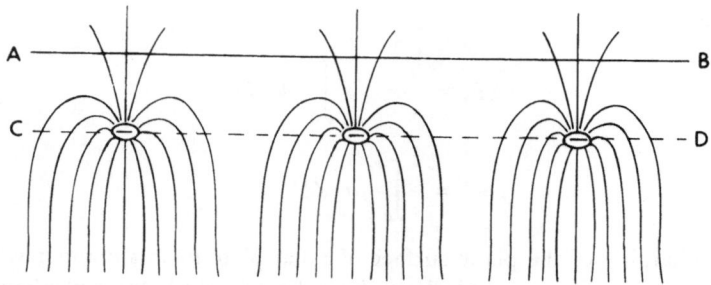

FIGURE 5.18. The discreteness of charge according to Bell, Levine, and Pethica.[47]

Here β_d is a constant whose values at 20°C and 35°C will be 1.34 and 1.37, respectively, if the area A per adsorbed molecule (or charged group for RNa) is expressed in nm^2. Here, c stands for $c_R + c_{Cl^-}$ as explained in Section 4.11. Combining equations (5.22), (5.23), and (5.24), therefore, yields

$$\pi_e^d = \alpha_d c^{1/2} \left[\cosh \sinh^{-1} \left(\frac{\beta_d}{A c^{1/2}} \right) - 1 \right]$$

$$= \alpha_d c^{1/2} \left[\left(\frac{\beta_d^2}{A^2 c} + 1 \right)^{1/2} - 1 \right] \tag{5.25}$$

Numerical values of α_d are 6.1 and 6.2 at 20°C and 35°C, respectively. Inserting the values of π_k and π_e from equations (5.20) and (5.25) in relation (5.19), the Davies equation of state[34] for the ionized monolayer may be explicitly given by

$$\pi^d = \frac{kT}{A - A_0} + \alpha_d c^{1/2} \left[\left(\frac{\beta_d^2}{A^2 c} + 1 \right)^{1/2} - 1 \right] \tag{5.26}$$

The superscript d in π^d and π_e^d is used to indicate the Davies model of calculation for the theoretical values of the total and electrical pressures.

In Fig. 5.15, π–A curves for the sodium octanoate monolayer in the presence and absence of neutral salts based on γ–c_R measurements have been compared with π^d–A curves constructed on the basis of equation (5.26). At a given value of A, π is always found to be less than π^d for octanoate, sebacate, azelate, and SDS monolayers adsorbed at various types of oil–water interfaces.[29,45,46]

5.9. THE DISCRETE-ION EFFECT

Bell, Levine, and Pethica (BLP)[47] have attempted to account for the ionized monolayer based on the discreteness of the charge at the plane CD near the interface (vide Fig. 5.18). The charge on the interface is not uniformly distributed but somewhat discrete in nature, as a result of which the electrical potential at a particular point near the interface will be less than the average electrostatic potential ψ_0 by an amount ϕ_f. ϕ_f is termed the self atmosphere or the fluctuation potential, the value of which can be calculated theoretically using only a few assumptions. The total electrical free energy change ΔF_e^{BLP} for charging the liquid interface in a discrete manner with subsequent formation of the electrical double layer is given by

$$\Delta F_e^{BLP} = \Delta F_e + \Delta F_{ic} \tag{5.27}$$

For a uniformly charged plane CD generating uniform lines of force at a short distance, the free energy contribution ΔF_{ic} due to the discreteness of the surface charge will be zero and $\Delta F_e^{\mathrm{BLP}}$ becomes equal to ΔF_e. For plane CD charged discretely, ΔF_{ic} makes a contribution to the electrical pressure by an amount π_f which is termed the fluctuation pressure. The total electrical pressure π_e^{BLP} according to the BLP theory is then expressed by the relation

$$\pi_e^{\mathrm{BLP}} = \pi_f + \pi_e^d \tag{5.28}$$

π_f exhibits a negative contribution to π_e. It can be calculated[47] with the help of the following equation:

$$\pi_f = -\frac{4\varepsilon^2}{D\pi}\left[\frac{1}{9r_0^3} + \frac{D - D_r}{D + D_r}f(r_0, 2\beta)\right]$$

$$+ \frac{4\varepsilon^2}{D\pi}\left[f(r_0, \kappa^{-1}) + \frac{D - D_r}{D + D_r}f(r_0, \overline{\kappa^{-1} + 2\beta})\right] \tag{5.29}$$

Here r_0 represents the radius of the circle on the surface plane at the center of which there exists an adsorbed ion leading to an uniform charge density to this limiting circular area. If A' represents the area per charged group (equal to A/z), it may be taken to be equal to πr_0^2. The distance of the head group β from the phase boundary plane is usually taken as 0.25 nm^2.[48–50] D stands for the uniform dielectric constant of the bulk aqueous medium. D_r is the dielectric constant of this medium at the interfacial region where the charged carboxyl groups exist. Its value is usually taken as 5 by Levine and co-workers.[47]

Values of the functions occurring in equation (5.29) are given by the general equation

$$f(r_0, x) = \left(\frac{2x^2}{9r_0^6} + \frac{5}{36r_0^4} + \frac{1}{24r_0^2x^2}\right)(r_0^2 + x^2)^{1/2}$$

$$+ \frac{1}{24x^3}\ln\frac{x + (x^2 + r_0^2)^{1/2}}{r_0} - \frac{x}{4r_0^4} - \frac{16x^3}{9r_0^6} \tag{5.30}$$

Here three sets of equations will be obtained by substituting x equal to 2β, κ^{-1}, and $(\kappa^{-1} + 2\beta)$ successively, so that three functions occurring in equation (5.29) can be solved from the experimental data.

The terms in equation (5.29) represent the fluctuation pressure arising from the effects of (i) the discreteness of surface charge, (ii) the forces due to the formation of the electrical image at the interface, and (iii) the screening of the double layer in the presence of neutral salts.

Using the complicated equation (5.29), π_f for a fully ionized monolayer can be calculated theoretically at a given value of A, so that π_e^{BLP} can be computed on the basis of relation (5.28). In Table 5.1 and also in Fig. 5.19, π_e^{BLP} for several monolayers calculated from the double layer theories have been compared with π_e.[29,46] A refined model of the electric double layer at a completely ionized monolayer with discreteness-of-charge effect has been recently examined by Feat and Levine.[51]

5.10 COUNTERION BINDING

In Fig. 5.15, the π–A curve for the neutral monolayer of nonionic octanoic acid is shown by the dashed line M. Here π_e will be zero, so that π becomes equal to π_k in the light of equation (5.19). The pressure–area curves of the octanoates at the oil–water interface in the presence and absence of excess neutral salt are presented in the same figure. At a given value of A, therefore,

TABLE 5.1. π_k, π_e, π_e^d, π_f, π_e^{BLP} for Sodium Salts of Dibasic Acids at the Heptane–Water Interface (NaCl concentration: 1 M)[a]

$C_R \times 10^3$ (mol/liter)	π (mN m^{-1})	A (nm^2/molecule)	π_k (mN m^{-1})	π_e (mN m^{-1})	π_e^d (mN m^{-1})	π_f (mN m^{-1})	π_e^{BLP} (mN m^{-1})
\multicolumn{8}{c}{Sodium salt of sebacic acid}							
5.00	7.9	0.94	7.1	0.8	12.9	−5.5	7.4
8.00	10.2	0.75	10.2	0.0	17.3	−7.1	10.2
10.00	11.6	0.68	12.5	−0.9	19.5	−8.8	10.7
15.00	14.6	0.56	19.3	−4.7	24.8	−14.5	10.3
20.00	17.1	0.47	32.7	−15.6	30.5	—	—
\multicolumn{8}{c}{Sodium salt of azelaic acid}							
2.0	7.0	1.13	6.9	0.1	10.1	−3.9	6.2
3.0	8.8	0.95	9.6	−0.8	12.7	−4.9	7.8
4.0	10.5	0.86	12.1	−1.6	14.5	−5.7	8.8
5.0	11.8	0.80	14.6	−2.8	15.9	−6.4	9.5
6.0	12.8	0.79	15.1	−2.3	16.2	−6.5	9.7
\multicolumn{8}{c}{Sodium salt of suberic acid}							
5.0	7.5	1.04	6.1	1.4	11.3	−4.4	6.9
8.0	9.6	0.82	8.9	0.7	15.6	−6.1	9.5
10.0	10.8	0.73	10.9	−0.1	18.0	−7.5	10.5
12.0	12.0	0.66	13.3	−1.3	20.3	−9.1	11.2
15.0	13.6	0.58	17.7	−4.1	23.7	−12.9	10.8

[a] R. P. Pal, Ph.D. thesis, Jadavpur University, 1973.

TABLE 5.2. Thickness of the Various Types of Monolayers at the Air–Water Interface[63]

Monolayer	$a_s \times 10^{28}$ $\left(\dfrac{\text{ergs} \cdot \text{cm}^2}{\text{molecule}^2}\right)$	$a \times 10^{34}$ $\left(\dfrac{\text{ergs} \cdot \text{cm}^3}{\text{molecule}^2}\right)$	τ_{nm}	τ_{nm}^{m}
Butyric acid	0.80	0.26	0.78	0.79
Valeric acid	1.29	0.38	0.99	0.91
Caproic acid	1.69	0.53	1.25	1.04
Caprylic acid	2.92	0.86	1.57	1.29
Na-octanoate (1 M NaCl)	1.15	0.86	2.33	1.29
Na-octanoate	4.1	0.86	1.29	1.29
NaLS	10.0	1.30	0.86	2.04
NaLS (0.1 M NaCl)	2.0	1.30	1.83	2.04
$C_{16}H_{33}N(CH_3)_3^+$	9.3	3.56	2.50	2.58
$C_{18}H_{37}N(CH_3)_3^+$ (0.01 M HCl)	9.0	4.21	3.01	2.83
$C_{26}H_{53}N(CH_3)_3^+$ (0.001 M NaCl)	10.4	7.34	4.78	3.85

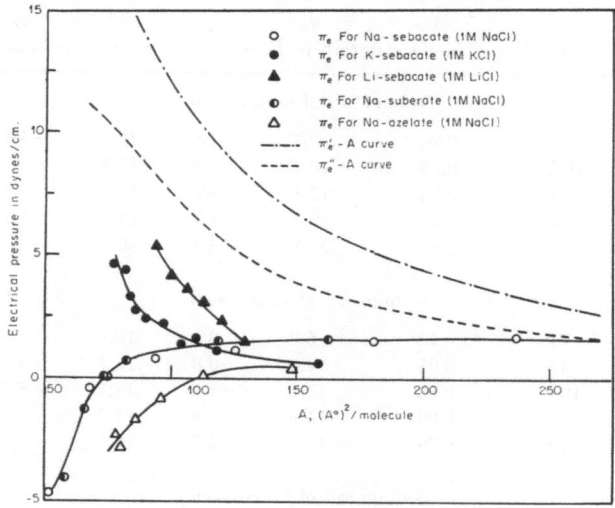

FIGURE 5.19. Electrical pressure vs. area per molecule for salts of dibasic acids at heptane–water interface (35°C)[46] π_e^d and π_e^{BLP} are written as π_e' and π_e''. (Reproduced with permission from Academic Press, New York.)

difference in π for salt and acid monolayers will lead to the evaluation of π_e from the experimental data using equation (5.19). The values of π_e for octanoate and SDS monolayers at the oil–water interface are considerably lower in magnitude than π_e^d and π_e^{BLP} (vide also Table 5.1). For salts of dibasic acids, π_e becomes even negative (vide Fig. 5.19). These results indicate that the Gouy model of the electrical double layer alone cannot explain the electrical free energy change associated with ionization of the monolayer even when the discrete-ion effect has been taken into account.

The wide variation between the experimental and theoretical values of π_e is suspected to be due to the neglect of the counterion binding in the charged film. Such binding of counterions will lead to the formation of the fixed part of the double layer as proposed by Stern (vide Section 4.13). It has been proposed recently[46] that

$$\pi_e = \pi_e^d + \pi_e^{BLP} + \pi_e^s \qquad (5.31)$$

where π_e^s is the contribution of the surface pressure due to the formation of the counterion binding in the Stern layer. If one subtracts the calculated value of π_e^{BLP} from π_e, then

$$\pi_e^c = \pi_e^d + \pi_e^s \qquad (5.32)$$

Then π_e^c is the electrical pressure generated at the smoothly charged interface and is equal to $(\pi_e - \pi_e^{BLP})$. π_e^c thus contains contributions only from the diffuse and the fixed double layers with elimination of the discrete-ion effect. Values of π_e^c for different sebacate monolayers at the heptane–water interface in the presence of excess neutral salt has been compared with π_e^d in Fig. 5.20. We may imagine that the π_e^d–A curve is valid for the diffuse double layer when the counterion binding is completely absent in the monolayer. For an identical value of π_e^c and π_e^d, respectively, if A_e^c and A_e^d be the values of the area per adsorbed boloform ion (two COO^- groups) in π_e^c–A and π_e^d–A curves in Fig. 5.20 then

$$f^c = 1 - A_e^c / A_e^d \qquad (5.33)$$

Here f^c stands for the fraction of the COO^- group per unit film area binding the counterions. Values of f^c for several sebacate monolayers are shown in Fig. 5.21.

Neglecting the discrete-ion effect, Matijevic and Pethica[52] have considered the aspects of counterion binding qualitatively for the octanoate monolayer formed at the air–water interface. One can calculate the fraction f of counterion

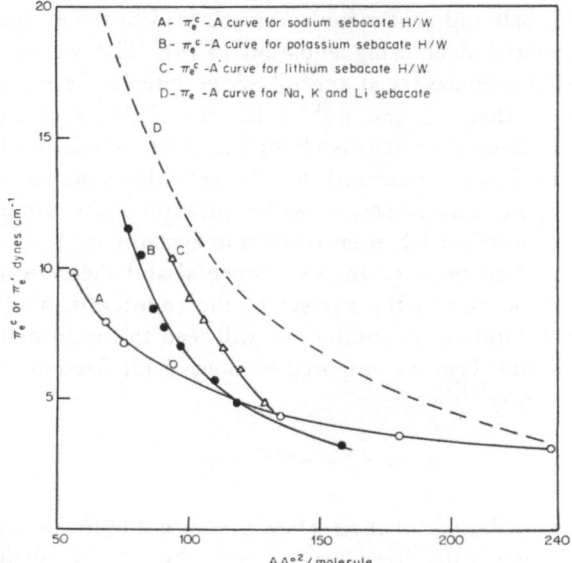

FIGURE 5.20. π_e^c-A curves (A, B, C) and π_e^d-A curve (D) for sodium, potassium, and lithium sebacate at heptane–water interface (35°C).[46] (Reproduced with permission from Academic Press, New York.)

binding for sebacate monolayers at the heptane–water interface after setting π_e^{BLP} equal to zero. In such a situation f will be higher than f^c (vide Fig. 5.21). It may be pointed out that calculation of f in such a case cannot be extended to the region A in Fig. 5.19, where π_e becomes negative. The observed negative electrical pressure can only be explained clearly if the discrete-ion effect is precisely taken into account.

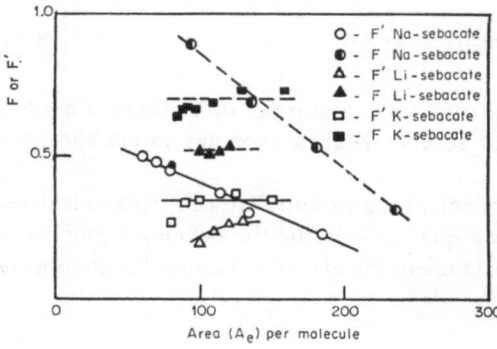

FIGURE 5.21. Fractions f and f^c (written as F and F') of adsorbed organic ions binding counterions as function A of the area per adsorbed molecule.[46] A may be either A_e or A_e^c. (Reproduced with permission from Academic Press, New York.)

5.11. DISCUSSION ON THE COMPUTATION OF π_e

From all these discussions, it is clear that the equation of state for the charged monolayer at the oil–water interface may be suitably represented by the general equation

$$\pi_{O/W} = \pi_k + \pi_e^d + \pi_e^{BLP} + \pi_e^s \qquad (5.34)$$

π_k can be computed with the help of equation (5.20) provided the value of A_0 for the adsorbed ions is known. Values of A_0 for octanoate or sebacate monolayers are assumed to be the same as those of corresponding organic acids. This may or may not be valid. For SDS and quaternary amine monolayers, values of A_0 are in fact uncertain. Computation of $\pi_{O/W}$ with the help of equation (5.19) will remain uncertain to some extent in the high-pressure region when π_k becomes sensitive to the magnitude of A_0.

Based mainly on the calculation of Verwey and Overbeek,[42] Davies has shown π_e^d to be equal to $\int_0^{\psi_0} \sigma' \, d\psi'$. Refined derivations of this relation either from theories of the double layer or from the integration of the Gibbs adsorption equation have been presented by Payens[53] and Hachisu.[40] Davies[4] derived an approximate expression for π_e^d on the basis of the Donnan effect for the charged monolayer. Theoretical computation of π_e^d for given values of A may be achieved with the help of equation (5.25).

Based on the theory of the discrete-ion effect, π_e^{BLP} can also be theoretically evaluated at given value of A using the relatively complicated equations (5.28), (5.29), and (5.30). In Fig. 5.19 and Table 5.1, π_e^{BLP} is observed to differ widely from π_e^d as the value of A decreases. Further, it has been shown that the negative value of π_e experimentally obtained in the high-pressure region can only be explained on the basis of the BLP theory, so that the existence of the discrete-ion effect in the ionized monolayer seems to be a reality.

Direct elaborate experimental studies on counterion binding in the spread and adsorbed monolayers have been made in recent years.[54–56] A statistical mechanical theory for the adsorption of metal ions onto the fatty acid monolayer has been developed by Tanaka and Fukutome.[57] Recently, on the basis of site-binding or surface complexation model for the ionized spread monolayer at the oil–water interface, James[58] has shown that

$$\pi = \frac{kT}{A - A_0} + \int_0^{\psi_0} \sigma' \, d\psi' + \int_0^{\psi_s} \sigma_s \, d\psi_s' \qquad (5.35)$$

Here ψ_0 and ψ_s (equal to $\psi_0 - \psi_\delta$) appear to have identities with the quantities discussed for the Stern model for the double layer (vide Section 4.13). It also seems that the last term on the right-hand side of equation (5.35) may be taken to be equal to π_e^s in equation (5.34), if the π_e^{BLP} term is neglected. In fact, in deriving relation (5.35) from the site-binding model as well as from the integration of the Gibbs equation, the discrete-ion effect has been neglected by James. He has also shown that on the basis of the computer calculation of π for different values of A, the theoretically obtained curve may be fitted with the experimentally obtained π–A curves for sodium octadecyl sulfate monolayer spread at oil–water interfaces at several ionic strengths of the medium. Since the first two terms on the right-hand side of equation (5.35) are definite for given values of A, the last term (representing π_e^c) needs adjustment in fitting the theoretical and experimental curves. From the results presented in Fig. 5.21, it has been shown that the fraction of counterion binding in this procedure will be overestimated at the higher-pressure region because of the neglect of the discrete ion effect.

5.12. ELECTRICAL DOUBLE LAYER AND ELECTRICAL FREE ENERGY

Models of the electrical double layer proposed, respectively, by Helmholtz, Gouy, and Stern have been discussed in detail in Chapter 4. Application of the Helmholtz model gives simple expression for the coefficient m of the Gibbs adsorption equation, although this model as such is not quite satisfactory. In dealing with the data on the adsorption at the charged interface, the Gouy model has been most widely used, although fit of the experimental data to this model is not always satisfactory. The Stern model of the double layer is used also qualitatively and the limit of its application lies in the lack of interest for the direct measurement of the binding of the counterions within the interfacial double layer.

From the precise experimental measurements of the interfacial tension of the mercury–water system in the presence of different salts for various values of applied potential, charge, or capacity of the double layer, the detailed structure of the double layer has been worked out by a number of workers.[59–61] According to Bockris, Devanathan, and Muller,[61] the structure of the double layer at the mercury–water interface in the presence of salts may be represented by Fig. 5.22. In this picture, the first layer adsorbed on the mercury interface may be composed of mostly water molecules of low dielectric constant 6. A few anions having the same charge as that of the metal interface may be specifically adsorbed on this layer situated between the metal plane ϕ_M and the (inner) Helmholtz plane ϕ_1. Further, solvated cations are adsorbed on the first adsorbed layer. The centers of the adsorbed cations lie on the Gouy plane ϕ_2 (outer Helmholtz layer). Formation of the diffuse double layer is believed to start close to the plane ϕ_2. The water molecules existing between the metal plane ϕ_M and planes ϕ_2 (and also a short distance away from it) are highly oriented and their dielectric constants and other properties are not uniform.

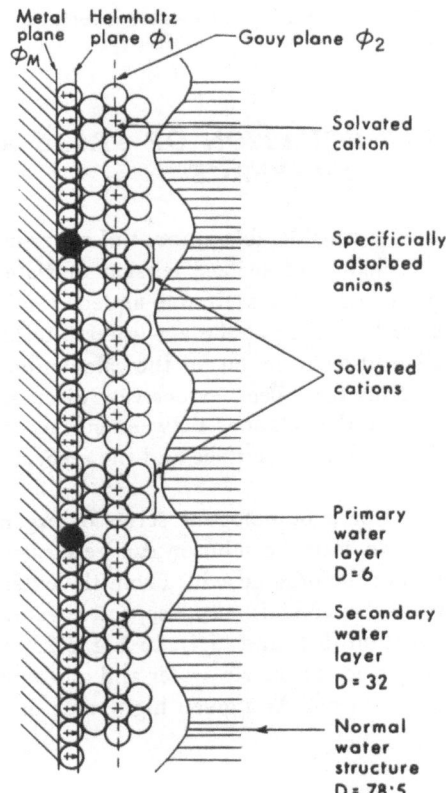

FIGURE 5.22. Structure of the double layer at the metal–water interface in the presence of salts.[61]

Whether the entire picture of the double layer depicted for the metal–solution interface remains valid for the ionized monolayer at the oil–water interface is an open question to be resolved in the future. But the most important feature is that the experimentally derived quantity π_e represents the actual change in the free energy due to the formation of the electrical double layer at the liquid interface. Although $\pi_e^d + \pi_e^{\mathrm{BLP}}$ have been theoretically estimated with some confidence, this sum is significantly different from π_e. The difference has been arbitrarily attributed to π_e^s via equation (5.31). In fact, based on the most refined pictures of the double layer, the above difference should be equated to $\pi_e^s + \pi_e^{\mathrm{hy}} + \pi_e^m$. Here π_e^{hy} is the effect of the change in the hydration of the interfacial phase due to the ionization of the monolayer. The fact that π_e^c in Fig. 5.20 depends upon the nature of the alkali metal ions indicates that the contribution of the hydration of the monolayer to π_e is not negligible. The other specific and nonspecific effects based on any refined model of the double layer may be included in the term π_e^m. An attempt should be made to make precise analysis of the structure of the electrical double layer at the oil–water interface in terms of the experimentally measured π_e.

5.13. EQUATION OF STATE AT THE AIR–WATER INTERFACE

A considerable amount of experimental data have so far accumulated on the change of surface tension of water due to the addition of organic and inorganic solutes to the aqueous solvent.[2,3,62] Theoretical aspects of the adsorbed layer at the air–water interface and the applicability of the Gibbs adsorption equation for the calculation of the surface excess in such aqueous systems have been extensively discussed in Chapters 3 and 4. We will now discuss the relations between surface pressure of the adsorbed monolayer at the air–water interface and the respective surface and bulk concentrations Γ_R and c_R.

For a homologous series of organic acids RH at low pH, the surface tension γ of the solution can be measured as a function of the concentration (c_R) of the organic acid. From the extent of surface tension lowering and also from the slope $d\gamma/dc_R$, respective values of Γ_R and A can be calculated using equation (5.1) and (5.4). In Fig. 5.23, π–A curves for a series of homologous fatty acids at the air–water and oil–water interfaces have been compared with each other.[63] At a given high value of A, $\pi_{A/w}$ is observed to decrease more

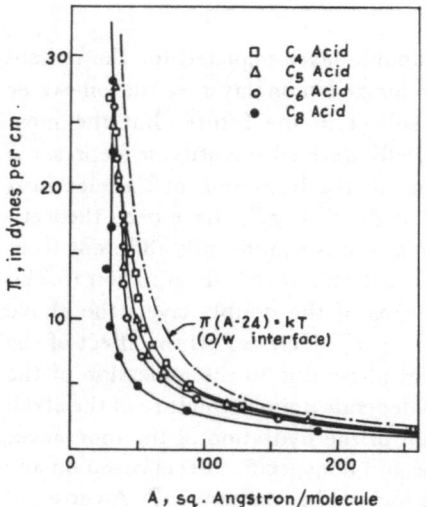

FIGURE 5.23. Pressure–area plots for the neutral monolayers (n-acid) at air–water interface (35°C). Dashed curve, general π–A curve for the acids (C_4–C_8) at oil–water interfaces (35°C). (Redrawn from Ref. 63.)

and more with increase of the length of the hydrocarbon chain attached to the COOH group of the acid. At a given value of A of an acid, $\pi_{A/W}$ of an acid is considerably lower than $\pi_{O/W}$ at the oil–water interface. Also in contrast to the behavior at the air–water interface, value of $\pi_{O/W}$ is independent of the length of the hydrocarbon chain of the acid. The equation of state for the neutral monolayer of the organic solute at the air–water interface is represented most suitably by the relation given by Davies[31]

$$\pi_{A/W} = \pi_k + \pi_c \tag{5.36}$$

Here π_k is the kinetic pressure whose value can be calculated using equation (5.7) or (5.20). π_c is the cohesional pressure due to the attraction between the hydrocarbon groups of the acid, which are oriented to the air side of the interfacial plane. Such cohesional pressure (π_c) at the oil–water interface is zero so that plot of $1/\pi_{O/W}$ against A becomes linear (vide Fig. 5.6) according to equation (5.7). One thus finds that a hydrocarbon chain of the adsorbed solute surrounded by molecules of oil will not be able to exert net attraction to other similar chains in the film. When oil in the system is replaced by air, cohesional interaction in the monolayer will be significant, so that for the same value of A in Fig. 5.23

$$\pi_{A/W} - \pi_{O/W} = \pi_c \tag{5.37}$$

This equation was obtained by Davies,[3,31] who also pointed out that π_c is negative since $\pi_{A/W} < \pi_{O/W}$.

Earlier from an empirical consideration of Guastala,[30] Davies[31] has suggested an equation for $A > 100 \text{ Å}^2$

$$\pi_c = \frac{400 n_c}{A^{3/2}} \tag{5.38}$$

Here n_c stands for the effective number of CH_2 groups present in the hydrocarbon chain. More recently from the analogy of the van der Waals equation, it has been suggested by Chattoraj and Chatterjee[64] that

$$\pi_c = \frac{a_s}{A^2} \tag{5.39}$$

where a_s is termed as the surface van der Waals constant. π_c for the fatty acids can be evaluated from the data using equation (5.36). Plot of π_C against $1/A^2$ in Fig. 5.24 has been found to be linear; from the slope of the curve, a_s for various acids has been calculated and presented in Table 5.2.

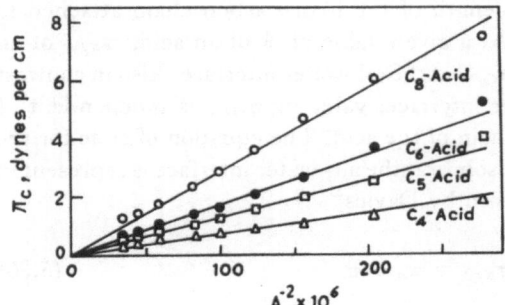

FIGURE 5.24. Cohesive surface pressure (π_c) of neutral monolayers against A^{-2}. (Redrawn from Ref. 64.)

Gershfeld[65] has pointed out that following the thermodynamic treatment of Defay and Prigogine,[66] it is legitimate to write

$$\pi_c = \left(\frac{\partial F_c}{\partial A}\right)_T$$

$$= \left(\frac{\partial E_c}{\partial A}\right)_T - T\left(\frac{\partial S_c}{\partial A}\right)_T \tag{5.40}$$

Here F_c, S_c, and E_c stand for the (Helmholtz) free energy, entropy and internal energy associated with cohesional interaction. If v^s is the volume of the surface phase per adsorbed molecule, then it is equal to τA, where A as before is equal to $\mathscr{A}/N$ [vide equation (5.4)].

We can then write

$$\pi_c = \tau\left(\frac{\partial E_c}{\partial v^s}\right)_T - T\left(\frac{\partial S_c}{\partial A}\right)_T$$

$$= \frac{a}{\tau}\frac{1}{A^2} - T\left(\frac{\partial S_c}{\partial A}\right)_T \tag{5.41}$$

This relation is obtained after inserting the thermodynamic equation $(\partial E_c/\partial v^s)_T$ equal to $a/(v^s)^2$ for van der Waals gases[67] enclosed in a three-dimensional container of volume v^s. Combining equations (5.39) and (5.41) therefore

$$\tau = \frac{a}{a_s + A^2 T(\partial S_c/\partial A)_T} \tag{5.42}$$

It is assumed that the hydrocarbon region in the monolayer is homogeneous. From the known values of critical temperature and pressure, a for a real gas having the same number of CH_2 groups as that present in the monolayer may be calculated. These are also presented in Table 5.2. Further, from thermodynamics, it can be shown that

$$\left(\frac{\partial S_c}{\partial A}\right)_T = \frac{k}{A} \qquad (5.43)$$

so that

$$\tau = \frac{a}{a_s + kTA} \qquad (5.44)$$

Gershfeld, in a sample calculation, has shown that τ is insensitive to the value of A. For A equal to $0.60 \, nm^2$, τ calculated for monobasic acids, SDS, and amines are shown in Table 5.2. These values are in satisfactory agreement with the length (τ_m) of the organic molecule within the monolayer.[63,65] In deriving equation (5.44), only the contribution due to the translational motion of the molecules has been taken into account but that due to the molecular rotation has been neglected.

Because of the wicket structure formed by the dibasic acid at the air–water interface (vide Fig. 5.7), the cohesive interaction between the chain in this film becomes considerably restricted, so that values of a_s become considerably less.[14] Possibly because of this, the interchain cohesion for the protein monolayer at the air–water interface is totally absent.[20]

For the ionized film at the air–water interface, the electrical double layer formed at the aqueous side will be similar to that existing at the oil–water interface provided that both cases are identical. Thus the equation of state for the charged monolayer at the air–water interface can be expressed by[31]

$$\pi_{A/W} = \pi_k + \pi_c + \pi_e \qquad (5.45)$$

For the adsorption of some organic electrolytes at the air–water and oil–water interfaces, therefore, equation (5.37) will remain valid for the charged monolayer also. Using this relation, π_c for several charged monolayers has been calculated[31,63] for various values of A. The linear plots of π_c against $1/A^2$ for quaternary amine monolayers are shown in Fig. 5.25. The surface van der Waals constant a_s and τ calculated from equations (5.39) and (5.44) are included in Table 5.2; τ is found to be very close to τ_m, the extended length of the molecules in the monolayer.

From Fig. 5.25, it appears that the plot of π_c deviates from linearity when A is less than $1.10 \, nm^2$. To explain this deviation, it has been suggested by Chattoraj and Chatterjee[64] that in the high-pressure region, a certain fraction, x, of the adsorbed molecules is undergoing aggregation, so that equation (5.39)

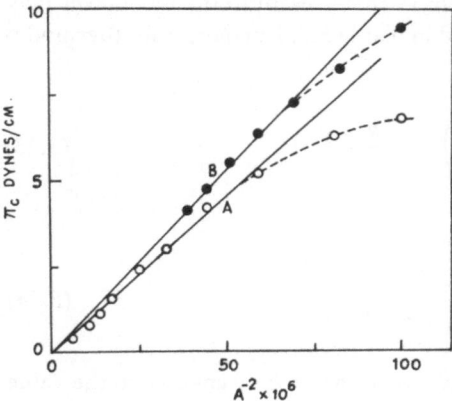

FIGURE 5.25. π_c as a function of the inverse square of the area per molecule. (A), $C_{18}H_{37}N(CH_3)_3^+$ monolayer in a substrate of $10^{-3}\,M$ HCl at 20°C. (B), $C_{26}H_{53}N(CH_3)_3^+$ monolayer in a substrate of $10^{-3}\,M$ NaCl at 21°C. (Redrawn from Ref. 64).

will assume the modified form

$$\pi_c = \frac{a_s^c}{A^2}(1 - nx + x^2) \qquad (5.46)$$

Here a_s^c is the average van der Waals constant for the monomer plus micelles. For $A < 1.0\ \text{nm}^2$, the aggregation numbers n calculated in this manner are 19 and 6 for $C_{18}H_{37}N(CH_3)_3^+$ and $C_{26}H_{53}N(CH_3)_3^+$ monolayers, respectively. The formation of surface micelles having honeycomb structure has been earlier proposed by Langmuir.[68]

The deviation from linearity in Fig. 5.25 may also be explained if we assume empirically

$$\pi_c = \frac{a_s}{A^2} + \frac{b_s}{A^q} \qquad (5.47)$$

Here b_s, like a_s, is constant, and q is a constant. The second term on the right-hand side of equation (5.47) may arise as crowded molecules begin to attract a particular molecule from different sides within the monolayer at high surface pressure. From the linear plot of $\log(\pi_c - a_s/A^2)$ against $\log A$ shown in Fig. 5.26, b_s and q are numerically found to be 180 and 3.3, respectively, for $C_{18}H_{37}N(CH_3)_3^+$ monolayers.

Recently, Tajima and co-workers[69,70] have presented $\pi\text{–}A$ curves for SDS monolayers at the air–water interface based on the $\gamma\text{–}c_R$ data experimentally obtained by them. They have also calculated the cohesive pressure (π_c') from the relation $\pi - \pi_e^d - \pi_k$ where π_e^d is the electrical pressure calculated from the Davies equation (5.25). They have shown from an empirical consideration

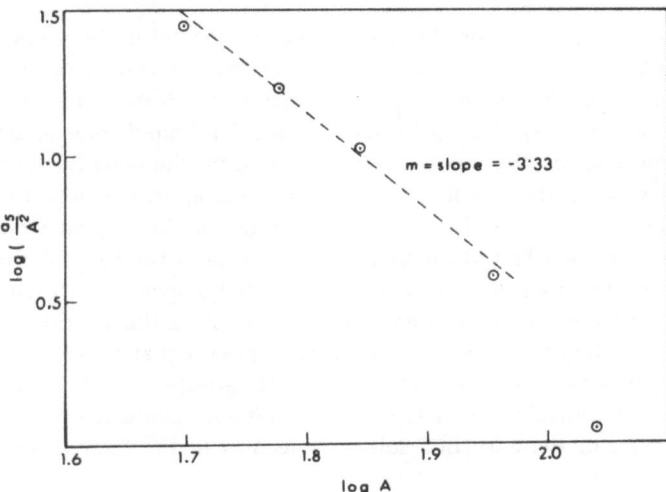

FIGURE 5.26. Plot of $\log(a_s/A^2 - \pi_s)$ against $\log A$ for $C_{18}H_{37}N(CH_3)_3^+$ monolayer. (Das and Chattoraj, unpublished.)

that

$$\pi_c' = \frac{0.43kT}{A} \qquad (5.48)$$

Since π_e in equation (5.31) is equal to $\pi_e^d + \pi_e^{BLP} + \pi_e^s$, π_c' is actually equal to $\pi_c - \pi_e^{BLP} - \pi_e^s$, so that equation (5.48) cannot be valid solely for the cohesional interaction. Similarly, Ghosh and co-workers[71,72] have calculated cohesional interaction between the hydrocarbon chains of the ionized monolayer at the oil–water interface having assumed that π_e and π_e^d are equal. Their calculation will not be correct because of the fact that π_e is not equal to π_e^d.

5.14. FREE ENERGIES OF ADSORPTION AT THE LIQUID INTERFACE

By differentiating equation (5.2) and then combining with the Gibbs equation (3.1), it can be shown that

$$\Gamma_R = \frac{b}{RT}\,c_R$$

$$= \alpha_R c_R \qquad (5.49)$$

This is the equation for the distribution of the solute between the surface and bulk phases being valid when the solution is considerably dilute. The distribution coefficient α_R is proportional to the initial slope b of the γ–c_R plot (vide Fig. 5.4). Traube[9] showed that for a homologous series of normal chain compounds, the concentration required for the same lowering of surface tension by a compound of n carbon atoms was approximately 3 times higher than that of the next higher homologue containing $(n + 1)$ carbon atoms. This regularity of the effect of the hydrocarbon group is the basis of Traube's rule. According to this rule, the ratio of the initial slopes of γ–c_R curve of acids with n and $n + 1$ carbon atoms will be 3. From a theoretical consideration, Langmuir[2] showed that the ratio of these slopes is equal to $\exp\left(-\Delta G^0_{CH_2}/RT\right)$, so that the free energy of transfer per CH_2 groups is 2670 J/mol at 25°C at the air–water interface. Ward[73] discussed different conformational and orientation factors of the adsorbed solute molecules in the light of the validity of Traube's rule.

Deviation from relation (5.49) for the ideal adsorption isotherm takes place at higher solute concentrations, when γ–c_R curves are no longer linear (vide Fig. 5.4). A number of empirical equations have been proposed to explain the relationship between Γ_R and c_R at moderate to high solute concentrations.[32,74,75] A brief review of these aspects has been reported by Davies.[3]

Based on the concept of monolayer adsorption of gases on solid interface, Langmuir[1,2] presented his equation for adsorption in the most convenient form. Following this concept, we will imagine that the liquid interface possesses S_0 total sites per unit area for adsorption of adsorbate molecules such that S_0 molecules of adsorbate per unit area may be accommodated at the state of monolayer saturation. If X_S and X_{RS} stand for the mole fractions on the free and occupied sites of the interface at equilibrium with the X_R mole fraction of the organic solute present in the bulk, then for the reaction

$$R + S \rightleftharpoons RS \qquad (5.50)$$

$$\frac{X_{RS}}{X_R \cdot X_S} = K$$

$$= e^{-\Delta G^0/RT} \qquad (5.51)$$

Here S stands for the surface and K is the equilibrium constant for the surface reaction. ΔG^0 is the standard free energy change in the mole fraction scale for such a reaction. Since at any given mole fraction X_R, the number of surface sites per unit area remaining in the bound state is $1/A$, it follows that

$$\frac{X_{RS}}{X_S} = \frac{1/A}{(1/A_0) - (1/A)}$$

$$= \frac{A_0}{A - A_0} \tag{5.52}$$

Combining equations (5.51) and (5.52), and writing $c_R/55.51$ for X_R,

$$\frac{A_0}{A - A_0} = \frac{c_R}{55.51} e^{-\Delta G^0/RT} \tag{5.53}$$

This is the Langmuir equation for monolayer adsorption on the surface having fixed sites for the attachment of the adsorbates. According to this concept of Langmuir, so long as a site at the interface remains occupied, it is not available for further adsorption. However, in the absence of cohesive interaction, the adsorbed molecules at the oil–water interface become perfectly mobile, as a result of which the site at the oil–water interface previously occupied may become free and hence available for adsorption, by the lateral displacement of the occupied molecule to another site. A statistical correction to equation (5.53) for this lateral motion has been introduced by de Boer[76]; the modified equation (5.54) may be written as follows:

$$\frac{A_0}{A - A_0} \exp \frac{A_0}{A - A_0} = \frac{c_R}{55.51} \exp\left(\frac{-\Delta G^0}{RT}\right) \tag{5.54}$$

A and A_0 for the organic acids at the oil–water interface have been calculated from γ–c_R data. The plot of the left-hand side of equation (5.54) against c_R is found to be linear (vide Fig. 5.27) so that ΔG^0 can be calculated[8,12,14] from the slope. Values of the free energy of transfer from the bulk to the oil–water interface evaluated in this manner are presented in Table 5.3. As first pointed out by Haydon,[12] equation (5.54) is most suitable for the calculation of ΔG^0 at the oil–water interface. The plot of the left-hand side of the Langmuir equation (5.53) against c_R in Fig. 5.28 is, however, found to be nonlinear.

For the oil–water interface, the free energy of transfer ΔG^0 for the monobasic acids depends on the chain length of the organic solute. From the difference in the values of free energy between two homologous acids, the average values of $\Delta G^0_{CH_2}$ per CH_2 group for benzene–water and petrol-ether–water interfaces are found to be -3180 and -3380 J/mol, respectively.[8] ΔG^0 of dibasic acid does not increase uniformly with increase of chain length (vide

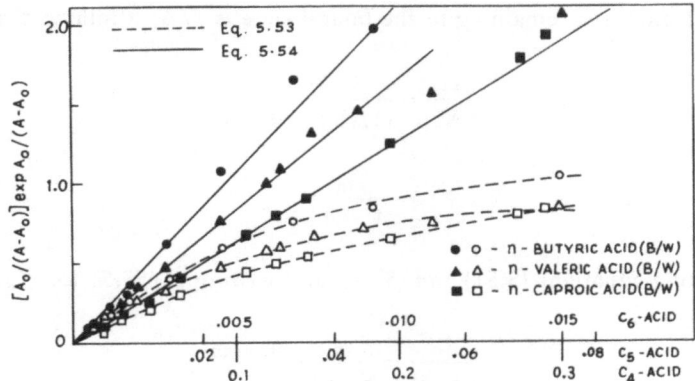

FIGURE 5.27. The adsorption isotherms for un-ionized acids at 35°C. (Redrawn from Ref. 8.)

Table 5.3). The uniform effect will, however, be observed if acids containing odd and even number of carbon atoms are grouped separately; then $\Delta G^0_{CH_2}$ is equal to -2980 J/mol[14] (vide Table 5.3). The lower magnitude for $\Delta G^0_{CH_2}$ for dibasic acid compared to that of monobasic acid is possibly due to the relative rigidity in the peculiar conformation of the former at the interface (vide Fig. 5.7). The difference in the behavior for the odd and even chain acids further indicates that the free energy of adsorption between two amphiphiles cannot be compared if their conformations at the interface are different. It appears that a correction of the standard state for the orientation effect is needed in this case if values of ΔG^0 for various solutes are to be compared with one another.

TABLE 5.3. ΔG^0 for Monobasic and Dibasic Acids at the Oil–Water Interface[a]

Name of the organic acid	Interface	ΔG^0 (kJ/mol)
n-Butyric acid	Benzene–water	16.2
n-Valeric acid	Benzene–water	19.1
n-Caproic acid	Benzene–water	22.6
n-Caprylic acid	Benzene–water	28.8
n-Caprylic acid	Petroleum ether–water	31.4
Pimelic acid	*n*-Heptane–water	23.5
Azelaic acid	*n*-Heptane–water	29.4
Suberic acid	*n*-Heptane–water	24.7
Sebacic acid	Benzene–water	27.8
Sebacic acid	*n*-Heptane–water	30.6
Sebacic acid	*n*-Heptane–1 M NaCl	33.9
Azelaic acid	*n*-Heptane–1 M NaCl	31.9

[a] Refs. 8 and 14.

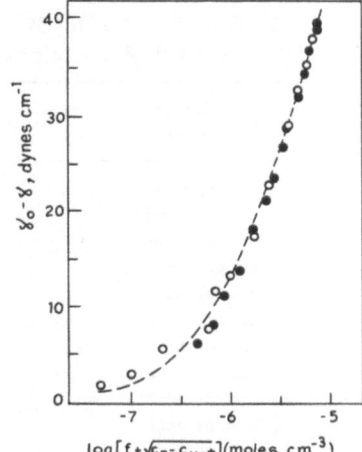

FIGURE 5.28. Interfacial pressures $\gamma_0 - \gamma$ of SDS–NaCl solutions against heptane as a function of mean ionic product of RNa. $\bigcirc$, $C_{NaCl} = 0$; $\bullet$, $C_{NaCl} + C_{RNa} = 8 \times 10^{-6}$ mol/ml. Line: ideal surface theory. (Redrawn from Ref. 91.)

Haydon and Taylor[77] have indicated that ΔG^0 for the ionized monolayer at the oil–water interface can be expressed by the relation

$$\Delta G^0 = \Delta G^0_{chem} + \varepsilon\psi_0 + \Delta G^0_{hy} + \Delta G^0_{sp} \tag{5.55}$$

Here ΔG^0_{chem} is the free energy of transfer of the ion in the hypothetical absence of the electrostatic and other effects. Haydon *et al.*[77] believe that the contributions of the dehydration and ion binding energies ΔG^0_{hy} and ΔG^0_{sp} for the monolayer to ΔG^0 are small. Combining equations (5.55) and (5.54),

$$\left(\frac{A_0}{A - A_0}\right)\exp\left(\frac{A_0}{A - A_0}\right)\exp\left(\frac{\varepsilon\psi_0}{kT}\right) = a\exp\left(\frac{-\Delta G^0_{chem}}{kT}\right) \tag{5.56}$$

Here a stands for the bulk activity of the surface-active ion in the mole-fraction scale. One weakness of this equation is that ψ_0 may have to be calculated using the Gouy model of the double layer. Using this equation, $\Delta G^0_{CH_2}$ for ionized monolayers or alkyl sulfates has been evaluated by Haydon and Taylor.[77] $\Delta G^0_{CH_2}$ for the ionized monolayer is considerably lower in magnitude than that obtained for the neutral monolayer.[8,78]

Neglecting the difference between the Helmholtz and Gibbs free energies at the interfacial phase, Chattoraj and Ghosh have pointed out (unpublished data) that the surface free energy change (ΔG_s) per unit area due to the change in the bulk mole fraction of the solute from zero to X_R can be obtained from the relation

$$\Delta G_s = \gamma - \gamma_0$$

$$= -\pi \tag{5.57}$$

TABLE 5.4. ΔG^0 for Monobasic and Dibasic Acids at the Air–Water and Oil–Water Interface Calculated Using Equation (5.59)[a]

Name of the acid	Interface	$-\Delta G^0$ (kJ/mol)
n-Butyric acid	Benzene–water	21.0
n-Valeric acid	Benzene–water	24.5
n-Caproic acid	Benzene–water	28.4
n-Caprylic acid	Benzene–water	35.0
Pimelic acid	Air–water	3.7
Suberic acid	Air–water	27.1
Azelaic acid	Air–water	28.8
Sebacic acid	Air–water	32.5
Pimelic acid	Air–1 M NaCl	23.3
Suberic acid	Air–1 M NaCl	26.7
Azelaic acid	Air–1 M NaCl	30.0
Sebacic acid	Air–1 M NaCl	33.4

[a] D. K. Chattoraj and L. N. Ghosh, unpublished.

This change results from the accumulation of Γ_R moles of solute per unit area at the interface. Keeping Γ_R constant, one can hypothetically dilute the bulk activity of the solute from $f_R X_R$ to unity (standard state of reference for solute in bulk) so that the free energy change ΔG_d for dilution per unit surface area may be calculated from the relation[20,79,80]

$$\Delta G_d = -\Gamma_R \ln \frac{1}{f_R X_R}$$

$$= \Gamma_R RT \ln f_R X_R \qquad (5.58)$$

Total free energy change per unit area due to the transfer of Γ_R moles of solute from the bulk to the interface of unit area for changing mole fraction of solute in bulk from zero to a hypothetical state of unity may be represented by equation (5.59)

$$\Delta G = \Delta G_s + \Delta G_d$$

$$= -\pi + \Gamma_R RT \ln f_R X_R \qquad (5.59)$$

ΔG for neutral or charged monolayers of a solute at the air–water and oil–water interfaces may be calculated with the help of equation (5.59) at a given value of π, f_R, and X_R (equal to $c_R/55.51$). Plots of ΔG against Γ_R for various monobasic acids and their salts are found to be linear. The slopes $\Delta G/\Gamma_R$ representing ΔG^0, the standard free energy change for the transfer of one mole of R from the bulk to the surface, have been presented in Table 5.4. Values of ΔG^0 thus evaluated are a little bit higher than those obtained from equations (5.54) and (5.56) but values of $\Delta G^0_{CH_2}$ in both cases remain the same. The difference is due to the implicit inclusion of the interaction terms in ΔG in (5.59) which is excluded in equations (5.54) and (5.56). Equation (5.59) is independent of any model of the interface. An advantage of equation (5.59) is its universal application for charged and uncharged oil–water and air–water interfaces. The application of this equation to biological and solid systems will be discussed later (vide Chapter 10).

Recently, Bull and Breese[81] measured surface tension of 0.1 M sodium chloride solution at 30°C as a function of increasing concentrations c_R of 20 different amino acids dissolved in the aqueous medium. The measurement was carried out at the isoelectric pH of each amino acid. The surface tension is found to decrease or increase with increase of c_R depending upon the nature of the amino acid. In Table 5.4, the amino acids are arranged in increasing

TABLE 5.5. Free Energy of Transfer (ΔG) and Free Energy of Transfer of Residue Relative to Glycine (Δg) of the Amino Acids[a]

Amino acids	pH	$d\gamma/dC_R$	ΔG (kJ/mol)	Δg (kJ/mol) B + B[b]	Δg (kJ/mol) N + T[c]
Ala	6.08	+0.96±0.08	2.53	−0.84	−2.08
Arg	11.12	+1.03±0.01	2.86	−0.50	—
Asu	4.15	+1.17±0.12	3.70	0.33	—
Asp	2.94	+0.96±0.24	2.53	−0.84	—
Cys	5.01	+0.69±0.26	14.97	−1.87	—
Gln	4.61	+1.21±0.14	4.03	0.66	—
Glu	3.23	+0.86±0.25	2.11	−1.24	—
Gly	6.07	+1.12±0.04	3.33	0	0
His	7.68	+1.03±0.13	2.86	0.50	−2.08
Ile	6.15	−15.16±0.33	−6.03	−9.40	−12.05
Leu	5.94	−21.91±0.99	−6.86	−10.24	−7.49
Lys	10.02	−0.92±0.14	1.92	−1.45	—
Met	5.73	−3.01±0.18	−2.74	−6.11	−5.41
Phe	5.83	−17.28±1.69	−6.32	−9.69	−10.39
Pro	6.11	−0.49±0.09	−0.70	−4.07	—
Ser	5.68	+0.76±0.11	1.75	−1.62	1.24
Thr	5.73	+0.59±0.11	1.20	−2.17	−1.66
Trp	5.73	−9.63±0.65	−5.0	−8.37	−14.15
Tyr	5.46	−15.09 ± 5.90	−5.94	−9.31	−9.57
Val	6.21	−3.74±0.20	−3.13	−6.48	−6.24

[a] Ref. 81.
[b] B+B=results of Bull and Breese.
[c] N+T=results of Nozaki and Tanford.

order of their hydrophobic character in terms of magnitude and sign of $d\gamma/dc_R$. Leucine, containing a large proportion of hydrophobic group in the side chain, seems to be most surface-active, whereas glycine and asparagine being surface-inactive, accumulate as negative surface excess, i.e., $d\gamma/dc_R$ is positive. Bull *et al.*[81] have also computed the free energy of transfer for all these amino acids using an empirical procedure [vide equation (8.50)]. This scale for free energy is found to be consistent with the hydrophobicity scale given for the amino acids from solubility measurements by Nozaki and Tanford[82]. The hydrophobicity scale for amino acid residues may be of importance in understanding the stabilizing influence of hydrophobic interactions on the structure of native proteins in solution. These results are presented in Table 5.5.

The standard free energy change for monolayer adsorption at the air–water interface is frequently calculated[84–86] using the following equation:

$$\Delta G^0 = -RTb \tag{5.60}$$

where b is given by equation (5.2). For the calculation of free energy in this manner, one needs measurement of γ at extremely dilute solution.

Szyszkowski[74] proposed an empirical equation

$$\frac{\gamma_0 - \gamma}{\gamma_0} = K_S \log\left(\frac{c_R}{B_S} + 1\right) \tag{5.61}$$

for surface tension lowering at the air–water interface. Here K_S and B_S are constants. This equation can be obtained from the combination of the adsorption equation given by Gibbs with that presented by Langmuir. The logarithmic term in equation (5.61) can be expanded so that using the experimental data of γ_0, γ, and c_R, the values of K_S and B_S can be obtained. Many workers[87–89] have calculated the standard free energy of adsorption using the following equation:

$$\Delta G^0 = -RT \ln B_S \tag{5.62}$$

Recently Rosen and Aronson[90] have proposed equation (5.63) for the calculation of ΔG^0 on thermodynamic grounds:

$$\Delta G^0 = -\pi A_0 + RT \ln c_R \tag{5.63}$$

ΔG^0 for long-chain alcohols and various other types of surfactants have been evaluated by these workers using this equation. There exist some striking similarities between equations (5.59) and (5.63) although thermodynamic treatments for their derivations appear to be different.

5.15. GENERALIZED FORM OF THE SURFACE EQUATION OF STATE

Lucassen-Reynders et al.[23] have derived several general relations between surfactant concentration in bulk, surface excesses, and surface tension on the basis of the thermodynamic approach of Gibbs without using any model for the interfacial phase. The chemical potential μ_i of the ith component in the bulk or surface phases in this treatment is expressed by equation (5.64) or (5.65), respectively:

$$\mu_i = \mu_i^0 + RT \ln f_i X_i \tag{5.64}$$

$$\mu_i = \mu_i^{0,\sigma} + RT \ln f_i^\sigma X_i^\sigma - \gamma A_i^\sigma \tag{5.65}$$

Here μ_i^0 and $\mu_i^{0,\sigma}$ are standard chemical potentials, f_i and f_i^σ the activity coefficients, and X_i and X_i^σ the mole fractions of the ith component in the bulk and the surface phases respectively. The partial molar area A_i^σ in the interfacial phase is also defined by the equation

$$A_i^\sigma = \left(\frac{\partial \bar{\mathscr{A}}}{\partial n_i^\sigma} \right)_{\gamma,\, T,\, n_j^\sigma} \tag{5.66}$$

Here $\bar{\mathscr{A}}$ stands for the total area of the interface formed due to the presence of n_1^σ, n_2^σ, ..., n_i^σ moles of components in the interfacial region as imagined by Gibbs. Components 1 and 2, respectively, stand for solvent and surfactant. It has further been assumed

$$n_1^\sigma + n_2^\sigma + \cdots + n_i^\sigma = \Gamma_2^m \bar{\mathscr{A}} \tag{5.67}$$

where Γ_2^m is the limiting value of surfactant adsorption. Its value can be evaluated from experiment by the application of the Gibbs adsorption equation. In equation (5.67), n_1^σ is the moles of solvent present in the interfacial region whose value is positive (and not zero as imagined by Gibbs for relative excess) so that Γ_2^m may be regarded as absolute and not relative excess. For strongly surface-active substance, however, Γ_2^1 is very close to $n_2^\sigma / \bar{\mathscr{A}}$ so that evaluation of Γ_2^m with the help of the Gibbs equation (3.1) may be justified. It may be pointed out here that Δn_i has close similarity with n_i^σ in many respects.

Differentiating equation (5.67) at constant values of γ, T, and n_j and then combining with equation (5.66) it may be shown that

$$A_2 = \frac{1}{\Gamma_2^m}$$

Lucassen–Reynders *et al.*[23] have further shown that

$$A_1 = A_2 = \cdots = A_i \tag{5.68}$$

and

$$X_2 = \frac{n_2^\sigma / \bar{\mathscr{A}}}{\Gamma_2^m} \tag{5.69}$$

and the whole treatment is independent of the monolayer or multilayer models of the interface. The dependence of this treatment on the position of the Gibbs dividing plane (vide Figs. 3.6 and 3.7) is not very significant.

Combining equations (5.64) and (5.65) for binary solution containing solvent component 1 and surfactant component 2, respectively, it can be shown that

$$\gamma_0 - \gamma = RT\Gamma_2^m \ln \left(f_2 X_2 \right) e^{\lambda_\sigma^0 / RT} + \frac{f_1^{0,\sigma}}{f_1^\sigma} \tag{5.70}$$

$$\Gamma_2 = \Gamma_2^m \left[\frac{f_2 X_2}{f_2 X_2 - (f_1^{0,\sigma} / f_1^\sigma) \, e^{-\lambda_\sigma^0 / RT}} \right] \tag{5.71}$$

and

$$\gamma_0 - \gamma = -RT\Gamma_2^m \ln \left(1 - \frac{\Gamma_2}{\Gamma_2^m} \right) - RT\Gamma_2^m \ln \frac{f_1^\sigma}{f_1^{0,\sigma}} \tag{5.72}$$

λ_σ^0 is regarded as molar free enthalpy of adsorption of surfactant when its concentration in the bulk remains infinitely dilute. Its value is given by the equation

$$\lambda_\sigma^0 = \mu_2^{0,\sigma} - \mu_2^0 - \gamma_0 / \Gamma_2^m \tag{5.73}$$

The activity coefficient of pure solvent in the interfacial phase is represented by $f_1^{0,\sigma}$.

For ideal bulk as well as surface solutions forming a liquid system, $f_1 \simeq f_1^{0,\sigma} \simeq f_2^\sigma = 1$ and in this circumstance, equations (5.70) and (5.71) assume the forms given by Szyszkowski and Langmuir, respectively.

Lucassen-Reynders *et al.*[23] have finally concluded that the behavior of surfactant at the interface is completely described by three parameters, f_i^σ, Γ_2^m, and λ_σ^0. Inserting these parameters in their equations (5.70), (5.71), and (5.72) with best fit of the experimental data of γ vs c_R, they have further noted that f_i^σ is close to unity even for higher values of c_R, and λ_σ^0 (related to c_R at the state of half saturation) is influenced by the same factors that affect the critical micellar concentration of the detergent in solution. From this observation it has been suggested that the processes of adsorption and micelle formation are markedly similar.

Lucassen-Reynders[91] has further extended the thermodynamic approach to include ionizing surfactants and non-surface-active electrolytes. The derived equations are similar in form to those of nonionizing surfactants except for the molar concentrations of the solutions being replaced by the geometric mean of the concentrations of long-chain ions and counterions. Thus for two electrolytes RNa and NaCl with a common ion, it can be shown that

$$\gamma_0 - \gamma = 2RT\Gamma_R^m \ln\left[2\frac{f_\pm}{f_\pm^\sigma}\left(\frac{X_{Na^+}\cdot X_R}{X_{RNa}}\right)^{1/2} + \frac{f_1^{0,\sigma}}{f_1^\sigma}\right] \tag{5.74}$$

where

$$RT\ln X_{RNa} = (\mu_R^{\sigma,0} + \mu_{Na^+}^{\sigma,0}) - (\mu_R^0 + \mu_{Na^+}^0) - \frac{2\gamma_0}{\Gamma_R^m} \tag{5.75}$$

Equation (5.74) predicts one value of γ corresponding to a given value of the ionic product $(c_R \cdot c_{Na^+})$ irrespective of the presence or absence of NaCl. Here $f_\pm$ and $f_\pm^\sigma$ are the mean activity coefficients of the ions in the bulk and surface phases whose values may be put equal to unity for the ideal behavior of the phases. In Fig. 5.28, $\gamma_0 - \gamma$ versus $\log[f_\pm(c_R c_{Na^+})^{1/2}]$ plot from experiments at the heptane–water interphase and that calculated on the ideal surface theory have been found to agree with each other most satisfactorily. At oil–water interfaces, the ionized surfactants thus form ideal surfaces so that $f_\pm^\sigma$ becomes unity. At the air–water surface, the long-chain surfactants do not form ideal surfaces. The extensive analysis of the experimental data proves that electrical interactions do not measurably affect the ideality of a surface, and it is the interaction between the hydrophobic chains which determines the ideality of a monolayer.

5.16 SUMMARY AND COMMENTS

From the extent of lowering of boundary tension of a liquid due to the addition of a surface-active solute component in the liquid phases, the surface pressure (π) for the soluble monolayer can be evaluated indirectly. The relation between π and the surface excess concentration Γ_R of the organic solute at the oil–water interface has been established on the basis of the ideal or van der Waals gas models for the adsorbate present in the monolayer. General theoretical approaches for the calculation of the excluded areas of the solute in the monolayer phase are not developed yet with precision.

In the surface solution model proposed alternatively for the interpretation of the π–A curves, the role of solvent in the monolayer phase has been considered but the inhomogeneous nature of the surface phase has been neglected. Motomura and co-workers have recently studied the surface pressure of soluble monolayers at the oil–water interface as functions of temperature, pressure, and bulk solute concentration (c_R). From the detailed theories developed by them, partial molar volume, partial molar entropy, partial molar internal energy of a component in the interfacial phase have also been evaluated by these workers so that thermodynamics of mixing of components in the interfacial phase are gradually being understood. How far this approach remains valid for more complex types of systems is yet to be seen.

From the comparison of the π–A curves for the unionized organic acid with that of the ionized monolayer of the corresponding salt of the fatty acid at the oil–water interface, the contribution of the electrical free energy change for the formation of the interfacial double layer has been directly calculated. This gives an opportunity to test the validity of the various theoretical models for the interfacial double layer frequently proposed by many workers. Based on these free energy data, it is also desirable to develop a generalized theory for calculating the activity coefficient of the solute in the interfacial phase which may include chain–chain interaction, solvation, inhomogeneous distribution of the solvent and the ions in the interfacial phase.

The cohesive pressure of the monolayer at the air–water interface can be calculated with the help of the Davies equation. Using these data, the thickness of the interfacial phase may be evaluated on the basis of the van der Waals gas model. The results are consistent with the extended orientation of the organic molecules in the interfacial phase. However, it is not clear yet (i) the nature of cohesional forces in the high-pressure region, (ii) decrease of cohesional pressure with increase in the proportion of hydrophilic groups attached to the organic molecule, (iii) disappearance of cohesional pressure when air in the system is replaced by oil.

The calculation of standard free energies ΔG^0 for adsorption of organic molecules at the oil–water interface on the basis of the Langmuir adsorption equation appears to be satisfactory if modification for the lateral motion of the molecules in the monolayer is taken into account. Several general methods of approach for the calculation of ΔG^0 applicable to monolayers at air–water and oil–water interfaces have been discussed, but it is not clear whether their values or the values of $\Delta G^0_{CH_2}$ for different class of compounds having different molecular orientation at the interface are strictly comparable to each other. It is highly desirable that ΔG^0 values should be compared with reference to a fixed (real or hypothetical) orientation of the molecule at the interface.

The generalized equations of state derived on the basis of the thermodynamic concept put forward by Lucassen-Reynders and van den Tempel appear to be extremely useful for the interpretation of γ, c_R, and Γ_R for ionic and nonionic systems in terms of f_i^σ, Γ_R^m and λ_σ^0. The standard chemical potentials $\mu^{0,\sigma}$ and $\mu^{0,s}$ used in equation (5.65) and the Butler equation (3.119) are different from each other so that f_i^σ and f_i^s refer to different standard states. The importance of the experimental evaluation of Γ_R^m for the thermodynamic interpretation of the monolayer properties has also been emphasized by Lucassen-Reynders. That treatment also leads to the evaluation of the entropy of adsorption in a simple manner. However, the exact physical concept of the surface excess Γ_i^σ and their dependence on the position of the Gibbs dividing plane particularly for ionized monolayer is not very clear. The validity of equation (5.67) for such an ionic system having both positive and negative values of Γ_i^σ may be questioned. Differences in the values of Γ_R^m for neutral and ionized monolayers in the presence and absence of inorganic salt may account for the insignificant contribution of the electrical free energy of the double layer to the surface activity coefficient.

SPREAD MONOLAYER

6.1. INTRODUCTION

If a very small quantity of a practically insoluble and nonvolatile organic substance is carefully placed on the surface of a liquid such as water, the latter having a relatively high surface tension, the following may take place: (1) the substance may either remain as a compact drop (or as a solid mass), leaving the rest of the liquid surface clean, or (2) it may spread out over the whole available surface of the liquid. The necessary and sufficient condition for that substance to spread as a monolayer is that its molecules must attract the water more than they attract each other. In other words, quantitatively, the work of adhesion between the substance and the water must be greater than the work of cohesion of the substance itself. If such a balance of forces exists, then as many as possible of the molecules of the spreading substance move into direct contact with the surface molecules of water (substrate), thus forming a unimolecular film, under given conditions. Similar considerations would apply to any other interface, e.g., oil–water, Hg–water.

The structural properties of these monomolecular films are found to be of much interest in the understanding of many other processes. From the study of these films one can easily determine the molecular arrangement, as regards the size, shape, and other physical properties of the individual molecules.

We have seen elsewhere (Chapter 5) that the monolayer film will also be formed when an amphiphile solute dissolved in bulk phase is adsorbed in excess at the interface (air–water or oil–water). Adsorbed monolayers have usually been treated by the Gibb's adsorption equation, which allows one to calculate the excess quantities of thermodynamic extensive properties except volume. In the case of spread monolayer, the interface is generally treated thermodynamically as a homogeneous phase. It is also seen that the spread monolayer may be assumed as a limiting state of adsorbed monolayer, and thus the same thermodynamic treatment could be applied to both types of monolayers. However, there are certain subtle differences which require the development of thermodynamic considerations which are separate for the spread monolayer case from those already described for the adsorbed system.

In the case of spread monolayers, the amount of amphiphile present at the surface is assumed to be equal to the amount applied (since the amount in bulk subphase, c_R, is assumed to be negligible). These spread films can be studied with the help of the Langmuir balance (as described in Chapter 2) whereby the surface pressure, π, and area, A (area per spread molecule), are measured.

It is thus obvious that the surface property of the liquid (subphase) is considerably affected in the presence of a spread film. However, besides the change in the surface tension, the other physical properties which have been extensively investigated are surface potential, surface viscosity, surface evaporation rates, surface waves, optical properties, and film–solid interaction (so-called Langmuir–Blodgett films). The quantity surface pressure is of fundamental interest since it is of thermodynamic significance. The arrangement of monomolecular layers at liquid surfaces, being a very useful model for many complicated systems such as emulsions, microemulsions, and biological cell membranes, has given rise to much of importance in the studies of these films.

HISTORICAL DEVELOPMENT

In ancient times it was well known that surface films of oil on rough seas had a protective action, since the oil films had a wave-breaking effect. It was reported[1] that in more quantitative terms a teaspoon of oil was sufficient to have a calming effect on the half-acre surface of a lake. In 1890 Rayleigh[2] reported that the erratic movement of a camphor crystal on the surface of water was stopped when a certain amount of oleic acid ($C_{17}H_{33}COOH$) sufficient to give a film of only about 1.6 nm (16 Å) thick was added.

The most important contribution came from the observations made by Pockels[3] on the surface films. It was reported that surface films can be handled by pushing them in front of strips or barriers extending the whole width of a trough of water filled to the brim, while the barriers made contact with the surface of liquid thus completely enclosing the film-forming substance. It was reported that the surface tension did not change much until the fatty acid films were compressed to about 0.20 nm^2/molecule. The properties of these films were studied in great detail for the first time by Langmuir.[4]

6.2. STATES OF MONOMOLECULAR FILMS

In the following we will discuss the various states of monomolecular films as analyzed from the π–A isotherms (vide Chapter 2) in the case of simple amphiphile molecules. Physical forces acting between the amphiphile molecules at the interface, that is, the forces acting between the alkyl–alkyl chains, and between the polar–polar groups through the substrate, should be considered. In the process where two amphiphile molecules are brought closer during the measurements of π as a function of A, the interaction forces would undergo certain changes which would be related to the packing of the molecules in the two-dimensional plane. The $\pi-A$ isotherm for a film is analogous, in many respects, to the three-dimensional $P-V$ isotherm of a gas. We know that as pressure, P, on a gas (e.g., CO_2), is increased, the molecules approach more closely and transition to the liquid phase takes place below the critical temperature. Further compression of the liquid results in the formation of a solid. In the case of alkanes, the distances between solid molecules is about 0.5 nm, while it is about 0.6 nm in the case of the liquid state. The distances in the gas phase are of the order 1000 times larger than in liquids or solids. It is found that the isotherms of the two-dimensional films also resemble the three-dimensional $P-V$ isotherms, and that one can use the same classical descriptions as regards the qualitative description in the analysis of the various states. However, it is obvious that there will not be a complete comparison between the three-dimensional and the two-dimensional structures, since there are very subtle differences in the structures. In the three-dimensional structural build-up, the molecules are in contact with near neighbors, as well as in interaction with molecules which may exist 5 to 10 molecular dimensions apart. This is apparent from the fact that in liquids there is a long-range order of interaction up to 5–10 molecular dimensions. On the other hand, in the two-dimensional films the state is much different. The amphiphile molecules are organized such that the polar groups are oriented towards the substrate, while the alkyl parts are placed away from the substrate. The structure is stabilized through the interactions between alkyl–alkyl chains, polar group–substrate, and polar–polar group, respectively.

As close packing in three-dimensional structures leads to the solid state, a similar kind of close packing would also give a "two-dimensional solid" like film at high surface pressures. The surface pressure (π) versus area (A) isotherms (schematic) are given in Fig. 6.1, as a function of temperature for simple amphiphiles.

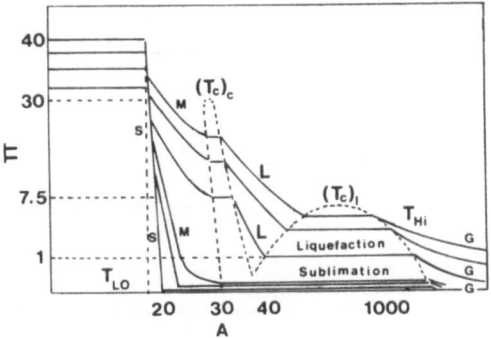

FIGURE 6.1. π in mN/m versus A in Å^2/molecule at different temperatures of spread monolayers at air–water interface (schematic). Low: LO; high: HI; T_c, critical temperature; G, gas; L, liquid; S, solid; M, mesomorphic. (Adapted from Ref. 22.)

6.2.1. GASEOUS FILMS

This film (G) would consist of amphiphiles which are at a sufficient distance apart from each other that the lateral adhesion (van der Waals forces) is negligible. However, there is sufficient interaction between the polar group and the substrate (e.g., water) that the molecules cannot be lost into the "vapor phase," and that the amphiphile is almost insoluble in water.

The molecule will have average kinetic energy, $\frac{1}{2}kT$ for each degree of freedom. The surface pressure measured arises from the collision between the film molecules and the float or the barrier from the translational kinetic energy having two degrees of freedom. The ideal film obeys the relation [vide equation (5.6)]

$$\pi A = kT \tag{6.1}$$

where A is the area/molecule. At 25°C, the product kT becomes equal to 411 if units of π are mN/m and of A are Å^2, respectively. Since ideal gas films are present at $\pi < 0.50$ mN/m where the area available per amphiphile molecule is very large, the studies of such films require Langmuir balances with very high sensitivity (± 0.001 mN/m). In Fig. 6.2 is given a typical πA–π plot for the monolayer of valinomycin (a dodecapeptide) which shows ideal gas film properties[5] (see Section 6.18.2).

In equation (6.1), we have assumed that the molecules are present as single entities in the film. However, if these are present as dimers or trimers, then the limiting value of πA would be less than the value of kT, indicating a certain degree of association.

6.2.2. LIQUID FILMS

In the case of long-chain amphiphiles, in several cases, transition phenomena have been found between the gaseous and coherent states of films which show a very striking resemblance to the condensation of vapors to

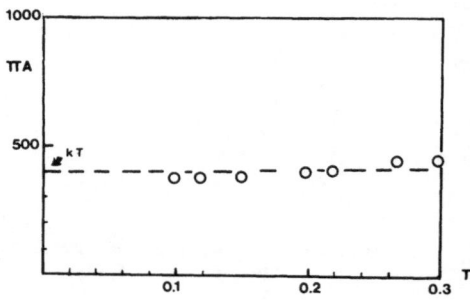

FIGURE 6.2. πA versus π plot for valinomycin monolayer[5] (25°C) at air–water interface. π is expressed in mN/m and A in Å^2/molecule (M.W. = 1100).

liquids in the three-dimensional systems. The liquid films show various states in the case of some amphiphiles, as shown schematically in Fig. 6.1, e.g., L_{ex} or $L_{co}(M)$.

6.2.3. LIQUID-EXPANDED FILMS (L_{ex})

In general, there are two distinguishable types of liquid films. The first state has been called the liquid expanded, L_{ex}.[6-8] If one extrapolates the π–A isotherm to zero π, the value of A obtained ($\simeq 0.5$ nm^2 per molecule) is much larger than that obtained for close-packed films. This shows that the distance between the molecules is much larger than one will find in solid films, as will be discussed later. If the surface pressure for L_{ex} is lowered, a first-order transition to gaseous film will occur.

6.2.4. LIQUID-CONDENSED FILMS (L_{co})

As the area per molecule (or the distance between molecules) is further decreased, there is a transition to a so-called liquid-condensed state L_{co}. These states have also been called "vapor-expanded" films. The effect of temperature, as shown in the films of n-pentadecylic acid in Fig. 6.3, is very typical for these films.

6.2.5. SOLID FILMS (S)

These films, also called condensed solid[6-8] films, S, are observed in systems where the molecules adhere to each other through the van der Waals forces very strongly. The π–A isotherm shows generally no change in π at high A, while at a rather low A value, one observes a sudden increase in π, as shown in Fig. 6.1. In the case of straight chain amphiphile molecules, like

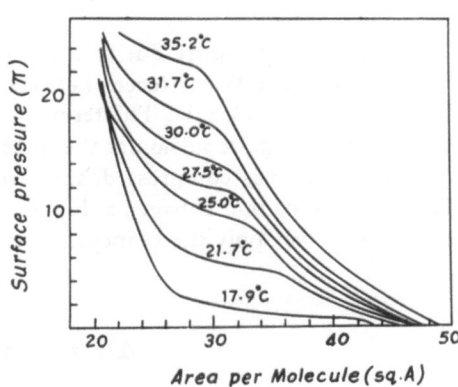

FIGURE 6.3. π–A isotherms[7,8] of pentadecylic acid for spread monolayers on aqueous subphase with pH = 2.

stearyl alcohol, the sudden increase in π is found to take place at $A \approx 0.20$ nm^2, at room temperature (that is much lower than the phase transition temperature of stearyl alcohol).

From the above discussion, it is thus seen that the films may under given experimental conditions show three different first-order transition points (Fig. 6.1), e.g., (i) transition from the gaseous film to the liquid-expanded L_{ex} state, (ii) transition from the liquid-expanded, L_{ex}, to the liquid-condensed, L_{co}, and (iii) transition from the liquid-expanded or the liquid-condensed to solid state, S. These various states exhibit very characteristic surface elasticity, E_S, which is defined by

$$E_S = \left(\frac{\partial \gamma}{\partial \ln A}\right)_T \qquad (6.2)$$

This is analogous to the elasticity of three-dimensional systems,[8] since the transition pressure, π_{tr}, varies continuously with the area, A. On the other hand, a true first-order transition requires that π_{tr} remains independent of A. A much intensive interest has been aroused in the past decade in the phase transition behavior of lipids of biological importance. We will consider this phase transition behavior of lipid films separately in this chapter.

6.3. EQUATION OF STATE FOR MONOMOLECULAR FILMS

In real films, we should also consider the interactions arising from other forces which will be expected to contribute to the measured surface pressure, π. The surface pressure of a film can be expected to arise mainly from the following different forces[9] (vide also Section 5.13):

$$\pi_{A/W} = \pi_k + \pi_c + \pi_e \qquad (6.3)$$

where π_k is the pressure due to the kinetic forces. The term π_c would arise from the van der Waals forces between the alkyl groups at the interface of the amphiphile molecule. The term π_e would arise in charged films through the charge–charge repulsion at the interface.

As the film is compressed, the transition for a gas–liquid film to (L_{ex}) becomes a transition between a dilute film to a more concentrated film at the interface. If we apply the Clapeyron analogy to these transitions, then we can write[6–8]

$$\Delta H = T \cdot \Delta A \left(\frac{d\pi}{dT}\right) \qquad (6.4)$$

where ΔA is the difference in A (cm^2/mole) for the L_{ex} state and gas phase and ΔH is then the latent heat of vaporization in erg/mole.

The heats of vaporization[6] are about 8, 13, and 40 kJ/mole for tridecylic acid, myristic acid, and pentadecylic acid, respectively. As discussed earlier, since the polar groups of the amphiphile molecules are interacting with the substrate, these values of enthalpy will be much different from those reported for the bulk materials.

6.3.1. LIQUID-EXPANDED STATE

The liquid-expanded state of the films is generally observed for the insoluble long-chain amphiphile molecules (e.g., fatty alcohols, fatty acids, lecithins) which have strong cohesion between the chains, and when the charge–charge repulsion forces are either absent or negligible (as for long-chain acids, alcohols, amides, and nitriles). Typical π–A isotherms are given in Fig. 6.3 in the case of pentadecylic acid. In these films the cohesive forces may be constant over a large range of areas and the equation of state is[9]

$$(\pi - \pi_c)(A - A_0) = kT \tag{6.5}$$

π_e is generally absent or negligible for such films. In another study,[10] the following equation of state was described:

$$\left(\pi + \frac{n_c \cdot \varepsilon_1 \cdot A_0}{2A^2}\right)\left[A\left(1 - \frac{A_0}{2A}\right)^2\right] = kT \tag{6.6}$$

where $\pi_c = -(n_c\varepsilon_1 A_0)/2A^2$.

In this equation it is assumed that film-forming molecules act as a stack of disks (n_c is the number of CH_3 and CH_2 groups) anchored to the subphase, water, by their polar head groups. The Lennard–Jones energy parameter, ε_1 for the disks is taken as that for methane (= 2.05×10^{-21} J). A_0 is the coarea and is twice the area defined by the distance, d, of closest approach, i.e., $A_0 = (\pi d^2/2)$. In Table 6.1 the estimated values of $(n_c\varepsilon_1 A_0)$ are given for various systems.

It is clear that the modified equation (6.6) is better than the relation given in equation (6.5). The interaction constant term varies rather widely in the case of single-chain amphiphiles. The variation of constant is much less in the case of a double-chain amphiphile, e.g., dimyrystyl lecithin, DML. The results of the interaction parameter, ε_1, further indicate that the interaction energy per carbon atom in the different films is: pentadecylic acid, 240; myristic acid, 410; DML, 76 (in DML there are two alkyl chains and accordingly the number of carbon atoms is twice the number in myrystyl chain, i.e., 13). Even though the variations are too large in the interaction constant term, the model

TABLE 6.1. The Estimated Values of the Interaction Energies in Liquid Expanded Films as Given by Equation (6.6)[a]

Substance	Range of A (Å^2/molecule)	Calculated value of $(n_c \varepsilon_1 A_0)$	t (°C)
Pentadecylic acid[b]	34–40	73 000–63 000	25
Myristic acid[c]	27–43	158 000–56 000	25
DML[d,e]	64–96	88 000–94 000	17

[a] K. S. Birdi (unpublished).
[b] Reference 7.
[c] Reference 11.
[d] DML = dimyristoyl-phosphatidyl choline.
[e] Reference 12.

does indicate that the measured surface pressure is lower due to the cohesive forces, which can be approximately estimated from the relation given in equation (6.6).

In earlier studies,[9] π for the liquid-expanded neutral films was given by another empirical relation (vide also Section 5.13)

$$(\pi + 400 n_c / A^{3/2})(A - A_0) = kT \qquad (6.7)$$

in the case of films with $A > 100 \text{ Å}^2$.

In many recent studies[13,14] the magnitude of π_c has been computed by using the two-dimensional van der Waals equation.[15,16] Using equations (5.39) and (6.5), Rakshit *et al.*[16b] have recently calculated the two-dimensional van der Waals constant a_s and limiting area A_0 from π–A data for the spread monolayers of a series of saturated, unsaturated, and hydroxy fatty acids (vide Table 6.2).

TABLE 6.2. van der Waals Constants a_s and A_0 for Spread Monolayers of Different Fatty Acids[a] at 25°C at A/W Interface

Fatty acids	$a_s \times 10^{27}$ (erg cm^2/molecule^2)	A_0 (nm^2/molecule)
Myristic acid	49.1	34.30
Palmitic acid	49.9	34.05
Cis-3-octadecanoic acid	55.5	35.75
Cis-9-octadecanoic acid	41.9	30.00
Trans-3-octadecanoic acid	59.0	39.07
2-Hydroxy stearic acid	52.7	34.50
3-Hydroxy stearic acid	46.5	33.49
6-Stearolic acid	48.5	34.62

[a] Reference 16b.

The cohesive forces between two linear long chains in two-dimensional arrays have recently been analyzed by quantum mechanical treatment.[17] The van der Waals attractive forces, W_d, were given as follows:

$$W_d = (Æn_c\pi)/(8L_dD_d^5) \tag{6.8}$$

where W_d is the interaction energy in kcal/mol per methylene group (or more correctly the number of carbon atoms, since the interaction between $-CH_3$ terminal groups would not be expected to be the same as for the $-CH_2$ groups), $Æ$ is the coefficient of dispersion interaction between the methylene groups, L_d the effective $-C-C$ distance along the chain, and D_d the chain separation distance. In the case of linear long-chain alkanes, the values of the various constants are[17,18]

$$Æ = -1.34 \times 10^3 \, kcal/mol \qquad L_d = 1.26 \, Å \tag{6.9}$$

The total attraction force arising from van der Waals interaction between two parallel and opposed paraffin chains at a distance D_d is (when the magnitude of D_d is much smaller than the mutual length)

$$W_d = -1.24 \times 10^3 \cdot n_c/D_d^5 \tag{6.10}$$

$$= -1.24 \times 10^3 \cdot n_c/[(2 \, Å/3^{1/2})^{1/2}]^5 \tag{6.11}$$

The relation in equation (6.11) is obtained on the assumption that in such monolayers a hexagonal close packing is present[17-20] which gives from the geometrical relations $A = (3^{1/2}D_d^2)/2$. The relation given in equation (6.11) was applied recently[18] in the estimation of protein penetration into lipid monolayers.

Let us assume that the van der Waals forces, W_d, are the most predominant in neutral lipids at moderate π values. Further, packing arrangements other than hexagonal can be considered in a more rigorous analysis. We will assume that the term arising from the cohesive forces contribution, i.e., π_c, is a function of the distance between the molecules as follows[20]:

$$\pi_c \approx W_d \propto 1/D_d^5 \propto 1/A^{2.5} \tag{6.12}$$

With modification of this equation, we have

$$\pi_c = C_1 + [C_2/(A)^{2.5}] \tag{6.13}$$

where C_1 is a constant and C_2 would be some function of W_d. However, in this analysis the one main parameter used is that the cohesive forces are related

TABLE 6.3. Equation of State for Lipid Monolayers; at A/W Interface[a]
[see equation (6.13)]

Lipid	Temperature (°C)	C_1	C_2	Correlation coefficient (r^2)
DML	17	−12.9	644.055	0.989
DPL	25	−2.83	394.294	0.905
	35	12.25	−419.685	0.996
Myristic acid	17	0.423	−10.048	
	25	1.96	−143.000	0.989
	35	9.305	−150.784	0.999

[a] Reference 20.

to the distance between the molecules (which in turn is determined from area/molecule).

The values of C_1 and C_2 of various lipid monolayers which were analyzed using equation (6.13) are given in Table 6.3. It is seen that in all cases the magnitude of π_c varies linearly with $1/(A)^{2.5}$ with a correlation coefficient of about 0.99. It is also seen that this linear relation is observed both for fatty acid and lecithin films, even when the properties of films of the latter differ in certain ways from those of the former (see Section 6.15.2).

6.3.2. EQUATION OF STATE FOR OTHER TYPES OF CONDENSED FILMS

For the solid films, Fig. 6.1, it has been found that the following relation holds[7]:

$$A = b - a\pi \tag{6.14}$$

where a and b are constants. The solid films, S, are very viscous due to strong hydrogen bonds between the substrate (water) and the polar group of the amphiphile.

Recently[21] it has been reported that in condensed films, alkyl chains are in tilted orientation in the case of different amphiphiles.

6.4. THERMODYNAMICS OF SPREAD MONOLAYERS

The insoluble monolayer in contact with the bulk phases of water and air, respectively, may be assumed[22–25] to be composed of water (solvent component 1) and amphiphile (solute component 2). Besides, air may also be present in the monolayer phase, but the effect of it is usually neglected in the thermodynamic treatment. Following the thermodynamic treatment of Butler and others (vide Sections 3.17 and 5.15), the chemical potential μ_1 of water in the monolayer phase at constant temperature and pressure can be expressed by the equation

$$\mu_1^m = \mu_1^{0,m} + RT \ln f_1^m X_1^m - \gamma A_1 \qquad (6.15)$$

Here $\mu_1^{0,m}$ is the standard chemical potential, X_1^m the mole fraction, and f_1^m the activity coefficient of water in the monolayer phase, respectively. A_1 is partial molar surface area. Using an equation similar to relations (3.111) and (5.66), A_1 can also be defined appropriately. At equilibrium, μ_1^m will be equal to μ_1^0, the chemical potential of pure solvent present in the subphase, so that

$$\gamma A_1 = (\mu_1^{0,m} - \mu_1^0) + RT \ln f_1^m X_1^m \qquad (6.16)$$

When X_1^m equals unity, γ becomes equal to γ_0, the surface tension of the pure solvent, so that

$$\gamma_0 A_1^0 = \mu_1^{0,m} - \mu_1^0 \qquad (6.17)$$

As per Rault's law convention, f_1^m under this situation for pure solvent is taken also as unity. The partial molar surface area A_1^0 of the pure solvent system is frequently taken to be equal to A_1. Subtracting (6.16) from (6.17), we find

$$(\gamma_0 - \gamma)A_1 = -RT \ln f_1^m X_1^m \qquad (6.18)$$

or

$$\pi = -\frac{RT}{A_1} \ln f_1^m X_1^m \qquad (6.19)$$

Equation (6.19) has a striking similarity with the equation for the osmotic pressure for solution in the bulk derived from the thermodynamic consideration (vide also Section 5.5). Replacing X_1^m by $1 - X_2^m$ and further replacing $\ln (1 - X_2^m)$ by $- X_2^m$ for highly dilute solution in the monolayer phase, we obtain

$$\pi = RT \frac{X_2^m}{A_1}$$

$$\simeq RT \frac{n_2^m}{n_1^m A_1} \qquad (6.20)$$

Here n_1^m and n_2^m are, respectively, the number of moles of components 1 and 2. For such a highly dilute solution, $n_1^m A_1$ may almost become equal to the total area of the liquid surface and $n_1^m A_1 / n_2^m$ becomes equal to $\mathscr{A}$, the area per mole of spread molecule. Equation (6.19) may now become equivalent to the ideal equation of state (6.1). The same equation for the adsorbed gaseous monolayer has already been derived on the basis of the validity of the empirical equation (5.2).

The thermodynamic analysis of the monolayer phase of nonideal type has been carried out based on the regular solution theory.[22] In such a case, equation (6.15) is to be replaced by equation (6.21):

$$\mu_1^m = \mu_1^{0,m} + RT \ln X_1^m + \alpha l (X_2^m)^2 - \gamma A_1 \qquad (6.21)$$

For a regular solution, $\alpha l (X_2^m)^2$ is the contribution of the extent of deviation from ideality to the chemical potential μ_1^m. In the statistical theories developed for the regular solution in the bulk and the interfacial phases, it is assumed that an individual molecule is surrounded by neighboring molecules of types 1 and 2 for a binary solution, thus forming a quasicrystalline lattice. α in this model is given by[22]

$$\alpha = NZ\left[\varepsilon_{12} - \frac{\varepsilon_{11} + \varepsilon_{22}}{2}\right] \qquad (6.22)$$

where the coordination number Z is the number of neighboring molecules in contact with the individual molecule thus considered. Let l stand for the fraction of these Z molecules which is in contact with the individual molecule in the same lattice plane and m be the fraction of Z remaining in the adjacent plane or planes. Also ε_{11} and ε_{22} and ε_{12} represent the respective potential energies of interaction between 1 and 1, 2 and 2, and 1 and 2 types of molecules in the binary solution.

Combining equation (6.21) with relation (6.17) in a similar manner, it can be shown[22] that

$$\pi = -\frac{RT}{A_1}\left[\ln(1 - X_2^m) + \frac{\alpha l}{RT}(X_2^m)^2\right] \qquad (6.23)$$

Further, putting $A_1 = A_2$, on the basis of the assumption that the sizes of the solvent and solvent molecules in the monolayer mixture are almost the same, the total monolayer surface area A_m of the liquid will be given by

$$A_m = (n_1^m + n_2^m)A_2 \qquad (6.24)$$

so that

$$X_2^m = \left(\frac{n_2^m}{n_1^m + n_2^m} \right) = \frac{n_2^m A_2}{A_m} \tag{6.25}$$

Combining relations (6.23), (6.24), and (6.25), we obtain

$$\pi = -\frac{RT}{A_2} \left[\ln \left(1 - \frac{n_2^m A_2}{A_m} \right) + \frac{\alpha l}{RT} \left(\frac{n_2^m A_2}{A_m} \right)^2 \right] \tag{6.26}$$

When $n_2^m A_2 / A_m$ is small, one can expand equation (6.26) and obtain the virial equation (6.27) of state for the insoluble monolayer

$$\pi = \frac{n_2^m RT}{A_m} + RT A_2 \left(\frac{n_2^m}{A_m} \right)^2 \left(\frac{1}{2} - \frac{\alpha l}{RT} \right) + \cdots \tag{6.27}$$

The second virial coefficient B^* in this virial equation may actually be written as[22] (see Section 6.8)

$$B^* = A_2 \left[\frac{1}{2} - \frac{\alpha l}{RT} \right] \tag{6.28}$$

Equations (6.27) and (6.28) have close similarities with the virial equations for van der Waals gases.

Detailed derivation of the ideal equation of state (6.1) based on statistical mechanics has also been considered by several workers.[22,26,27]

6.5. SURFACE ACTIVITY COEFFICIENT

For a spread film containing an un-ionized surface-active solute i, the chemical potential μ_i of the film has been defined by Gershfeld *et al.*[28] by the relation

$$\mu_i = \mu_i^0 + RT \ln \Gamma_i + RT \ln f_1^m \tag{6.29}$$

Here μ_i^0 is the standard chemical potential, Γ_i the Gibbs surface excess, and f_i^m the surface activity coefficient of component i in the appropriate reference scale. From the Gibbs equation,

$$d\pi = \Gamma_i \, d\mu_i$$

$$= \frac{1}{\mathscr{A}} \, d\mu_i \tag{6.30}$$

$\mathscr{A}$ stands for area per mole of the spread film. Differentiating equation (6.29) after replacement of Γ_i by $1/\mathscr{A}$ and then combining with equation (6.30), it can be shown

$$\frac{d \ln f_i^m}{d\pi} = \frac{\mathscr{A}}{RT} - C_m \qquad (6.31)$$

where the surface compressibility $C_m [= E_s^{-1}$; equation (6.2)] is given by

$$C_m = -\frac{1}{\mathscr{A}} \left(\frac{\partial \mathscr{A}}{\partial \pi}\right)^T \qquad (6.32)$$

From the slopes of π–$\mathscr{A}$, isotherms of the spread film, and hence $\mathscr{A}/RT - C_m$, can be evaluated for various values of π. Integrating equation (6.32) graphically

$$\ln f_i^m = \int_0^\pi \left(\frac{\mathscr{A}}{RT} - C_m\right) d\pi \qquad (6.33)$$

one can estimate f_i^m, the surface-activity coefficient. Following this procedure, surface-activity coefficients of lauric acid, hexadecanol, and m-octadecyl phosphate have been calculated by Gershfeld and Patlak.[28] They have also compared these values with those obtained from an alternative procedure based on desorption kinetics.[28]

Lucassen–Reynders[29] has also discussed the surface-activity coefficient of the monolayer phase with reference to the limiting value of the saturation of the monolayer by the individual surfactant molecules. The outline of this concept has already been discussed in Section 5.15.

On the basis of the concept of the surface osmotic pressure (vide Section 5.5), equation (6.19) has been derived also by Fowkes.[30] Assuming surface activity coefficient to be unity, he has also obtained the following relation [vide equation (6.24)]:

$$A_m = n_2^m A_1(n_1^m / n_2^m) + n_2^m A_2 \qquad (6.34)$$

In agreement with this equation, plot of A_m against n_1^m / n_2^m becomes linear for many monolayer systems even at a relatively high-pressure region in the absence of a phase transition. The surface activity coefficient of the solvent in the monolayer phase is very close to unity in this situation. The solvent activity coefficient will, however, significantly alter when phase transition occurs in the monolayer phase.

6.6. PROTEIN AND POLYMER FILMS

Many proteins and other types of polymers under suitable conditions can be spread on the aqueous solution of salts. The surface pressure of proteins can be suitably measured as a function of the total area of the surface using a highly sensitive film balance (vide Section 2.3). The pressure–area curves for spread protein monolayers have been extensively studied by various workers in detail because of the importance of such investigations in the biomembrane system. π–A curves for the protein monolayer do not exhibit characteristics of phase transition within the monolayer. At high-pressure regions, the surface protein separates out as gels.

At extremely low surface pressure, many protein films spread extensively on the surface of liquid, thus forming a gaseous film. For expansion of this film, water has to be transferred from the bulk of the subphase to the surface phase. Free energy change ΔG for the transfer of one mole of water from the subphase to the clean interface (of surface free energy γ_0) at constant temperature and pressure, according to Bull,[31,32] may be expressed by the relation

$$\Delta G = A_1^0 \gamma_0 = RT \ln \frac{X_1^{m,0}}{X_1} \tag{6.35}$$

Here A_1^0 and $X_1^{m,0}$ are the molar surface area and mole fraction of water in the monolayer phase, respectively. For the transfer of one mole of H_2O from the substrate to the monolayer phase (of surface free energy γ) containing n_2^m moles of surface-active substance, the free energy change ΔG will similarly be given by

$$\Delta G_1 = A_1 \gamma = RT \ln (X_1^m / X_1) \tag{6.36}$$

Here A_1 and X_1^m are the molar surface area and mole fraction of component 1 at the interfacial phase, respectively. Subtracting equation (6.36) from (6.35) and inserting further $A_1^0 \simeq A_1$ and $X_1^{m,0} = 1$, one finds

$$A_1(\gamma_0 - \gamma) = \pi A_1 = RT \ln (1/X_1^m) \tag{6.37}$$

Identity of equation (6.37) with (6.19) will be noted with interest. Replacing $\ln X_1^m$ or $\ln (1 - X_2^m)$ by $-X_2^m$ for highly dilute solution in the monolayer phase, it can be shown that

$$\pi n_1^m A_1 = n_2^m RT \tag{6.38}$$

since X_2^m is almost equal to n_2^m / n_1^m. Also the total interfacial area A_m is equal

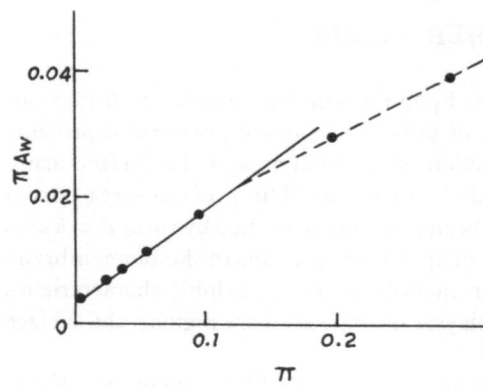

FIGURE 6.4. πA_w versus π plot[32] in low-pressure region for BSA on 5% $(NH_4)_2SO_4$. Area expressed in m^2/mg. (Redrawn from Ref. 32.)

to $n_1^m A_1 + n_2^m A_2$ [vide equation (6.24)] so that

$$\pi A_m = n_2^m (RT + A_2 \pi) \tag{6.39}$$

If w_2^m is the constant weight of the biopolymer spread on the total surface area A_m (vide Chapter 2) then n_2^m is equal to w_2^m / M_2, where M_2 is the molecular weight of the polymer in the monolayer phase. One can then write

$$\pi A_w = \frac{RT}{M_2} + \left(\frac{A_2}{M_2}\right)\pi \tag{6.40}$$

Here A_w stands for the surface area per gram of the polymer at the interface.

For extremely low values of π (below 0.5 or 0.2 mN/m) for many proteins, plots of πA_w versus π are satisfactorily linear (vide Fig. 6.4), so that from the intercept and slope of the curve, the molecular weight M_2 and effective

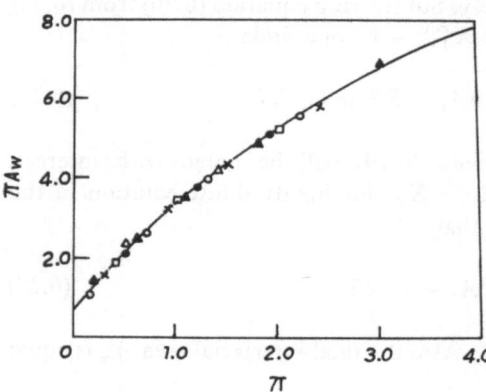

FIGURE 6.5. Plots of πA_w against π for a series of fractions of polyvinyl acetate. ($\pi =$ mN/m; $A_w = m^2/mg$.) (Redrawn from Ref. 33.)

molar area A_2 of the biopolymer in the monolayer phase may be calculated[31,32] (vide Section 6.9 also).

However, with increase of π, the plot of πA_w against π for bovine serum albumin and other proteins will deviate from linearity (vide Fig. 6.4). A similar deviation from linearity in Fig. 6.5 has also been observed in the high-pressure region for the monolayers of a series of polyvinyl acetate fractions[33] spread at the air–water interface. The departure from ideality is usually related to the neglect of the effect of entropy and enthalpies of mixing of the polymer and solvent components in the monolayer phase in deriving equation (6.40).

6.7. RIGOROUS EQUATIONS OF STATE FOR POLYMER FILMS

In the statistical treatment presented by Singer[34] for the equation of state of the spread film, the model for the surface used is a two-dimensional quasilattice in which the linear macromolecule may assume various types of orientations. This picture is analogous to the Flory–Huggins[35,36] model proposed for the polymer solution in bulk. The total surface area here is assumed to be composed of N_0 cells each of area A_0^c. In this theory, the interaction between the solvent and the macromolecules is neglected. It is also assumed that each solvent molecule occupies the same area as the monomeric unit of the macromolecule.

The relation derived by Singer[34] from a statistical approach can be represented by the following equation:

$$\frac{\pi \mathscr{A}_0}{RT} = \left(\frac{\nu - 1}{\nu}\right)\left(\frac{Z}{2}\right) \ln\left(1 - \frac{2\mathscr{A}_0}{Z\mathscr{A}}\right) - \ln\left(1 - \frac{\mathscr{A}_0}{\mathscr{A}}\right) \qquad (6.41)$$

Here ν stands for the number of flexible units in the polymer chain. $\mathscr{A}$ and $\mathscr{A}_0$ are the area and the limiting area, respectively, per mole of the macromolecule. Z is the coordination number of the lattice imagined to be present at the interface. Its value for a rigid macromolecule is 2 and in that case equation (6.41) can be converted to the form (6.40) for extremely dilute solution. The deviation of πA_w (or $\pi \mathscr{A}$) versus π plots in Figs. 6.4 and 6.5 indicates flexibility of the polymer chain in the monolayer phase so that values of Z becomes larger than 2. From the Singer equation actually one can estimate Z using the experimental data. A quantity $Z-2$ in the Singer equation is assumed to be a measure of the chain flexibility due to the unfolding of the chain at the interface against cohesive forces between the segments.[37,38] On the basis of this approach, the percentage flexibilities of gliadin, ovalbumin, and hemoglobin (in the presence of acid) at the oil–water interface are found to be 65, 22, and 38, respectively, whereas at the air–water interface the corresponding values are 3, 0.8, and 82. In spite of the approximation that the enthalpy of the surface is negligible, relation (6.41) agrees satisfactorily with the experimental data when π is not too high. The range of the Singer equation can be extended if the value of $\mathscr{A}_0$ used is that of collapse point $\mathscr{A}_{collapse}$ (K. S. Birdi, unpublished).

Motomura et al.[39] have pointed out that the assumption about the value of Z equal to 2 for a rigid macromolecule in the Singer equation is indeed too small. An alternative equation based on the combinatory theory of Guggenheim has also been presented by these workers in which the cohesive forces W between the polymer segments in the monolayer have been taken into account. The cohesive forces for the monolayers of polyvinyl acetate, polymethyl acetate, and polymethyl methacrylate are found to be significant in the region of high pressure. Their equation also approaches the Singer equation when the cohesive force tends to zero.

Flexible macromolecules present as insoluble films may form loops which can be compressed so that values of $\mathscr{A}_0$ may be altered. This has been taken into account in the statistical derivation of the equation of state,[40] which in simplified form reads

$$\frac{\pi\sigma}{N_s kT} = -\ln\left(1 - \frac{\nu N_1'}{N_s}\right) + \left(\frac{s}{2}\right)\left(\frac{\nu - 1}{\nu}\right)\left[\frac{1}{(1 - 1/t)}\right]$$

$$\times \ln\left[1 - \frac{2\nu}{s}\left(1 - \frac{1}{t}\right)\frac{N_1}{N_s}\right] \tag{6.42}$$

where σ is the total area, ν is the number of segments of polymer, N_s is the total number of sites in the surface, $\nu N_1'$ is the number of sites occupied by polymer segments, s is the coordination number in surface, and t is the number of segments adsorbed (complete adsorption means that $\nu = t$). Since, $\sigma = N_s A_0$, $A = \sigma/tN_1' = N_s A_0/tN_1'$, we get

$$\frac{\pi A_0}{kT} = \frac{s}{2}\frac{(\nu - 1)}{\nu(1 - t^{-1})}\left\{\ln\left[1 - \frac{2}{s}(1 - t^{-1})p\frac{A_0}{A}\right]\right\}$$

$$- \ln\left(1 - \frac{pA_0}{A}\right) \tag{6.43}$$

where $p = \nu/t$.

For $s \gg 1$ and $\nu \to t$ this relation reduces to the equation of state as derived by Singer [see equation (6.41)] for a completely deposited polymer.

6.8. VIRIAL EQUATION OF STATE FOR TWO-DIMENSIONAL POLYMER FILMS

Analogous to the three-dimensional virial equation of state for gases, the following kinds of expressions have been reported for two-dimensional films[41-45]:

$$\pi A_m = \frac{W_2^m RT}{M_2} + b_1 \pi + b_2 \pi^2 + b_3 \pi^3 \qquad (6.44)$$

Here b_1, b_2, and b_3 are virial coefficients. The Huggins virial coefficients[41] B_1 and B_2 are related to b_1 and b_2 by the equations

$$b_1 = B_1/\rho \qquad (6.45a)$$

and

$$b_2 = \frac{M(B_2 - B_1^2)}{RT\rho^2} \qquad (6.45b)$$

where ρ is the surface density (equal to M_2/A_0). In the case of several synthetic polymers[42,43] and biopolymers[44,45] it was shown that from πA_m versus π plots the value of M_2 could be determined as $\pi \to 0$. The virial coefficients of different biopolymers are given in Table 6.4.

In further analyses[44,45] it was given that

$$B_1 = \frac{n f_m^2}{2} \left(1 - \frac{2}{Z} \right) \qquad (6.46a)$$

$$B_2 = \frac{n}{3} f_m^3 \left(1 - \frac{4}{Z^2} \right) \qquad (6.46b)$$

where n is the number of residues, Z the coordination number of the two-dimensional quasilattice, f_m the average fraction of polymer in the surface

TABLE 6.4. Values of Virial Coefficients in Equation (6.67) for Proteins[a]

Polymer	$w_2^m RT/M$	b_1	b_2	b_3
BSA	0.0364	1.504	0.1384	0.01348
Ovalbumin	0.0551	1.818	0.2598	0.0539
Transferrin	0.0275	0.1628	0.00816	0.000427
Myoglobin	0.1457	0.0722	0.022	−0.0028

[a] References 44 and 45.

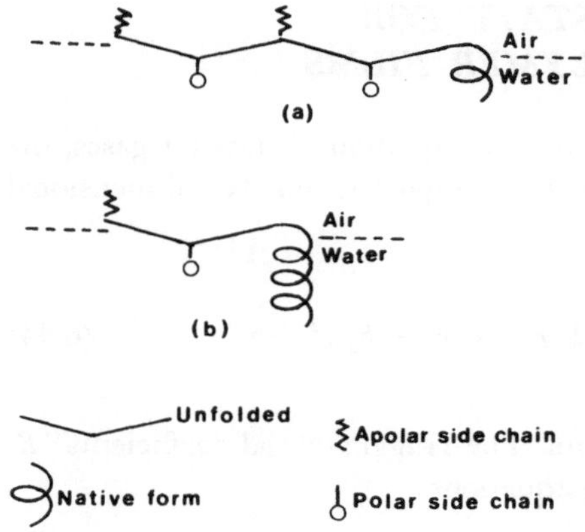

FIGURE 6.6. Orientation of protein molecules at air–water interface (schematic).

monolayer. The orientation of the biopolymers at the interface is depicted in Fig. 6.6. The values of A_0, B_1, B_2, and f_m are given in Tables 6.4 and 6.5 for proteins and polyamino acids. Values of f_m are higher for proteins which are more apolar (BSA, hemoglobin) than for polar proteins (transferin) as determined from their amino acid composition. The polyamino acids in α-helical conformation exhibit larger values of f_m than those for the β-form.[44,45] As regards the value of Z, the relations given in equations (6.46a) and (6.46b) are subjected to criticism (K. S. Birdi, unpublished).

6.9. CHARGED MONOLAYERS OF AMPHIPHILES AND BIOPOLYMERS

The ionized monolayers of linear amphiphiles have been extensively analyzed.[46-53] The treatments developed for the adsorbed ionized monolayer and discussed in detail in Chapter 5 are also expected to remain valid for the

TABLE 6.5. Values of A_0, B_1, B_2, and f_m for Biopolymers[a]

Polymer	$A_0(m^2/mg)$	B_1	B_2	f_m
BSA	1.16	1.296	1.683	0.976
Ovalbumin	1.47	1.229	1.516	0.933
Hemoglobin	0.975	1.264	1.601	0.952
Transferrin	0.206	1.115	1.252	0.844
Myoglobin	0.173	0.117	0.280	0.510
PLA	1.18	1.45	2.12	1.0
PBLG	0.64	1.14	1.30	0.9
PALS	0.87	1.03	1.10	0.8

[a] References 44 and 45.

charged spread film. The presence of an electric charge on the film-forming molecule gives rise to an appreciable effect on the π–A isotherm.

In the low-surface-pressure region, the contribution from π_c for a spread protein film can be considered to be negligible while π_e will be of larger magnitude.[54,55] We can thus write the equation of state

$$(\pi - \pi_e)(A - A_0) = kT \qquad (6.47)$$

so that $\pi = \pi_k + \pi_e$. The presence of charges would give an increase in π which is equivalent to π_e, at any given A. The surface model as described by Davies (see Chapter 5) in detail for the charged monolayers assumes that the extra free energy is required to bring changes from infinity to a state when surface potential is ψ_0 [vide equations (5.21) and (5.23)]. The quantitative expression (5.25) for the term π_e (equal to π_e^d) was derived (see Chapter 5) based on the Gouy diffuse double-layer model. Equation (5.25) was derived for a theoretical model of a charged monolayer formed due to the presence of uni-univalent organic salt at the interface. It is not easy to reformulate this relation to use it for the presence of an organic polyelectrolyte at the interface. However, as a first approximation, it has been useful to apply the relation given in equation (5.25) to the data for the charged protein monolayer spread at the air–water interface. The πA–π plots for various proteins[55] (e.g., insulin, BSA, and hemoglobin) are given in Fig. 6.7. From all these data, it is observed that πA becomes equal to kT when $\pi \to 0$ and pH of the subphase is the isoelectric point of the protein. This indicates that at isoelectric pH due to the absence of any net charge in the protein molecule, the magnitude of $\pi_e \to 0$, and thus $(\pi A)_{\pi \to 0} = kT = 411$. In other words, the molecular weights found were as one should expect at the isoelectric pH of the respective protein[55]: insulin as 6000, BSA as 68,000, and hemoglobin 68,000.

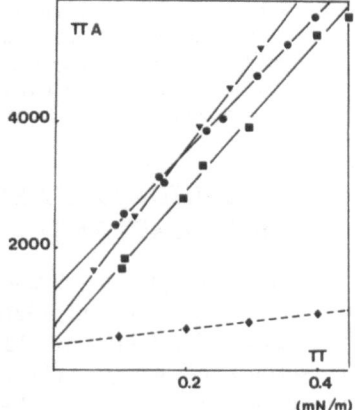

FIGURE 6.7. πA vs. π plots of monolayers of various proteins at the air–water interface: Insulin, pH = 7.4 (◆); hemoglobin, pH = 7.4 (■), pH = 5.1 (●): BSA, pH = 5.1 (▼).[55] [π = mN/m (dyn/cm); A = Å^2 per molecule ($\pi A = kT = 411$, at 25°C).]

TABLE 6.6. Variation of Charges per
Molecule z_p, of BSA with pH, as Determined
from Bulk and Monolayer Studies[a]

pH	z_p (bulk)	z_p (monolayer)
4	25	15–25
5.5	0	0
7.4	10	15–25

[a] Reference 55.

The number of charges per molecule in the monolayer at different pH calculated from equation (6.47) has been compared with that for protein in bulk obtained by other physicochemical methods (vide Table 6.6). After π_e (equal to π_e^d) is estimated, the value of A_e is found from equation (5.25), and thus the magnitude of $z_p = A/A_e$ = number of charges/protein molecule. The data for BSA have been found to fit[55] the following relation in the low π region:

$$\left(\frac{\pi A}{kT}\right)_{\pi \to 0} = 1 + 0.00413 z_p^2 \tag{6.48}$$

This result is consistent with bulk measurements, while the other literature reports[56,57] are in disagreement, due to their neglect of the effect of π_e in equation (6.47).

NOTE ADDED IN PROOF: Recent monolayer studies of ionophore peptides, e.g., melittin [K. S. Birdi *et al.*, *Colloid Polym. Sci.* **261**, 767 (1983)] and valinomycin (K. S. Birdi *et al.*, to be published) show that these peptides exhibit "discrete charges" on an aqueous subphase with added electrolytes. The magnitude of π_e as calculated by equations (6.47) and (5.25) gave very low values for charges per peptide, z. An equation of state was derived which was based on the discrete charge interactions (through a low dielectric medium) and which gave satisfactory analyses of both melittin (a maximum of six charges) and valinomycin (a maximum of one charge) monolayers. The reason the Gouy–Chapman model gave low values for z for these charged monolayers arose from the fact that this model is insensitive to variation of the dielectric medium and is only useful for charges as found in the aqueous subphase.

6.10 HELMHOLTZ FREE ENERGY FOR PROTEIN MONOLAYERS

When an aqueous solution of a protein is very carefully applied to the surface of an aqueous salt solution, the protein molecule tends to unfold.[55,58] The degree of unfolding would be consistent with the maximum lowering of the surface free energy. Thus, monolayer studies should be very useful in providing information about the forces operative at the surface (e.g., hydrogen bonds, electrostatic interactions, hydrophobic effects). A general method of

TABLE 6.7. Magnitudes of W_c for Different Proteins[a]

Protein	W_c (J/residue)
BSA	760
Ovalbumin	787
Transferrin	85.3
Myoglobin	150

[a] Reference 54.

evaluating these forces at the surface would be the determination of the work of compression, W_c (Helmholtz free energy) of the monolayer[54,58]:

$$W_c = -\int_{A_{initial}}^{A_{final}} \pi \, dA \qquad (6.49)$$

The values of W_c for different proteins were determined (Table 6.7). A linear relation between W_c and the amount at the surface has been reported. It was also found that the values of W_c were identical under the two different rates of compression for BSA and myoglobin.

From the π–A isotherm the limiting area/residue were determined, as a function of the amount of protein at the surface. These results are summarized in Table 6.8. As reported by various investigators, the area/residue for a completely unfolded protein at the surface is 0.15–0.17 nm² residue. Further it has been suggested[58] that for each protein, this value will be dependent also

TABLE 6.8. Magnitudes of Limiting Area/Residue (as Determined from π–A Isotherms) of Various Proteins Spread at the A/W Interface[a]

Protein	Limiting area (nm²/residue)
BSA	0.17
Hemoglobin	0.18
Ovalbumin	0.14
Insulin	0.17
Transferrin	0.04
Myoglobin	0.04
Cytochrome c	0.04
Ribonuclease	0.04
Lysozyme	0.04
β-Lactoglobulin	0.04

[a] References 54 and 58.

to some degree on the different charge effect at a given pH. From the results in Table 6.8, it can thus be concluded that BSA, hemoglobin, insulin, and ovalbumin are in a completely unfolded state, while transferrin, myoglobin, cytochrome c, lysozyme are incompletely unfolded. The data for ovalbumin and transferrin show that the Singer equation (6.41) could only be applied in the low-surface-pressure region, i.e., below 5 mN/m. It was also found that the coordination number Z in these calculations decreased with increasing amount of protein at the surface.

The protein molecule is thus adsorbed at the surface, such that the underside of the film is predominantly hydrophilic while the upperside is predominantly hydrophobic.[58] This model is shown schematically in Fig. 6.6. The polypeptide backbone is shown at the interface, and the nonpolar and polar side chains are depicted. The incompletely unfolded protein molecule is also shown in Fig. 6.6. In this case, the part of the molecule adsorbs and unfolds at the surface, while the rest of the molecule in its native form is dipping inside the aqueous subphase. It is obvious that the part of the molecule which is inside the subphase does not contribute to the surface pressure.

In the literature, very few studies on the monolayers at the oil–water interface have been reported. In the absence of cohesive forces, the area per residue for various proteins at the oil–water interfaces are somewhat larger than those values found at the air–water interface. Comparatively strong polar proteins like lysozyme give very stable[9,58] films at the oil–water interface, whereas the monolayer of this protein at the A/W interface is unstable.

6.11. ANALYSIS OF PROTEIN UNFOLDING AT INTERFACES

There have been a great number of studies reported in the literature on the prediction of the three–dimensional structure of a globular protein from the amino acid sequence of its polypeptide chains.[58–62] These studies have been able to predict with variable success the α-helix content, β-structure, and random coil[62] present in the protein dissolved in the aqueous medium.

In a recent study,[18,58] it was shown how the degree of unfolding of a protein at the interface was related to its amino acid composition with some exceptions[18] (Table 6.9). We can assume as a first approximation that the energy of adsorption at air–water interfaces is mainly determined by the nonpolar residues. There is a contribution to the adsorption energy from some of the polar side chains, for example lysine. We will assume that the nonpolar residues are ala, leu, val, met, phe, trp, tyr.[18,58] Further, analogous to the bulk folding algorithms, we will also argue that it is the short-range interactions[63] which mainly contribute to the adsorption at the interface. The long-range interactions are of much less importance in the unfolded state.

TABLE 6.9 Degree of Unfolding of Different Proteins as a Function of Various Parameters[a]

Protein	Degree of unfolding	Polar: apolar ratio	Helix content (%)	Number of S–S bonds	Number of chains	Molecular weight
Insulin	Complete	1.08	37	3	2	6 000
Zinc–insulin	Complete	1.08	37	3	2	6 000
Hemoglobin	Complete	1.03	75	0	4	68 000
Ovalbumin	Complete	1.26	45	0	1	40 000
Bovine serum albumin	Complete	1.56	50	17	1	68 000
Cytochrome c	Incomplete	2.1	0	0	1	14 000
Lysozyme	Incomplete	1.74	35	5	1	12 000
Transferrin	Incomplete	1.62			1	90 000
Myoglobin	Incomplete	1.29	75	0	1	17 000

[a] Reference 18a.

In the various algorithms given in the literature on the prediction of the three-dimensional conformation of a protein molecule, some have been based on the principle that hydrophobic side chains will be expected to interact if present at the i, $i + 1$, $i + 2$, $i + 3$, and $i + 4$ positions.[63] We will now examine how the hydrophobic character of the polypeptide chain varies in a sequence, for each segment of 5 amino acids from i to $i + 4$. In Table 6.10 we give the number of nonpolar amino acids for each segment of five residues for various proteins from their sequence data.

There are large differences in the nonpolar character of various proteins as determined by their sequence of nonpolar residues. It can also be concluded that the mere consideration that the nonpolar segments are responsible for the adsorption does not fully explain the degree of unfolding, as given in Table 6.9. It is useful to compare the plots of two typical proteins, e.g., hemoglobin and cytochrome. We find that in hemoglobin the adsorbing segments are fairly evenly distributed throughout the chain, thus giving

TABLE 6.10. Number of Nonpolar Residues in Each Segment of Five Residues in the Sequence of Various Proteins (see text)[a]

Hemoglobin (bovine, α-chain):	41332333413212143333122332433433321
Myoglobin (whale):	22443224221223213330132133232212232322
Lysozyme (egg):	2241223311101312011231212143 0131
Ribonuclease (bovine):	23112122102223120012102311 41432
Cytochrome c (bovine):	11311002202211223132410431

[a] Reference 58.

complete unfolding as also found by experiments (area/residue 0.17 nm^2). On the other hand, the cytochrome molecule shows rather predominant patches of polar features, which would give no adsorption energy, and thus an incomplete unfolding at the interface will be predicted, as also found by experiments (area/residue 0.04 nm^2). Thus it can be concluded that if the nonpolar character in a chain is distributed more or less evenly throughout the chain, then the molecule should be expected to adsorb and unfold completely, whereas if this distribution takes place more unevenly and is concentrated in some parts of the molecule, then the adsorption will be incomplete.

However, we would like to add some quantitative dimensions to these postulates. The hydrophobic energy required to transfer a hydrocarbon chain from water to the surface, i.e., air–water interface, according to the Traubes rule is of the order of 2900 J/mol CH$_2$ group (vide Section 5.14). It is also now known that the hydrophobic energy is most correctly defined by considering the surface area of the alkyl chain in contact with water.[58,64,65] The surface area per amino acid side chain for the apolar residues was calculated from the values reported in the literature.

The standard free energy of transfer from water to air for various amino acids, $\Delta G^0_{W \to A}$, was calculated from the relationship[58]

$$-\Delta G^0_{W \to A} = 107 \, \text{(surface area of molecule)}/(\text{J/mol}) \qquad (6.50)$$

These values are given in Table 6.11. The measured values for the $\Delta G^0_{o \to w}$, i.e., standard free energy of transfer from oil to water are also given for comparison. It is seen that the ratio $\Delta G^0_{A \to w}/\Delta G^0_{O \to w}$ differ by a factor 4.3 ± 2.0.

TABLE 6.11. Calculation of Hydrophobic Energy of Transfer from Water to Air $\Delta G^0_{A \to W}$ for Apolar Amino Acid Residues[a]

Amino acid	Area of apolar side chain (nm^2)	$\Delta G^0_{A \to w}$ kJ/mol (calculated)	$\Delta G^0_{O \to w}$ (measured)[b] (kJ/mol)
Proline	1.40	30.0	11.1
Phenylalanine	2.79	15.0	10.9
Methionine	2.13	22.6	5.43
Valine	1.66	17.6	7.12
Iso-leucine	2.53	26.7	12.3
Leucine	2.11	22.6	10.0
Alanine	0.84–0.90	9.2	3.14
Tyrosine	2.78	29.7	11.9
Tryptophan	2.94	31.4	12.5

[a] Reference 58.
[b] Reference 125.

In the following we will assume that the energy of transfer from bulk to the surface is approximately the same as that of the transfer of an alkyl chain from water to the surface phase. In the adsorption process we are only going to calculate by this procedure the free energy of adsorption, $\Delta G^0_{w \to a}$, for the nonpolar group for each residue, e.g., ala, val, leu, ile, met, pro, tyr, trp. This procedure thus gives the desired magnitudes of free energy (ΔG^0_ϕ) for each 5-residue segment from i to $i + 4$ residue. These calculated values of various proteins with known sequence are given in Fig. 6.8.

The following proteins have been found to unfold partially, viz., lysozyme, ribonuclease, transferrin, myoglobin, cytochrome. The data in Fig. 6.8 shows that in the case of ribonuclease, lysozyme, and cytochrome, there are rather extensive patches in the chain where there are predominantly polar residues. A value of zero hydrophobic energy indicates that nonpolar residues in the segment of 5-residue of the chain are absent. Comparing these proteins with hemoglobin, we find that in the latter there are extensive patches with hydrophobic character. This observation then would suggest that hemoglobin should

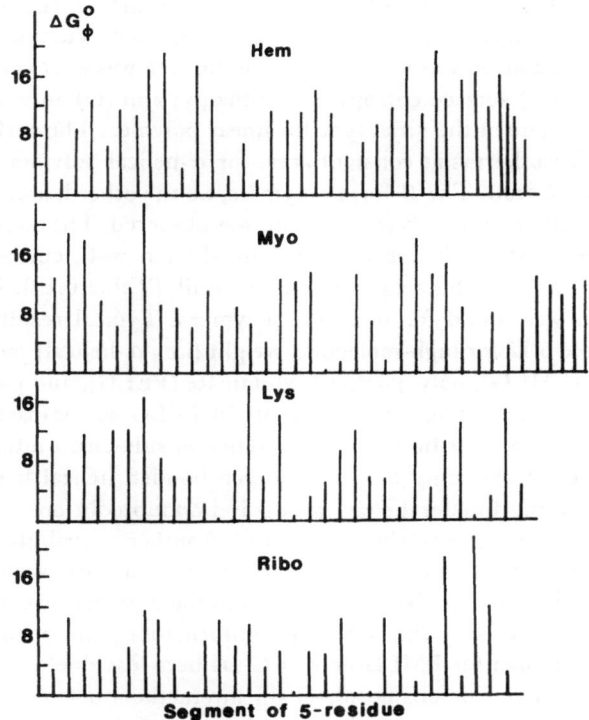

FIGURE 6.8. Change in free energy, ΔG^0_ϕ (kcal/mol) arising from only the transfer of hydrophobic groups from water to surface in each segment of 5-residues in the sequence of various proteins[58] (Hemoglobin: Hem; myoglobin: Myo; lysozyme: Lys; ribonuclease: Ribo.).

adsorb differently than the other proteins, which is in agreement with the experimental data.

The data in Fig. 6.8 indicate that myoglobin exhibits a greater number of regions with hydrophobic energy less than 40 kJ as compared to the hemoglobin molecule. Another property which is of vital interest in the adsorption process is that the low-energy regions in myoglobin are more concentrated near each other, whereas in the hemoglobin molecule the low-energy regions are more evenly distributed near the high hydrophobic energy regions.

6.12. MONOLAYERS OF SYNTHETIC POLYAMINO ACIDS

It is well established that synthetic polypeptides and proteins tend to form α-helical or β-structures in the solid state or in solutions, depending on the particular amino acid and the molecular weight.[58] The following properties of the polypeptide monolayers have been reported[45]: (i) the relationship between the cohesive forces in the bulk and at the surface; (ii) comparison of the compression surface pressure (π) versus area (A) curves with surface pressure versus concentration curves where the area was kept constant (called π–C_s curves); (iii) surface entropy and enthalpy; and (iv) an equation of the bidimensional state of the virial type for linear polymer. The surface pressure (π) versus C_s isotherms at constant area, for α-helical polymers gave a very characteristic plateau. The β-form polymers, on the other hand, did not yield a flat plateau, but instead only an inflection was observed. The collapse pressure (π_{col}) for the β-form polymers, as determined from π–C_s curves, was found to be related to the stabilizing forces in the bulk (Table 6.12). However, no such relation was found for α-helical polymers. Typical results of π_{col} are given in Table 6.12 for high-molecular-weight (i.e., α-helical) poly-γ-methyl-L-glutamate (PMLG), poly-γ-ethyl-L-glutamate (PELG), and poly-γ-benzyl-L-glutamate (PBLG). The greater stability of PBLG as compared to PMLG in solution has been attributed to a difference in side-chain interactions.

The efficient shielding of the hydrogen-bonded helical backbone from the solvent by the bulkier benzyl group is undoubtedly one of the factors responsible for the greater stability of PBLG. Another contribution to stability would be the nonbonded (hydrophobic) interactions between the phenyl moieties, which would be bonding at the interface (water–backbone interaction), when PBLG is in the α-helical conformation; the collapse pressure would be lower than for PMLG or PELG (as here less shielding is possible).

Intramolecular side-chain–side-chain interactions between helical polymers would be expected to contribute to the magnitude of π_{col}. Thus PMLG and PELG show higher π_{col}, since the intramolecular side-chain packing of the methyl and ethyl groups is efficient, while the bulky benzyl groups hinder

TABLE 6.12. Comparison of Solvent Composition for Destabilization of β- or α-Structure and Collapse Pressure, π_{col}, for Various Polypeptides[a]

	Solvent composition[b]	π_{col} (mN/m)
β-structure		
Poly-acetyl-serine	40%DCA + 60%CHCl$_3$	10.0
Poly-carbobenzoxymethyl-cysteine	3%DCA + 97%CHCl$_3$	3.5
Poly-methyl-cysteine	—	11.0
Poly-test-butyl-serine	—	22.5
α-helical		
Poly-methyl-glutamate	60%DCA + 40%CHCl$_3$	20.0
Poly-ethyl-glutamate	—	22.0
Poly-benzyl-glutamate	69%DCA + 31%CHCl$_3$	7.0

[a] Reference 59.
[b] DCA: dichloroacetic acid.

compact packing. On the other hand, in the case of the β-forms, where intermolecular side-chain–side-chain interactions are predominant, the magnitude of π_{col} would be more directly related to these interactions.

6.13. MONOLAYERS AT THE OIL–WATER INTERFACE

The surface pressure π at the oil–water interface is taken as $\pi_k + \pi_e$ since the cohesive pressure π_c in this case is taken as zero in equation (6.3). However, it has been reported[66] that there exists a residual interchain cohesion at the oil–water interface as evaluated from π–A isotherms.

One of the difficulties for the study of the spread films at the oil–water interface is the fact that suitable procedures for obtaining π–A isotherms are not easily realized. Further π is measured as a function of the variable concentrations of amphiphiles at the interface. Only in a few cases has π been measured as a function of A_m directly. This causes a great deal of inconsistency in terms of the comparison of the various experimental data.[67]

Recently the π–A curves for the monolayers of sodium octadecyl sulfate and octadecyl trimethyl ammonium bromide spread at the heptane–water and air–water interfaces, respectively, have been compared by Mingins *et al.*[68] at 5°C and 20°C. These workers have also computed the Helmholtz free change ΔF for a monolayer in going from area A_1 to A_2 by the equation

$$\Delta F = -\int_{A_1}^{A_2} \pi \, dA \tag{6.49a}$$

A common reference area 200 nm^2 per molecule was taken as the area of the monolayer at the initial state (A_1) of the monolayer compression process. The entropy change ΔS_c^m for the monolayer compression from this initial area to a chosen area A_2 can be calculated using the following equation:

$$\Delta S_c = -\frac{d(\Delta F)}{dT} \qquad (6.51)$$

ΔS_c in all cases is negative and further its magnitude varies linearly iwth $\log_{10} A$ (vide Fig. 6.9). ΔS_c is found to be independent of the ionic strength and nature of the head group but its magnitude at the O/W interface is relatively higher. This indicates that the conformations of the hydrocarbon chains at the oil–water and air–water interfaces are different in nature.

In a recent study,[69,70] π–A data for 1,2-distearoyl-sn-glycerol-3-phosphoryl choline (SPL) at the n-heptane–water interface at 0 to 1 molar sodium chloride concentration were analyzed at different temperatures. The two-dimensional first-order phase transition enthalpy ΔH calculated[69] with the help of equation (6.4) was nearly equal to 59 kJ/mol at 25°C.

6.14. PHASE RULE FOR TWO-DIMENSIONAL FILMS

In order to describe equilibrium in multicomponent systems, it is always necessary to use the Gibbs phase rule principles in general for either three- or two-dimensional systems. According to this rule, the variables of the system

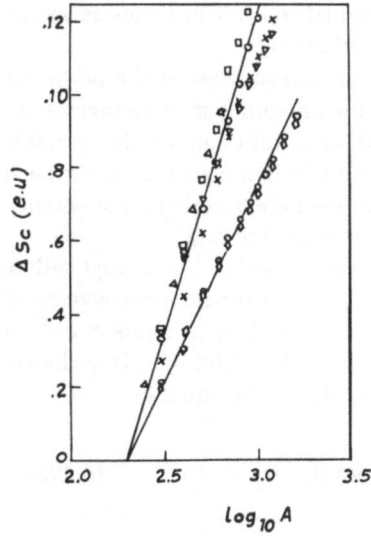

FIGURE 6.9. Entropies of compression from experiment A/W: C_{18} sulfate $\odot$, 0.01 M NaCl; C_{18} TAB $\diamondsuit$, 0.1 M NaCl. O/W: C_{18} sulfate $\times$, 0.1 M, $\square$, 0.01 M, $\bullet$, 0.001 M NaCl; C_{18} TAB $\triangledown$, 0.1 M NaCl (C_{18} sulfate + C_{18} TAB)$_{\text{mixture}}$ $\triangle$, 0.001 M NaCl. (Redrawn with changes from Ref. 68.)

are (i) concentration of each component in each phase (or the composition), (ii) other intensive properties of the system, e.g., temperature and pressure. In addition to the chemical potential of each component, the temperature and pressure, any further intensive factors may be included such as gravitational or electrical potential; i.e., another degree of freedom being allowed for each such intensive property included. Hereafter, the interfacial tension or energy per square centimeter of surface will be considered as such a property.

For a system at constant temperature and external pressure,[71] the phase rule can be written as

$$F = C - P \tag{6.52}$$

where F is the number of degrees of freedom, C is the number of components and P is the number of phases. Since external pressure is assumed constant, only plane surfaces are considered in this case.

If one considers that each surface is a separate phase, then for P^s surface phases there would be $(C - 1)P^s$ new variables, since all components except one may be varied independently in each surface phase. There will also be P^s terms for surface. For each surface phase, there will be P^s terms for surface energies, thus giving a total of $(C - 1)P^s + P^s$ or CP^s additional variables. Further, we would have equilibrium constraints: $\mu_i^S = \mu_i^B$, where B and S indicate bulk phase and surface phase, respectively. Similarly, if there is more than one phase in a surface, we will also have $\mu_i^{S_i} = \mu_j^{S_j}$.

If one or more of the components, designated C^S, is only confined to a fluid interface, and not equilibrated with the rest, this interface becomes equivalent to a phase separated by a membrane permeable to all the components except C^S. The interface then possesses an independently variable energy term due to the two-dimensional osmotic pressure π.

Further, there will be isolated "surface phases" which are quite distinct from one another and are completely delimited by mechanical constraints; other physically distinct surface phases are intimately contiguous and form a part of the same surface. The former types are called "surfaces," while the latter are called "surface phases." For example: oil–water and oil–air interface separated by a column of bulk phase, or two air–water interfaces separated by a mechanical barrier, will be accordingly separate "surfaces." On the other hand, a condensed fatty acid film in equilibrium with its vapor film will be "surface phases" in the same surface.

Without going into extensive derivations for the general case, the case of a single surface, containing q surface phases, is written as

$$F = C^S + C^B - P^B - (q - 1) \tag{6.53}$$

$$= C - P^B - (q - 1) \tag{6.54}$$

where the total number of components is $C = C^B + C^S$.

Let us apply this phase rule to some of monolayer systems:

   i. Monocomponent monolayer at air–water (A/W) or oil–water (O/W) interface:

$$\text{A/W: } C^B = 1, \quad C^S = 1; \quad P^B = 1, \quad q = 1; \quad F = 1 \tag{6.55}$$

$$\text{O/W: } C^B = 2, \quad C^S = 1; \quad P^B = 2, \quad q = 1; \quad F = 1 \tag{6.56}$$

   ii. If two surface phases are present, such as solid or liquid in equilibrium with two-dimensional vapor, $q = 2$ and $F = 0$. This result implies that a flat curve would be observed in a π–A isotherm.

  iii. If three surface phases are present, $q = 3$, analogous to a triple point, $F = -1$. This result implies that this system can only exist at a unique temperature.

    The thermodynamics of mixed monolayers have been described by various investigators[22,25,71–76] in many of which the phase rule approach has been found to be useful.

6.15. LIPID PHASE TRANSITION IN MONOLAYERS

    It is now well established that the lipids play a very crucial role in the function and the structural characteristics of biological membranes.[58,77] The structure of the biological membrane is very complex and it is useful to study the physical properties of lipids in model systems. The various model systems which have been used are bilayers, vesicles, and monolayers,[58,78] structures of which are given in Fig. 6.10.

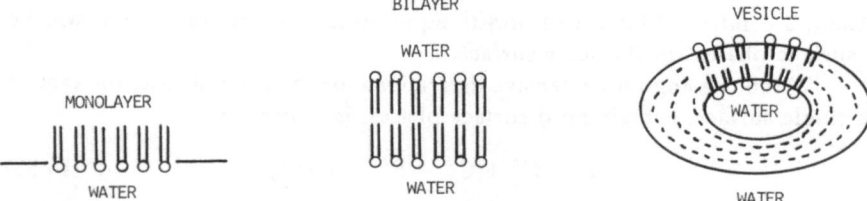

FIGURE 6.10. Various model membrane systems: monolayers; bilayer; vesicles (schematic). O=, lipid molecule.

The bilayers represent the biological membrane structure very closely, since a bilayer of lipid separates two aqueous solutions, as one has in biological cell membranes. The vesicle also accordingly resembles the cell membrane, due to the presence of the inside and outside nature of these structures. On the other hand, the monolayer at the air–water interface is not expected to resemble the biological membranes completely for the simple reason that the system is placed at the air–water interface. The lipid monolayers studied at an oil–water interface should be expected to resemble the membrane behavior more closely.[78] But the dispersion forces are almost absent between the alkyl chains of the lipids at oil–water interface, while these are present both at the air–water interface and in biological membrane (and also in bilayers and vesicles).

The phase transition of lipid bilayers is an example of a two-dimensional cooperative transition. When the pressure of the dispersion is increased, the phase transition temperature is shifted to higher values, indicating an increase in molecular volume during the transition.[79-81] The phase transition characteristics, such as temperature and enthalpy, are dependent on the chain length of the alkyl part, and the nature of the polar head group of the lipid, pH, ionic strength, and the nature of the electrolytes in the aqueous medium.

All the lipids found in the biological world possess such a hydrophobic–hydrophilic balance that they readily spread to form monolayers at the air–water interface (and with some exceptions also at the oil–water interface). This observation suggests that the amphiphile character of the lipid molecules must be a requirement for their function in the biological processes. The lipid monolayers have been extensively studied under various experimental conditions. In this chapter, we will discuss only certain isothermal properties of the lipid monolayers.[82-87]

6.15.1. THEORETICAL ANALYSIS OF LIPID PHASE TRANSITION IN MONOLAYERS

Based on the description of state of lipid monolayers (vide Fig. 6.1.), it is noted that for very large values of A, π varies slowly, with a hyperbolic shape. The first phase transition occurs at very low surface pressures. The second transition occurs at the "liquid expanded" to "liquid condensed" states and finally the liquid condensed state to the "solid" monolayer. The "liquid expanded" to "liquid condensed" transition has been argued to be a second-order phase transition. A second-order phase transition may be observed merely due to an orientational effect of asymmetrical molecules.[88-90] The vesicles of pure lipids dipalmitoyl phosphatidylcholine (DPL) have been reported to show true isothermal first-order transitions.[91]

The two-dimensional gas–liquid condensation of monolayers (of *n*-pentadecanoic acid) has been reported in the literature.[92,93] The density in monolayer of the liquid phase, ρ_L, and the density of the gas phase, ρ_v, were obtained by fitting the π–A data with a polynominal in ρ near each side of the coexistence curve and taking the intersection of these fitted curves. These analyses for the variation of compressibility (K_T) are given in power equations for T close to the critical temperature T_c:

$$(K_T)_v = 2.73 \times 10^3 (T_c - T)^{-0.97} \, \text{cm/dyn} \tag{6.57}$$

$$(K_T)_{L_{ex}} = 1.25 \times 10^3 (T_c - T)^{-0.98} \, \text{cm/dyn} \tag{6.58}$$

$$(K_T)_{T > T_c} = 6.73 \times 10^2 (T - T_c)^{-0.98} \, \text{cm/dyn} \tag{6.59}$$

$$(\rho)_{L_{ex}} - \rho_V = 41.9 (T_c - T)^{0.5} \, \text{molecules}/10^4 \, \text{Å}^2 \tag{6.60}$$

At critical temperature, ρ_c and π_c are equal to 41.7 molecules/10^4(Å^2) and 174 mdyn/cm, respectively. K_T is the two-dimensional compressibility coefficient equal to ($d \ln \rho / d\pi$). The π–A isotherms are most characteristic with respect to the dependence of compressibility, K_T, on the state of the film, e.g., gas, liquid (or gel), or solid. It is of interest to determine how the above-mentioned gas-to-liquid expanded transition could be extended to the $L_{ex} \rightarrow L_{co}$ transition data.

The π–A data for the liquid-expanded to -condensed transition for DML (dimyristoylphosphatidyl choline) and DPL isotherms (Fig. 6.11) was recently analyzed[94] using the above relations, and the following was found (by a curve-fitting procedure):

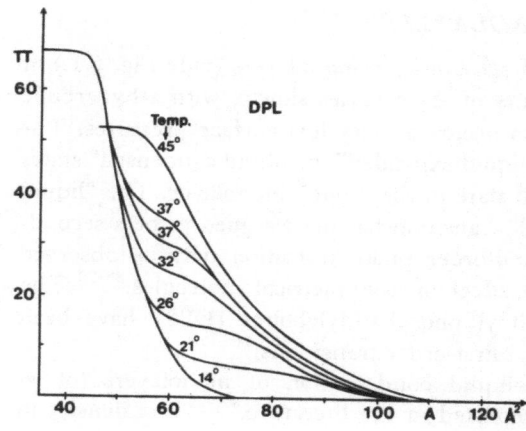

FIGURE 6.11. π–A isotherm for DPL. π is in dyn/cm^2 and A is in Å^2 per spread molecule.

For DML:

$$\Delta\rho = \rho_{co} - \rho_{ex} = 0.000591(T_{tr} - T)^{1/2} \qquad (6.61)$$

For DPL:

$$\Delta\rho = \rho_{co} - \rho_{ex} = 0.00081(T_{tr} - T)^{1/2} \qquad (6.62)$$

The magnitudes of T_{tr} were found from these equations (plots of $\Delta\rho$ versus T are given in Fig. 6.12) as follows: T_{tr} for DML = 22.3°C (from vesicles 25°C) and for DPL = 41.2°C (from vesicles 41°C). The limiting value of $(\Delta\rho)^2$ approaches zero as $T \to T_{tr}$ in π–A isotherms at the same temperature where vesicles melt. This further indicates that these monolayer structures are equilibrium isotherm systems. Further, plots of $\Delta\rho$ (or $A_{ex} - A_{co}$) versus T also gave the same T_{tr} values at $\Delta\rho = 0$ (Fig. 6.12).

The π–A isotherms at the oil–water interface have been reported at different temperatures. The plot of ΔA (or $A_{ex} - A_{co}$) versus T for DSPL[69] at an oil–water interface leads to the value of T_{tr} close to 42°C. This value is lower than that found for A/W monolayers or vesicles ($\simeq$ 60°C). More investigations are indeed required at O/W for such systems in order to determine why T_{tr} for monolayers at A/W and vesicles (or membranes) are almost the same, while T_r is lower at the O/W interface.

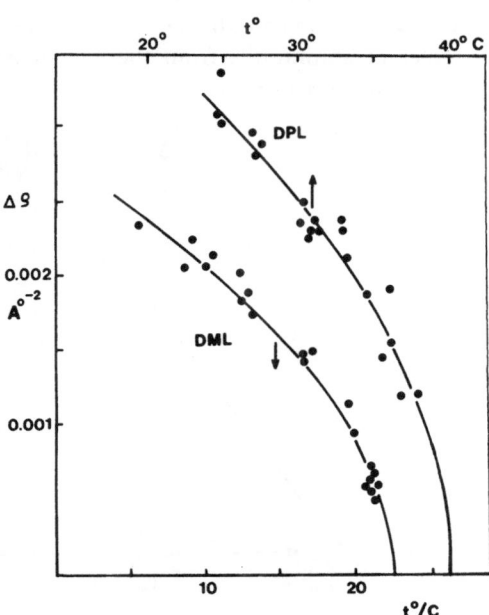

FIGURE 6.12. Variation of $\Delta\rho$ versus t (°C) [equation (6.61)] for DML and DPL.[94]

6.15.2. EQUILIBRIUM SURFACE PRESSURE OF LIPID FILMS

If a pure amphiphile in the form of a solid or liquid is placed on the surface of water, it spreads with the formation of monolayer which will remain in equilibrium with condensed solid or liquid bulk phases. Various types of equilibria[95] thus attained are shown in Fig. 6.13. The equilibria $B_S \rightleftharpoons V$ and $B_L \rightleftharpoons V$ represent the processes of sublimation and vaporization, respectively, for bulk lipid phase. The equilibrium surface pressure π_{eq} determined from experiments is assumed to remain constant in the whole range of the transition process.

There exists a certain confusion, however, in the description of the above processes, such as that the melting temperature of the bulk lipid thus spread may or may not agree with the conversion of liquid expanded to a solid film or the transition temperature for the change of gaseous film to the liquid film (as discussed below).

Analogous to the Clausius–Clapyron equation derived for phase transition in the bulk phase, the heat of spreading (Q_S) of a monolayer for an insoluble amphiphile can be written as[96-99]

$$Q_S = TA_e \left(\frac{\partial \pi_{eq}}{\partial T} \right) \tag{6.63}$$

where A_e is the molecular area of the amphiphile in the monolayer state.

From the thermodynamic analysis, it can be shown that the enthalpy change ΔH_i involved in bringing n_1 moles of solvent from the bulk phase and n_2 moles of amphiphile from the condensed phase to the interface may be given by the equation

$$\Delta H_i = T \cdot A \left(\frac{\partial \gamma_{eq}}{\partial T} \right) \tag{6.64}$$

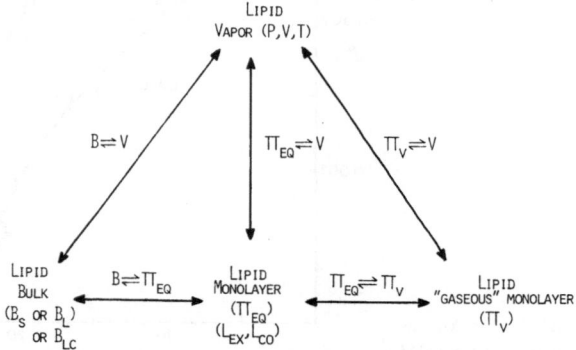

FIGURE 6.13. Various equilibria[95] between bulk phases (liquid B_l; solid, B_s, monolayers (π_{eq} or gaseous, π_v) and vapor (v).

Here γ_{eq} is the equilibrium free energy at the interface in the presence of amphiphile. In the analogous manner, the enthalpy change ΔH_0 due to the formation of the pure interface in the absence of the amphiphile can be written as

$$\Delta H_0 = TA\left(\frac{\partial \gamma_0}{\partial T}\right) \tag{6.65}$$

One can then write

$$\Delta H_0 - \Delta H_i = TA\left(\frac{\partial \pi_{eq}}{\partial T}\right) \tag{6.66}$$

The heat of fusion ΔH_f can be written as

$$\Delta H_f = T_m A_m^0 \left[\frac{\partial \pi_{eq}^c}{\partial T} - \frac{\partial \pi_{eq}^1}{\partial T}\right] \tag{6.67}$$

Here T_m is the melting point, A_m^0 the molecular area at T_m so that A_m becomes equal to $(Nn_2^m A_m^0)$. $\partial \pi_{eq}^c / \partial T$ and $\partial \pi_{eq}^1 / \partial T$ are the changes in π_{eq} of the monolayer for T below and above T_m, respectively. In Fig. 6.14, the equilibrium spreading pressures of various types of fatty acids has been plotted against temperature.[100] As seen from this figure, π_{eq} changes linearly with temperature exhibiting a sharp break at the bulk melting temperature T_m. Values of A_m^0 at T_m estimated from π–A curves obtained from separate surface balance experiments at different temperatures are also given in Table 6.13.

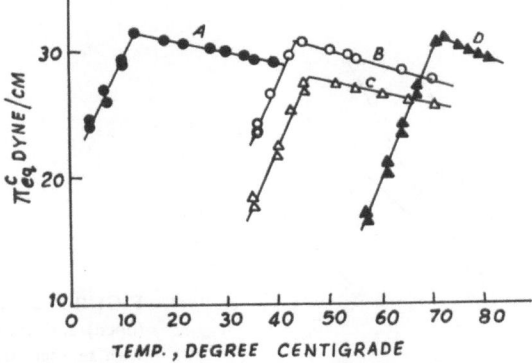

FIGURE 6.14. Equilibrium spreading pressure vs. temperature for ^{18}C fatty acids.[100] A, Oleic acid; B, elaidic acid; C, 9-stearolic acid; D, stearic acid. (Redrawn with permission from Academic Press, New York.)

TABLE 6.13. Comparison of Heats of Fusion (ΔH) Calculated from the Data on Spreading Pressure to those Determined Calorimetrically[a]

Fatty acid	T_m (°C)	A_m^0 (nm²/ molecule)	ΔH_f (kJ/mol) [equation (6.67)]	ΔH_f (kJ/mol) (calorimetry)
Stearic acid	71	0.26	55.0	62.2
Palmitic acid	63	0.26	54.0	52.6
Myristic acid	55	0.26	44.7	44.7
Erucic acid	32	0.30	48.4	47.7
Cis-6-octa- decenoic acid	31	0.30	47.7	48.0
Linoelaidic acid	30	0.27	48.4	44.7

[a] References 11 and 100.

In Table 6.13, values of ΔH_f calculated on the basis of equation (6.67) for several systems have been reported. These values in the table agree exceedingly well[100] with the values obtained from the direct thermal analysis used in the calorimetric experiments.

There are, however, various criticisms raised on the value of ΔH_f for phase transitions observed in π–A isotherms at $T < T_m$. The π_{eq}–T data reported on methyl palmitate and ethyl palmitate both showed[101] that heats of melting as calculated by using equation (6.67) have agreed with the calorimetric results. It is worth mentioning that π_{eq}–T and π_{co}–T curves were the same at $T > T_m$. It is also seen that $(d\pi_{tr}/dT)$ obtained from π–A isotherms at various temperatures for various fatty acid monolayers compare well with $(d\pi_{eq}/dT)$ as shown in Fig. 6.15.

In principle, measurements of π_{eq} as a function of T indicates the existence of all types of phase transitions in two-dimensional monolayers. Thus

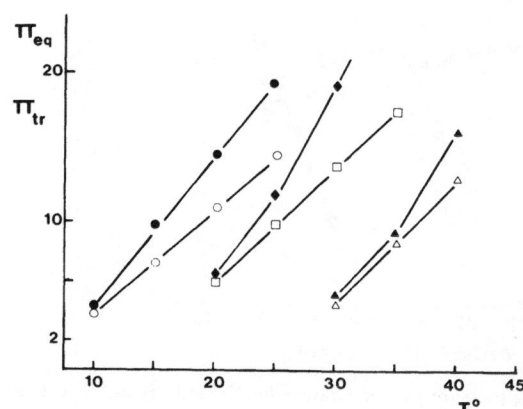

FIGURE 6.15. Variation of π_{eq} (open) and π_{tr} (filled) (from π–A) with temperature for myristic acid (●, ○); palmitic acid (▲, △); cis-3-decenoic acid (◆, □).[11]

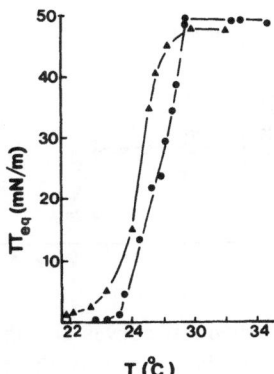

FIGURE 6.16. Variation of π_{eq} with temperature for DML: ●, Ref. 102; ×, Ref. 103.

$\pi_{eq}-T$ data for dimyristoyl lecithin (DML) spread at the air–water interface[102,103] has been presented in Fig. 6.16. The temperature study for gel–liquid transition indicates in this case that $T_m \simeq 25°C$. Using equation (6.67), for the DML system, ΔH_f is observed to be 2500 kJ/mol. This value is absurd since the calorimetric value estimated for the system is nearly 40 kJ/mol. Thus, the equilibria represented in Fig. 6.13 appear to be valid for fatty acids but do not hold[103] good for the DML or lecithin systems. These analyses allow us to conclude that the equilibria shown in Fig. 6.13 need reevaluation, since these are obviously not general. Further in the $\pi-A$ isotherms, the transition from L_{ex} to L_{co} has not been extensively analyzed in the current literature.[99,103] Primarily there exists a disagreement as regards whether the phase transition is of first order or second order. Thus, for a first-order transition

$$\left(\frac{\partial \mu_i^m}{\partial \pi}\right)_T = A \qquad (6.68)$$

and for a second-order transition

$$\left(\frac{\partial^2 \mu_i^m}{\partial \pi^2}\right)_T = \left(\frac{\partial A}{\partial \pi}\right)_T \qquad (6.69)$$

Here μ_i^m stands for the chemical potential of the ith component in the interfacial phase.

If the system undergoes first-order transition, then the $\pi-A$ isotherm should exhibit a horizontal curve near the transition region. The second-order transition, on the other hand, requires that there will be no horizontal region. All the $\pi-A$ isotherms reported in the literature for lipid monolayers generally show no horizontal regions. It has been argued[82,94,104] that the monolayer is a complex system as compared to pure three-dimensional systems where

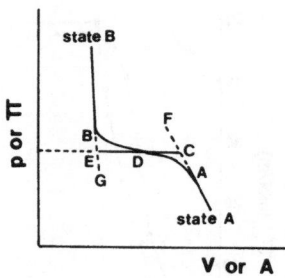

FIGURE 6.17. Schematic pressure (p or π) versus volume (v) or area (A) isotherm for small systems at a temperature where a first-order phase transition takes place. BDA is a typical experimental curve.[105]

horizontal curves are observed. This difference arises from the fact that the monolayer π–A isotherm is related not only to the lipid–lipid interactions, but the subphase also determines the shape of the π–A curve in the transition region.

Analogous to three-dimensional systems,[105] the π–A isotherm would be expected to show a sharp transition, ACDEB, below the critical temperature only if an infinite number of molecules is involved in the phase transition. On the other hand, if say only 20, 100, or 1000 molecules are involved, then the phase transition region would follow the curve ADB instead of ACDEB (Fig. 6.17). It has been shown that in such transition regions with finite molecules taking part in the phase transition from state A to B the latter is valid, so that

$$ED = DC \qquad (6.70a)$$

and

$$\text{Area (BED)} = \text{area (ACD)} \qquad (6.70b)$$

Preliminary analysis of π–A isotherms of DML and DPL[103] indeed shows that the data agree with the relations in equations (6.70a) and (6.70b). The Gibbs phase rule at constant atmospheric pressure for this system is applicable. With bulk lipid and at least one surface phase always present, F according to the phase rule may be unity or zero. Thus, when π_{eq} varies with temperature, $F = 1$, and the chemical potential of lipid in each equilibrium phase also varies with temperature.

6.16. MIXED MONOLAYERS

Extensive research has recently been carried out on the experimental and theoretical aspects of the mixed films of two different amphiphiles spread at the air–water interface.[25,72–76,95,106–114] In these studies, a definite amount of the mixture of the two substances is spread in the usual manner on the liquid surface formed in a surface balance so that average area per molecule can be calculated. The surface pressure π of the mixed monolayer thus formed is then measured with the help of the highly sensitive surface balance. Earlier π–A curves of mixed L-type of films of alcohols, amines, and acids have been examined by Harkins and Florence.[115,116] In the absence of the interaction between the spread components, ideal mixed monolayers will be formed. Strong interaction between the components may frequently occur within the nonideal monolayer. Such interaction may sometimes lead to the formation of the surface complexes. In many cases, the experimental investigations will indicate that two components in the monolayer are partially or completely immiscible with each other.

The pressure–area curves of the monolayers of individual components are frequently compared with that of the mixed monolayers, whereby some informations about the interaction in the monolayer phase can be obtained. In the complete absence of such interaction, the experimental data should fit the equation

$$\pi_{1,2} = \pi_1 X_1^m + \pi_2 X_2^m \qquad (6.71)$$

Here $\pi_{1,2}$ is the measured surface pressure of the mixed monolayer at a given value A. π_1 and π_2 are the respective pressures for the spread monolayers of individual components also for the same value of A. X_1^m and X_2^m are, respectively, the mole fractions of components in the mixed monolayer whose values during π–A measurement remain constant for a particular isotherm.

In Fig. 6.18, typical π–A plots for mixture of myristic acid and elaidic acid[110] at fixed mole fractions of the acids are presented. Here the range of A varies from 30 nm^2 per molecule down to a point of compression where the monolayer is in the liquid state. From the point of discontinuity of this

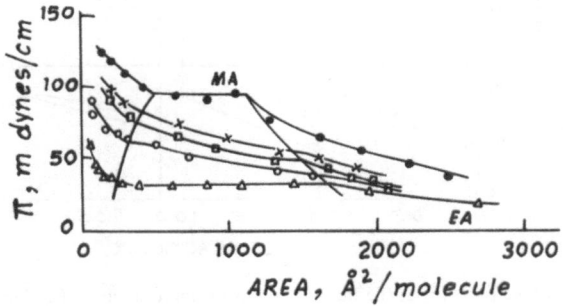

FIGURE 6.18. Surface pressure vs. area per molecule for a mixture of myristic acid (MA) and elaidic acid (EA) at 25°C. (Redrawn from Ref. 110.)

curve, the surface vapor pressure at the liquidus point can be estimated. In Fig. 6.19, the surface pressures at the liquidus point for various acid mixtures[110] have been plotted against X_1^m, the mole fraction of one of the components in the mixture. It is of interest to note that oleic acid, ^{13}C cicosenoic acid, and linoleic acid each exhibit ideal mixing behavior with myristic acid since π in all these cases varies linearly with mole fraction of one of these components in agreement with equation (6.71). The mixtures of monosaturated *trans* isomer of oleic, elaidic, and linoleic acids with myristic acid show positive deviation from an ideal behavior since the π–X_2 plot is not linear.

As in the case of the bulk system, the deviation from ideality for a mixed monolayer system can be expressed in terms of the activity coefficients of the components in the condensed phase. It is also noted from the experiments that one of the components in the condensed phase of the mixed monolayer is relatively more volatile than the other. The activity coefficient f_2^m of this volatile component can be calculated using the relation[110,117]

$$f_1^m = \frac{X_1^{(v)}}{X_1^{(c)}} \cdot \frac{\pi_V^{(c)}}{\pi_V^{(1,0)}} \tag{6.72}$$

Here $X_1^{(v)}$ and $X_1^{(c)}$ are the mole fractions of components 1 in the vapor and surface condensed states. $\pi_V^{(c)}$ and $\pi_V^{(1,0)}$ are the surface pressures at the liquid point and of pure component 1, respectively. Using π–A data (vide Fig. 6.18) for mixtures, values of all the quantities occurring on the right-hand side of equation (6.72) can be calculated so that f_1^m may be estimated for a given value of X_1^m. These values for several fatty acid mixtures[110] are given in Table 6.14. Following the same types of procedures, Tajima and Gershfeld[117] have

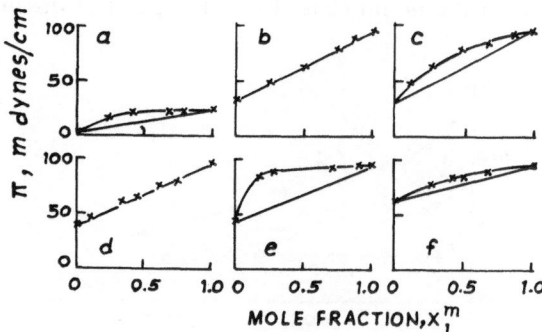

FIGURE 6.19. Surface pressure at the liquidus point vs. mole fraction. (a) Palmitic acid (1) and stearic acid (2); (b) myristic acid (1) and oleic acid (2); (c) myristic acid (1) and elaidic acid (2); (d) myristic acid (1) and linoelaidic acid (2); (e) myristic acid (1) and linoleic acid (2); (f) myristic acid (1) and 9-stearolic acid (2). (Redrawn from Ref. 110.)

estimated the activity coefficients of the component present in cholesterol–dimyristoyl lecithin and cholesterol–dioleyl lecithin mixtures.

As in the case of binary solution in the bulk phase, f_1^m can be related to f_2^m by the Gibbs–Duhem equation so that the value of the latter can be estimated[110] from the known values of f_1^m using the analytical equation put forward by Margules. Values of f_2^m for several mixtures are also presented in Table 6.14.

From a detailed thermodynamic analysis, Goodrich[72] has shown that the free energy change ΔG_{mix}^m due to the mixing of X_1^m moles of component 1 with X_2^m moles of component 2 at a given pressure π in ideal state can be represented by

$$\Delta G_{\text{mix}}^m = RT(X_1^m \ln X_1^m + X_2^m \ln X_2^m) \tag{6.73}$$

when the mixing is not ideal, the excess free energy of mixing ΔG_{ex}^m will be defined as

$$\Delta G_{\text{ex}}^m = \Delta G_{\text{mix}}^m - RT(X_1^m \ln X_1^m + X_2^m \ln X_2^m) \tag{6.74}$$

Unlike the case of an ideal mixture, ΔG_{ex}^m for nonideal mixed film is not zero. Goodrich[72] has further shown that

$$\Delta G_{\text{ex}}^m = \int_{\pi^*}^{\pi} (\mathscr{A} - X_1^m \mathscr{A}_1 - X_2^m \mathscr{A}_2)\, d\pi \tag{6.75}$$

Here $\mathscr{A}$, $\mathscr{A}_1$, and $\mathscr{A}_2$ refer to areas per mole for the mixed monolayer and for individual monolayers of components 1 and 2, respectively. Here π^* is the measured low surface pressure very close to zero. Using this equation, ΔG_{ex}^m for various types of mixed monolayers have been measured by several workers.[18,72,118,119,114]

TABLE 6.14. Activity Coefficients[a] of the Components in the Binary Mixture at 25°C

Myristic acid (1) and palmitic acid (2)			Oleic acid (1) and palmitic acid (2)			Myristic acid (1) and linoleic acid (2)		
X_1	f_1^m	f_2^m	X_1	f_1^m	f_2^m	X_1	f_1^m	f_2^m
0.10	1.57	1.03	0.364	2.35	2.97	0.347	2.02	5.50
0.30	1.43	1.09	0.500	1.78	3.86	0.444	1.83	7.50
0.50	1.28	1.00	0.616	1.49	4.33	0.711	1.30	8.08
0.70	1.17	1.00	0.856	1.15	10.38	0.900	1.10	23.51
0.90	1.04	2.13						

[a] Reference 110.

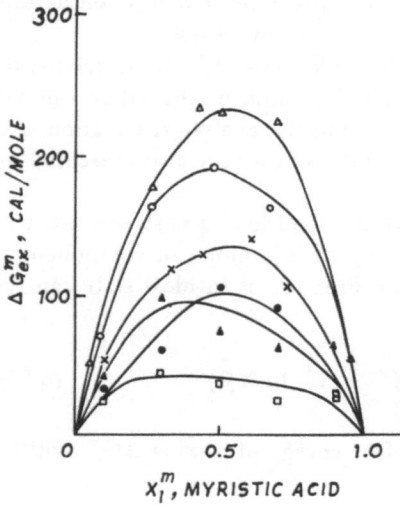

FIGURE 6.20. Excess free energy-composition diagram for different two-component systems: ▲, palmitic acid and myristic acid; △, 9-stearolic acid and myristic acid; ○, elaidic acid and myristic acid; □, oleic acid and myristic acid; ×, linoelaidic acid and myristic acid; ●, ^{13}C eicosenic acid and myristic acid. (Redrawn from Ref. 110.)

When the activity coefficients f_1^m and f_2^m of the mixed condensed films are known from experiments (vide Table 6.14), ΔG_{ex}^m can be calculated using the following relation directly:

$$\Delta G_{ex}^m = RT(X_1 \ln f_1^m + X_2 \ln f_2^m) \tag{6.76}$$

In Fig. 6.20, values of ΔG_{ex}^m for various types of fatty acid mixtures[110] have been plotted as functions of X_1^m.

6.17. MISCIBILITY IN MIXED MONOLAYERS

In the theoretical treatment with mixed monolayers, it is usually assumed that the two components present are completely miscible. To test the miscibility, various experimental approaches are used. In a particular approach, the area A of the mixed monolayer at a fixed value of π is plotted against mole fractions X_1^m or X_2^m. If the plot is linear, indicating the additive properties of areas, then

$$A = X_1^m A_1 + X_2^m A_2 \tag{6.77}$$

where A_1 and A_2 are the areas of the respective monolayers of pure components for the same value of π. The linear plot[106,120] may indicate a situation of either some type of ideal mixing or complete immiscibility. Deviation from this type of additivity is the indication of miscibility.[106]

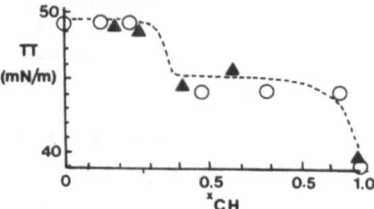

FIGURE 6.21. Comparative values of π_{eq} (O)[121a] and π_{col} (▲)[103] as a function of mole fraction of cholesterol, x_{CH}, in mixed films of DML + CH (29.5°C).

Change in the collapse pressure for the mixed film may serve as a test for the miscibility.[106] If the two components are immiscible[18b,23] then as the value of π reaches the value equivalent to $\pi_{col,1}$, one will observe the collapse of the component with the lower value. The second component (at higher $\pi_{col,2}$) would then be ejected at the corresponding surface pressure. The monolayers of polyalanine (PLA) and lipid (e.g., cholesterol, CH, or stearyl alcohol, ST) were studied[18b] in which the plots of A versus X_{PLA} were linear at a definite pressure. On the other hand, the π–A isotherms of both systems showed that besides two collapse states, corresponding to pure components, a new collapse state was observed in the mixed film. This shows that the criterion of immiscibility as found from the validity of equation (6.77) is not very reliable.

The miscibility can also be studied by measuring the equilibrium surface pressure, π_{eq}, of mixed films.[103,121] The π_{eq} of DML–cholesterol and DOL(dioleoyl-lecithin)–cholesterol (CH) at 29.5°C indicate interactions only at certain mole ratios. π_{eq} was constant and equal to the value of pure lecithin (DOL or DML) when the mole fraction of CH was 0–0.33.[121a] The surface phase rule predicts that two bulk lipid phases (i.e., pure lecithin and lecithin-CH complex) can coexist. Through rigorous thermodynamic analysis it was established that lecithin and cholesterol form a stable 2:1 complex at 29.5°C. The magnitudes of π_{eq} as given in Fig. 6.21 are compared with collapse pressures, π_{col} of the same system at identical temperature.[103] It is seen that $\pi_{eq}^{l} = \pi_{col}$ when the experimental temperature is greater than T_m. It thus suffices to conclude that in mixed monolayers both $(\pi_{eq} - T)$ and $(\pi_{col} - T)$ isotherms provide the same information when $T > T_{tr}$. The differences observed at $T < T_{tr}$ where $\pi_{col} > \pi_{eq}^{s}$ arise from the incorrect thermodynamic considerations in the literature.[103] The same thermodynamics have been applied to lipid–biopolymer and mixed polymer films.[106]

6.18. LIPID–PROTEIN MONOLAYERS (A MEMBRANE MODEL SYSTEM)

From electrical and permeability studies it was concluded that the biological membrane consists of a thin lipid-containing layer surrounding the cell.

TABLE 6.15. Ratio of Protein to Lipid in Various Membranes[a]

Membranes	Ratio of protein:lipid (w/w)
Myelin	0.25
Erythrocyte	1.25
Mitochondria	3.0

[a] Reference 121b.

A small portion of the membrane consists of very narrow pores that permits polar molecules and ions of small dimensions to leak through. The large proportion of the lipid content controls the movement of solutes through their lipid-solubility characteristics. In addition to these two types of unfacilitated transport across the membranes, the membranes also exhibit highly specific carrier systems which give rise to a greater degree of selectibility of the membrane, allowing certain substances to penetrate at highly enhanced rates. The analysis of membranes has shown that the protein:lipid ratio varies a great deal in various systems, as given in Table 6.15. The composition of the lipid component as found in various membranes (Table 6.16) shows that while phospholipids are always present in the membranes, the neutral and nonlinear lipid, cholesterol, is found only in some of the membranes.

The membrane consists of a lipid bilayer with the protein molecules located in a fixed orientation across the membrane and some protein molecules associated with the inner surface of the bilayer.[122,123] This structural description has led to the conclusions that lipid–lipid, lipid–protein, and protein–protein interactions are of major importance as regards the structural stability and the biological function (Fig. 6.22).[124–126]

TABLE 6.16. Lipid Composition of a Variety of Membranes[a]

Lipid	Myelin	Chloroplasts	Erythrocytes	Mitochondria
Phospholipids	32	10	55	95
Cholesterol	25	0	25	9
Sphingolipids	31	0	18	0
Glycolipids	0	41	0	0
Others	12	50	2	0

[a] Reference 121b.

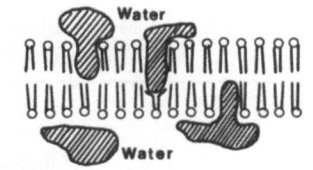

Biological membrane structure

FIGURE 6.22. Biological membrane structure (schematic).

α≔ **Lecithin,** **Protein.**

6.18.1 LIPID AND PROTEIN INTERACTIONS IN THE MONOLAYERS

The interaction of lipids with proteins by the monolayer method is studied in the following procedure. A film of lipid is first spread at the surface of aqueous salt solution. After equilibrium has been reached, as regards the surface pressure, π, the protein is carefully injected under this lipid film, which gives rise to a change in surface pressure, $\Delta\pi$. The interaction between cholesterol and gliadin was studied by this method, which indicated that the protein was ejected from the lipid–protein mixed monolayer at high surface pressures.[127] In other studies[128] it was proposed that a layer of spread protein was formed beneath the lipid film, followed by a second layer of unspread (that is in native conformation) beneath the first layer. These results were alternatively interpreted to show[129] that protein unfolded at the interface and formed a mixed film with the lipid, giving rise to a change in surface pressure, $\Delta\pi$. Other workers[58] have discussed the role of the hydrophobic side-chains of the protein molecule as regards the magnitude of $\Delta\pi$. The results clearly indicated that charged interactions take place between proteins and lipids under these conditions. The interaction between stearic acid and poly-L-lysine was reported in another study.[130]

Insulin and various other proteins[58] have been reported to be able to penetrate the monolayers of various lipids. At a certain lipid pressure called the "limiting π," no further penetration by the protein molecule is observed, i.e., $\Delta\pi$ is zero. In Fig. 6.23 the relation between π and $\Delta\pi$ is given for the interaction between insulin and various lipids, e.g., stearyl alcohol, stearic

FIGURE 6.23. $\Delta\pi$ for several lipid films as a function of lipid π. Stearyl alcohol, O; stearic acid, O; DPL, ▲; cholesterol, ●.[58]

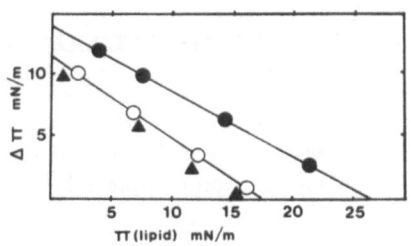

TABLE 6.17. The Magnitudes of van der Waal Forces, W_d [see equation (6.11)] at Limiting π for Lipid and Insulin Films[a]

Lipid	Limiting π (mN/m)	Area per molecule of lipid (nm^2)	Distance between lipid molecule (nm)	W_d (kJ/mol)
Stearyl alcohol	17.0	0.34	0.63	−9.6
Stearic acid	17.0	0.34	0.63	−9.6
DPL	17.0	0.58	0.58	−11.9
Cholesterol (assuming 32 CH$_2$ groups)	27.0	0.40	0.68	−11.3

[a] Reference 58.

acid, lecithin, and cholesterol.[58] It is seen that in all cases $\Delta\pi$ decreases linearly over the entire range of the lipid π. In the lipid monolayers the van der Waals forces, which are known to play a very important role in the stability of such films, can be estimated from the relationship given by equation (6.11). It is reasonable to expect that when the value of lipid π is equal to the "limiting π," the van der Waals interactions between the hydrocarbon chains are of such magnitude that no penetration by protein molecule is possible, especially in the case of stearyl alcohol and cholesterol. The value of W_d^{st} for stearyl alcohol is found to be -9.61 kJ/mol by applying equation (6.11). It is found that $W_d^{st} \approx W_d^{ch}$ at the limiting π, if one uses $n_c = 32$, when actually there are 27 carbon atoms in the nonlinear cholesterol molecule (vide Table 6.17). The results given in Fig. 6.21 indicate that the "limiting π" for stearic acid is the same as that of stearyl alcohol. At the "limiting π" for lecithin, A is equal to 0.58 nm^2 or 0.29 nM^2 for each hydrocarbon chain. Using these values and since n_c is 15 for lecithin in the present case, $W_d^{el} = -11.9$ kJ/mol. It is safe to conclude that the magnitude of "limiting π" in the various lipid films is related to the magnitude of the dispersion forces in these films. The differences observed between the plots of $\Delta\pi$ versus lipid π for cholesterol and DPL are general.

Monolayer techniques have thus been found to be useful for the study of interaction of surfactants with various proteins.[131–134]

6.18.2. MONOMOLECULAR FILMS OF VALINOMYCIN AND LIPIDS

It is well known that the ion transporting antibiotics fall into two categories: (1) the diffusion carriers which act by diffusing through the membrane with the ion complexed in the central cavity of the antibiotic molecule (e.g., nonactin, valinomycin, and nigericin) and (2) the channel carriers which act by forming channels or pores of well-defined size through the membranes (e.g., gramicidin). The neutral diffusional carriers form a metal complex which has a net positive charge. On the other hand, the carboxylic diffusional carriers (e.g., nigericin, monesin) are usually linear compounds and their metal complex has no net charge. The antibiotics, especially valinomycin and gramicidin, exhibit highly polar–apolar character such that stable monomolecular films are formed at the surface of water. The formulas of these two compounds are given below[58]:

Gramicidin A:

HCO-L-VAL-D-GLY-L-ALA-D-LEU-L-ALA-D-VAL-L-VAL-D-
VAL-L-TRP-D-LEU-L-TRP-D-LEU-L-TRP-D-LEU-L-TRP-NCH$_2$CH$_{11}$OH

Valinomycin:

-L-VAL-D-HYV-D-VAK-L-LAC-L · VAL-D-HYV-
      -L-LAC-D-VAL-D-HYV-L-VAL-L-LAC-D-VAL

It is obvious that due to the high content of the apolar side chains of the residues, VAL, LEU, and TRP, we should expect rather low solubility in water, as is also the case. The formation of stable monolayers[58,135,136] thus indicates that the hydrophobic–hydrophilic character is of appropriate balance for formation of an insoluble monolayer. The πA–π plot of valinomycin is given in Fig. 6.2 (substrate = H$_2$O).

The surface pressure of valinomycin spread on water with various concentrations of KCl showed that the K$^+$ ion was giving rise to a significant effect on the π versus A isotherm.[58,135,136] From monolayer studies the cation selectivity for valinomycin was found to be in complete agreement with similar measurements using other methods, e.g., the solvent extraction method and bioionic potentials.[58,135,136] The π–A isotherm of valinomycin on water shows a collapse at $\pi = 24$ mN/m. As the concentration of KCl increases, there appear two collapse states, as shown in Fig. 6.24 for 1 M KCl. In Table 6.18 are given the values of π_{c1}, A_{c1}, π_{c2}, and A_{c2}, where subscripts $c1$ and $c2$ refer to the first and second collapse states.

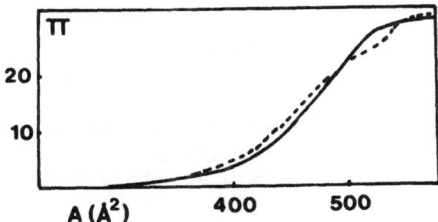

FIGURE 6.24. Surface pressure (π) versus A isotherms[58] of valinomycin on water (——) and 1 M KCl (- - - -) subphase, 25°C.

TABLE 6.18. Effect of KCl Concentration on Valinomycin Films[a]

KCl (M)	π_{c1} (mN/m)	A_{c1} (Å^2)	π_{c2} (mN/m)	A_{c2} (Å^2)
0	25	205	No second collapse	
0.2	24	201	No second collapse	
1.0	21	219	ca. 25	ca. 160
1.5	17.3	230.5	ca. 25	ca. 160
2	15.6	246	ca. 25	ca. 160
3	14.4	242	ca. 25	ca. 160

[a] Reference 58.

The π–A curves of valinomycin spread over water and 1 M KCl were studied at various temperatures, e.g., 15, 25, and 28.5°C. The data fit the equation $\pi(A - A_0) = RT$.

Plots of πA versus π were linear for low π values for water and 1 M KCl subphases at various temperatures. We find that πA is equal to $kT (= 400)$ as π approaches zero; in other words, the valinomycin films behave as ideal films (with molecular weight = 1100) at low π values (Fig. 6.2). These plots can be described by the linear equations given in Table 6.19.

The magnitude of A_0 increases significantly with temperature (Table 6.19) in the KCl systems, while it is less sensitive to temperature in the absence of KCl. The mixed films of valinomycin + DPL and valinomycin + cholesterol were studied under similar conditions. The πA versus π plots of pure DPL

TABLE 6.19. Equation of State for Films of Valinomycin[a]

Subphase	Temperature (°C)	Equation of state[b]
Water	15	$\pi(A - 286) = 400 = kT$
	25	$\pi(A - 304) = 400 = kT$
	28.5	$\pi(A - 292) = 400 = kT$
1 M KCl	15	$\pi(A - 308) = 370 = kT$
	25	$\pi(A - 324) = 380 = kT$
	28.5	$\pi(A - 340) = 390 = kT$

[a] Reference 58.
[b] π = mN/m; $A = 10^{-20}$ m^2/molecule = Å^2/molecule.

TABLE 6.20. Data for Valinomycin Monolayer^{a,b}

	Water			1 M KCl		
	15°C	25°C	28.5°C	15°C	25°C	28.5°C
A (limiting area/molecule)	371	378	377	392	429	452
A_{c1}	186	196	203	239	253	263
A_{c2}		Absent		163	183	188
π_{c1}	32	31	31	24	24	24
π_{c2}		Absent		33	32	31

^{a} Reference 58.
^{b} $A = \text{Å}^2/\text{molecule} = 10^{-20} \, \text{m}^2/\text{molecule}$; $\pi = \text{mN/m} = \text{dyn/cm}$.

and cholesterol crossed through the origin. On the other hand, all the mixed films gave πA versus π plots which intersected the A axis at 200 (i.e., $\frac{1}{2}kT$).

The data of mixed films thus allow one to conclude that the valinomycin and lipids (e.g., DPL and cholesterol) form ideal mixed films. The effect of temperature on the A_0 in the mixed films is of further interest. The value of A_0 does not change with temperature, both at water and KCl subphase, in the case of valinomycin + cholesterol. However, in the case of valinomycin + DPL, the value of A_0 changes somewhat with temperature, even though the values in water and KCl subphase remain of the same magnitude.

It is of interest to compare the area per molecule values from monolayer data to that with the space-filling models. It has been reported[58] that a molecular area of 1.50 nm^2 for the valinomycin plus K$^+$ complex can be estimated from the space-filling model, when projected on a surface such that the plane of the ring is parallel to the plane of the surface. Similarly the space-filling model of uncomplexed valinomycin gave an area per molecule of magnitude 1.70–1.75 nm^2. From these data it was concluded that the uncomplexed model occupies an area 0.20–0.25 nm^2 larger than the model of the complexed state.

In analyzing these data, the authors[136] compared the A_{c1} of water (1.86 nm^2) and A_{c2} over KCl (1.63 nm^2) with those obtained from the space-filling models. It was concluded that there was an increase of area of about 0.20–0.25 nm^2 in the uncomplexed molecule, which agreed with the space-filling models. The data at 25°C give a difference of 0.09 nm^2, while at 28.5°C it is of magnitude 0.15 nm^2. Since the experimental error in A is $\approx 10\%$, we would conclude that these differences are about 0.15–0.20 nm^2 in the range of 15–28.5°C. It is obvious that the monolayer model system offers a very useful method in studying the function and mechanisms of such antibiotics in the membranes, and in the design of such molecular configurations.

6.19. THE INSOLUBLE MONOLAYER AND THE GIBBS SURFACE EXCESS

For the system of the insoluble film, the presence of the lipids and fatty acids in the bulk liquid or bulk nonaqueous phases (e.g., air) are assumed to be negligibly small. Under this circumstance, the amount n_2^m (or Δn_2) of this substance thus spread may be safely taken to be equal to Γ_2^1 according to the Gibbs equation (Chapter 3) since $n_1^m X_2 / X_1$ is zero for all practical purposes. Because of this and also for other possible reasons, the pressure–area curves (vide Fig. 6.1) for the insoluble monolayer have frequently been explained on the basis of the properties of the lipids with complete disregard of the role of the water and air components forming subphases. The analysis based on this direct study gives a reasonable picture of the behavior of the insoluble lipid films such as phase transition, intermolecular attraction and repulsion at the interfacial phase, formation of island and solid films at the interface, etc. Such a direct conclusion is difficult to achieve from experimental data in the case of the adsorbed monolayer of solute at the liquid interface or adsorbate at the solid interface, based on the concept of the Gibbs surface excesses (vide Chapters 5 and 8).

However, Motomura and co-workers[25,137–140] have pointed out only recently that many properties measured from the study of the π–A curves of the insoluble monolayers can be explained quantitatively if the thermodynamic treatment includes the contribution made by water and air in the monolayer phase. These treatments are highly rigorous and are particuarly useful for the interpretation of the π–A data of the mixed monolayer systems. The surface phase rule has been further clarified on the basis of this treatment. The expression for ΔG_{ex}^m obtained by Motomura is different from that obtained by Goodrich, who has in fact neglected the role of water in shaping the monolayer properties. The entropy change for the transfer of film-forming materials from the bulk to the monolayer phase has also been calculated by Motomura based on these derived equations. The theoretical expressions are also consistent with the experimental data for phase transitions in two-component monolayers. The phase diagrams for the mixed monolayers have also been constructed by these workers. The entropy and enthalpy changes for the phase transitions in mixed monolayers calculated from the experimental data are observed to be negative. The features of the azeotropic transformation in mixed monolayers of ethyl heptadecanoate and tetradecanoic acid have been elaborated in the light of this theory.

It may be pointed out here that in deriving equation (6.40) of state for insoluble protein films, Bull has assumed that the interfacial phase is a monomolecular layer composed of both biopolymer and water. Fowkes[141–143] also had assumed that the properties of the insoluble films are controlled both by the organic substance and water forming a monolayer phase. The penetration of the hydrocarbons, soluble surfactants, proteins, and polymers to the insoluble film can be satisfactorily explained by the theories proposed by Fowkes.[141-143] It may be pointed out here that although both solute and solvent molecules are involved in the Gibbs concept of the interfacial phase, it does not state with precision that the interfaical phase is a monolayer.

Lucassen-Reynders[144,145] developed thermodynamic treatment for the calculations of the activity coefficients of the mixed monolayers. In this treatment, the Gibbs concept for the interface has been used, although the position of the Gibbs dividing surface remains uncertain.

It is highly desirable that a unified treatment should be developed for the soluble and insoluble films based on the Gibbs picture of the interface which will fit all kinds of experimental data most satisfactorily. The position of the dividing plane in such a picture should remain well defined also so that the properties of the interfacial phase are understood unambiguously.

6.20. SUMMARY AND COMMENTS

The studies carried out on spread monolayers, as described above, clearly show that these systems provide much useful information as regards the stability and function of various other important structures, as one finds in industrial and biological systems. The measurement of surface pressure, π, further, provides a quantitative analysis of the different forces, e.g., kinetic, van der Waals, and electrostatic charge repulsion, which are known to be present in these organized structures. This kind of information, as otherwise, is not easily measured from most of the other methods. The lipid monolayers thus allow one to obtain the necessary information as regards the emulsion (or microemulsion) stability. The vesicles and monolayers are shown to give the same transition temperatures. This allows one to use monolayers as a suitable model to understand the physics of phase transition which takes place in cell membranes and vesicles. The latter study is further supported by the various physical theoretical analyses as carried out on the monolayer data in the past decade.

The protein and biopolymer (and synthetic polymer) monolayers provide information which is useful as regards the unfolding of these polymers at interfaces. It is shown how the sequence of amino acid and the monolayer characteristics are interrelated. This observation needs be studied in more detail, due to the fact that amino acid sequence is now known to determine also the biological function of protein molecules. The charged monolayers of membrane proteins, e.g., melittin and valinomycin, with ionophore properties need to be investigated in more detail. The thermodynamics of these systems has been extensively studied. However, there remains much to be studied, due to the fact that as more molecular description becomes available about the emulsions and membranes, the monolayers need to be investigated in the same detail. In fact, it is already seen that the monolayer behavior of single-alkyl-chain lipids is different from that of two-alkyl-chain lipids.

In general, the number of studies where the sensitivity was very high (less than 0.1 mN/m) has been very small. However, with the help of modern apparatus it is shown that this goal can be easily achieved.

WETTABILITY AND CONTACT ANGLES

7.1. INTRODUCTION

In everyday life one encounters many different systems where it is apparent that various processes take place at the line of contact between a liquid phase and a solid phase. The molecules in the liquid phase can move larger distances than those in the solid phase. Therefore, since the molecules in a solid phase are well fixed, this gives rise to the fact that it is in general not possible to study the surface forces of the solid phase in the same way we have described for the liquid phase elsewhere in this book. In spite of this, there exist many examples where we have to attribute a definite value of surface tension to the solid surface, in order to be able to describe the different reactions taking place at the solid surface in contact with liquid phase.

There exist surface defects and irregularities on a solid surface which accounts for such properties as catalysis. Further, the adsorption of a solute at the liquid–solid interface is a very important process in present day life and must also have played a crucial role in the biotic and prebiotic age in the evolution process on earth where we have had gas–solid–liquid contacts. We might also mention the tertiary oil recovery process as being another typical example of such a system.

7.2. SURFACE TENSION OF A SOLID AND THE CONTACT ANGLE

The work of adhesion,[1] W_{AB} (per unit area), between two different liquids, A and B, was given by the following equation (vide Fig. 7.1):

$$W_{AB} = \gamma_A + \gamma_B - \gamma_{AB} \tag{7.1}$$

where γ_A and γ_B are the surface tensions of the two phases. Analogously to this description we can write the work of the adhesion, W_{SL}, of a solid and liquid, as being the work necessary to separate the liquid from the solid

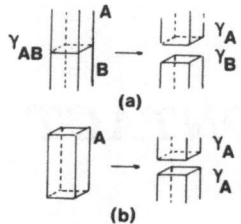

FIGURE 7.1. Schematic description of work of adhesion (a) W_{AB}; and work of cohesion, W_{AA} (b).

perpendicularly from each other, against the adhesive forces existing between them:

$$W_{SL} = \gamma_{SV} + \gamma_{LV} - \gamma_{SL} \tag{7.2}$$

Here γ_{SV} and γ_{SL} are the surface tensions of solid against air and liquid, respectively, and γ_{LV} is the surface tension of the liquid. The work of cohesion W_{AA} (per unit area) is defined as the work of adhesion when both the phases are the same, so that from equation (7.1)

$$W_{AA} = 2\gamma_A \tag{7.3}$$

A liquid drop when placed on a solid surface would strive to reach a shape such as to achieve a minimum free energy. For the solid–liquid–gas system shown in Fig. 7.2, this condition would require that the different surface forces balance each other at equilibrium so that

$$\gamma_{SV} = \gamma_{SL} + \gamma_{LV} \cdot \cos \theta \tag{7.4}$$

where θ is the contact angle (measured in the liquid). This equation is known as the Young's equation.[2] The proof of Young's equation from energy considerations has only been recently given by Adam,[3] McNutt and Andeo.[4] If a

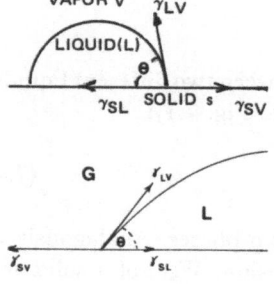

FIGURE 7.2. Equilibrium of forces, e.g., γ_{SV}, γ_{LV}, and γ_{SL} at the line of contact at vapor–liquid–solid interface (schematic).

small change in areas takes place in a system with vapor–liquid interface of area A_{LV}, vapor–solid interface of area A_{SV}, and a liquid–solid interface of area A_{SL}, then the free energy of system will change as

$$dG = \gamma_{LV}\, dA_{LV} + \gamma_{SV}\, dA_{SV} + \gamma_{SL}\, dA_{SL} \qquad (7.5)$$

If only the drop slides, then

$$dA_{SL} = -dA_{SV} \qquad (7.6)$$

Assuming that

$$dA_{LV} = \cos\theta\, dA_{SL} \qquad (7.7)$$

Combining equations (7.5), (7.6), and (7.7), we obtain

$$dG = \gamma_{LV} \cos\theta\, dA_{SL} - \gamma_{SV}\, dA_{SL} + \gamma_{SL}\, dA_{SL} \qquad (7.8)$$

Since at equilibrium $dG = 0$, this relation gives Young's equation (7.4). The term $\gamma_{LV} \cdot \cos\theta$ has been defined as the "adhesion tension" and is equal to the difference between the solid–air and solid–liquid surface tensions, i.e., $W_{SL} - \gamma_{LV}$. The meaning of the term "wetting" of a solid by a liquid is only a matter of definition. The most convenient procedure would be to describe[5] the degree of wetting in terms of contact angle, θ. Thus it is convenient to relate a contact angle of value zero to complete wetting, and a finite value of contact angle to an incomplete wetting system.

One of the main criticisms which has been raised by Bikerman[6,7] is that in the derivation of Young's equation the vertical force component of γ_{LV} as given by $\gamma_{LV} \sin\theta$ has been ignored (see Fig. 7.2). In some studies it has been indeed reported that on soft surfaces this component is actually not negligible since a circular ridge is seen at the periphery of a drop.[8,9] On the other hand, it is quite reasonable to expect that the vertical component would produce a very negligible effect on the contact angle measured on rigid solid surfaces. It has thus been explained that in an actual sense Young's equation, besides these needful considerations, provides us with an equilibrium contact angle between a liquid and a solid which relates the three surface tension forces.[5] In other words, the contact angle as measured includes the drawbacks in Young's equation to varying degree so that it applies to smooth solid surfaces with rigidity such that strains developed due to the placement of the mass of the liquid are small and negligible.[5,10] Before describing the solid–liquid contact angle further, we need at this stage to discuss the interfacial tension between two phases.

7.3. MOLECULAR INTERPRETATION OF INTERFACIAL TENSION

At any interface formed between two phases, I and II, the adjacent interfacial macromolecular layers are subjected to an unsymmetric force field and consequently the interfacial tension is the net result of these forces (vide Fig. 7.3).

The surface molecules of phase II which were in contact with gas phase before are instead now in interaction with the molecules of phase I, and the same is true for molecules of phase I. A general comment would be in place at this stage to consider what kind of intermolecular forces one would need to describe when two dissimilar phases interact. The scope of this comment would be concise since extensive theoretical treatments on the intermolecular forces, both in the bulk phase[11] and at interphases,[12,13] are given in current literature.

The state of the molecules at liquid interface are first of all best described by Stefan's law,[14] according to which the surface molecules are under greater tension due to the structural asymmetry as compared to the bulk molecules. Hence, the ratio surface tension: heat of evaporation should be ~0.5 according to Stefan's law for symmetrical molecules. In practice, the value of this ratio is found to be 0.5 for the case of H_2, N_2, etc., while it is <0.5 for such nonspherical molecules as H_2O. This observation thus clearly indicates that the molecules at the interface are under different intermolecular forces than in the bulk. Thus it is easily seen that if one considers a water–vapor system in comparison to a water–heptane system, we will expect that H_2O molecules at the interfaces are situated under different force fields in the two systems.

The free energy of interaction between dissimilar phases can be given as work of adhesion, W_A (energy per unit area)[15–17]:

$$W_A = W_A^D + W_A^H + W_A^{DD} + W_A^I + W_A^\pi + W_A^{DA} + W_A^E + \cdots \quad (7.9)$$

where W_A is expressed as sum of different intermolecular forces, e.g., London dispersion forces (D), hydrogen bonds (H), dipole–dipole interactions (DD),

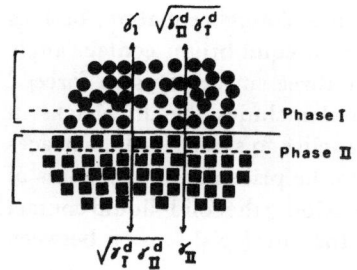

FIGURE 7.3. Various intermolecular forces at an interface between dissimilar phases, I and II.

dipole-induced dipole interactions (I), π-bonds (π), donor–acceptor bonds (DA), and electrostatic interactions (E). It is also easily seen that the W_A^D term will be always present in all systems, i.e., liquids and solids, while other contributions would be present to varying degree determined by the magnitude and nature of the dipole associated with the molecules. We will compile all the intermolecular forces arising from the dipolar nature to W_A^P, i.e.,

$$W_A = W_A^D + W_A^P \tag{7.10}$$

where

$$W_A^P = W_A^H + W_A^{DD} + W_A^I + W_A^\pi + W_A^{DA} + W_A^E + \cdots \tag{7.11}$$

The molecular description of dispersion forces has been given in detail in current literature.[11,15-19] The expression for dispersion forces between two molecules as a function of the distance R center-to-center is, according to London,[20]

$$E_{12}^a = \frac{K_1 \alpha_2}{R} \tag{7.12}$$

where

$$K_1 = \frac{3\alpha_1 I_1 I_2}{2(I_1 + I_2)} + \mu_1^2 \tag{7.13}$$

and α_i, I_i, and μ_i are polarizability, ionization potential, and dipole moment, respectively. In order to determine how theoretical treatment agrees with experimental data, the surface tension of n-octane is found to be given as[15-17]

$$\gamma_{\text{octane}} = \gamma_L^D \tag{7.14}$$

$$= 0.3\pi N_i^2 C_i f_i \lambda_i / \varepsilon_i d_i^2 \tag{7.15}$$

Since $\gamma_L^P = 0$ for a nonpolar molecule, and where N_i is the density of CH_2 or CH_3 groups ($= 3.1 \times 10^{22}$ groups/cm^2), C_i is the interaction energy, $f_i \approx 0.90$, $\lambda_i = 0.87$, $\alpha_i = 2.12 \times 10^{-24}$ cm^3, $I_i = 10.55$ V, $d_i = 4.6$ Å, and $\varepsilon_i = 1.05$. The calculated value for γ_{octane} is 19.0 mN/m, while the measured value is 21.5 mN/m at 20°C. The real outcome of this analysis is clearly to show that such theoretical analyses do indeed predict the surface dispersion forces, γ^D, as measured experimentally. In a further analysis the Hamaker constant A_i for liquid alkanes is found to be related to γ_L^D as[15-17]

$$A_i = 3 \times 10^{-14} (\gamma_L^D)^{11/12} \tag{7.16}$$

This was further expanded to include two components at an interface, 1 and 2:

$$A_{12} = \frac{3 \times 10^{-14}}{D_2} \left(\sqrt{\gamma_1^D} - \sqrt{\gamma_2^D} \right)^{11/6} \tag{7.17}$$

where D_2 is the dielectric constant of phase 2. However, in some cases, forces other than dispersion forces would also be present at the interface between two phases.

The manifestation of intermolecular forces in the direct measurement of any interfacial property thus requires some general picture of the different forces[15-19] responsible for bond formation, as discussed in the following:

(i) Ionic bonds: The force of attraction between two ions is given as

$$F_{ionic} = g^+ g^- / r^2 \tag{7.18}$$

and the energy U_{ionic} between two ions is related to r by the equation

$$U_{ionic} = g^+ g^- / r \tag{7.19}$$

where two charges g^+ and g^- are at a distance r.

(ii) Hydrogen bonds: In terms of molecular structure, those conditions under which hydrogen bonds might be formed are (a) presence of a highly electronegative atom, such as O, Cl, F, N, or a strongly electronegative group such as $-CCl_3$ or $-CN$, with a hydrogen atom attached; (b) in the case of water the electrons in two unshared sp^3 orbitals are able to form hydrogen bonds; (c) two molecules like $CHCl_3$ and acetone may form hydrogen bonds when mixed with each other, which is of much importance in interfacial phenomena.

(iii) Weak-electron sharing bonding: In magnitude this is of the same value as the hydrogen bond. It is also the Lewis acid–Lewis base bond (comparable to Bronsted acids and bases). Such forces might contribute appreciably to cohesiveness at interfaces. A typical example is the weak association of iodine (I_2) with benzene or any polyaromatic compound. The interaction is the donation of the electrons of I_2 to the electron-deficient aromatic molecules (π electrons).

(iv) Dipole-induced dipole forces: In a symmetrical molecule, such as CCl_4 or N_2, there is no dipole ($\mu_a = 0$); through overlapping of electron clouds from another molecule with dipole, μ_b, it can interact by induction.

The typical magnitudes of the different forces are given for comparison in Table 7.1. It will thus be clear that various kinds of interactions would have to be taken into consideration whenever we discuss surface tensions of solids (γ_S), liquids (γ_L) or solid–liquid (γ_{SL}) interfaces.

TABLE 7.1. Intermolecular and Interatomic Forces
between Molecules

Force	Energy (kJ/mol)	Ref.
Chemical bonds		
Ionic	590–1050	a
Covalent	60–700	b
Metallic	100–350	c
Intermolecular forces		
Hydrogen bonds	50	d
Dipole–dipole	20	
Dispersion	42	
Dipole–induced-dipole	2.1	

[a] Y. K. Syrkin and M. E. Dyatkina, *Structure of Molecules and the Chemical Bond* (Wiley-Interscience, New York, 1950).
[b] L. Pauling, *Nature of the Chemical Bond*, 3rd. ed. (Cornell University Press, Ithaca, 1960).
[c] N. F. Mott and H. Jones, *Theory of the Properties of Metals and Alloys* (Oxford, New York, 1936).
[d] G. C. Pimentel and A. L. McClellan, *The Hydrogen Bond* (Freeman, San Francisco, 1960).

7.4. LIQUID–LIQUID INTERFACIAL TENSION

It is known that various types of molecules exhibit different intermolecular forces; and their different force and potential-energy functions can be found.[11] If the potential-energy functions were known for all the atoms or molecules in a system, and the spatial distribution of all the atoms, it could in principle be then possible to add up all the forces acting across an interface. Further, this would allow one to estimate the adhesion and wetting character of interfaces. Because of certain limitations in the force and potential-energy functions this is not quite easily attained in practice. Further, the microscopic structure at molecular level is not known at the present state. For example, in order to calculate surface tensions of liquids one needs the knowledge of radial pair-distribution functions. However, for the complex molecules, this would be highly difficult to measure, although data for simple liquids (such as argon) have been found to give the desired results. The intermolecular forces in saturated alkanes arise only from London dispersion forces. Now, at the interface, the hydrocarbon molecules are subject to forces from the bulk molecules, which is equal to γ (we denote phase I as hydrocarbon in Fig. 7.3). Also, the hydrocarbon molecules are under the influence of London forces due to the molecules in phase II. As described by different investigators[15-19,21] the geometric means of the force due to the dispersion attraction should predict the magnitude of the interaction between any dissimilar phases.

As described in Section 7.3, the surface tension, γ, of any system, solid or liquid, may be described by the following equation:

$$\gamma = \gamma^D + \gamma^P \tag{7.20}$$

Here γ^D is the surface tensional froce due to the dispersion interaction and γ^P that due to polar interaction. For example, the interfacial tension between a hydrocarbon and water can be written as[15-17] (Fig. 7.3)

$$\gamma_{12} = \gamma_1 + \gamma_2 - 2(\gamma_1^D \gamma_2^D)^{1/2} \tag{7.21}$$

where subscripts (1) and (2) refer to alkane and water, respectively; and D indicates the dispersion force component of surface tension. Considering the solubility parameter analysis of mixed liquid systems,[22] we find that there the geometric mean of attraction force gives the most useful analysis. Analogous to that analysis in the bulk phase, the geometric mean should also be preferred for the estimation of intermolecular forces at interfaces. The geometric mean term needs to be multiplied by a factor of 2 since the interface experiences this amount of force exerted by each phase. However, relation (7.22) was alternatively proposed by Antonow[23]

$$\gamma_{12} = \gamma_1 + \gamma_2 - 2(\gamma_1 \gamma_2)^{1/2}$$

$$= (\gamma_1^{1/2} - \gamma_2^{1/2})^2 \tag{7.22}$$

This relation is found to be only approximately valid for such systems as fluorocarbon- or hydrocarbon–water interfaces, while not applicable to organic liquids–water interfaces. A modification of this relation has been given by Good[18]

$$\gamma_{12} = \gamma_1 + \gamma_2 - 2\phi(\gamma_1 \gamma_2)^{1/2} \tag{7.23}$$

where values of ϕ vary between 0.5 to 1.15.

For alkanes where only dispersion forces exist, $\gamma_L = \gamma_L^D$. The value of $\gamma_{H_2O}^D$ was found to be 21.8 mN m^{-1}, against different alkanes, using equation (7.21). This has been questioned by various authors.[24,25] It has been argued that plots of $(\gamma_2 - \gamma_{12})/\gamma_1$ versus γ_1^{-1}, although linear, do not intercept 0, -1, as required by equation (7.21), where subscripts 1 and 2 are for alkane and water, respectively. Another criticism is that γ_2^D calculated by using equation (7.22) varies smoothly with alkane chain length. These criticisms are found to be incorrect on the basis of recent studies made by Aveyard.[26] The plot of $W_A = 2(\gamma_1^D \gamma_2^D)^{1/2}$ versus $\gamma_1^{1/2}$ is linear, so that

$$W_A = 6.6\gamma_1^{1/2} + 12.0 \tag{7.24}$$

where γ_1 is in mN/m units. The observations are consistent with a value of $\gamma_2^D = 10.9$ mN/m, and there being a residual interaction over the interface resulting possibly from the Debye forces of 12.0 mN/m. However, this appears unlikely since theoretical calculations[15-17,27] clearly give a value of 19.2 mN/m for $\gamma_2^D(= \gamma_{H_2O}^D)$ and Debye forces could only contribute about 2 mN/m.

It has been assumed that if alkane molecules lie flat at the interface, then the additive contributions from $-(CH_3)$ and $-(CH_2)-$ groups to W_A can be given as

$$W_A = \frac{2W_{CH_3} - \sigma_{CH_3} + (N-2)\sigma_{CH_2} \cdot W_{CH_2}}{(N-2)\sigma_{CH_2} + 2\sigma_{CH_3}} \tag{7.25}$$

where N is the number of carbon atoms in the alkane chain and σ denotes the surface area for $-CH_3(= 0.11$ nm$^2)$ or $-CH_2-$ $(0.05$ nm$^2)$ group. The values for works of adhesion for $-CH_3$ and $-CH_2-$ groups are estimated as 30 mN m$^{-1} = W_{CH_3}$ and 52 mN m$^{-1} = W_{CH_2'}$ respectively. The plots of γ_{12} versus N by using equation (7.21) show that the relation given in equation (7.23) for flat alkane orientation model at interface is in agreement with experimental data. It is thus concluded that $\gamma_{H_2O}^D$ of magnitude 19.5 mN m^{-1} is in agreement from experimental data with the calculated values. We would, however, like to suggest that in future analysis it should be borne in mind that the surface areas of alkanes be used as given elsewhere,[28,29] for instance in an linear alkane chain, $CH_3CH_2CH_2CH_2CH_3$; surface area is equal to $[2 \times 84.9 + 2 \times 45 + (n_C - 4)31.8]$Å^2/molecule, where n_C is the number of carbon atoms in the alkane (Å$^2 = 100$ nm^2).

7.5. SOLID–LIQUID CONTACT ANGLE

The solid–liquid interface is classified for the sake of convenience into low-energy and high-energy solid surfaces.[30] In the case of high-energy solid surfaces, the surface attracts the liquid molecules more strongly than they attract one another, while in low-energy surfaces the solid surface attracts the liquid molecules weakly.

By combining equations (7.4) and (7.2), we get

$$\cos \theta = -1 + \frac{2(\gamma_S^D \gamma_L^D)^{1/2}}{\gamma_L} \tag{7.26}$$

CHAPTER 7

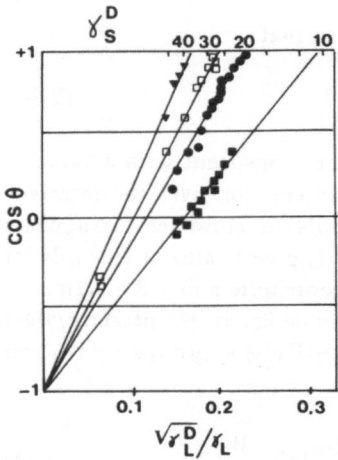

FIGURE 7.4. Plots of $\cos\theta$ versus $(\gamma_L^D)^{1/2}/\gamma_L$ for various solid surfaces: polyethylene (▼), paraffin wax (□), $C_{36}H_{74}$, Decanoic acid (●), dodecanoic acid (■).[16]

(where subscript V has been dropped, i.e., $\gamma_S \approx \gamma_{SV}$), for the case where $\gamma_S = \gamma_S^D$, i.e., the solid surface exhibits only dispersion attraction force, and the polar forces are absent, i.e., $\gamma_S^P = 0$, such as paraffin, polyethylene, Teflon. A plot of $\cos\theta$ versus $\sqrt{\gamma_L^D}/\gamma_L$ would give a straight line with origin at $\cos\theta = -1$ and with slope $= 2\sqrt{\gamma_S^D}$. This method of contact angle analysis has proved far more useful than the earlier empirical method of analysis suggested by other investigators,[30] Fig. 7.4.

The empirical procedure suggested[30] was based on plotting $\cos\theta$ versus γ_L, which would give the intercept at $\theta = 0$, with γ_C the critical surface tension of solid. However, this procedure has the limitations that the $\cos\theta$ versus γ_L plots are not necessarily linear if liquids used contain large variation in the values of the component γ_L^P. It gives rise to large inaccuracies, as discussed by Birdi.[31-34]

In the case that equilibrium film pressure of adsorbed vapor on the solid surface, π_e^{sp}, is not negligible, equation (7.26) can be modified to the form

$$\cos\theta = -1 + \frac{2(\gamma_S^D \gamma_L^D)^{1/2}}{\gamma_L} - \frac{\pi_e^{sp}}{\gamma_L} \qquad (7.27)$$

π_e^{sp} is the spreading film pressure, which is given by the relation

$$\pi_e^{sp} = \gamma_{LV} - \gamma_{LF} \qquad (7.28)$$

Here γ_{LF} refers to the film covered surface of liquid. In the present case, the Gibbs adsorption equation can be used to derive the relation[21,35]

$$\pi_e^{sp} = \gamma_S - \gamma_{SV} \qquad (7.29)$$

where γ_S is the surface free energy of the solid (in vacuum) and γ_{SV} is the surface free energy of solid in equilibrium with liquid vapor. Thus the magnitude of π_e^{sp} can be determined from adsorption isotherms of the vapor. It has also been pointed out that in a system with contact angle, the adsorption isotherm must cross P_0 (saturated vapor pressure) line and that there must be an unstable region.[35] The first part of the curve is given as

$$\pi_e^{sp} = \gamma_S - \gamma_{SV}$$

$$= kT \int_{\Gamma=0}^{\Gamma} \Gamma d \ln p \qquad (7.30)$$

where Γ is surface excess. For the case of infinitely thick duplex film,

$$\gamma_S - \gamma_{SL} - \gamma_{LV} = kT \int_{\Gamma=0}^{\infty} \Gamma d \ln P$$

$$= \pi_{duplex} \qquad (7.31)$$

However, it has been reported that in the case of contact angle system, the equilibrium adsorbed film will necessarily be different from the bulk liquid.[21]

π_e^{sp} is negligible if θ is large, so that $\gamma_S \approx \gamma_{SV}$. Further, in hydrophobic surfaces $\gamma_S^P \sim 0$, π_e^{sp} is negligible since the plots agree with equation (7.26). (Fig. 7.4.) All these extensive data confirm the validity of the relation given in equation (7.28) in the case of nonpolar solid surfaces.

7.6. ESTIMATION OF THE POLAR FORCES AT SOLID–LIQUID INTERFACES

The surface forces were classified as arising from dispersion and polar forces, i.e., for any liquid or solid [vide equation (7.20)]. In spite of the fact that the theoretical basis of this procedure in splitting γ is extensively described in the literature, there has not been reported any systematic investigation on the studies of the role of γ_S^P on the solid–liquid interfacial behavior. It was shown[15–17,21] that the work of adhesion between a solid and liquid in the absence of polar forces is given by

$$W_{adhesion} = 2(\gamma_S^D \gamma_L^D)^{1/2} \qquad (7.32)$$

Now, using the expression for γ_{SL} which includes the polar forces, we get the expression[18,19]

$$\gamma_{SL} = \gamma_S + \gamma_L - 2(\gamma_S^D \gamma_L^D)^{1/2} - 2(\gamma_S^P \gamma_L^P)^{1/2} \qquad (7.33)$$

Using the geometric mean, we can write the expression for the work of adhesion in the presence of both dispersion and polar forces as

$$W_{\text{adhesion}} = 2[(\gamma_S^D \gamma_L^D)^{1/2} + (\gamma_S^P \gamma_L^P)^{1/2}] \qquad (7.34)$$

Combining equations (7.4) and (7.33) we get the following relation:

$$\cos \theta = -1 + \frac{2(\gamma_S^D \gamma_L^D)^{1/2}}{\gamma_L} + \frac{2(\gamma_S^P \gamma_L^P)^{1/2}}{\gamma_L} \qquad (7.35)$$

which is the expression[15–17,31–33] relating θ to γ_S^D and γ_S^P.

According to Fowkes,[15–17] the magnitude of the dispersion force contribution to the surface free energy can be determined by using a solid hydrocarbon (paraffin) as the reference solid phase with known value of $\gamma_S = \gamma_S^D$, and by determining the value of γ_L^D from contact angle measurements and equation (7.26). This procedure has been used by various workers for different liquids (Table 7.2) or mixtures (Fig. 7.5). In Fig. 7.5 is given a plot of $\sqrt{\gamma_L^D}/\gamma_L$ vs. molar concentration of methanol or n-propanol, where the values of $\sqrt{\gamma_L^D}/\gamma_L$ were determined from paraffin film contact angles for various alcohol–water mixtures.[31]

The surface properties of fibrinogen films were investigated by measuring contact angles of n-propanol–water mixtures. The plot of $\cos \theta$ vs. $\sqrt{\gamma_L^D}/\gamma_L$ in Fig. 7.6 gives a straight line with origin at $\cos \theta = -1$. These data show that only dispersion forces are acting across the interface since the data agree with equation (7.26). It is reasonable to expect[31] that addition of alcohol to water gives rise to heterogeneous hydrogen bonding, thus reducing the capacity of the water molecules to create large contributions at the surface due to γ_L^P. However, this behavior is only observed at very high alcohol concentrations, where no dependence on concentration is observed for the different alcohols. It is thus seen that the alcohol–water mixtures are very useful systems for studying the surface properties of fibrinogen. From the data in Fig. 7.6 we

TABLE 7.2. Contact Angle for Various Liquids on Fibrinogen and Value of γ_S^D as Determined from Equation (7.28)[a]

Liquid	Contact angle (degree)	γ_S^D (mN/m)
Symm. tetrabromoethane	39.0	37.5
1-Bromo-naphthalene	31.6	38.0
1-Methyl-naphthalene	25.0	34.5

[a] Reference 31.

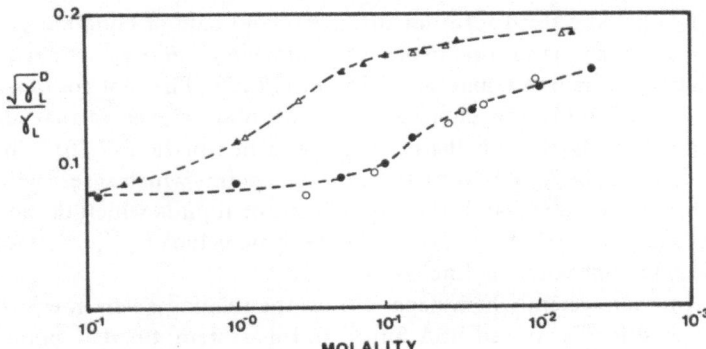

FIGURE 7.5. Variation of $(\gamma_L^D)^{1/2}/\gamma_L$ with molar concentration of methanol ($\bullet$) or n-propanol (Δ) and water mixtures.[31]

find that the value of γ_S^D is 37.0 dyn/cm. A few studies on fibrinogen have been reported by other workers,[31] but none of these results were analyzed by Fowkes relations. Contact angles were also measured of various organic liquids, Table 7.2, which indicate that the value of γ_S^D is around 37.0 dyn/cm as calculated from equation (7.26). It can thus be concluded that accurate determination of γ_S^D of fibrinogen is possible from contact angle measurements of n-propanol–water mixtures and by using equation (7.26). The precise knowledge of the surface behavior of fibrinogen is of considerable importance, since fibrinogen plays an important role in the blood coagulation at synthetic surfaces.

In the course of different studies of various solid surfaces,[31-34] it has become apparent that plots of $\cos \theta$ versus $\sqrt{\gamma_L^D}/\gamma_L$ in many cases show agreement with equation (7.26), but in quite a few cases the plots of $\cos \theta$ versus $\sqrt{\gamma_L^D}/\gamma_L$ do not cross the ordinate at -1. This could only be due to

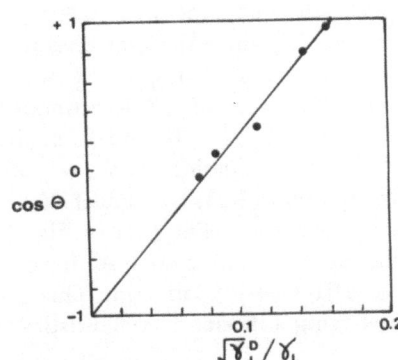

FIGURE 7.6. Plot of $\cos \theta$ versus $(\gamma_L^D)^{1/2}/\gamma_L$ for dried film of fibrinogen (liquids used were water and n-propanol mixtures).[31]

the fact that the third term on the right-hand side of equation (7.38) is not negligible. We thus see that in order to estimate γ_S^D and γ_S^P we need to analyze the data with help of equations (7.26) and (7.35). The plot $\cos \theta$ vs. $\sqrt{\gamma_L^D}/\gamma_L$ for various liquids and cholesterol laurate film[31-34] shows that all the data points do not agree with the relation given in equation (7.26). The value of γ_S^D is found to be 35.0 dyn/cm from the data points which agrees with equation (7.26). On the other hand, the data points for liquids which do not cross the origin at $\cos \theta = -1$ are found to give the same value of γ_S^D, i.e., 35.0 dyn/cm, at the intercept with the line $\cos \theta = -1$.

Another system which departs from the relation given in equation (7.26) is reported for denatured BSA film.[31] In this system, the data points give sets of straight lines, which both intercept at the same point on the line $\cos \theta = -1$. The value of γ_S^D determined at the intercept of $\cos \theta = +1$ is 37.0 dyn/cm.

The surface properties of BSA and Lysozyme films,[31] in the native state, have also been studied in detail. The plots of $\cos \theta$ vs. $\sqrt{\gamma_L^D}/\gamma_L$ are exactly similar for these two proteins. The data points do not agree with equation (7.26), and the value of γ_S^D as determined from the intercept with line $\cos \theta = +1$ is found to be 35.0 dyn/cm for both the proteins. The value γ_S^D for these proteins is of reasonable magnitude, as compared with the value of fibrinogen. It was reported that BSA was more apolar than lysozyme.[37] The contact angle measurements, however, indicate that the hydrophobic effect of these two proteins is the same, since the magnitude of γ_S^D is of the same order of magnitude. On the other hand, the polar forces, i.e., γ_S^P, might not be of the same order of magnitude in these two proteins.

It can thus be concluded that Fowkes' relation as given in equation (7.26) is very useful, when only dispersion forces are acting across the liquid–solid interface, as also concluded by other workers.[15-17,31-33,37-39] We have further shown that in systems where data points do not agree with equation (7.26), and rather the modified Fowkes equation (7.35) should be expected to be valid, the plot of $\cos \theta$ vs. $\sqrt{\gamma_L^D}/\gamma_L$ is still useful insofar as the determination of γ_S^D is concerned. As one should expect, the solid glass surface would exhibit polar surface forces, i.e., $\gamma_S = \gamma_S^D + \gamma_S^P$, which is also obvious from the data in Fig. 7.7, since the plot does not intercept the ordinate at -1. In order to solve equation (7.35) for γ_S^P we need to know the quantities γ_L, γ_L^D, γ_L^P, γ_S^D. The value of γ_S^D is estimated from $\cos \theta$ versus $\sqrt{\gamma_L^D}/\gamma_L$ plots using liquids with $\gamma_L^P \approx 0$, and from the intercept with abscissa when $\cos \theta = 1$. After the calculation of γ_S^D, we can then estimate the value of $2(\gamma_S^D\gamma_L^P)^{1/2}/\gamma_L$ in equation (7.35). A plot of $2(\gamma_S^P\gamma_L^P)^{1/2}$ versus $\sqrt{\gamma_L^P}$ gives the values of γ_S^P from the slope. The data of Fig. 7.7 are replotted in Fig. 7.8, as $2(\gamma_S^P\gamma_L^P)^{1/2}$ versus $\sqrt{\gamma_L^P}$. These analyses for glass gives $\gamma_S^D = 31$ mN/m, $\gamma_S^P = 64$ mN/m, $\gamma_S = 31 + 64 = 95$ mN/m. This procedure has been found to be useful for analyzing adhesion characteristics for a variety of polar solid surfaces.[32-34] In

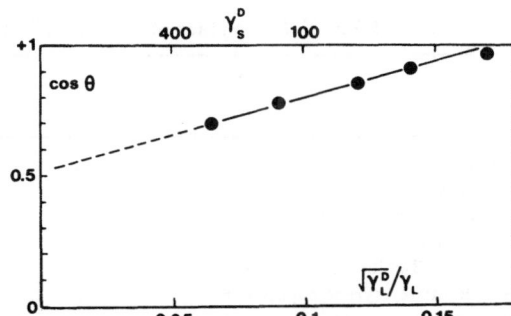

FIGURE 7.7. Plot of $\cos \theta$ versus $(\gamma_L^D)^{1/2}/\gamma_L$ on glass plate.

another interpretation[18] the interfacial tension is given by the equation

$$\gamma_{LS} = \gamma_L + \gamma_S - 2\phi(\gamma_L \gamma_S)^{1/2} \tag{7.36}$$

where ϕ is an adjustable parameter and can be determined from the molar volumes of the two phases

$$\phi = \frac{4(V_L V_S)^{1/3}}{(V_L^{1/3} + V_S^{1/3})^2} \tag{7.37}$$

where V_L and V_S are molar volumes of liquid and solid, respectively. From this, one obtains the relation

$$\cos \theta = -1 + 2\phi \left(\frac{\gamma_S}{\gamma_L} - \frac{\pi_e^{sp}}{\gamma_L} \right)^{1/2} \tag{7.38}$$

As described above, it is reasonable to assume that π_e^{sp} vanishes for $\gamma_L > \gamma_S$. In this treatment, γ is not expressed in the dispersion and polar components, as suggested by earlier theory.[15-17] However, the main drawback in the relation given in equation (7.23) arises from the fact that the molar volumes generally

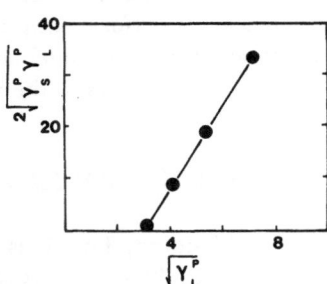

FIGURE 7.8. Plot of $(\gamma_S^P \gamma_L^P)^{1/2}$ versus $(\gamma_L^P)^{1/2}$ for glass surface (data from Fig. 7.7).

TABLE 7.3. Values of γ_S^D and γ_S^P for Synthetic Polymers[a]

Polymer	γ_S^D (mN/m)	γ_S^P (mN/m)	$\gamma_S = \gamma_S^D + \gamma_S^P$ (mN/m)
Polystyrene	41	1	42
Polyethylene	25	0	25
Poly(methylmetacrylate)	27	13	40
Polyformaldehyde	28	10	38
Poly(cis-3,3-chlormethoxetane)	29	2	31
Acrylonitrite/styrene copolymer	33	8	41

[a] Reference 40.

are not known, nor are these available to calculate for high molecular weight polymers.

Studies on contact angles have been recently reported[40] on plasma protein layers absorbed on selected polymer surfaces at 37°C. The γ_S^D and γ_S^P values reported for polymers are given in Table 7.3. The contact angle data have been analyzed by calculating the work of adhesion, $W_{adhesion}$ with the help of equation (7.34). The works of adhesion of either protein were in order polyethylene < poly(bis,3,3,chloromethyloxetane) < polystyrene < polyformaldehyde < acrylonitrile < polymethylmethacrylate.

The polar (γ_S^P) and apolar (γ_S^D) surface tensions of wood have been studied by various investigators[41,42] by using the relations given in equations (7.26) and (7.35). These investigations have also clearly shown that γ_c measurements are not useful as regards the characterization of wood surface energy. On the other hand, the measurements of γ_S^D and γ_S^P have given a useful analysis of different surfaces (γ_S^D varying between 39.5 and 45.2 mN/m and γ_S^P between 3.8 and 11.2 mN/m). The values of γ_c for the same surfaces have varied between 43.0 and 51.8 mN/m. A few reports on the surface tension of human skin as measured by using contact angle have been given in the literature.[43-47] The analysis of skin surface tension hitherto given has been found to be very unsatisfactory and conflicting. A systematic study is urgently needed at this stage, since the role of human skin is very important at the biological level (for example the body temperature is mainly related to the evaporation rate of water at skin, which in turn is related to the rate of penetration and thus to the magnitudes of γ_S^D and γ_S^P of skin).

7.7. LIQUID₁–SOLID–LIQUID₂ SYSTEM

Solid–liquid₁–liquid₂ is a very different system, and hitherto has been very little studied. The procedure followed has been to use liquid₁, n-octane, which has $\gamma_{L_1} \simeq \gamma_{L_1}^D = 21.8 \, mN/m$, and liquid₂, L_2, as water, which has

$\gamma_{L_2}^D = 21.8$ mN/m. Since only those forces which are common to both interfaces are able to act across the interface, this shows that when $\gamma_{L_1}^D \simeq \gamma_{L_2}^D$, Young's equation can be easily solved (see Fig. 7.9).

Applying the same arguments as followed for the liquid–solid system, we can write[48,49]

$$\gamma_{L_1 L_2} \cos\theta + \gamma_{SL_1} = \gamma_{SL_2} \qquad (7.39)$$

If L_1 is n-octane and L_2 is water, then from experiments, $\gamma_{WO} = \gamma_{L_1 L_2} = 48.3$ dyn/cm. Combining expressions from the Fowkes equation for γ_{SO} and γ_{SW}, we get

$$\cos\theta = [\gamma_W' - \gamma_0 - 2(\gamma_S^D \gamma_W^D)^{1/2} + 2(\gamma_S^D \gamma_O^D)^{1/2}]/48.3 \qquad (7.40)$$

The subscript S, O, and W denote solid, octane, and water, respectively. Here γ_W' is the surface tension of octane-saturated water ($= 51.6$ dyn/cm), γ_0 is the surface tension of water-saturated octane ($= 21.8$ dyn/cm). The values of $\gamma_W^D \simeq \gamma_O^D = \gamma_0 = 21.8$ dyn/cm, so that equation (7.40) will assume the simple form

$$\cos\theta = [\gamma_W' - \gamma_0 - 2(\gamma_S^P \gamma_W^P)^{1/2}]/48.3 \qquad (7.41)$$

If γ_S^P becomes zero then the equation reduces to the form

$$\cos\theta = 0.6169 \qquad (7.42)$$

since $\gamma_0^D \cong \gamma_W^D$. However, if $\gamma_S^P \neq 0$ then the third term on the right-hand side of equation (7.41) will be needed, as described earlier. The effect of the adsorption of water at the octane–solid surface should be considered. However, this term can be neglected since its magnitude has been found to be small[50,51] in most cases.

The procedure employed is to place a drop of octane on the solid surface placed inside water (Fig. 7.9). This setup is not easily manageable, and it has

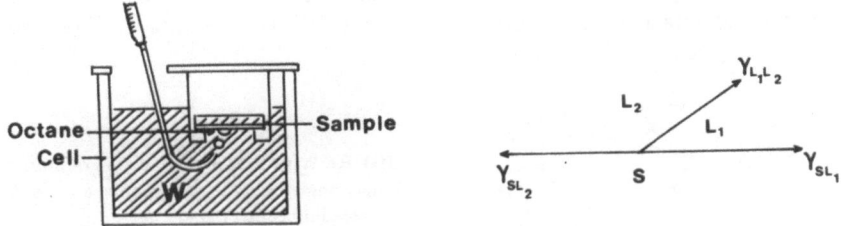

FIGURE 7.9. Contact angle, θ, measuring setup at liquid (L_1)–solid (S)–liquid (L_2).

been found[50] that if the solid is fixed at the bottom of the liquid octane and a drop of water is placed on the solid then one should have little or no technical difficulties. The difference between these two procedures is that the value of θ would be different by 180°.

For the case like paraffin–H_2O–octane we get from the equation (7.41), $\theta = 50°$. The results reported in the literature prove that the derivations given by Fowkes and these calculations for solid–L_1–L_2 are correct, since one indeed measures $\theta = 50°$ as predicted, when $\gamma_S^P = 0$. In other words the arguments used for derivations based on Young's equation for liquid–solid, and further extension to liquid$_1$–solid–liquid$_2$ are all correct.

The basic advantage in the procedure of S–L_1–L_2 is based on the fact that only *one* contact angle measurement is sufficient to determine both γ_S^D and γ_S^P values. However, it has one disadvantage, namely that the contact angle measurement must be free of any artifact like the adsorption of L_1 at the S–L_2 interface. Since recent studies show that[50] γ_S^D and γ_S^P measured by methods S–L_1 and S–L_1–L_2 were in agreement, this allows us to conclude that both methods should be comparatively used.

7.8. CONTACT ANGLE HYSTERESIS

If a liquid drop is moved back and forth, the state of equilibrium as shown in Fig. 7.10 should not change on a smooth, horizontal solid surface, so long as the values of γ_{SV}, γ_{LV}, and γ_{SL} remain unchanged. However, in literature many reports are found which describe the existence of advancing angle, θ_a and a receding angle, θ_r, when a drop of liquid is pushed on a smooth solid surface as shown in Fig. 7.10. The difference $\theta_a - \theta_r$ is termed the hysteresis.

The existence of hysteresis on rough surfaces is easy to understand.[9,39,52,53] The same is true for the case where the drop is placed on a tilted solid surface.[52] Let θ_e be the contact angle for a liquid–solid–vapor system at equilibrium without any gravitational effects. We have mentioned earlier in this section that for any rigorous analysis of contact angle, θ, measurements, a plot of $\cos \theta$ versus $\sqrt{\gamma_L^D}/\gamma_L$ gives much useful information.

Accordingly, the analysis[39] of the contact angle (both advancing, θ_a, and receding, θ_r) data as reported on polymeric solids[54] has been made by plotting

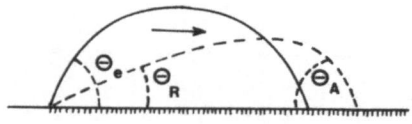

FIGURE 7.10. Schematic state of equilibrium contact angle, θ_e, advancing contact angle, θ_a, and receding contact angle, θ_r.

TABLE 7.4. Surface Tensions, γ_S, γ_S^D, and γ_S^P of Polymeric Solids Estimated from Plots of cos θ_a versus $\sqrt{\gamma_L^D/\gamma_L}$ [a]

Solid	γ_S^D (mN/m)	γ_S^P (mN/m)	γ_S (mN/m)
Teflon	18.9 (17.3)	0	18.9
Polypropylene	27.7	0	27.7
Polycarbonate	27.7	6.5	34.2
Nylon 6	35.4 (31.9)	5.8 (ca. 12)	41.2
Polystyrene	34.6	0.5	35.1
Polyvinylchloride	39.1 (32.4)	1.5 (8.5)	40.6
Kevlar 49	25.0	14.4	39.4
Graphite	43.3	0.15	43.8

[a] Reference 39; γ_S^D and γ_S^P values from literature are given within brackets.

cos θ versus $(\sqrt{\gamma_L^D/\gamma_L})$. These plots are linear for both cases. The value of γ_S^D has been evaluated in all cases from the extrapolation of plot to cos $\theta = 1$. In Table 7.4 the values of γ_S^D for θ_a and θ_r data are given. As one should expect, Young's equation [equation (7.4)] is valid for a given value of γ_S and γ_L, by a fixed value of θ_e when the gravitational force effect is absent. The plot of cos θ_e versus $\sqrt{\gamma_L^D/\gamma_L}$ will then agree with the relation as given in equation (7.26) if $\gamma_S^P = 0$. If any gravitational force effects are present, as would be the case on a tilted solid surface,[52] then one needs additional term arising from this force in equation (7.8). The analysis for the case of Teflon and polypropylene surfaces thus shows that since the data of θ_a fit the relation in equation (7.4) then $\theta_a = \theta_e$ and $\gamma_S^P = 0$ as expected.

One can also present another explanation of the hysteresis. If one places a very small volume (10–20 μl) of liquid on a solid surface, the contact angle measured is always equal to θ_e. However, if the volume of liquid used is much larger than 20 μl, then the contact angle measured may differ from θ_e under certain conditions. This may arise from some lateral movement during the process which gives rise to a small displacement of the liquid such that the change in area of solid covered by liquid, ΔA, is found to be related to $\Delta G_{\text{surface}}$ by the equation

$$\text{Change in surface energy} = \Delta G_{\text{surface}} = \Delta A \cdot (\gamma_{SL} - \gamma_S)$$

$$+ \Delta A \cdot \gamma_L \cdot \cos(\theta_e - \Delta\theta) \qquad (7.43)$$

The hysteresis thus arises from the fact that $(\theta_e - \Delta\theta) = \theta_r$, which would require a restoring force of sufficient magnitude to overcome the adhesion work term, i.e., $\Delta A(\gamma_{SL} - \gamma_S)$. If the restoring force is lesser in magnitude than $\Delta A(\gamma_{SL} - \gamma_S)$, then one would observe hysteresis, while if it is greater, then one would measure θ_e. From this description, as well as from our analysis given above, we feel that θ_r is nothing but a state of nonequilibrium of quantities in equation (7.8), which are imposed upon a system during the experimental procedure.

If the $\cos \theta$ versus $\sqrt{\gamma_L^D}/\gamma_L$ plot does not intersect at -1 on the ordinate axis, then the term $(\gamma_S^P \gamma_L^P)^{1/2}/\gamma_L$ in equation (7.35) is not negligible. The magnitude of the term $(\gamma_S^P \gamma_L^P)^{1/2}$ on the right-hand side of equation (7.35) is calculated and plotted against $\sqrt{\gamma_L^P}$. From such plots the values of γ_S^P are calculated from the slopes and their values are given in Table 7.4. It is clear that such polymeric solids as Kevlar and polycarbonate would be expected to exhibit high γ_S^P values, as is also observed.[39]

This analysis shows convincingly that: (a) $\cos \theta$ versus $\sqrt{\gamma_L^D}/\gamma_L$ plots for θ_a and θ_r give useful analysis of γ_S^D. In fact, when $\gamma_S^P = 0$, it is shown that $\theta_a = \theta_e$. (b) Since $\theta_a = \theta_e$ then θ_r is observed due to the term related to the adhesion work, and has no effect on the determination of γ_S^D or γ_S^P. (c) Since in systems where $\gamma_S^P = 0$ (like Teflon, Polypropylene), $\theta_e = \theta_a$, indicates that θ_r provides no information of thermodynamic basis in the absence of gravitational effect.

7.9. CONTACT ANGLES AND HEATS OF IMMERSION

The contact angle variation with temperature can be used to estimate heat of immersion. Since heat of immersion can be measured directly with the help of calorimetry, then it offers a unique combination to verify the two methods.

The evacuated solid is allowed to interact with the liquid and the heat evolved is determined as follows. The change in surface free energy is

$$g_i = \gamma_{SL} - \gamma_S \qquad (7.44)$$

Combining this relation with equation (7.33), we obtain

$$g_i = \gamma_{LV} - 2(\gamma_S^D \gamma_L^D)^{1/2} \qquad (7.45)$$

if $\gamma_S^P = 0$. The enthalpy change per unit area, h_i, can be written as

$$h_i = h_{SL} - h_S$$

$$= g_i - T\left(\frac{\partial g_i}{\partial T}\right) \qquad (7.46)$$

We thus obtain the relationship

$$h_i = \gamma_{LV} - 2(\gamma_S^D \gamma_L^D)^{1/2}$$

$$- T\left(\frac{\partial \gamma_{LV}}{\partial T}\right) - 2\sqrt{\gamma_{LV}^D}\left(\frac{\partial \sqrt{\gamma_S^D}}{\partial T}\right)$$

$$- 2\sqrt{\gamma_S^D}\left(\frac{\partial \sqrt{\gamma_{LV}^D}}{\partial T}\right) \tag{7.47}$$

In the case of hydrocarbons and low-energy solids, it has been shown that the following relation is valid[21]:

$$h_i = -\gamma_S^D + T\left(\frac{\partial \gamma_S^D}{\partial T}\right)$$

$$\approx -\gamma_S^D - 21 \tag{7.48}$$

where $\partial \gamma_S^D / \partial T \approx -0.07$ has been used as found for various systems. The heats of immersion of graphon into n-hexane, n-heptane, and n-octane have been calculated with the help of this equation, and one finds that by using $\gamma_S^D = 70$ mN/m for graphon $h_i = -90$ erg/cm$^2 \approx -82$ to -102 erg/cm^2 (measured).

If we write $h_{i(SV)}$ for the case where the solid surface is covered by vapor, then we have

$$h_{i(SV)} = h_{SL} - h_{SV} \tag{7.49}$$

and

$$g_{i(SV)} = \gamma_{SL} - \gamma_{SV}$$

$$= -\gamma_{LV} \cos \theta \tag{7.50}$$

Combining with Young's equation, we get

$$h_{SL} - h_{SV} = \gamma_{SL} - \gamma_{SV} - T\left(\frac{\partial g_i}{\partial T}\right)$$

$$= -\gamma_{LV} \cos \theta$$

$$+ \left[T\gamma_{LV}\left(\frac{\partial \cos \theta}{\partial T}\right) + \cos \theta \left(\frac{\partial \gamma_{LV}}{\partial T}\right) \right]$$

$$= \left[-\gamma_{LV} - \frac{T \partial \gamma_{LV}}{\partial T} \cos \theta \right] + \gamma_{LV} T \frac{\partial \cos \theta}{\partial T} \tag{7.51}$$

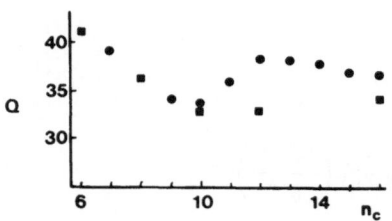

FIGURE 7.11. Comparative heats of wetting (■) and calorimetric heats of immersion (●) of Teflon/n-alkanes.[53] ($Q = \text{erg/cm}^2$.)

or

$$-h_{i(\text{SV})} = h_{\text{LV}} \cos\theta - \gamma_{\text{LV}}\, T\left(\frac{\partial \cos\theta}{\partial T}\right) \tag{7.52}$$

This means that by measuring θ as a function of temperature the heat of immersion can be estimated. At this stage in literature there are not many examples where the magnitudes of $h_{i(\text{SV})}$ estimated from equation (7.52) have been compared with direct immersion calorimetry,[20,52,53] Fig. 7.11. This kind of comparison is thus of urgent need and must be a challenge to any beginner in this field of surface chemistry.

7.10. ROLE OF POLAR SOLID SURFACE TENSION AND CELL ADHESION

The understanding of the surface forces which determine the adhesion characteristics between a cell and solid, obviously is of much importance in order to be able to explain the mechanism by which certain bacteriological phenomena develop. For example, the oral cavity is one of the many biological systems where the bacterial adhesion to tooth enamel is a very vital phenomenon. At this stage, the analysis of the cell adhesion in the oral cavity on tooth surface is far from completely understood. Another example one might mention is cancer cells selectivity behavior in metathesis, etc.

The cell surface is mainly made up of glycoprotein and glycolipid, which together with the phospholipid bilayer of cell membranes would determine the adhesive properties, as well as the glycoproteins attached to the cell membrane.[33,34] It is easily seen that in the approach of a cell towards the solid, the forces involved in adhesion would be related to the adsorption properties of both the glycoproteins and lipid layer. The forces which would have to be taken into consideration, which would be related to the components of the cell membranes, would be hydrophobic, hydrophilic interactions, and electrostatic interactions (as described in Section 7.3). On the other hand, one should expect that the solid surface would be easily described, as regards the physicochemical forces, i.e., the solid would be able to exhibit hydrophobic interactions (γ_S^D) and polar forces (γ_S^P).

The well-known colloidal theory, the so-called DLVO theory[52] have been used by various authors[55] to describe the cell adhesion. While DLVO theory would have predicted that negatively charged cells should adhere weakly on negatively charged surfaces, the opposite has been observed[32,33,34,55] in many cases. Many of these criticisms raised by various investigators[55] on the deficiency of the DLVO theory in explaining the cell adhesion on solid surfaces, may not be valid. Two surfaces with similar charges would experience no charge–charge repulsion due to the effect of counter ions in the electrical double layer. This leads us to conclude that in cell adhesion, the counter ions would be of importance in determination of the adhesion process.

Various investigators have attempted to describe the cell adhesion characteristics on the basis of the surface forces acting across the cell–solid interface.[55] This approach was considered a less rigorous alternative, arising from the close-range intermolecular forces (which were considered to be hydrophobic forces, and polar forces (such as hydrogen bonds, dipole, and electrostatic interactions)). In one of the studies the surface[55] tension of solids is characterized by the so-called "critical surface tension" (γ_c) for the solid. In one case it has been found that adhesion of cells to solids is determined by the value of γ_c. It has been reported that both blood platelets and fibroplasts have exhibited very low adhesion to hydrophobic solid interfaces (as characterized by γ_c analysis). These investigators also attempted to conclude that depending on whether the two phases (here the cell and the solid) approach closer than a certain distance, then one could apply either the DLVO theory or the surface force approach. This differentiation is not fully acceptable to many authors.[34]

It has been reported that negatively charged cell adhesion is very low on untreated polystyrene surface,[55] while after the polystyrene surface was treated with sulfuric acid, which gave polar surface, the adhesion increased. In this

TABLE 7.5. The Magnitudes of Dispersion (γ_S^D) and Polar (γ_S^P) Forces of Treated and Untreated Surfaces of Polystyrene with Sulfuric Acid[a]

Solid surface	γ_S^D (mN/m)	γ_S^P (mN/m)
Untreated	35	0
5 min[b]	35	50
15 min[b]	35	89

[a] Reference 33.
[b] Time of contact of polystyrene with conc. H_2SO_4.

analysis, it is shown, rather empirically, that γ_c is different for sulfuric-acid-treated polystyrene, and therefore this increase in γ_c value can be related to the increase in cell adhesion.

The results of $2(\gamma_S^P \gamma_L^P)^{1/2}$ versus $\sqrt{\gamma_L^P}$ for the various samples of polystyrene are given in Table 7.5. The most characteristic conclusion[33,34] one obtains is that while the dispersion forces, γ_S^D, remain the same (i.e., within the experimental accuracy) for the different solid samples, the polar force component, γ_S^P, changes appreciably (Table 7.5). This leads us to conclude that cell adhesion is related to the polar forces (i.e., γ_S^P).[33,34]

7.11. SUMMARY AND COMMENTS

In this chapter we have discussed some of the most important interfacial systems, i.e., gas–solid–liquid or liquid–solid–liquid$_2$. Such interfacial systems are found both in various industrial and biological phenomena. In spite of this, the number of investigations reported in the literature on these interfaces is much fewer than studies carried out on other interfaces, e.g., liquid–gas, liquid$_1$–liquid$_2$. Accordingly, the theoretical analyses are not as satisfactory at the molecular level for the gas–solid–liquid interfaces. Further, there are also some subtle reasons for the latter observation arising from the technical aspects. One of these is that while one can purify the liquids to ~100% grade, the solid surface always remains ambiguous. This obviously leads to much data scatter as reported in the literature by different investigators on the same solid. On the other hand, the main theme of the research, as actively carried out by various institutes around the world, has clearly shown that these problems can be solved. Future research should be more oriented toward the approach where molecular forces acting at the gas–solid or liquid–solid interface are analyzed. This requires another parameter which has not been extensively investigated, i.e., the enthalpy of adsorption on solids. This is important since the entropy of adsorption is large and therefore the effects of temperature are generally very large.

CHAPTER 8

ADSORPTION AT
SOLID–LIQUID INTERFACES

8.1. INTRODUCTION

In Chapters 3, 4, and 5, we have discussed extensively the detailed thermodynamic aspects of the adsorption of solute and solvent at the interface formed by the contact of either liquid–gas or liquid–liquid phases. The surfaces in all these cases are regarded as homogeneous across the plane of contact, although perpendicular to the surface plane the surface phase is considered as inhomogeneous. Surface free energy (or interfacial tension) in such systems involving liquid interface can be accurately measured in the presence of solute component in the bulk solution so that excess adsorption at the interface can be calculated using the Gibbs adsorption equation.

When a significant amount of a powdered solid is added to a solution, one of the components can preferentially accumulate at the solid–liquid interface. This excess accumulation (or adsorption) per gram of the solid powder can be measured directly using various analytical techniques as shown in Section 2.6. If the surface area per gram of the solid powder is known from a separate independent experiment, then one can express the results in terms of the grams (or moles) of component adsorbed per unit surface area of the solid provided the surface area of the particle itself does not alter with the progress of adsorption. Particles of solid like carbon black, silica, alumina, barium, sulfate, etc. will maintain their rigid surface structure when many soluble substances like acids, dyes, polymers are accumulated at their inter- faces. On the other hand, many polymers such as polyacrylic acid, gums, gelatin, casein, etc. may remain insoluble in the presence of binary solution at suitable pH. Excess accumulation at the interface of the polymer may lead to the expansion or contraction of the surface area of the solid–liquid interface. In such a situation the adsorption per gram of solid rather than per unit surface area becomes more meaningful. In contrast to the liquid interfaces, different regions of the surfaces of both rigid and flexible solids in contact with the solution may be highly heterogeneous across the surface plane.

The adsorption at the solid–liquid interface is of major importance in both industrial processes and many biophysical phenomena. Just for the sake of examples, one can consider chromatographic methods of separation, soil phenomena, tertiary oil recovery, adsorption in biomembranes, liposomes, and fibrous proteins.

Experimental investigations on the adsorption from solution by powdered particles are extensive. This was reviewed extensively very early by Kipling[1]. The physicochemical and the thermodynamic aspects of the adsorption process at the solid–liquid interface on the basis of the treatment given by Gibbs for liquid interface has been developed only in recent years.[1-5] In this chapter, we shall limit our discussions mainly to this aspect of adsorption at the solid–liquid interface.

The adsorption at the liquid interface is in general reversible so that thermodynamic principles are applicable. In the case of adsorption of many substances at the solid–liquid interface, it is generally assumed that the process is a physical adsorption. In a few cases, the reversible nature of the adsorption and desorption has been verified from direct experiments. In a few other cases, however, physical adsorption becomes mixed up with chemisorption, so that physicochemical interpretation of the data becomes relatively complex.

8.2. POSITIVE AND NEGATIVE EXCESSES

For clarification of certain basic concepts of adsorption, we shall first discuss the adsorption of alcohols and glycols from their aqueous mixture by gelatin powder. Gelatin itself is a denatured protein which forms a turbid colloidal solution in water at ordinary temperature. On addition of alcohols and glycols, the colloid precipitates out in solid form. Gelatin in contact with pure glycols and alcohols forms solid agglomerate.

Let us first consider that a known weight of gelatin remains in contact with a definite weight of a pure nonaqueous component ethylene glycol. If a definite weight of water is added to this system, the insoluble protein will adsorb water solute from the mixture. When equilibrium is attained in the system, one can calculate the amount of water (Γ_1^2) adsorbed per kilogram of gelatin powder with the help of the analytical equation

$$\Gamma_1^2 = \frac{w_2^t}{1000}(m_1^t - m_1) \tag{8.1}$$

Here m_1^t and m_1 are the molal concentrations of water solute before and after adsorption, respectively; w_2^t is the total weight of component 2 in the system per kilogram of the dry protein in the mixture.

In Fig. 8.1, values of Γ_1^2 for different nonaqueous solvents have been plotted as functions of mole ratio compositions X_1/X_2. It is found that, in all cases, Γ_1^2 is positive, which means that m_1^t is greater than m_1. Γ_1^2 increases

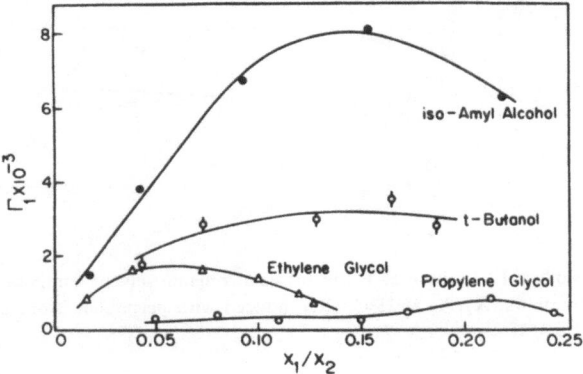

FIGURE 8.1. Plot[6] of Γ_1^2 (written as Γ_1) in moles of water (solute) adsorbed per 100 kg of dry gelatin vs. X_1/X_2 at 25°C. (Reproduced with permission from *Indian Journal of Biochemistry and Biophysics.*)

with increase of X_1/X_2. At a given value of X_1/X_2, its magnitude depends also on the nature of component 2. This indirectly indicates that component 2 involves itself in a competitive manner to the phenomena of adsorption.

If in equation (8.1), m_1', m_1 and w_2' are replaced by $1000n_1'/M_2n_2'$, $1000n_1/M_2n_2$, and M_2n_2', respectively, then it can easily be shown[6] that

$$\Gamma_1^2 = n_1' - n_2'X_1/X_2 \qquad (8.2)$$

where M_2 is the molecular weight of component 2 in the system; n_1' and n_2' are respective moles of components 1 and 2 present in the system before adsorption and n_1 and n_2 are their values at adsorption equilibrium. The ratio n_1/n_2 is therefore equal to X_1/X_2, where X_1 and X_2 are the mole fractions of the two components at equilibrium. The right-hand side of equation (8.2) is obviously the Gibbs surface excess of component 1 per kilogram of solid powder which can thus be computed directly from the experimental data. Equation (8.2) can be written in the alternative form

$$-\Gamma_1^2X_2/X_1 = n_2' - n_1'X_2/X_1 \qquad (8.3)$$

If we define the left-hand side of the equation to be equal to the relative surface excess Γ_2^1 of component 2 in the mixture per kilogram of solid, then

$$\Gamma_2^1 = n_2' - n_1'X_2/X_1 \qquad (8.4)$$

Values of Γ_2^1 computed from the experimental data are found in Fig. 8.2 to be negative, as expected.

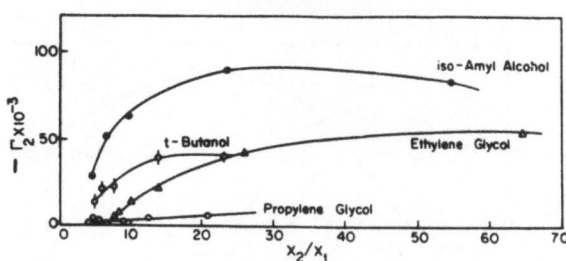

FIGURE 8.2. Plot[6] of Γ_2^1 (written as Γ_2) in moles of organic solvent component adsorbed per 100 kg of dry gelatin vs. X_2/X_1 at 25°C. (Reproduced with permission from *Indian Journal of Biochemistry and Biophysics*.)

Chatterjee and Chattoraj[6] have also measured the extent of adsorption of the organic solute component Γ_2^1 in aqueous solvent by insoluble gelatin using the analytical equation

$$\Gamma_2^1 = \frac{w_1^t}{1000}(m_2^t - m_2) \tag{8.5}$$

Here w_1^t is the amount of solvent component per kilogram of gelatin. From the chemical analysis, the initial concentration m_2^t of component 2 before adsorption is always found to be less than the concentration m_2 at adsorption equilibrium so that Γ_2^1 becomes negative. The negative excesses for the glycols and alcohols are found to increase with increase of the mole ratio composition X_2/X_1. At a given value of X_2/X_1 also, the magnitude of Γ_2^1 depends on the nature of the solute (vide Fig. 8.3).

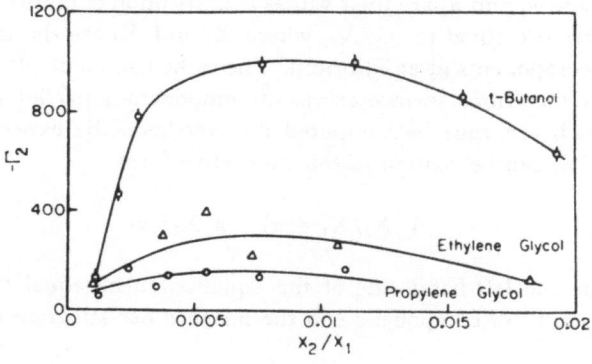

FIGURE 8.3. Plot[6] of Γ_2^1 (written as Γ_2) in moles of organic solute component adsorbed per 100 kg of dry gelatin vs. X_2/X_1 at 25°C. (Reproduced with permission from *Indian Journal of Biochemistry and Biophysics*.)

Following a similar procedure for the replacement of m_2^i, m_2, and w_1^i in equation (8.5) by $1000n_2^i/M_1n_1^i$, $1000n_2/M_1n_1$, and $M_1n_1^i$, respectively, we can directly obtain equation (8.4). By algebraic operation, equation (8.4) can subsequently be converted to the form (8.2) if one puts $(-\Gamma_2^1X_1/X_2)$ to be equal to Γ_1^2 in conformity with the concepts of the Gibbs positive and negative excesses. Equations (8.2) and (8.4) will then remain valid irrespective of the nature of the solvent or solute of either of the two components, and in general, the following relation will be applicable to these systems:

$$\Gamma_2^1X_1 + \Gamma_1^2X_2 = 0 \tag{8.6}$$

Since Γ_2^1 are all negative in Fig. 8.3, it is obvious that Γ_1^2 are all positive and their magnitudes for various values of X_1/X_2 can be calculated with the help of equation (8.6). In Fig. 8.4, Γ_1^2 for various organic solutes have been plotted against X_1/X_2. The results in Figs. 8.1 and 8.4 further indicate that water existing either as a solute or as a solvent possesses more affinities for attachment with the surface of the gelatin powder than those of various organic components present in the system, which one should also expect.

In analogy to equations (3.50) and (3.50a), one may also write the following equations for the adsorption at the solid–liquid interface[6,7]:

$$n_1^i = n_1 + \Delta n_1 \tag{8.7}$$

$$n_2^i = n_2 + \Delta n_2 \tag{8.8}$$

Here Δn_1 and Δn_2 are moles of components 1 and 2, respectively, present in the inhomogeneous surface bound phase per kilogram of solid powder. Also n_1 and n_2 are moles of these components in the bulk phase of the system containing 1 kg of gelatin.

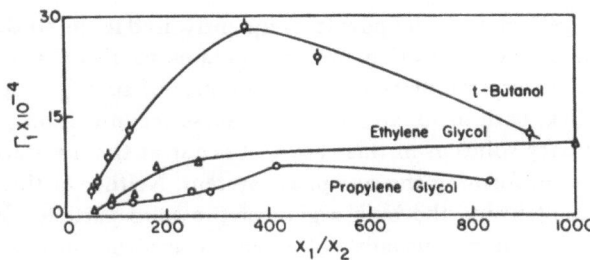

FIGURE 8.4. Plot[6] of Γ_1^2 (written as Γ_1) in moles of water (solvent) component adsorbed per 100 kg of dry gelatin vs. X_1/X_2 at 25°C. (Reproduced with permission from *Indian Journal of Biochemistry and Biophysics*.)

Combining equations (8.7) and (8.8) with relations (8.2) and (8.4), respectively, it can easily be shown[6] that

$$\Gamma_1^2 = \Delta n_1 - \Delta n_2 (X_1/X_2) \tag{8.9}$$

$$\Gamma_2^1 = \Delta n_2 - \Delta n_1 (X_2/X_1) \tag{8.10}$$

These two equations can explain explicitly the competitive nature of the adsorption since either Γ_1^2 or Γ_2^1 depend on Δn_1 as well as on Δn_2. Plots of Γ_1^2 or Γ_2^1 against X_1/X_2 or X_2/X_1, respectively, in Figs. 8.1–8.4 are not linear in any range of concentration studied by Chatterjee *et al.*[6] This means that both Δn_1 and Δn_2 themselves do not remain constant but they vary with change of composition of the binary solution.

In the analytical equations (8.1) and (8.5) and also in equations (8.7) and (8.8), it has been implicitly assumed that components 1 and 2 are not present in the interior bulk phase of the solid so that Γ_1^2 may be identified with the extent of positive adsorption of water at the interfacial region of the solid powder. This is, in all probability, not true for denatured protein, which may alter its conformation in the gel with its gradual hydration. The gel may swell extensively by hydration so that Γ_1^2 may represent extent of "sorption" (vide Chapter 1) rather than "adsorption." The protein gel in an alternative approach may be regarded as a "single phase" for which the validity of equations (8.9) and (8.10) based on the "two-phase" model may be questionable. Equations (8.1), (8.2), (8.4), and (8.5) will, however, remain operationally valid when adsorption is replaced by sorption. The thermodynamic aspects of hydration and solute interaction with protein have been further discussed more extensively in Chapters 9 and 10.

8.3. ADSORPTION FROM A BINARY LIQUID MIXTURE BY A RIGID SOLID

Unlike gelatin, the solid powder frequently used in the study and technical applications of the adsorption process contains particles which are rigid in respect to their crystalline structure. Examples of such powders are carbon, alumina, silica, barium sulfate, etc. The excess accumulation of a component from the binary solution in this case will occur at the interface between the solid and liquid forming the two-phase system. Neither of these components will be present inside the bulk region of the solid particle. Swelling of the system by adsorption is usually not extensive and the surface area per gram of the powder is expected to remain unchanged by gradual adsorption of a component from the liquid mixture. For adsorption in such systems, equations (8.9) and (8.10) will remain undoubtedly valid.

For the study of adsorption in binary solution of completely miscible liquids, calculations of Γ_2^1 and Γ_1^2 with the help of analytical equations (8.1) and (8.5) in the whole range of liquid composition will lead to confusion since the definition of molality (or molarity) will be invalid in this case. In such a case, Γ_2^1 and Γ_1^2 could be calculated easily from the experimental data using equations (8.2) and (8.4), respectively. However, this convenient procedure has never been followed. In dealing with this situation, a new operational quantity Γ_2^n, sometimes referred to as apparent excess adsorption, is estimated from the experimental data using the following equation:

$$\Gamma_2^n = (n_1^i + n_2^i)(X_2^i - X_2) \tag{8.11}$$

Here X_2^i and X_2 stand for the respective mole fractions of component 2 before and after adsorption. Values of Γ_1^n can also be defined in a similar manner.

Values of Γ_1^n and Γ_2^n for different powdered materials in the presence of various types of liquid mixtures have been evaluated by many workers[1,4,7] using equation (8.11). Such typical isotherms[8] are shown in Fig. 8.5 at several temperatures.

The extent of apparent adsorption Γ_2^w in a binary solution has been occasionally calculated on the basis of the difference in the weight fractions W_2^i and W_2, respectively, found before and after adsorption using the following

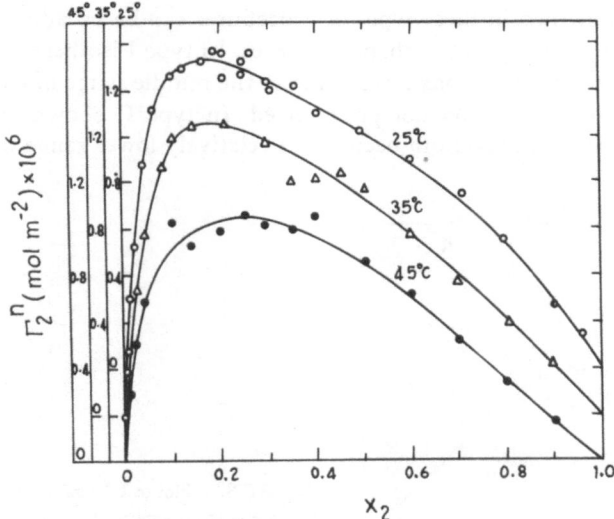

FIGURE 8.5. Plot Γ_2^n (in mol m^{-2}) vs. X_2 for the adsorption of n-heptane (component 1) + n-hexadecane (component 2) mixtures. (Redrawn from Ref. 8.)

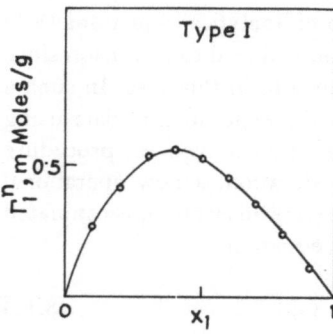

FIGURE 8.6. Plot of Γ_1^n against X_1 for adsorption of 1,2-dichloroethane (component 1) + benzene (component 2) on alumina gel at 25°C. (Redrawn from Ref. 7.)

equation:

$$\Gamma_2^w = (w_1^t + w_2^t)(W_2^t - W_2) \tag{8.12}$$

Here w_1^t and w_2^t are the total weights of the two components before adsorption and w_1 and w_2 their respective values in solution at adsorption equilibrium. The weight fractions W_2^t and W_2 are, respectively, equal to $w_1^t/(w_1^t + w_2^t)$ and $w_2/(w_1 + w_2)$. The apparent adsorption of component 1 can also be similarly defined. According to Schay,[7] the theoretical basis for the interpretations of the adsorption in terms of equations (8.11) and (8.12) stand on the same footing.

Schay and co-workers[9–11] have pointed out that the different adsorption isotherms obtained in these types of procedures can be broadly grouped into five types depending on the shape of curves. In type I isotherm, presented in Fig. 8.6, there occurs a maximum lying in the middle range of concentration. The maximum is usually not pronounced. In type II shown in Fig. 8.7, a more pronounced maximum occurs at a relatively lower range of concentra-

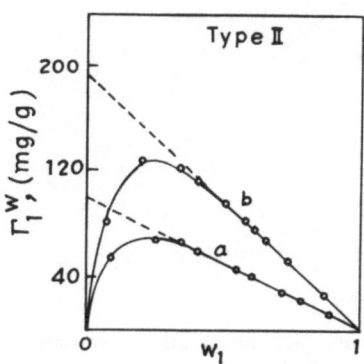

FIGURE 8.7. Plot of Γ_1^w against W_1 for adsorption of benzene (component 1) + n-heptane (component 2) on (a) alumina gel; (b) silica gel at 25°C. (Redrawn from Ref. 7.)

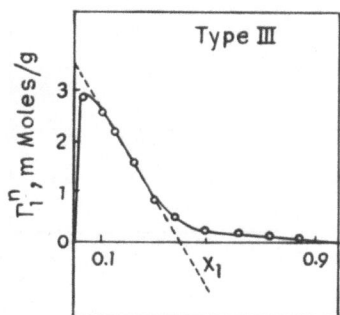

FIGURE 8.8. Plot of Γ_1^n against X_1 for adsorption of ethanol (component 1) + water (component 2) on charcoal at 25°C. (Redrawn from Ref. 7.)

tions. In the higher range of concentrations, Γ_1^w falls off sharply and linearly with X_1. Adsorption isotherms observed in molecular sieves are often of this type. In type III, shown in Fig. 8.8, there exists an inflection and the maximum occurs at relatively lower range of concentration. Beyond the maximum, the curve is linear in a certain range of concentration. At higher values of X_1 at the end, the isotherm tends asymptotically to zero. In type IV, presented in Fig. 8.9 also, there appears a maximum which is followed by a linear portion valid in wide regions of X_1. However, the value of Γ_1^n becomes zero and even negative when X_1 is quite high and finally it becomes zero again when X_1 becomes unity. A zero value thus appearing in the curve across the X_1 axis corresponds to an adsorption azeotrope. Type V is similar to type IV except that in the former the isotherm is nonlinear everywhere (vide Fig. 8.10). There is no straight portion near the azeotropic point.

The adsorption of a binary mixture of alcohol and benzene by solid adsorbent montmorillonite yields type II isotherm.[12] If the surface of this solid is modified by treatment with hexadecyl pyridinium ions, the adsorption isotherm for the alcohol–benzene mixture becomes of the type IV. Theoretical aspects of the change in shape have been discussed by Dekany *et al.*[13]

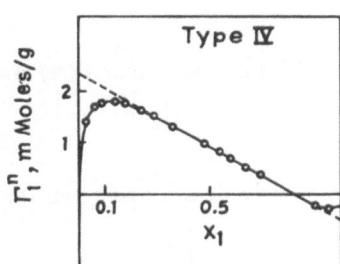

FIGURE 8.9. Plot of Γ_1^n against X_1 for adsorption of benzene (component 1) + ethanol (component 2) on charcoal at 25°C. (Redrawn from Ref. 7.)

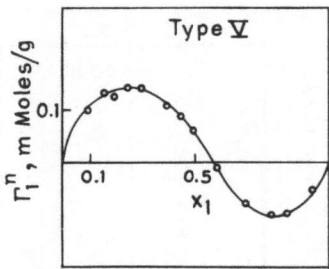

FIGURE 8.10. Plot of Γ_1^n against X_1 for adsorption of 1,2-dichloroethane (component 1) + benzene (component 2) on charcoal at 25°C. (Redrawn from Ref. 7.)

8.4. APPARENT EXCESS AND THE GIBBS EXCESS

Apparent excess Γ_2^n is measured from the difference in the mole fraction of component 2 before and after adsorption. Equation (8.11) defining Γ_2^n operationally can be written in the form

$$\Gamma_2^n = n_1^t X_1 \left(\frac{X_2^t - X_2}{X_1 X_1^t} \right)$$

$$= n_1^t X_1 \left(\frac{X_1 - X_1^t}{X_1 X_1^t} \right) \tag{8.13}$$

On furher algebraic simplification,

$$\Gamma_2^n = n_1^t X_1 \left[\left(\frac{1}{X_1^t} - 1 \right) - \left(\frac{1}{X_1} - 1 \right) \right]$$

$$= n_1^t X_1 \left(\frac{n_2^t}{n_1^t} - \frac{n_2}{n_1} \right)$$

$$= X_1 \left(n_2^t - n_1^t \frac{n_2}{n_1} \right) \tag{8.14}$$

Combining equations (8.4) and (8.14),

$$\Gamma_2^n = X_1 \Gamma_2^1 \tag{8.15}$$

Similarly, it can be shown that

$$\Gamma_1^n = X_2 \Gamma_1^2 \tag{8.16}$$

Combining equations (8.6), (8.15), and (8.16) one obtains

$$\Gamma_1^n + \Gamma_2^n = 0 \tag{8.17}$$

Apparent excesses are thus related to the Gibbs excesses through equations (8.15) and (8.16). The positive or negative sign of the Gibbs excess and the apparent excess is also the same. The magnitudes of Γ_1^n and Γ_2^n are, however, the same. The magnitudes of Γ_1^2 and Γ_2^1 are obviously different on the basis of equation (8.6). Thus Γ_1^n versus X_1 plots become the same as $(-\Gamma_2^n)$ vs. $(1 - X_2)$ plots in Fig. 8.5. Equations (8.15), (8.16), and (8l17) have been deduced by Schay[7] rigorously using the algebraic method of analysis of the Gibbs adsorption equation.

Combining equations (8.9) and (8.10) with equations (8.15) and (8.16), one finds

$$\Gamma_1^n = \Delta n_1 X_2 - \Delta n_2 X_1 \tag{8.18}$$

Similarly,

$$\Gamma_2^n = \Delta n_2 X_1 - \Delta n_1 X_2 \tag{8.19}$$

An alternative algebraic derivation of these equations has earlier been presented by others.[1]

These two equations can further be written in the linear form

$$\Gamma_1^n = \Delta n_1 - (\Delta n_1 + \Delta n_2)X_1 \tag{8.20}$$

$$\Gamma_2^n = \Delta n_2 - (\Delta n_1 + \Delta n_2)X_2 \tag{8.21}$$

Equation (8.20) further indicates that the two unknown quantities Δn_1 and Δn_2 can be evaluated by this single equation, if by chance, the plot of Γ_1^n versus X_1 becomes linear in a certain range of concentrations. Such types of linear plots have been actually observed,[7] in Figs. 8.8 and 8.9. The absolute composition of the interfacial phase in terms of Δn_1 and Δn_2 thus remains unchanged when X_1 is varied in this linear region. All these approaches are thus consistent with those presented for the adsorption at the liquid interface and discussed in Chapter 3.

Using similar types of algebraic operations, equation (8.12) can be converted to the form

$$\Gamma_2^1 = \frac{\Gamma_2^w}{W_1 M_2} \tag{8.22}$$

Thus the Γ_1^n versus X_1 plot in Fig. 8.5 is the same as the $-\Gamma_1^n$ versus $(1 - X_2)$. The value of Γ_2^w based on the experimental data is related to the Gibbs excess Γ_2^1 through equation (8.22). Combining equations (8.10) and (8.22), it can be further shown that

$$\Gamma_2^w = M_2 \Delta n_2 - (M_1 \Delta n_1 + M_2 \Delta n_2) W_2 \qquad (8.23)$$

Similar equation may be obtained for Γ_1^w. In Fig. 8.7, plots of Γ_1^w are observed to be linear when W_1 is high so that values of Δn_1 and Δn_2 can be calculated from the slopes and the intercepts.

8.5. ADSORPTION AZEOTROPE AND MONOLAYER ADSORPTION

At the point of the formation of the adsorption azeotrope, Γ_1^n becomes zero so that according to equation (8.19)

$$\frac{\Delta n_2}{\Delta n_1} = \frac{X_2}{X_1} \qquad (8.24)$$

Although the overall mole ratio composition of the interfacial phase at this state becomes equal to that of the bulk phase, the different regions of the surface phase may not be regarded as homogeneous (vide Section 3.3). In type IV isotherm shown in Fig. 8.9, the straight line portion of the curve includes the azeotropic point. This means that $\Delta n_2/\Delta n_1$ is constant but X_2/X_1 varies with change of X_1 until (X_2/X_1) becomes equal to the mole ratio composition of the interfacial phase. In Fig. 8.10, the azeotropic point lies on the nonlinear curve. This means that $\Delta n_2/\Delta n_1$ as well as X_2/X_1 vary with change of X_1 until they become equal to each other.

If the molecules of components 1 and 2 together can form a monolayer at the interfacial phase of the solid, then one can write an independent equation (8.25) simply from a geometric consideration:

$$A_m = N(\sigma_1 \Delta n_1 + \sigma_2 \Delta n_2) \qquad (8.25)$$

Here A_m is the surface area per gram of the solid powder. σ_1 and σ_2 are the effective areas of the surface of the solid covered by molecules of components 1 and 2, respectively. Estimates of σ_1 and σ_2 can be made from the cross-sectional area of the molecule and also on the basis of certain assumptions about the molecular orientations and packing in the interfacial phase.[7] A_m for charcoal thus estimated from equation (8.25) using the values of Δn_1 and Δn_2 obtained in the region of the linear plot in Fig. 8.8 is 756 m^2/g. This compares very well with the BET value $760 \, \text{m}^2/\text{g}$ for the same charcoal used in these experiments. From the linear plot in Fig. 8.9, the surface area of another charcoal sample has been calculated to be $492 \, \text{m}^2/\text{g}$ whereas the BET surface area of the respective charcoal was $480 \, \text{m}^2/\text{g}$. Similar agreements are reported by Schay[2] for many other solid powders giving isotherms of various types in the presence of different liquid mixtures (vide Table 8.1).

One can logically conclude that even in the nonlinear regions of the plots shown in Figs. 8.6–8.10, equation (8.25) based on the monolayer concept of the interfacial phase may remain valid. As pointed out earlier, in this nonlinear region, Δn_1 and Δn_2 are not constant but they alter with variation of X_1 or X_2. However, solving two simultaneous and independent equations (8.18) and (8.25) for Δn_1 and Δn_2, it can be shown that

$$\Delta n_1 = \frac{(X_1 A_m - N\sigma_2\Gamma_2^n)}{N(\sigma_2 X_1 + \sigma_2 X_2)} \qquad (8.26)$$

and

$$\Delta n_2 = \frac{(X_2 A_m - N\sigma_1\Gamma_1^n)}{N(\sigma_1 X_1 + \sigma_2 X_2)} \qquad (8.27)$$

Mole fractions X_1^s and X_2^s at the interfacial phase can be expressed by

$$X_1^s = \frac{(X_1 A_m - N\sigma_2\Gamma_2^n)}{A_m - N(\sigma_1\Gamma_1^n + \sigma_2\Gamma_2^n)} \qquad (8.28)$$

and

$$X_2^s = \frac{(X_2 A_m - N\sigma_1\Gamma_1^n)}{A_m - N(\sigma_1\Gamma_1^n + \sigma_2\Gamma_2^n)} \qquad (8.29)$$

TABLE 8.1. Surface Area of the Solid by the BET Method and by Adsorption from Solution[a]

Adsorbent	Solution mixture	A_m (BET) (m^2/g)	A_m (adsorption) (m^2/g)
Silica gel	Toluene + n heptane	540	560
Silica gel	Benzene + n heptane	450	460
Alumina gel	Benzene + n heptane	240	236
Charcoal	Ethanol + benzene	612	587
Charcoal black	Ethanol + benzene	68	72

[a] Reference 2.

One can calculate values of X_1^s and X_2^s from the knowledge of Γ_1^n, Γ_2^n, and A_m obtained from experiments. Values of σ_1 and σ_2 are to be selected judiciously, and further one has to assume that orientations of the molecules in the interfacial phase do not alter with change of X_1 or X_2. Other ways of calculating X_1^s and X_2^s have also been discussed.[1,7]

In several cases of the studies of the adsorption at solid interfaces, values of A_m evaluated from equation (8.25) do not agree with that obatined from the BET plot.[2,7] In some of these cases, chemisorption mixed with physical adsorption appears to be the cause of such discrepancies. The limitations of the monolayer model as a concept has also been discussed.[2,7,14-15] The short-comings of the monolayer model and its inconsistency with the Gibbs adsorption isotherm at the liquid interface have been examined by Defay and Prigogine.[14] They have shown that the discrepancy may be eliminated if the surface is imagined to possess two layers of molecules.

8.6. ADSORPTION AND ADSORPTION ISOTHERM

In the case of adsorption of a single gas by a solid powder, only one component is accumulated at the interface in absolute amount at a constant temperature and pressure. This accumulation may be regarded as the absolute adsorption and it is directly measurable from the experiments. When adsorption by solids in the presence of binary liquid mixture is experimentally measured using equations (8.5) or (8.11), evaluation of the extent of such absolute adsorption defined by Δn_1 and Δn_2 is difficult indeed. Such evaluation can be obtained from the linear plot of the data in accordance with equations (8.10) or (8.21).

In the nonlinear region, such evaluation is difficult since we then have only a single independent equation containing two unknown and variable quantities. If one assumes the validity of the monolayer model, then another independent equation (8.25) relating Δn_1 and Δn_2 and a measurable quantity A_m will be available so that combining this with equation (8.21), values of absolute adsorption for the individual component can be evaluated.

Williams[1,16] originally suggested a direct method through which experimental data can be related to Δn_1 and Δn_2 with the help of a second independent relation. This has been further developed by Kipling *et al.*[1,17,18] and others.[19,20] In the experimental setup followed by these workers (vide Section 2.6.1), the solid powder is suspended in the vapor of the liquid mixture until adsorption equilibrium is reached. The total weight w of the mixture of the two components per gram of the solid is measured at this stage so that

$$w = M_1 \Delta n_1 + M_2 \Delta n_2 \tag{8.30}$$

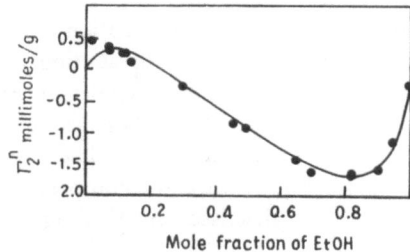

FIGURE 8.11. Composite isotherm for adsorption by carbon from mixtures of benzene and ethyl alcohol; points are experimental. (Redrawn from Ref. 17.)

Neglecting the presence of small amounts of the components in the vapor phase, another independent relation of the type (8.18) can also be obtained from the Gibbs treatment. Combining this equation with relation (8.30) the absolute adsorption of ethanol and benzene by charcoal at different mole fraction of ethanol have been determined. These are shown in Fig. 8.11. Kipling and Tester[17] further showed that this kind of adsorption occurring in the vapor phase is consistent with the Langmuir adsorption isotherm for mixed gases and also with the isotherm obtained when the adsorption by the solid takes place in direct contact with the binary liquid mixture.

Since experimental determinations of Δn_1 and Δn_2 are difficult and inconvenient, adsorption from all kinds of binary liquid mixtures are in general expressed by the Gibbs excesses Γ_1^2 and Γ_2^1, or the Gibbs apparent excesses Γ_1^n and Γ_2^n. These excesses may become positive or negative following equations (8.6) and (8.17), respectively. The absolute adsorptions Δn_1 and Δn_2 as defined by the respective equations (8.7) and (8.8) can never be negative. The different terms for excess adsorption are consistent with the thermodynamic concept involved in the derivation of the Gibbs adsorption equation. Earlier, positive excess adsorption from binary liquid mixture was termed preferential or selective adsorption.[21]

The curve obtained from the plot of absolute adsorption of a gas against pressure varying between zero and higher values at a constant temperature is called an adsorption isotherm at that temperature. The adsorption isotherms for individual components in the binary vapor mixture are shown in Fig. 8.12.

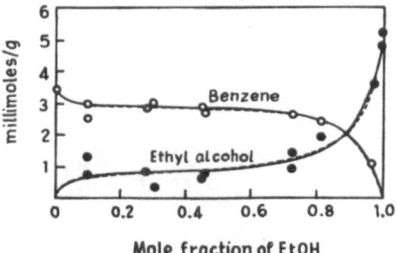
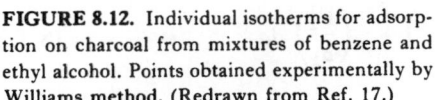

FIGURE 8.12. Individual isotherms for adsorption on charcoal from mixtures of benzene and ethyl alcohol. Points obtained experimentally by Williams method. (Redrawn from Ref. 17.)

In the case of direct adsorption of one of the components from a binary liquid mixture, the curves representing variation of Γ_2^1 or Γ_2^n against X_1 or m_2 (or c_2) at constant temperature (vide Figs. 8.6–8.10) are also frequently termed adsorption isotherms. These are referred to sometimes as composite isotherms since the excess adsorption is composed of both Δn_1 and Δn_2 [vide equation (8.19)]. The features of composite isotherms for liquid mixtures have been extensively discussed by Kipling and others.[1]

8.7. SURFACE ACTIVITY COEFFICIENT AT THE SOLID–LIQUID INTERFACE

In Section 3.17, we discussed the procedure of Schay and co-workers for the calculation of the surface activity coefficients of binary liquid mixture at the air–liquid interface with the help of equations (3.122) and (3.123). Here the surface mole fractions X_1^s and X_2^s have been calculated using equations (3.103) and (3.104) from Γ_1^2 and Γ_2^1, which in turn can be indirectly calculated from the measurement of surface tension at different partial vapor pressures of the liquid mixtures.

In the case of the adsorption of the liquid mixture by powdered solids, excess moles Γ_1^2 and Γ_2^1 adsorbed per gram of the solid are directly measurable. Since specific surface area A_m per gram of the solid powder can also be measured by an independent experiment, the surface excesses of components 1 and 2 per unit interfacial area will be equal to Γ_1^2/A_m and Γ_2^1/A_m, respectively. X_1^s and X_2^s can be estimated using equations (8.28) and (8.29). The equations (3.122) and (3.123) of Schay can also be used for the calculation of f_1^s and f_2^s at the solid–liquid interface provided one can estimate $(\gamma - \gamma_1^0)$ and $(\gamma - \gamma_2^0)$. Direct measurement of these quantities for the solid–liquid interface is difficult to achieve, but their values can be indirectly obtained from the integration of the Gibbs equation, so that[7]

$$\gamma - \gamma_1^0 = \int_{\gamma_1^0}^{\gamma} d\gamma$$

$$= -\left(\frac{RT}{A_m}\right) \int_1^{a_1} \left(\frac{\Gamma_1^2}{a_1}\right) da_1 \tag{8.31}$$

and similarly,

$$\gamma - \gamma_2^0 = -\left(\frac{RT}{A_m}\right) \int_1^{a_2} \left(\frac{\Gamma_2^1}{a_2}\right) da_2 \tag{8.32}$$

The graphical integrations may be carried out on the basis of the measured

TABLE 8.2. Adsorption of Water (Component 1) and Ethanol (Component 2) on Charcoal at 25°C; $A_m = 750 \text{ m}^2/\text{g}^a$

X_2	$\gamma - \gamma_2^0$ (mN/m)	f_2^s	$\gamma - \gamma_1^0$ (mN/m)	f_1^0
0.05	9.96	0.815	13.9	1.20
0.10	4.84	0.974	19.2	1.11
0.20	1.61	1.310	22.6	0.97
0.30	0.84	1.411	23.2	0.97
0.40	0.60	1.371	23.7	0.93
0.50	0.49	1.260	23.8	1.02
0.70	0.26	1.053	23.8	1.34
0.80	0.15	1.020	23.8	1.50

^{a} Reference 7.

values Γ_1^2 and Γ_2^1 for various values of the bulk activities, a_1 and a_2 so that $(\gamma - \gamma_1^0)$ and $(\gamma - \gamma_2^0)$ in equations (8.31) and (8.32), respectively may be computed. Values of f_1^s and f_2^s calculated by Schay *et al.*[7] for the alcohol–water mixtures in the presence of solid alumina powder are presented in Table 8.2.

One may note with interest in Tables 3.7 and 8.2 the significant differences in the values of f_1^s and f_2^s for the air–water and alumina–water interfaces. It seems that the nonaqueous phase can influence the interaction between the two liquid components at the interfacial phase to different extents. Comparisons of the activity coefficients in a quantitative manner will be questioned unless the standard chemical potentials appearing in equations (3.118) and (3.119) are proved to be the same, for two types of surfaces.

Everett[22–24] has made a calculation of the surface activity coefficient on the assumption that the exchange reaction between interfacial monolayer and the bulk phase takes place according to the chemical reaction

Component 1 (liquid) + component 2 (surface) $\rightleftharpoons$

component 1 (surface) + component 2 (liquid)   (8.33)

if $\sigma_1 \simeq \sigma_2$, the equilibrium constant for such reaction can be calculated from the equation

$$K' = \frac{(f_1^s X_1^s)(f_2 X_2)}{(f_2^s X_2^s)(f_1 X_1)} \tag{8.34}$$

If one further assumes that $(f_2 f_1^s / f_1 f_2^s)$ is close to unity then

$$X_1^s = \frac{K' X_1}{X_2 + K' X_1} \tag{8.35}$$

Further, equation (8.20) can be written in the form

$$\Gamma_1^n = (X_1^s - X_1)(\Delta n_1 + \Delta n_2) \tag{8.36}$$

Eliminating X_1^s between equations (8.35) and (8.36)

$$\left(\frac{X_1 X_2}{\Gamma_1^n}\right) = \frac{1}{\Delta n_1 + \Delta n_2}\left(X_1 - \frac{1}{1 - K'}\right) \tag{8.37}$$

so that from the linear plot, both $(\Delta n_1 + \Delta n_2)$ and K' can be calculated.

Also from the statistical mechanical approach, Everett[22-24] has shown that

$$\ln f_1 = X_2^s \ln\left[\frac{X_2^s}{X_1^s} \cdot \frac{a_1}{a_2}\right] - \int_0^{X_2^s}\left[\frac{X_2^s}{X_1^s} \cdot \frac{a_1}{a_2}\right] dX_2^s \tag{8.38}$$

From the known values of X_1^s, X_2^s, a_1, and a_2, f_1^s can be calculated. One can calculate f_2^s in a similar manner. It is also mentioned that the equation is consistent with the Butler equation as well as the Gibbs concepts of the surface phase. The values of f_1^s and f_2^s calculated by the models proposed by Schay and Everett, respectively, are expected to differ since the standard states for these coefficients are possibly not the same. More detailed experimental work is needed to clarify the position of the standard reference states of activities. An attempt should also be made to calculate the surface excess entropy for solid–liquid interfaces using equation (3.125).

8.8. ADSORPTION FROM DILUTE SOLUTION

We shall now consider the adsorption in the presence of powdered solid when the binary solution is relatively dilute with respect to one of the components. The density ρ of the dilute binary solution may be regarded as being the same as that of the pure solvent. The nonionic solutes used in many of these investigations are stearic acid and other long-chain fatty acids, ethyl stearate, decyl amine, malonic and sebacic acids, phenol, resorcinol, iodine, etc. These solutes may be dissolved in various solvents such as water, carbon tetrachloride, alcohol, cyclo-hexane, benzene, dioxan, and ether. Solid adsorbents may be charcoal, metal powders, alumina, resin, silica, etc. The study of adsorption in dilute solution has been extensively reviewed by Kipling.[1]

The extent of adsorption of the solute from dilute solution is calculated from the experimental data using the equation

$$\Gamma_2^l = \frac{V^l}{1000}(c_2^l - c_2) \tag{8.39}$$

where c_2^i and c_2 are molar concentrations of the solute in solution before and after adsorption. Here V^t is the total volume of the liquid in contact with a unit weight of the solid powder. Since the solution is dilute, c_2^i, c_2, and V^t may be replaced by $m_2^i \rho$, $m_2 \rho$, and w_1^i / ρ, respectively, so that equation (8.39) will assume the form of equation (8.5). Γ_2^l is therefore consistent with the concept of the Gibbs excess as expressed by equations (8.4) and (8.10). For a dilute solution, X_1 is close to unity so that $\Gamma_2^n = \Gamma_2^l X_1 \simeq \Gamma_2^l$.

Most of the solutes used in the studies of adsorption in dilute solution possess relatively strong affinities for the solid surface, whereas the affinities of the solvents for the surface are relatively weak. Under these circumstances, $(\Delta n_2 / \Delta n_1) \gg (X_2 / X_1)$ so that from equation (8.10)

$$\Gamma_2^l \simeq \Delta n_2 \qquad (8.40)$$

In most of the studies of adsorption in dilute solution, Γ_2^l increases with increase of c_2 at low solute concentration. At relatively higher values of c_2, Γ_2^l in many cases tends to reach a limiting steady value Γ_2^m which does not increase with further increase of c_2. In a few exceptional cases, the extent of adsorption may decrease or increase from Γ_2^m when c_2 is quite high.[1]

The interfacial phase of the solid in the presence of a dilute solution is usually regarded as monomolecular so that equations (8.25) and (8.33) will remain valid. With increase of c_2 from zero, Δn_2 (or Γ_2^l) will increase due to the positive accumulation of the solute component in the monolayer by displacement of the solvent molecules from the interfacial phase. When Γ_2^l reaches the value Γ_1^m, it is believed that the solvent is completely displaced from the interfacial phase so that Δn_1 is equal to zero. Then from equation (8.25), we obtain

$$A_m = N\sigma_2 \Gamma_2^m \qquad (8.41)$$

Inserting a reasonable value of σ_2, the magnitude of the specific surface area A_m of solid powder can be easily calculated from the value of Γ_2^m obtained directly from the adsorption experiments. The specific surface areas of powdered alumina and many other metallic oxides have been determined[25–27] in this manner from the adsorption of stearic or palmitic acids dissolved in the benzene medium assuming value of σ_2 to be 0.20 nM^2/molecule. A_m calculated in this manner for alumina agrees with that obtained from the BET method based on the gas adsorption.

Kipling *et al.*[28] and others have pointed out that for many systems, definite conclusions about the value of A_m on the basis of equation (8.41) cannot be achieved because of the difficulties of the interpretation of Γ_2^m and σ_2. The adsorbed long-chain fatty acid may remain in parallel, perpendicular, or inclined orientation with respect to the surface plane so that the fixation of correct value of σ_2 may become difficult.[1] In the case of adsorption of malonic acid in aqueous solution[1,29] by spheron 6, Γ_2^1 reaches a maximum value Γ_2^m and then it falls linearly with further increase of the solute concentration. This decrease indicates that Δn_1 is not zero when Γ_2^1 is equal to Γ_2^m so that A_m cannot be calculated using equation (8.41). The slope and intercept of the linear portion of the adsorption isotherm may lead to the evaluation of the values of Δn_1 and Δn_2 according to equation (8.21).

Stearic acid from benzene solution can be adsorbed on the surface of powdered steel[30] forming two layers. The first layer in contact with the surface is believed to be formed by the chemisorption, whereas the second layer on the top of the first layer is accumulated by the physical adsorption. Equation (8.41) will not remain valid in such situation. In the case of adsorption of iodine dissolved in dioxan or benzene by powdered magnesia, there exists more than one apparent state of saturation indicated by the presence of several well-defined steps in the isotherm. The different parts of the isotherm may correspond to excess accumulation on the chemically different sites on the solid surface which differ from each other in terms of adsorption energy.[1,31] In the case of adsorption of benzoic acid dissolved in water by graphon powder,[32] Γ_2^1 increases with increase of c_2 without reaching the limiting value even near the state of bulk saturation of the liquid. The sigmoid shape of the isotherm suggests the formation of multilayers of the acid at the interface. According to Kipling and Wright,[33] this corresponds to incipient crystallization of the solute at the interface.

8.9. LANGMUIR ADSORPTION ISOTHERM

For relatively dilute solution, the surface activity coefficient in equation (8.34) based on the monolayer concept of the interfacial phase may be put equal to unity so that K (equal to $1/K'$) for the reverse reaction in equation (8.33) reads

$$K = \frac{X_2^s}{X_1^s} \cdot \frac{a_1}{a_2} \tag{8.42}$$

Since the amount of solvent component 1 is larger, a_1 may be taken constant even when a_2 is varied. We shall put (K/a_1) to be equal to b, where b is a constant. Further X_1^s is equal to $(1 - X_2^s)$. Inserting all these in equation (8.42), we obtain

$$X_2^s = \frac{ba_2}{1 + ba_2} \tag{8.43}$$

If one assumes that the respective areas occupied by the molecules of the components 1 and 2 are equal, then σ_1 is equal to σ_2. After replacement of Δn_2 by Γ_2^1 in equation (8.25), therefore,

$$A_m = (\sigma_2 N \Gamma_2^1)/X_2^s \tag{8.44}$$

Combining equations (8.43) and (8.44), one obtains the following equation:

$$\Gamma_2^1 = \frac{(A_m b/N\sigma_2)a_2}{(1 + ba_2)} \tag{8.45}$$

In dilute solution, a_2 may be replaced by c_2. This is the appropriate form of the Langmuir equation for the adsorption of solute in solution by solid particles derived by Everett and others.[4,22–24] In the linear form, equation (8.45) can be written as

$$\frac{1}{\Gamma_2^1} = \left(\frac{N\sigma_2}{A_m b}\right)\frac{1}{c_2} + \left(\frac{N\sigma_2}{A_m}\right) \tag{8.46}$$

Thus if the plot of the experimental data of $(1/\Gamma_2^1)$ against $1/c_2$ is linear, then b and A_m can be calculated from its slope and the intercept provided σ_2 is known . Without the knowledge of Γ_2^m, therefore, A_m for a solid can be estimated provided the adsorption data fit the Langmuir equation. Further, assuming the activity of the solvent to be unity, b thus evaluated from the Langmuir plot can be identified with the equilibrium constant (K) for the reaction (8.33) for the monolayer adsorption. Since, the standard free energy change ΔG^0 for a reaction is equal to $(-RT \ln K)$, the value of the free energy of adsorption ΔG^0 can also be calculated from the Langmuir plot. This feature will be further discussed in a later section of this chapter.

When the concentration of the solute in solution is very low, $ba_2 \ll 1$ so that the plot of Γ_2^1 versus c_2 (or a_2) according to equation (8.45) will be linear (vide Fig. 8.13, curves L-1 and L-2). At higher concentrations, $(ba_2) \gg 1$, so that Γ_2^1 (sometimes referred as Γ_2^m) becomes equal to a constant value $(A_m/N\sigma_2)$. (Fig. 8.13, L-2). In the intermediate region of concentration, Γ_2^1 varies with c_2 in nonlinear fashion.

The shape of the adsorption isotherm obtained from experiments may deviate from that expected from the Langmuir equation for various reasons. Different shapes of the isotherms have been classified appropriately by Giles et al.[1,34] which are presented in Fig. 8.13. The Langmuir class in this figure is represented by L-types of curves. In L-1 curve the saturation state is not achieved from experimental measurement. In the L-3 type, there is indication that the interfacial phase tends to become polymolecular at higher values of c_2. In L-4 isotherm, two (or even more) sharp discontinuities are observed, the reason for which has been described in an earlier section. In mx-L type,

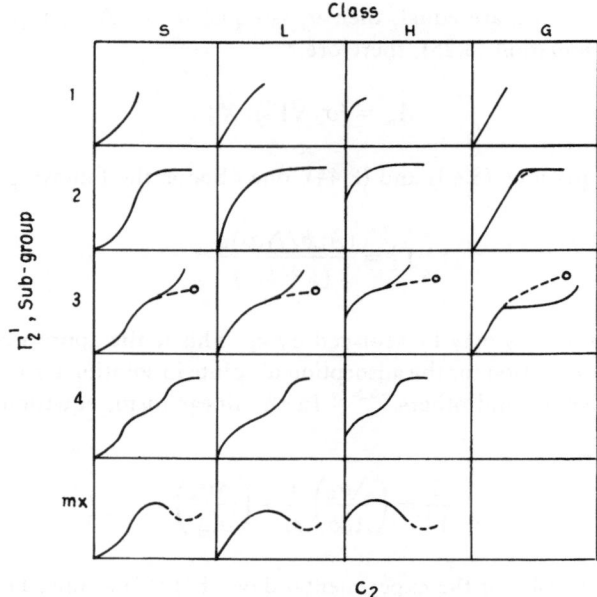

FIGURE 8.13. Classification of the different systems of isotherms. (Redrawn from Ref. 34.)

the isotherm exhibits a maximum sometimes followed by a minimum. One of the reasons for this is the incomplete displacement of water in the interfacial phase by the adsorbate at high value of c_2.

The adsorption isotherm forming the S-curve in Fig. 8.13 may be obtained as a result of the strong intermolecular attraction within the adsorbed layer. Strong adsorption of the solvent may be another reason for this kind of shape. The shape of the H-curve is caused from the high affinity between the adsorbate and adsorbent exhibited in very low range of solute concentrations. The shape may occur when solute is accumulated at the interface by chemisorption. The G-curve indicates that the solute is partitioned between the bulk solution and the interfacial phase in constant proportion so that (Γ_2^1/c_2) remains constant. This occurs in the adsorption phenomena in the presence of textile fibers.

8.10. ADSORPTION OF INORGANIC IONS AND ORGANIC DYES

Many inorganic precipitates are known to adsorb preferentially inorganic cations and anions.[35] Thus precipitates of silver iodide adsorb extensively silver or iodide ions from aqueous solutions containing either silver or iodide ions, respectively, whereby positively or negatively charged colloids of AgI are prepared. The colloid will be stable if the surface charge of the solid acquired by adsorption is high and the concentration of the neutral salts in the solution is reduced by exhaustive dialysis. Electrical double layer of the Stern–Gouy type is believed to be formed at the interfacial region of the solid (vide Chapter 4). The stability of the silver iodide and other similar colloidal systems has been explained widely on the basis of the theory of the interaction of double layer formulated by Deryaguin and Landeau[36] and also by Verwey and Overbeek.[37] A simple picture of the colloid stability on the basis of the critical zetapotential theory has also been proposed.[38,39]

The study of adsorption of cationic and anionic organic dyes is of considerable commercial importance in the textile industry. The subject has been extensively reviewed.[1,40–42] In Fig. 8.14, adsorption isotherms for the anionic dye chlorazal Sky Blue FF (CSBFF) on negatively charged viscose rayon[43] at a fixed concentration of NaCl and KCl are presented. At a given value of c_2, the difference in Γ_2^1 for various concentrations of NaCl and KCl indicates the significant role of hydration of adsorbents and adsorbates in shaping their interactions with each other at the interface. Iyer *et al.*[43] have also studied adsorption of this dye on rigid and hydrophobic solid graphon. The isotherms in all cases (with correction for the Donnan effect for electrical interaction)

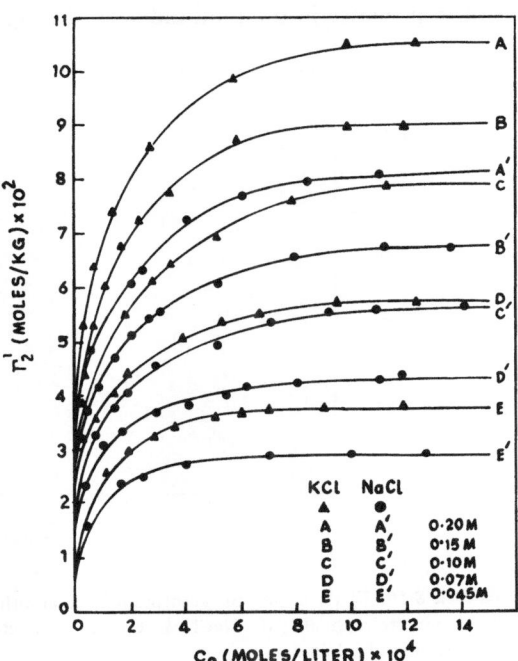

FIGURE 8.14. Plot of Γ_2^1 vs. c_2 for adsorption of CSBFF on viscose at different concentrations of NaCl and KCl at 50°C. (Redrawn from Ref. 43.)

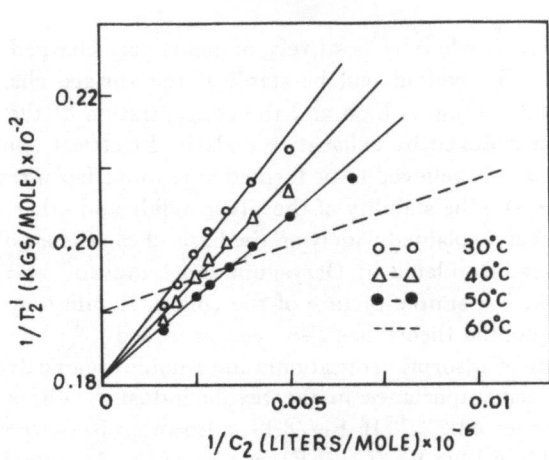

FIGURE 8.15. Langmuir plots of $1/\Gamma_2^l$ vs. $1/c_2$ for the adsorption of CSBFF on graphon at different temperatures. NaCl 0.13 M. (Redrawn from Ref. 43.)

are found to fit the Langmuir adsorption equation. The linear double reciprocal plots according to equation (8.46) are shown in Fig. 8.15. This leads to the evaluation of A_m and K from the slope and intercept of the plots.

Adsorption of cationic and anionic dyes by insoluble proteins like casein is of importance in the food industry. In Fig. 8.16, the adsorption isotherms

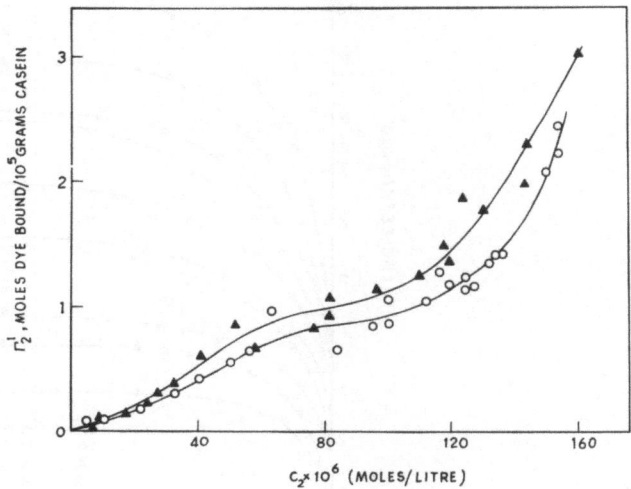

FIGURE 8.16. Γ_2^l vs. c_2 plot for adsorption of amaranth dye on casein surface. pH, 6.0; temp. 27°C, ionic strength, 0.1(M) NaCl. ▲, Casein with fat; ○, Defatted casein. (Das, Roy, and Chattoraj, unpublished data.)

for cationic and anionic dyes on caseins indicates deviation from the shape expected from the Langmuir adsorption isotherm.

8.11. ADSORPTION OF IONIC DETERGENTS

Adsorption of cationic and anionic detergents by powdered solid particles such as alumina, $BaSO_4$, bioglas, silica, polystyrene, etc. has been extensively studied as functions of the concentrations of the amphiphiles, pH, ionic strength and temperature below and above cmc.[44–53] Such investigations are useful in the field of detergency and tertiary oil recovery processes.[54–56] In Fig. 8.17, isotherms for the adsorption of dodecyl ammonium chloride (DAC), dodecyl pyridinium bromide (DPB) and sodium dodecyl sulfate (SDS) by powdered alumina suspended in the aqueous media have been presented. These isotherms are S-shaped, a feature that has been discussed in an earlier section. Close to cmc, Γ_2^1 reaches a limiting value whose magnitude depends on pH, ionic strength, nature and charge of the adsorbents and adsorbates.[49] Adsorption is thus significantly controlled by the electrostatic effect.

From the study of adsorption and electrokinetic potentials of the quartz particles in the presence of a homologous series of long-chain quaternary ammonium salts, Somasundaran *et al.*[52,53] have demonstrated the formation of hemi-micelles or two-dimensional aggregates of the amphiphiles in the

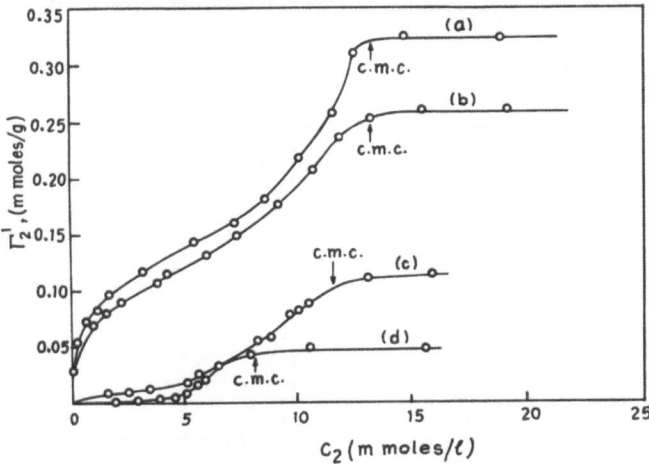

FIGURE 8.17. Adsorption of detergents from aqueous solution by solid alumina powder, (a) dodecyl ammonium chloride at 20°C, (b) dodecyl ammonium chloride at 40°C, (c) dodecyl pyridinium bromide at 20°C, (d) sodium dodecyl sulfate at 20°C. (Redrawn from Ref. 49.)

interfacial phase of the solid–liquid system. The thermodynamic analysis of the data in terms of the formation of the Stern–Gouy type of the electrical double layer at the interface leads to a value 580 J (140 calories) per mole for the transfer of CH_2 group from monomer to the hemi-micellar regions in the interfacial phase. This is consistent with the thermodynamic concept of the micelle formation in the bulk phase.

Mukherjee and Anavil[51] have studied adsorption of cationic detergent hexadecyl pyridinium bromide (HDPB) on powdered bioglas in the presence or absence of neutral salt. Here the adsorbent is a particular variety of glass having a network of rigid pores so that A_m is always very high. The extent of adsorption increases with increase of the concentration of the amphiphile in the solution until it reaches a maximum value near cmc. When the concentration of HDPB is increased far beyond cmc, Γ_2^1 decreases gradually. The decrease of Γ_2^1 beyond cmc is very pronounced when cationic detergent is replaced by anionic detergent such as SDS (sodium dodecyl sulfate) or STDS (sodium tetradecyl sulfate). These will be evident from Fig. 8.18. It is noted with considerable interest that decrease of the adsorption in the case of SDS is so pronounced that Γ_2^1 even becomes negative. Mukherjee *et al.*[51] have suggested that the micelles or hemimicelles remaining in the adsorbed state are gradually desorbed when detergent concentration in the bulk is significantly higher than cmc. This desorption is large when the net charge of the adsorbent and the adsorbate is same in sign.

The maximum in the isotherm (vide Fig. 8.18) followed sometimes by a minimum are frequently observed when ionic amphiphiles are adsorbed on the solid surface.[45–47,51] A detailed theory for the occurrence of these kinds

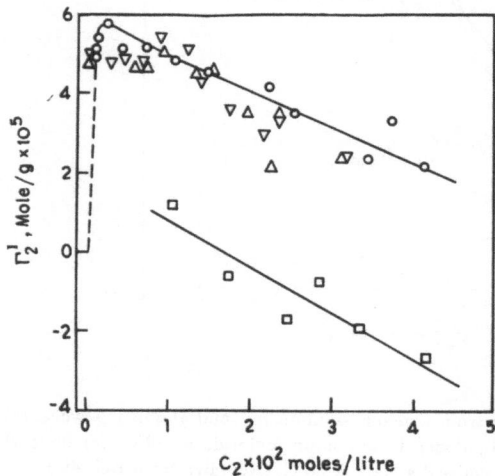

FIGURE 8.18. Adsorption of STDS and SDS to Bioglas 200 (1) at high concentrations and 35°C. O, STDS, △, STDS in 0.01 M NaCl, ▽, STDS in 0.03 M NaCl, and □, SDS. (Redrawn from Ref. 51.)

of inflections based on the presence of trace impurities in the detergent has been presented by Trogus, Schechter, and Wade.[57]

The adsorption of cationic and anionic detergents to protein gels has also been studied extensively by Sadhukhan and Chattoraj.[57a] The surfaces of the solids in these gels are not rigid but flexible. The features of these isotherms of casein gels for the adsorption of quaternary amines are discussed extensively in Section 10.11. The shape and the limiting value of these isotherms are dependent on the pH, ionic strength, and temperature. The general features of the adsorption of amphiphiles on the surface of rigid solid and flexible protein gels are similar.

8.12. ADSORPTION OF NONIONIC POLYMERS

Study of adsorption of nonionic synthetic or natural polymer by powdered solids is of considerable practical importance in pigment dispersion, adhesion, lubrication, surface treatment, membrane technology, etc. Many of these neutral polymers (e.g., synthetic rubber, polystyrene, polyisobutane, etc.) are soluble in nonaqueous solvents like benzene but are practically insoluble in water. The subject has been extensively discussed elsewhere[1,58–62] and only a short discussion on this topic will be presented in this section.

Synthetic polymers are polydispersed in nature. In the precise study of adsorption, the polymers are carefully fractionated. The number and weight average molecular weights of the monodisperse polymer thus prepared can be checked[61] from the osmometric, light-scattering, or other types of physicochemical experiments. At the solid–liquid interface, the adsorbed polymer may assume a large number of conformations, and for this possible reason, a long time (sometimes even a week) is required for the attainment of the adsorption equilibrium.

The extent of adsorption is most conveniently expressed by Γ_2^w [vide equation (8.12)] when the molecular weight of the polymer is not known. For dilute polymer solution, W_1 is always close to unity so that Γ_2^w becomes equal to $\Gamma_2^1 M_2$ according to equation (8.22). In Fig. 8.19, Γ_2^w for several fractions of polystyrene adsorbed on the surface of graphon have been plotted against weight concentration of the polymer in toluene solvent.[61] The shape of the isotherm is similar to that depicted in the H-2 curve in Fig. 8.13. An isotherm of this type is frequently obtained for the polymer adsorption.[1] For the adsorption of polystyrene on graphon, Dunn and Vold[61] have further proposed an empirical equation

$$(\Gamma_2^w)_{max} = 3.0 \times 10^{-2}\sqrt{M_2} + 28 \qquad (8.47)$$

Here $(\Gamma_2^w)_{max}$ is the limiting value of adsorption (vide Fig. 8.19) which can

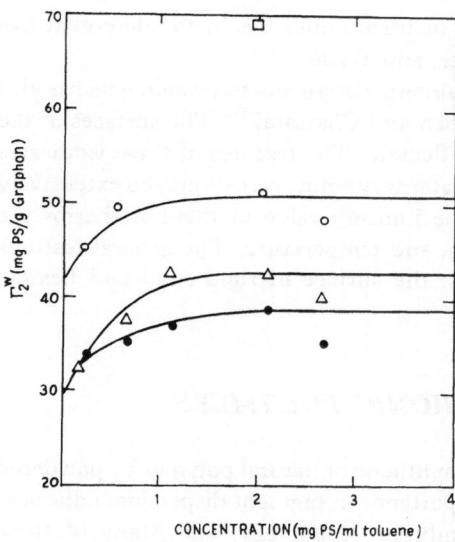

FIGURE 8.19. Plot of Γ_2^w vs. weight concentration for the adsorption of polystyrene (PS) from toluene onto graphon. ●, M.W. 110,000; △, M.W. 200,000; ○, M.W. 498,000; □, 1,800,000. Temperature: 24°C. (Redrawn from Ref. 61.)

be evaluated from the experimental data. $(\Gamma_2^w)_{max}$ thus increases with increase of the average molecular weight of the monodisperse polymer following equation (8.47).

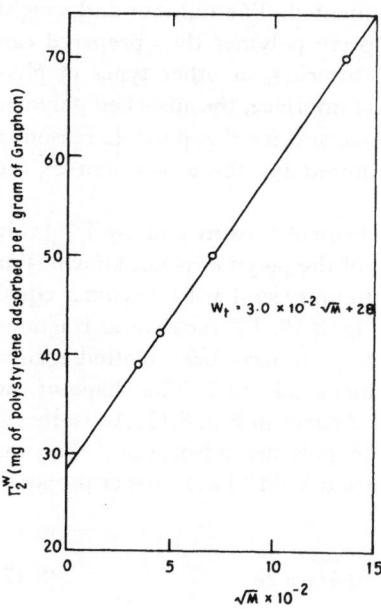

$$W_t = 3.0 \times 10^{-2} \sqrt{M} + 28$$

FIGURE 8.20. Plot of Γ_2^w vs. $\sqrt{M}$ for adsorption of polystyrene on graphon. (Redrawn from Ref. 61.)

If polystyrene lies flat on the surface of graphon, then from the determination of A_m of graphon by the BET method, one can theoretically estimate $(\Gamma_2^w)_{max}$ to be nearly equal to 28 mg per gram assuming an appropriate value of σ_2 for styrene monomer. On the other hand, if the adsorbed molecules of polystyrene assume a randomly coiled conformation, then its radius of gyration R_G becomes proportional to $\sqrt{M_2}$. The linear plot of $(\Gamma_2^w)_{max}$ against $\sqrt{M_2}$ should not have any intercept at all. If the whole polymer chain assumes perpendicular configuration then $(\Gamma_2^W)_{max}$ should vary linearly with M_2 with zero intercept. At the saturated state, therefore, it is proposed[61] that the surface is covered by a fraction of polystyrene molecule lying flat on the solid surface and the remaining part of the same adsorbed molecule may assume the random-coiled conformations (vide Fig. 8.21) so that equation (8.47) will be valid. This dual nature of conformations of various biopolymers at the interface has been advocated by many workers.[1,60,61,63,64] Various physicochemical techniques have been used to determine directly the conformation of the flexible polymer at the solid interface.[60]

The individual isotherms for the adsorption of polymers by powdered solid materials are frequently found to fit[64–66] the Langmuir equation (8.46) in linear form. However, the dependence of $(\Gamma_2^w)_{max}$ on the molecular weight of the polymer obviously indicates that the slope and intercept of this plot cannot give a unique value of A_m unless the Langmuir equation is appropriately modified. The initial slope of each isotherm in Fig. 8.19 is unusually sharp. This represents strong affinities between polymer and solid powder so that equation (8.40) remains valid.

Attempts have initially been made by Simha, Frisch, and Eirich[67–69] to derive from statistical consideration equations relating the extent of adsorption of polymer to its concentration in the liquid in bulk. These equations are found to be suitable for the interpretation of the experimental data to a certain extent.

In an alternative derivation based on the mass-action concept, Frisch *et al.*[70] considered that the interfacial monolayer is composed of solvent and polymer segments of equal size and the transfer of one molecule of polymer (component 2) from the bulk to the surface releases ν number of solvent

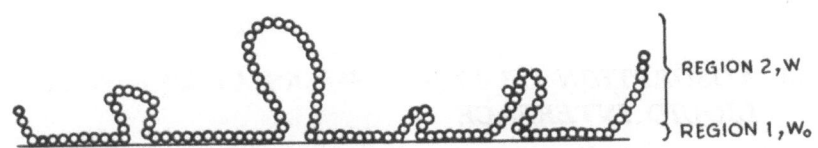

REGION 2, W

REGION 1, W_0

FIGURE 8.21. Hypothetical conformation of adsorbed polystyrene monomer segment. (Redrawn from Ref. 61.)

molecules from the monolayer. The reaction is represented thus:

Component 1 (surface) + component 2 (liquid) $\rightleftharpoons$
    (ν moles   1 mole

$$\text{component 1 (liquid) + component 2 (surface)} \quad (8.48)$$
    (ν moles)   (one mole)

so that

$$K = \left(\frac{X_2^s}{(X_1^s)^\nu}\right) \cdot \left(\frac{(X_1)^\nu}{X_2}\right) \tag{8.49}$$

Here X_1^s and X_2^s are mole fractions of the components at the interfacial phase. If θ stands for the fraction of the surface sites covered by ν polymer segments, then it will be proportional to (νX_2^s) mole fraction of the segments present in the interfacial phase. X_1^s is also proportional to $(1 - \theta)$. $(X_1)^\nu$ is nearly constant for the dilute solution and X_2 is proportional to the (weight) concentration c_2 of the polymer. Equation (8.49) can then be arranged in the form

$$\frac{\theta}{\nu(1 - \theta)^\nu} = bc_2 \tag{8.50}$$

where b is a constant. This equation is found to fit the adsorption data of polystyrene from toluene by charcoal with $\nu = 50$.

All these derived equations involve the assumption that ν is independent of the surface coverage θ. This is not expected to be strictly valid. More complicated equations for polymer adsorption have been derived by various workers[71-76] which include many (empirical) parameters and hence are difficult to use.

The magnitude of adsorption of polymer usually decreases with increase of temperature. Sometimes, however, it increases with increase of temperature and the process becomes controlled by increase of entropy. The adsorption also depends significantly on the nature, surface area, and porosity of the adsorbent. The extent of adsorption is strongly dependent on the nature of the solvent medium. The effects of solvent on adsorption of polyvinyl acetate on the cellulose fibers[62] are shown in Fig. 8.22.

8.13. ADSORPTION OF BIOPOLYMERS AT THE SOLID-LIQUID INTERFACE

Interactions of proteins at the solid–liquid and liquid–liquid interfaces are of considerable importance in the field of membrane biology and membrane transport phenomena.[77-79] Studies of adsorption of proteins on blood vessel

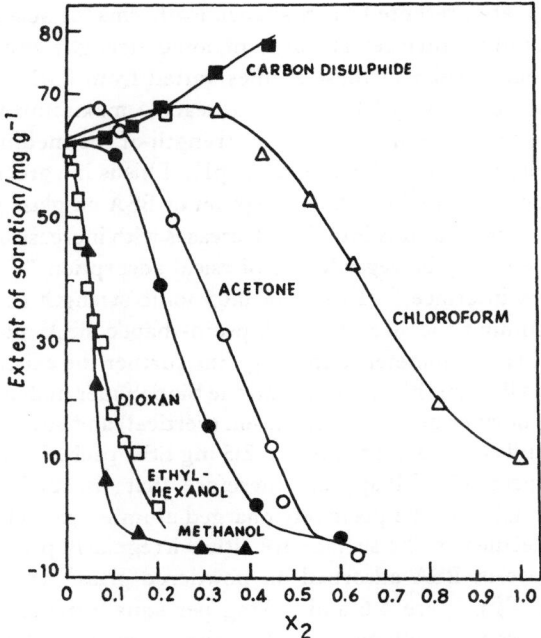

FIGURE 8.22. Adsorption of polyvinylacetate on the cellulose fibers from binary solvents based on benzene (component 1). The abscissa shows the mole fraction of the second solvent (component 2); the ordinate, the adsorption at an equilibrium concentration of 4 g/liter. (Redrawn from Ref. 62.)

walls,[80] interaction of proteins with synthetic biomedical materials,[81] and adsorption of immune proteins[82] on solid surfaces are very important in the biomedical field of research. In soil systems, the activities of the extracellular enzymes are altered as a result of their adsorption on clay particles containing alumina and other metallic oxides.[83] Phenomena of adsorption become important for the industrial production of many biochemicals using immobilized enzymes.[84]

Adsorption of bovine serum albumin and other proteins on the surfaces of glass powders, paraffins, resins, polystyrene latex, and hydroxyapatite has been studied extensively as functions of protein concentrations, pH, ionic strength, and temperature.[85–92] Using direct techniques such as chemical analysis of proteins (vide Chapter 2), microelectrophoresis, calorimetry, etc., Norde and Lyklema[93–95] have recently studied adsorption of human serum albumin and ribonuclease on the surface of polystyrene particles most extensively. These studies give an extensive analysis of the various physicochemical aspects of adsorbed proteins, which are highly interesting.

In Fig. 8.23, are presented the S-shaped isotherms for adsorption of BSA on powdered alumina-surfaces at constant ionic strength and temperature. The protein concentration in these studies varied from 0.01% to 0.20%. At high ranges of concentration Γ_2^w appears to reach a maximum value $(\Gamma_2^w)_{max}$. $(\Gamma_2^w)_{max}$ are found to depend on the ionic strength of the medium. The value of $(\Gamma_2^w)_{max}$ is quite high near the isoelectric pH. This is in agreement with the earlier observation of Bull[85] for the adsorption of BSA on glass particles (vide Fig. 8.24). $(\Gamma_2^w)_{max}$ for alumina interface decreases with increase of temperature so that the process may be regarded as physical adsorption.[96]

For alumina interface,[96] $(\Gamma_2^w)_{max}$ of 0.05 ionic strength (in Fig. 8.23) is 1.8 mg/m^2. Assuming the size of the ellipsoid-shaped BSA molecule[97] to be $(11.6 \times 2.7 \times 2.7)$ in nanometers and imagining further the existence of a layer of bound water of 0.5 nm thickness around the biopolymer molecule, formation of a compact monolayer of BSA with end-on (vertical) and side-on (horizontal) orientations would correspond to 8.4 and 2.5 mg BSA packed per square meter surface area, respectively.[95] It appears, therefore, that adsorbed BSA molecules are lying horizontally on the positively charged alumina particles and further asymmetric molecules on the surface are either irregularly packed or laterally expanded. In case of BSA adsorbed on powdered glass[85] and polystyrene[95] latex, values of $(\Gamma_2^w)_{max}$ are 4.6 and 3.0 mg per square meter of the surface respectively. Molecular packing of BSA at the interface is thus significantly dependent on the nature of the solid.

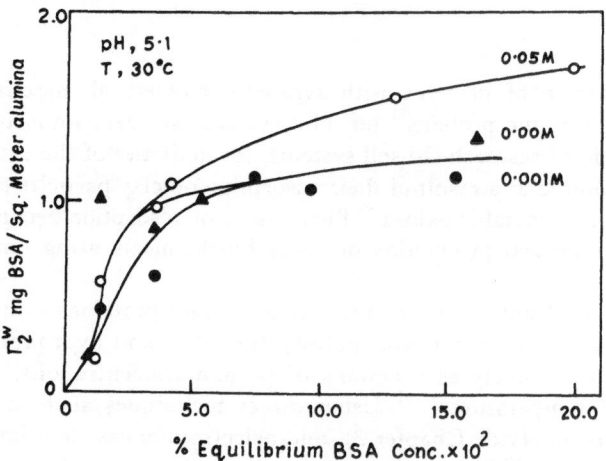

FIGURE 8.23. Plot[96] of Γ_2^w in mg of BSA adsorbed per square meter alumina vs. c_2 (percent protein concentration) at different molar concentrations of NaCl at 30°C. (Reproduced with permission of the *Indian Journal of Biochemistry and Biophysics*.)

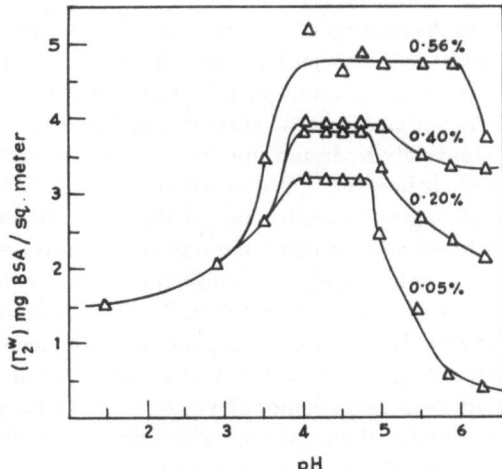

FIGURE 8.24. BSA adsorption in mg/m^2 on Pyrex glass as a function of pH at 0.05%, 0.20%, 0.40%, and 0.56% protein concentrations. (Redrawn from Ref. 85.)

In Fig. 8.25, each isotherm for adsorption of BSA on the alumina surface[96] at a fixed ionic strength exhibits a sharp maximum when the range of protein concentration exists between 0.001% and 0.003%. This is in agreement with the early works of Bull[86] with glass powder. Bull[85,86] has clearly demonstrated that a protein adsorbed on glass powder is extensively surface-denatured if

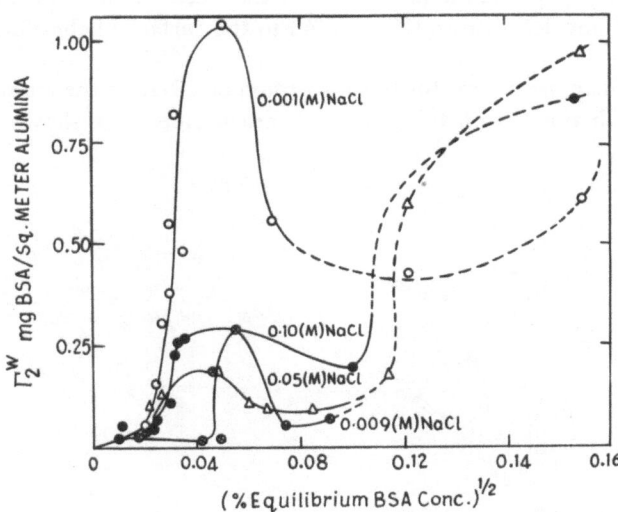

FIGURE 8.25. Plot[96] of Γ_2^w in mg of BSA adsorbed per square meter of alumina as a function of equilibrium percent concentration of BSA at different ionic strengths at 30°C. (Reproduced with permission from the *Indian Journal of Biochemistry and Biophysics*.)

its bulk concentration is very low. Here Γ_2^w increases with increase of concentration as more surface denatured BSA is accumulated on the alumina surface. Beyond the maximum, Γ_2^w decreases with increase of protein concentration. It is anticipated that surface denatured and dehydrated BSA at this state is extensively hydrated due to the refolding of the biopolymer so that the value of Γ_2^w is lowered.[96] With further increase of concentration folded and hydrated BSA begins to accumulate at the interface directly so that Γ_2^w increases again in the region of relatively higher concentration of the biopolymer.

Nucleic acids in combination with proteins are suspected to remain as gels in many biological systems. Also, adsorption chromatography has been frequently used for the separation of nucleic acids from each other. Electron-microscopic study of DNA structure is usually carried out by spreading nucleic acid over protein monolayer or solid surfaces.[98,99] There is now evidence[100] that bacterial chromosomes containing DNA are in attachment to invaginated portions of the cell membrane called the mesosome. It has also been suggested that both the origin and termini of DNA replication are preferentially bound to cell-surface membrane.

For some time, considerable interest has been shown for investigating the behavior of nucleic acids adsorbed on the solid–liquid and liquid–liquid interfaces. The techniques used for such study are differential capacity measurement, microelectrophoresis, electron microscopy, direct analysis, and radiotracer methods.[101–106] From the measurement of the particle mobility by the microelectrophoretic method, evidence has been obtained for the adsorption of DNA and RNA from the solution to the surface of the solid alumina powder.[104,105]

In Fig. 8.26, isotherms for the adsorption of DNA on the alumina–water interface[107] obtained from the analytical measurement are shown at several

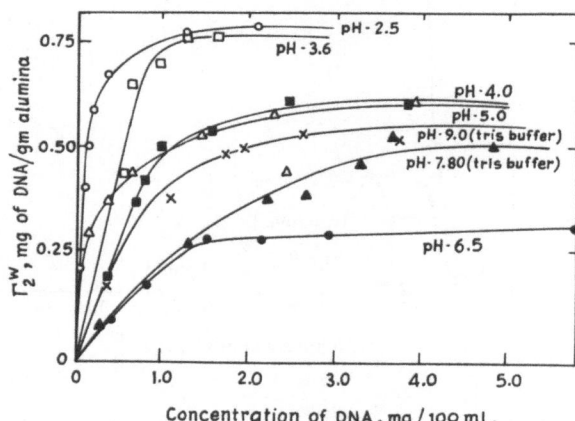

FIGURE 8.26. Adsorption of DNA onto alumina powder at 25°C, ionic strength 0.05. (Redrawn from Ref. 107.)

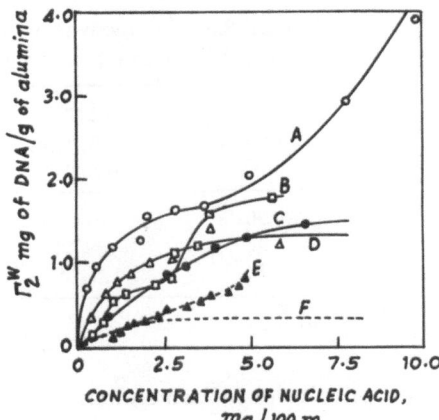

FIGURE 8.27. Adsorption of native and denatured DNA and RNA onto alumina powder at 25°C, pH 6.4–6.5, and ionic strength 0.05 with NaCl. A, Heat-denatured DNA; B, heat-denatured RNA; C, acid-denatured DNA; D, alkali-denatured DNA; E, native RNA; F, native DNA. (Redrawn from Ref. 107.)

pH values. The shape of the isotherm indicates monolayer adsorption and further the adsorption data fit the Langmuir equation satisfactorily. It is suggested that DNA on this solid surface remains as spherical gel bead of the radius of gyration of the order of a micron.[107] In Fig. 8.27, the adsorption isotherms of native and denatured DNA are compared with each other. At higher biopolymer concentrations also, heat-denatured DNA shows characteristics of multimolecular adsorption. Unlike rigid DNA, the structure of RNA is relatively flexible so that its isotherm also gives indications of multilayer adsorption.[107]

8.14. STANDARD FREE ENERGY OF ADSORPTION

Attempts have been made by many workers[108,109] to calculate the standard free energy change for the adsorption of a solute from binary solution by the solid particles. When the solute is strongly adsorbed, Γ_2^1 becomes equal to Δn_2 [vide equation (8.40)] and molar concentration (c_2^s) of the solute at the interface becomes equal to (Γ_2^1/τ) where τ is the thickness of the interfacial phase. Assuming a plausible value of τ, c_2^s can be calculated from the experimental data at a given value of the bulk solute concentration c_2. Since the partition constant for a system containing interfacial and bulk phases is equal to (c_2^s/c_2), the standard free energy change for adsorption can be calculated from the equation

$$\Delta G^0 = -RT \ln \frac{\Gamma_2^1}{\tau c_2} \qquad (8.51)$$

For the adsorption of n-octyl alcohol from solutions in benzene on a number of metal powders, ΔG^0 is found to be -12.6 kJ per mole. For the

TABLE 8.3. Thermodynamic Parameters for the Adsorption of CSBFF on Graphon in the Presence of 0.13 M NaCl[a]

Temperature (°C)	ΔG^0 (Langmuir) (kJ/mol)	ΔH^0 (kJ/mol)	ΔS^0 (kJ/deg/mol)
30	−26.0		
40	−27.4	+19.3	+149
50	−29.0		

[a] Reference 43.

adsorption of phenol on alumina from the benzene solution, ΔG^0 is −25.2 kJ per mole. In the calculation of ΔG^0 in this manner, values of τ remains uncertain.

When the adsorption data fit the linear plot according to the Langmuir equation (8.46), the value of b or K ($b = K/a_1 \simeq K$) for the monolayer reaction (8.33) can be calculated from the slope and intercept of the line. The linear plots for the adsorption of chlorazal sky blue FF on graphon at various temperatures are shown in Fig. 8.15. ΔG^0 for this adsorption process at three different temperatures are given in Table 8.3. ΔH^0 and ΔS^0 calculated from the temperature coefficient of ΔG^0 are also presented in this table. The results indicate that the adsorption is mostly entropy controlled due to the hydrophobic interaction of the dye with the hydrophobic graphon surface.

In the study of the adsorption of proteins on the surface of powdered glass, Bull[85,110] has deduced an explicit expression for the free energy of adsorption using the basic concept of the Gibbs adsorption equation. According to the Gibbs equation, given in Chapter 3, for the binary solution, the free energy a change $\Delta G'_a$ per square cm of the surface can be obtained from the equation

$$\Delta G'_a = \int_{\gamma_0}^{\gamma} d\gamma = -RT \int_0^{c_2} \left[\frac{(\Gamma_2^1)_{cm^2}}{c_2} \right] dc_2 \qquad (8.52)$$

Here $(\Gamma_2^1)_{cm^2}$ is the surface excess per unit surface area of the solid. γ and γ_0 are the surface free energies in the presence and absence of the solute of bulk concentration c_2. Multiplying $\Delta G'_a$ by A_m one obtains the value of the free energy (ΔG_a) of adsorption per gram of solid powder. Since $[(\Gamma_2^1)_{cm^2} A_m]$ is equal to the moles (Γ_2^1) of the solute adsorbed per gram of solid powder, one can write equation (8.52) in the form[55,110]

$$\Delta G_a = -RT \int_0^{c_2} \left(\frac{\Gamma_2^1}{c_2} \right) dc_2 \qquad (8.53)$$

For the comparison of ΔG_a for various systems, the free energy of adsorption must refer to the bulk concentration of solute at a standard state of reference. A suitable standard state is a hypothetical one molar solution. The bulk solution is thus hypothetically diluted from c_2 to one molar concentration keeping Γ_2^1 also hypothetically constant. At the state of saturation, Γ_2^1 actually remains constant during the hypothetical bulk dilution process. The free energy of dilution ΔG_d is given by

$$\Delta G_d = -\Gamma_2^1 RT \ln (1/c_2)$$

$$= (\Gamma_2^1 RT) \ln c_2 \qquad (8.54)$$

The total free energy change ΔG_B is equal to $\Delta G_a + \Delta G_d$ so that

$$\Delta G_B = -RT \int_0^{c_2} \left(\frac{\Gamma_2^1}{c_2}\right) dc_2 + \Gamma_2^1 RT(\ln c_2) \qquad (8.55)$$

ΔG_B can be calculated from the graphical integration using experimental values of Γ_2^1 for various values of c_2. ΔG_B is the free energy change for the transfer of Γ_2^1 moles of adsorbate in excess from the bulk to the interfacial phase when the bulk solute concentration is altered from zero to one molar concentration. Standard free energy change ΔG_B^0 is given by

$$\Delta G_B^0 = (\Delta G_B / \Gamma_2^1) \qquad (8.56)$$

ΔG_B is always found to vary linearly with increase of Γ_2^1 so that the slope of the curve represents ΔG_B^0. ΔG_B^0 for BSA and egg albumin adsorbed on glass powder at pH 4.66 is -34.5 and -28.5 kJ/mol.

Values of ΔG_B^0 for the adsorption of BSA on the surface of alumina calculated by Mitra and Chattoraj[96] with the help of equations (8.55) and (8.56) are given in Table 8.4. These values are found to be dependent on the ionic strength of the medium. From the results given in the table, it is concluded that about 9.0 kJ of free energy will be further reduced when native BSA transferred from the bulk to the surface is subsequently surface denatured.

Bull's equation (8.55) based on the Gibbs adsorption equation does not depend upon the thickness or the monolayer nature of the interface so that its application appears to be perfectly general. Electrostatic and all other types of interactions are included in the value of ΔG_B^0. For the conformational alterations of the adsorbents such as proteins, cellulose, etc. there is a need for further standard state corrections for the evaluation of the standard free energy of adsorption. This has been extensively discussed in Sections 10.3, 10.7, and 10.15. Using this approach, the standard free energy change ΔG^0 for the positive adsorption of either solute or solvent (vide Section 8.2) to

TABLE 8.4. Free Energy Change due to the Adsorption BSA at the Alumina–Water Interface[a]

Concentration of NaCl (M)	pH	Temperature (°C)	BSA concentration	$-\Delta G_B^0$ (kJ/mol)
0.0	5.1	30	High	32.6
0.05	5.1	30	High	31.8
0.001	5.1	30	High	34.6
0.001	5.1	40	High	34.6
0.001	5.1	25	High	36.3
0.0	5.1	30	Low	43.1
0.001	5.1	30	Low	40.7
0.05	5.1	30	Low	42.9

[a] Reference 96.

gelatin powders has been calculated by equations (10.10), (10.24), and (10.39). These values are given in Table 8.5. ΔG^0 for solute adsorption in positive excess represents the free energy change in bringing the solute mole fraction from zero to unit mole fraction in the bulk whereby a state of adsorption saturation is attained in the system. For solvent adsorption in positive excess, ΔG^0 stands for the change in free energy in bringing the solvent mole fraction in bulk from unity to 0.5 (a practical scale) whereby a state of fixed value of positive adsorption is reached. Γ_1^m in the table indicates the values of Γ_1 for the adsorption of water as a solute or as a solvent, respectively, at the final state of constant accumulation of water in excess in the interfacial phase.

8.15. SURFACE HETEROGENEITIES

The surfaces of the solid powders in many cases are heterogeneous in structure. For this reason the heat change Q_a due to the adsorption is not

TABLE 8.5. Standard Free Energies ΔG^0 of Binding of Gelatin at 25°C

Solute	Solvent	Γ_1^m moles H_2O adsorbed per kg of gelatin	ΔG^0 (kJ/kg of gelatin)
Water	*Iso*-amyl alcohol	790	−733
Water	*t*-butanol	320	−366
Water	Propelyneglycol	80	−63
Water	Ethyleneglycol	180	−192
t-Butanol	Water	2830	+4890
Propelyneglycol	Water	710	+1170
Ethyleneglycol	Water	1130	+1870

uniform throughout the surface. The standard free energy change ΔG^0 for the adsorption process, expressed by equations (8.33) and (8.34), is equal to $(-RT \ln K)$, which in its turn is nearly equal to $(-RT \ln b)$. Since ΔG^0 is equal to $(\Delta H^0 - T\Delta S^0)$, one can write[4]

$$b = e^{(\Delta S^0/R)} \, e^{-(\Delta H^0/RT)}$$

$$= b' \, e^{(Q_a/RT)} \tag{8.57}$$

Here $e^{\Delta S^0/R}$ is put equal to b' and $-\Delta H^0$ is taken to be equal to Q_a. Thus for a heterogeneous surface, the relation between b and Q_a will not be uniform. If Γ_2^m are the total moles of adsorption sites per gram of the solid powder out of which Γ_2^1 are occupied by the adsorbates as determined from the experiments, then one can write for the heterogeneous surface[4]

$$\left(\frac{\Gamma_2^1}{\Gamma_2^m}\right) = \int_0^{\infty} f(b)\theta \, db \tag{8.58}$$

Here $f(b)$ is the distribution function of b, which in its turn is a function of Q_a. θ depending upon b must be different for different regions of the surface having different values of Q_a.

With certain assumptions, equation (8.58) may be solved to a simplified form

$$\Gamma_2^1 = (\Gamma_2^m a_f)c_2^{1/n_f} \tag{8.59}$$

$$= K_f c_2^{1/n_f}$$

Here a_f, n_f, and K_f are constants. This equation is similar in form to the Freundlich equation,[111] which, however, was empirically proposed by Kuster. According to this equation, the plot of $\log \Gamma_2^1$ against $\log c_2$ should be linear so that the constants K_f and n_f representing roughly the adsorption capacity and intensity of adsorption can be estimated. For many systems, adsorption data in solution fit the Freundlich equation in the middle range of concentration possibly giving support that the surface is heterogeneous.

Henry[112] had shown that the Freundlich adsorption equation (8.59) can be derived also with the help of the Gibbs adsorption equation. If γ_0 is the average free energy (per unit area) of the surface in contact with the pure solvent and γ_1 that of the surface completely covered with a monolayer of the adsorbed solute, then the net surface free energy γ can be calculated using the following equation:

$$\gamma = \gamma_0(1 - \theta) + \gamma_1\theta$$

$$= \gamma_0 - (\gamma_0 - \gamma_1)\theta \tag{8.60}$$

Since θ is equal to Γ_2^l/Γ_2^m,

$$\gamma = \gamma_0 - (\gamma_0 - \gamma_1)\frac{\Gamma_2^l}{\Gamma_2^m} \tag{8.61}$$

Differenting this equation with respect to $d \ln c_2$, it can be shown that

$$\frac{1}{RT}\left(\frac{d\gamma}{d \ln c_2}\right) = -\left(\frac{\gamma_0 - \gamma_1}{RT\Gamma_2^m}\right) \cdot \left(\frac{d\Gamma_2^l}{d \ln c_2}\right) \tag{8.62}$$

The left-hand side of this relation is equal to $(-\Gamma_2^l)$ according to the Gibbs adsorption equation for the dilute solution so that, inserting this, we obtain

$$d \ln \Gamma_2^l = \left(\frac{RT\Gamma_2^m}{\gamma_0 - \gamma_1}\right) d \ln c_2 \tag{8.63}$$

Integrating this equation, we get

$$\ln \Gamma_2^l = \left(\frac{RT\Gamma_2^m}{\gamma_0 - \gamma_1}\right) \ln c_2 + \ln K_I \tag{8.64}$$

where $\ln K_I$ is the integration constant. Equation (8.64) agrees quantitatively with the logarithmic form of the Freundlich adsorption equation (8.59). This agreement further indicates that the Gibbs adsorption equation can quantitatively estimate the experimental value of Γ_2^l when the adsorption takes place from the solution to the heterogeneous surface of the solid powder.

Quantitative estimation of the extent of surface heterogeneities present in solid powder and also detailed theoretical analysis of the adsorption of liquid mixtures on the heterogeneous surface of the solids are indeed very difficult although efforts have been made in these directions by many workers.[113-118] Recently Dabrowski and co-workers[119] have developed quantitative theories for the studies of adsorption from binary solutions of nonelectrolytes on heterogeneous solid surfaces. Applying this method of approach, it has been confirmed that heterogeneity of solids is one of the major factors which can control the type, magnitude, and range of preferential adsorption from the solution. The treatment has been extended to cover up adsorption from ternary solution[120] by solid powder containing heterogeneous surfaces.

8.16. SUMMARY AND COMMENTS

In this chapter, the adsorption from binary solution by various types of solid powders has been mainly discussed in the light of the concept of the

Gibbs positive and negative excesses. In agreement with this concept, it has been demonstrated that the experimental data of adsorption from alcohol–water and glycol–water mixtures by solid gelatin powder may be positive or negative. From the detailed experimental studies of the adsorption from binary mixtures by various types of powders of rigid solids, the shapes of the isotherms have been broadly divided into five types. From the linear portion of these composite isotherms, the absolute amounts of the two liquid components present in the interfacial phase can be estimated. Based on this estimation, the surface area of many solids can be determined with confidence.

Using the adsorption data for the liquid mixture, the surface activity coefficients of the components present in the interfacial phase have been evaluated. The values of these coefficients calculated by Schay on the basis of the Gibbs adsorption equation do not always agree with those based on the statistical approach put forward by Everett, possibly because of the difference in the standard states. The effects of the nature of solids, modification of the solid interface, surface heterogeneity, etc. on the surface activity coefficients should be critically studied for proper understanding of the nature of the interaction phenomena in the interfacial phase.

The importance of the competitive nature of the interactions of solute and solvent in dilute solution with powdered solid surface has also been explained in terms of the concept of the Gibbs surface excess. The affinity of solute for the solid surface is predominant in this case. The form of the Langmuir equation for such a process can be derived only with the assumption that the interfacial phase containing the solute and solvent components is a monolayer. The dependence of the extent of positive adsorption of the organic dyes and ionic detergents on different solution parameters has been discussed objectively. Negative adsorption of the detergents by solids beyond CMC is possibly due to the large desorption of the surface micelles to the bulk phase. More experimental data will be necessary for complete theoretical analysis of the complex processes involved in these types of ionic adsorption.

The adsorption of neutral polymer from the nonaqueous solution to the surface of a solid powder depends on the average molecular weight of the macromolecule, temperature, nature of the solvent etc. The conformation of the flexible polymer at the solid–liquid interface can be qualitatively predicted from the study of adsorption isotherms determined under varying conditions of these solution parameters. It will be useful to carry out more experiments in this direction so that methodology for determining the average conformations of the polymer at the interface may be predicted quantitatively. Extensive studies of the adsorption of rigid and flexible proteins, DNA, and RNA on the surface of the solid in contact with the aqueous solution has been investigated intensively in recent years at various values of biopolymer concentrations, pH, temperature, ionic strength, etc. The conformations of the biopolymers and their possible aggregation at the interface can be predicted from such investigations.

Standard free energy change, ΔG^0, for adsorption calculated assuming arbitrary depth of the interfacial phase is convenient for ready computation but not useful for comparisons of the values for various systems. ΔG^0 can be calculated from the monolayer model for the interface provided the experimental data fit the linear Langmuir plot. A more general method for the calculation of ΔG^0 as proposed originally by Bull is based on the integrated form of the Gibbs adsorption equation combined with the free energy change for dilution to a final state of reference. The approach has been used for the biopolymer adsorption, but it seems to be theoretically valid for the simultaneous adsorption of small solute and solvent molecules at the interface. A universal scale for the comparison of the adsorption energies for different types of liquid systems undergoing interactions with solid surfaces of various types is really a need at the moment for clear interpretation of the experimental data.

CHAPTER 9

ADSORPTION OF WATER VAPOR BY BIOPOLYMERS

9.1. INTRODUCTION

Biological macromolecules usually exhibit strong affinities for water. Protein crystals contain a considerable amount of bound water.[1] The folded structure of a protein is known to be grossly altered if the bound water is completely removed from protein. Many globular proteins dissolved in the aqueous phase are found to be extensively hydrated.[2] Fibrous proteins and other types of protein powders are known to adsorb a considerable amount of water vapor from the moist atmosphere forming protein gels. If the adsorption is extensive, many proteins may dissolve in the aqueous phase forming a concentrated solution. It may be pointed out here that the biological functions of protein such as enzyme activities, membrane transport, muscle action, water evaporation from skin, etc. are exhibited in the living system only in the presence of large amounts of aqueous fluid. It is believed that these functions are guided by complicated water–protein interactions, the mechanism of which is not yet clearly understood.

Another group of biopolymers called nucleic acids, occurring within the living cell, control important genetic properties also in an aqueous environment. Water bound to deoxyribonucleic acid is known to maintain its double-helix structure in the correct manner.[3–5] Like proteins, nucleic acids remain extensively hydrated thus forming biogels or biofluids (e.g., chromosomes, cytoplasmic fluid). Polysaccharides present in many plant and animal systems become extensively hydrated in the presence of water.[6,6a] The information about hydration of proteins, nucleic acids, and carbohydrates has been recently reviewed.[2,6,6a]

The primary reason for the hydration of biopolymer is the existence of a large number of hydrophilic groups such as $-OH$, $-COOH$, $-NH_2$, phosphates which have strong affinities for water. The biopolymers at the same time possess a considerable number of water-repelling, i.e., hydrophobic groups. The latter control the properties of water in the close vicinity of the boundary region by exhibiting hydrophobic interaction.[7,8] In living systems occur lipids, in which the proportion of hydrophobic groups is considerably large compared to the hydrophilic groups, so that they are usually insoluble in water. The lipids may interact with each other or with neighboring biopolymer molecules which involve special arrangement of surrounding water molecules as a result of the hydrophobic interactions. Animal and plant systems also contain large amounts of neutral fats and oils (triglycerides) which are highly hydrophobic and completely insoluble in water. They may under proper conditions modify the structure of surrounding water molecules.

In the present chapter, we will discuss the adsorption of water vapor by polymers both in the presence and absence of inorganic and organic solutes, many of which occur in the living system and control the biological functions of the biopolymers. Only the thermodynamic aspects of interactions between water and biopolymers will be considered in the light of the concept of the Gibbs surface excess, as has been dealt with in Chapters 3 and 4.

9.2. INTERACTION OF PROTEIN WITH WATER MOLECULES AT LOW RELATIVE HUMIDITY

Using the isopiestic techniques discussed earlier (vide Chapter 2), Bull[9] studied extensively the uptake of water vapor by several proteins, at relative humidities ranging from zero to unity. In Fig. 9.1, water vapor adsorption curves have been presented for bovine serum albumin, hemoglobin, and egg-albumin.[10] The terms n_1 and p/p_0 (equal to a_1) in these plots have been explained in Sections 1.8 and 2.8. By linear extrapolation of the curve at p/p_0 equal to unity, the maximum number of moles (Δn_1^0) of water adsorbed per mole of protein may be calculated. Some of these values are stated in Table 9.1. Δn_1^0 for several other proteins have also been given by Bull.[9]

The S-shaped isotherms in Fig. 9.1 appear[11] to be of the BET type II. Accordingly, it may be thought that water adsorbed on the boundary region of the protein powder forms multilayers. In the lower range of p/p_0, the BET equation in the linear form

$$\frac{p/p_0}{V_g(1 - p/p_0)} = \left(\frac{1}{C_b V_m}\right) + \frac{C_b - 1}{C_b V_m}\frac{p}{p_0} \tag{9.1}$$

is found to be obeyed (vide Fig. 9.2) as shown by Bull,[9] Kuntz and Kauzmann.[2] Here V_g is the volume of the vapor adsorbed per gram of protein at NTP and may be taken to be equal to $n_1 RT/p$. From the slope and intercept of such plots, the BET constants V_m and C_b for various proteins have been calculated by Pauling.[12] Since V_m is equal to $\Delta n_m RT/p$, at NTP the maximum number Δn_m of moles of water attached per mole (or per gram) of protein directly in the first layer of adsorption may be calculated for various proteins. These are shown in Table 9.2. Δn_m has been found by Pauling[12] to be equal

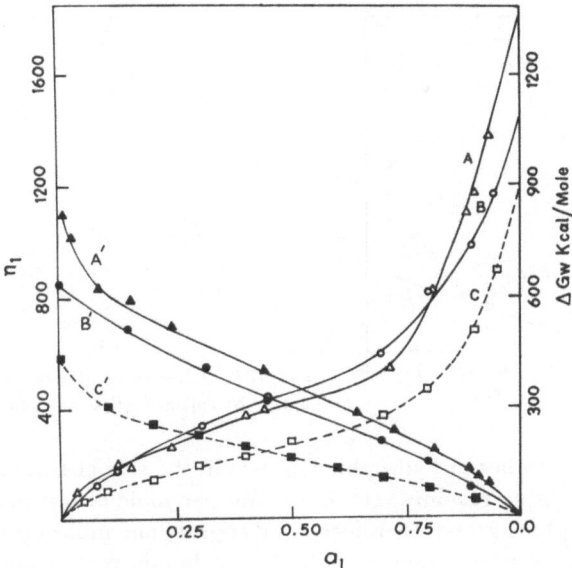

FIGURE 9.1. Plots of n_1 (left coordinate) against a_1 (A, BSA; B, bovine hemoglobin; C, egg albumin). Plot of ΔG_w (right coordinate) against a_1 (A', BSA; B', bovine hemoglobin; C', egg albumin[10]). (Reproduced with permission from the *Indian Journal of Biochemistry and Biophysics*.)

TABLE 9.1. Change of Free Surface Energy (ΔG_w) due to Water–Protein Interaction at 25°C between a_1 Equal to Unity and Zero[a]

Protein	Mol. wt.	Δn_1^0 (mol/mol)	ΔG_w (kJ/mol)	A_b (nm²/molecule)	$\gamma - \gamma_0$ (mN/m)
Serum albumin (bovine)	68,000	1800	3490	102	+57.2
Hemoglobin (bovine)	68,000	1600	2660	87.1	+51.3
Egg albumin (native)	45,000	1200	1790	78.4	+38.5
β-Lactoglobulin (lyophilized)[b]	40,000		1530	68.1	+37.8

[a] Reference 10.
[b] Reference 9.

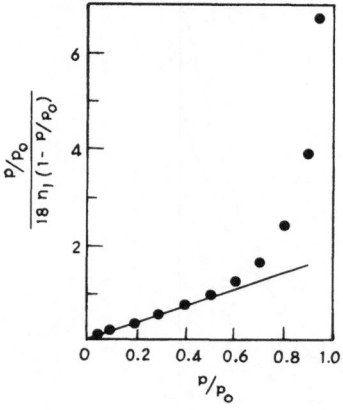

FIGURE 9.2. Representative isotherm for water uptake by collagen. (Redrawn from Ref. 2.)

to the total number of hydrophilic groups in the side chains of glutamic acid, arginine, tyrosine residues, etc. occurring per mole of a protein (Table 9.2). Each hydrophilic group, therefore, binds tightly one molecule of water. These are the groups which may form hydrogen bonds with water. According to Pauling, —CO— and —NH— groups in the peptide bonds occurring in a protein are incapable of forming hydrogen bonds with water for steric reasons. This has, however, been criticized by Mellon, Korn, and Hoover,[13] who suggested that —NHCO— group may also play an important role in the hydration of proteins.

TABLE 9.2. Water Uptake by Different Proteins at 25°C[a]

Protein	Water adsorbed in first layer (mol/10^5 g)	No. of polar groups (mol/10^5 g)
Silk	226	219–228
Ovalbumin		
Crystallized	329–344	277–313
Lyophilized	314	
Heat denatured	276	
Wool	366	303–341
Gelatin, collagen	485–529	328–609
C-Zein, B-Zein	210–228	305–390
Salmin	592	611–707
Serum albumin	374	420–424
β-Lactoglobulin		
Crystallized	370	472–508
Lyophilized	329	

[a] Reference 12.

9.3. WATER–PROTEIN INTERACTION AT HIGH RELATIVE HUMIDITY

In recent years, renewed interest has been shown on the water adsorption by proteins at high relative humidity, where the BET equation is not obeyed at all (Fig. 9.2). Thus Klotz[14] suggested that water bound to protein at high humidity region may assume an icelike structure, induced by the nonpolar parts of the amino acid residues. Bull and Breese[15] have suggested that all the water adsorbed by a protein at 92% relative humidity strongly interact with the biopolymer. The interaction above this relative humidity is regarded as weak. The weakly held water can be very easily squeezed out of the protein glass formed at 92% relative humidity; this excess water may be regarded as water of crystallization of protein crystals. Bull and Breese[15] have measured water uptake by several proteins at 92% relative humidity. A plot of the uptake against the effective number (x_n) of hydrophilic groups of the protein is shown in Fig. 9.3. The slope of this curve is found to be equal to 6. From this, it has been concluded that each hydrophilic group of a protein (e.g., hydroxyl, carboxyl, amino, imino, etc.) may be complexed with six molecules of water. The amido groups of asparagine and glutamine occurring in protein, however, do not complex in this manner. Further, each amido group present in the interfacial region of protein may break up the structure of six molecules of water bound or complexed with neighboring carboxyl, amino, or imino groups of the polypeptide chain.

From all these observations, it appears that water adsorbed by protein may be divided into strongly bound water, complexed water, water of crystallization, etc. Adsorbed water may thus be regarded as highly inhomogeneous and hence the concept of the Gibbs surface excess may be extremely useful in dealing with thermodynamics of such systems. The inhomogeneous property of water may become more pronounced since the protein surface either at

FIGURE 9.3. Water binding of the proteins as a function of the moles of polar residues. ○, sum of all polar residues; ●, sum of hydroxyl, carboxyl, and basic residues minus the amide residues. Moving up the y axis: insulin, lysozyme, chymotrypsinogen, egg albumin, β-lactoglobulin, ribonuclease, hemoglobin, cytochrome c, myoglobin. (Redrawn from Ref. 15.)

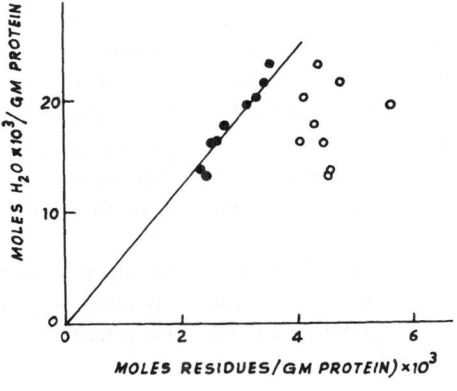

molecular or at powder levels is highly heterogeneous in character. There exist different kinds of hydrophilic groups on the protein surface, which render the neighboring region different. Besides, a considerable fraction of the protein surface is composed of hydrophobic groups of varied structures. It is no wonder that the BET equation fails for the case of water uptake by proteins at high humidity, since according to the BET concept, bound water beyond the first layer of adsorption should have uniform physical properties. In that context, the thermodynamic treatment of Gibbs for the inhomogeneous surface phase seems to be the most appropriate.

The water vapor adsorption isotherms for proteins are found to obey the equation of Frenkel,[16] Halsey,[17] and Hill,[18] according to which $\ln Vg$ should vary linearly with $\ln \ln p/p_0$. Such linear variation for collagen has been shown[2] throughout the whole range of p/p_0. The FHH equation is based on the potential theory for adsorption first proposed by Polanyi,[11] according to which the adsorbed gas is regarded as inhomogeneous so that the physical properties in this boundary region vary in gradient.

In the case of water vapor adsorption, n_1 does not depend on the degree of fineness of the protein powder.[19] This means that protein as a molecular solid interacts with water; the process continues as vapor pressure increases, and one may assume that water–protein interaction becomes complete when protein dissolves in water at p/p_0 equal to unity. In fact, in an alternative procedure, Cutler and McLaren[20] have assumed that protein gel in the whole range of relative humidity from zero to unity may behave as a single phase so that the Flory–Huggins lattice model[21] for polymer solution is valid for such system. According to this approach

$$\ln \frac{p}{p_0} = \ln V_1 + V_2 + \mu_f V_2^2 \qquad (9.2)$$

where V_1 and V_2 are volume fractions of water and polymer in the gel, respectively, and μ_f is a parameter which can be estimated from the regular solution theory. The equation has been found to fit the water vapor adsorption data of Bull[9] in the whole range of vapor pressure. For various reasons, the solution theory for water uptake by proteins has been recommended by Kauzmann and others.[2,22] In the lattice theory of polymers, a segment of a flexible polymer is comparable to the size of a solvent molecule, whereas a compact molecule of protein possesses enormously large size compared to the size of a water molecule, so that the lattice model for it may become very complex.

From all these discussions it appears that any thermodynamic treatment used for the hydration of protein may be accepted with confidence only when it is independent of the "two-phase" surface model and the "one-phase" solution model.

9.4. WATER ACTIVITY AND PROTEIN HYDRATION

The vapor pressure p of an aqueous solution of solute may be given by Raoult's law. According to this law [vide equation (1.23)]

$$a_1 = \frac{p}{p_0} \tag{9.3}$$

where a_1 is the activity of water in the solution, and p_0 is the vapor pressure of the pure solvent at a given temperature. For pure water a_1 becomes unity in mole fraction units and the liquid is in the standard state. This standard state is, however, based on Raoult's law.

In the isopiestic experiment (vide Section 2.8), the vapor pressure of the protein–water system is altered by using sulfuric acid solutions of different concentrations in the reference container. By decreasing p slightly from p_0, all the free water associated with the protein sample will evaporate and condense in the reference system at isopiestic equilibrium. The reason is simply that free water cannot remain in equilibrium with its vapor at pressure less than p_0 at a given temperature. The whole amount (Δn_1^0) of water associated with protein just below water activity equal to unity may be regarded as solvent bound to the protein.[10] The values of Δn_1^0 calculated from the extrapolation of the values of n_1 to a_1 equal to unity are shown in Table 9.1. This bound water associated with the protein sample has activity less than unity because of the water–protein interaction. Further, the value of the activity of the bound water decreases with the decrease in the value of n_1, when water–protein interaction becomes stronger.

From the experimental values of n_1 for different values of a_1 (equal to p/p_0), the gradient $(\partial a_1 / \partial n_1)_T$ may be easily calculated. Values of this gradient for three proteins, bovine serum albumin, bovine hemoglobin, and egg albumin are shown[10] in Fig. 9.4 for different values of n_1. It is of interest to note that this activity gradient sharply increases when n_1 is slightly increased from zero, reaches a maximum value, and then gradually decreases and finally tends to zero again when n_1 approaches Δn_1^0. The activity gradient of water is, therefore, related exclusively to the properties of the bound water. The existence of this gradient at the protein–water interface may indicate that energies associated with various layers of bound water are not identical in agreement with the picture given in the potential theory of adsorption presented originally by Polyani[11] and quantitatively modified by Frenkel, Halshey, and Hill.[16-18] Further, the gradient supports the experimental observations of Bull and Breese[15] who have demonstrated the presence of various types of water in the bound layer.

Bull and Breese[15] have indicated that above 92% relative humidity, water accumulated in excess interacts insignificantly with protein. As n_1 approaches Δn_1^0, i.e., as p/p_0 tends to unity, the activity gradient vanishes due to the

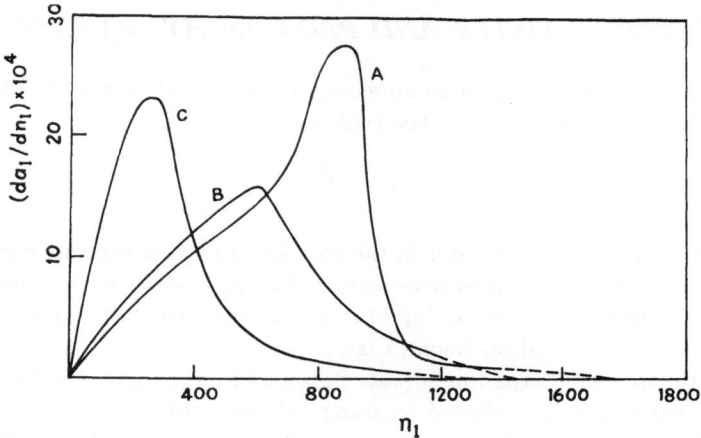

FIGURE 9.4. Plot of da_1/dn_1 against n_1 (A, BSA; B, bovine hemoglobin, C, egg albumin[10]). (Reproduced with permission from *Indian Journal of Biochemistry and Biophysics*.)

accumulation of nearly pure water in the sample. In the presence of pure water in the reference vessel, p/p_0 will be unity. The protein in the sample will then be saturated with all kinds of bound water; in addition, free water will also accumulate. So long as hydrated protein is able to exert osmotic pressure in an effective manner, vapor pressure exerted by the sample will be less than p_0, and hence more free water will be transferred to the sample. At infinite dilution of the sample, we may imagine equilibrium between bound and free water when p/p_0 is equal to unity. Both types of water will reach the standard state of chemical potential μ_1^0, when water activity becomes unity in Raoult's law scale.[10]

9.5. WATER–PROTEIN INTERACTION AND FREE ENERGY CHANGE

It is of considerable interest to calculate change in free energy due to the isopiestic hydration of proteins. For the transfer of dn_1 moles of water from unit activity to its lower value p/p_0 due to presence of protein in the sample, the free energy change dG_i is written by Bull[9,23] thus:

$$dG_i = dn_1 RT \ln \frac{p}{p_0} \tag{9.4}$$

Integration of this, following certain mathematical manipulation, leads to

$$\Delta G_i = n_1 RT \ln \frac{p}{p_0} - RT \int_0^p \frac{n_1 \, dp}{p} \tag{9.5}$$

Here ΔG_i is the free energy change due to the transfer of n_1 moles of water to the protein sample. The process of hydration may be imagined to involve two steps: (1) the vapor pressure of water is first altered from p_0 to p, so that the change of free energy becomes equal to $n_1 RT \ln p/p_0$. The second integral term in equation (9.5) indicates the free energy change when p is decreased gradually to zero when the reference state of dry protein is obtained. When p becomes equal to p_0, equation (9.5) will become

$$\Delta G_i = -RT \int_0^{p_0} \frac{n_1 \, dp}{p} \tag{9.6}$$

This represents the free energy of hydration when dry protein having p equal to zero is brought to the state p equal to p_0, i.e., the state of complete hydration of the protein. Values of ΔG_i for a large number of proteins at 25°C have been given in Table 9.3. The integration has been performed by plotting $n_1/(p/p_0)$ against p/p_0 and evaluating the graphical area between the limits p/p_0 equal to zero and unity.

Recently, the thermodynamic grounds for the derivation of Bull's equation have been criticized by Bryan,[24] who has assumed that calculation of the free energy change may become rational if the hydration of protein is imagined to consist of four steps instead of two considered originally by Bull. The detailed treatment based on this approach, however, leads to the same expression [equation (9.6)] for the calculation of ΔG_i.

Assuming the two-phase model for the protein–water system at isopiestic equilibrium, one can apply[10] the Gibbs–Duhem equation and obtain the

TABLE 9.3. Free Energies (ΔG_i) and Heats (ΔH_i) of Hydration[a] in Joules per 100 Grams of Dry Protein at 25°C

Protein	ΔG_i	ΔH_i
Silk	−2730	−5280
Wool	−4060	−8120
Salmin	−6040	−7490
Elastin	−4160	−9360
Collagen	−6740	−15,100
Egg albumin, native	−3970	−5620
Egg albumin, heat coagulated	−3380	−5830
β-Lactoglobulin, lyophilized	−3810	−6700
β-Lactoglobulin, crystallized	−4230	−3900
Serum albumin, horse	−4305	−6040
Calf thymus deoxyribonucleic acid[b]	−10,400	

[a] Reference 23.
[b] Calculated from data in Ref. 52.

following relation:

$$n_1 d\mu_1 + N A_p d\gamma = 0 \qquad (9.7a)$$

Here N stands for the Avogadro number and A_p the boundary area per molecule of globular protein present in the protein powder. As usual, γ here stands for the net change of free energy due to the formation of one square centimeter of the protein surface in contact with water.

The chemical potential μ_1 of water may be expressed by the equation

$$\mu_1 = \mu_1^0 + RT \ln (p/p_0) \qquad (9.7b)$$

Inserting equation (9.7b) in (9.7a), we obtain

$$n_1 RT\, d \ln (p/p_0) + n_1 d\mu_1^0 + N A_p d\gamma = 0 \qquad (9.8)$$

Here μ_1^0 is the chemical potential of pure water at vapor pressure p_0. If n_1 moles of this water are brought in contact with a mole of protein at p_0, reversibly, the value of $n_1 d\mu_1^0$ still remains zero since μ_1^0 depends on T and p_0 only. However, the protein–water system at p_0 will not remain in equilibrium; p_0 is assumed to alter reversibly to pressure p whereby equilibrium will be reached. The free energy change due to this is represented by the first term in equation (9.8). It represents the term $V_1 dP$ in the generalized Gibbs–Duhem equation. V_1 is the partial molal volume of water and P is the (swelling) pressure developed in the protein gel due to the dehydration of protein as a result of the change of vapor pressure from p_0 to p. Assuming water–protein binding to be an interfacial phenomenon, the total change in free energy $-\Delta G_w$ may be calculated[10] from the integration of equation (9.8):

$$-\Delta G_w = -N \int_{\gamma_0}^{\gamma} A_p d\gamma$$

$$= RT \int_{a_1=1}^{a_1} \frac{n_1}{a_1} da_1 \qquad (9.9)$$

Here γ_0 is the interfacial tension of protein–water system at water activity equal to unity. ΔG_w can be evaluated from the graphical area of the plot of n_1/a_1 against a_1. It will be noted that ΔG_w is equal in magnitude but opposite in sign to ΔG_i. In calculating ΔG_w by means of equation (9.9), it is assumed that protein in the fully hydrated state in contact with a large amount of bulk water possesses minimum free energy, which is arbitrarily set equal to zero so that it serves as the standard state. The characteristics of this standard state have been discussed in the previous section. In bringing water activity from

1 to 0.999, the graphical area of the plot of n_1/a_1 against a_1 will be negligibly small, that is, the free energy change of evaporation of a large proportion of accumulated free water from the protein sample will be negligible. Further decrease of a_1 will give significant and plausible values of ΔG_w. ΔG_w thus represents the change in free energy due to loss of water of the hydrated protein. Values of ΔG_w for several proteins in bringing water activity from unity to zero are given in Table 9.1. In Fig. 9.1 are included ΔG_w for several proteins at different values of water activity a_1. ΔG_w for various proteins are found to be different from one another as expected; the differences may be related to the amino acid composition, structure and shape, etc. of proteins. The results further indicate that hydration and not dehydration of the protein is a spontaneous process. Further, the right-hand side of equation (9.9) represents the increase in the osmotic work due to the change of water activity of swollen protein gel from unity to a_1. To attain equilibrium, the work due to interfacial dehydration has to be solely balanced by the osmotic work of swelling of the protein gel.

9.6. INTERFACIAL ENERGY OF BIOCOLLOID–WATER INTERFACE

The volume of a globular protein molecule is known to be at least a thousand times larger than that of the surrounding water. Let us assume that A_p stands for the surface area of a globular protein molecule which does not alter as n_1 varies. Equation (9.9) in such a situation will assume the form[10]

$$-\Delta G_w = -NA_p(\gamma - \gamma_0)$$

$$= RT \int_1^{a_1} \frac{n_1}{a_1} \, da_1 \qquad (9.10)$$

Here γ_0 is the interfacial tension of protein–water system at water activity equal to unity. Usually protein molecules are prolate ellipsoids whose axial ratios a/b are known from experimental measurements. Using a suitable geometrical relation between A_p and a/b, the surface area of a protein molecule can be calculated (vide Table 9.2). $\gamma - \gamma_0$ computed from equation (9.10) has also been given in this table for the complete dehydration of several proteins. A positive value of $\gamma - \gamma_0$ indicates that the interfacial tension of the protein biocolloid concerned may increase significantly when the protein–water interface is replaced by a protein–vacuum interface. The calculation of $\gamma - \gamma_0$ is, however, qualitative since with hydration A_p may change leading to alteration of surface curvature, internal structures, etc. The corresponding free energy change will be included in γ and hence in $-\Delta G_w$. The solution of the right-hand side of equation (9.9) takes into account all kinds of such changes in parameter due to the dehydration of protein.

9.7. ONE-PHASE MODEL FOR PROTEIN GEL

Protein in contact with adsorbed water may alternatively be considered to remain in one phase so that there exists no interface between protein and water and the properties of water are uniform throughout the phase. The activity a_1 (and chemical potential μ_1) of water in this single phase changes with increase of protein fraction in the gel, whereby swelling (osmotic) pressure of the system will increase. This includes protein–solvent and other types of interactions. The Gibbs–Duhem equation for such a single phase system containing n_1 moles of water and one mole of protein can be written in the form[10]

$$n_1 d\mu_1 + d\mu_p = 0 \qquad (9.11)$$

Here $d\mu_p$ is the free energy change of a mole of protein interacting with n_1 moles of water. The total free energy change ΔG_w may then be given by

$$-\Delta G_w = -\int_1^{a_1} d\mu_p$$

$$= RT \int_1^{a_1} \frac{n_1}{a_1} da_1 \qquad (9.12)$$

Comparing equations (9.9) and (9.12), we may write

$$N \int_{\gamma_0}^{\gamma} A_p d\gamma = \int_1^{a_1} d\mu_p$$

$$= \mu_p - \mu_p' \qquad (9.13)$$

Here μ_p and μ_p' are the chemical potentials of the protein in the gel at water activity equal to a_1 and unity, respectively.

The relation (9.13) shows the equivalence between the one-phase and two-phase models for the calculation of free energy change of protein hydration. The left side of equation (9.13) represents the increase in the interfacial energy due to protein dehydration which is attended with change of curvature, interfacial ionization, swelling, etc. The right-hand side of this equation represents the free energy change for segment–segment, segment–water, and water–water interactions according to the Flory–Huggins theory. Calculation of free energy change is, therefore, independent of the model. However, the physical interpretations of n_1 in the two models are entirely different. In the "two-phase" models n_1 moles of water remain physically bound inhomogeneously at the protein–water boundary. For such a system, the Gibbs method of studying inhomogeneous boundary phase will be very useful. In the "one-phase" model, however, the physical division between bound and free water is not allowed.

9.8. ENTHALPY CHANGE FOR PROTEIN HYDRATION

The isotherms for adsorption of water vapor by several proteins have been obtained at two or more temperatures[9] (vide Fig. 9.5). Here n_1 is found to decrease with increase of temperature. This reveals the physical nature of adsorption. The experimental results are useful for the calculation of enthalpies and entropies of adsorption.

The adsorption process may be expressed by the equation[11,18,25]

$$n_1 \text{ (vapor)} \underset{\text{at } p,\, T}{\rightleftharpoons} n_1 \text{ (adsorbate)} \atop \text{at } T \qquad\qquad (9.14)$$

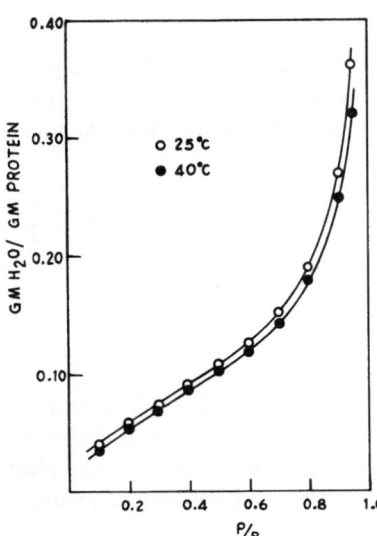

FIGURE 9.5. Water vapor adsorption isotherm of egg albumin at two temperatures. (Drawn from data in Ref. 9.)

If $\bar{\mu}_1$ is the chemical potential of the vapor behaving as an ideal gas, then

$$\bar{\mu}_1 = \bar{\mu}_1^0 + RT \ln p \qquad (9.15)$$

$\bar{\mu}_1^0$ is the chemical potential at a new standard state of vapor at atmospheric pressure.

At a given value of n_1, p depends on temperature (vide Fig. 9.5), so that

$$\left(\frac{\partial \ln p}{\partial T}\right)_{n_1} = \left[-\frac{1}{R}\left(\frac{d\bar{\mu}_1^0/T}{dT}\right) - \left(\frac{\partial \bar{\mu}_1/T}{\partial T}\right)_{n_1}\right]$$

$$= -\frac{\bar{H}_1 - \bar{H}_1^0}{RT^2}$$

$$= -\frac{\Delta \bar{H}_1}{RT^2} \qquad (9.16)$$

Here $\Delta \bar{H}_1$ is the partial molal enthalpy of hydration of protein. $\bar{H}_1$ and $\bar{H}_1^0$ are the partial molal enthalpies of adsorbed gas at pressure p and one atmospheric pressure, respectively. The differential isosteric enthalpy of hydration ($\Delta \bar{H}_1$) may thus be conveniently calculated from the following relation:

$$\Delta \bar{H}_1 = -RT^2\left(\frac{\partial \ln p}{\partial T}\right)_{n_1} \qquad (9.17)$$

The differential of $\ln p$ may be calculated using the isotherms in Fig. 9.5, whereby $\Delta \bar{H}_1$ for various values of T may be obtained. The differential partial molal entropy $\Delta \bar{S}_1$ of adsorption of water may be calculated from the relation

$$\Delta \bar{S}_1 = \frac{1}{T}(\Delta \bar{H}_1 - \Delta \bar{G}_1)$$

$$= -RT\left(\frac{\partial \ln p}{\partial T}\right)_{n_1} - R \ln p \qquad (9.18)$$

Here $\Delta \bar{G}_1$ or $\bar{\mu}_1 - \bar{\mu}_1^0$ is equal to $RT \ln p$ according to equation (9.15). In Fig. 9.6, differential enthalpy and entropy of adsorption of water vapor by egg albumin[24,26] have been plotted against n_1. It has been found that both enthalpy and entropy of protein hydration are significant.

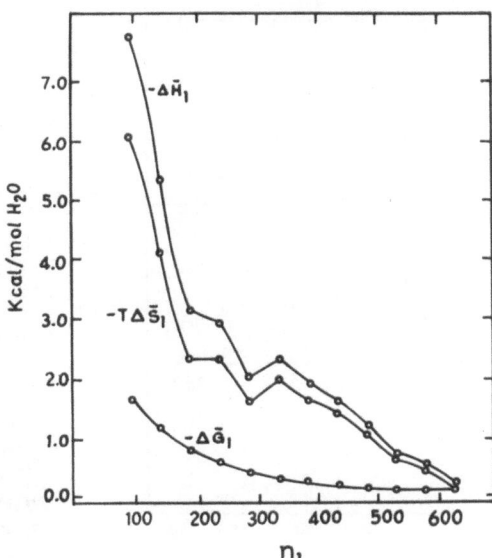

FIGURE 9.6. Calculated values of differential thermodynamic quantities of sorption for native egg albumin, desorption isotherm, calculated from data of Altman and Benson at 25°C and 40°C. (Redrawn from Ref. 24.)

The integral enthalpy of hydration of various proteins has been calculated by Bull[9,23] using the equation,

$$\Delta \bar{H}_i = \int_0^{\Delta n_1^0} \Delta \bar{H}_1 \, dn_1 \qquad (9.19)$$

This is the total change in the heat content for the transfer of Δn_1^0 moles of water to a unit amount of protein. The integration may be performed graphically. Values of $\Delta \bar{H}_i$ are also given in Table 9.3 together with those of integral free energies ΔG_i. The results here indicate that the hydration processes of proteins studied are largely controlled by the enthalpy change in the system. The average values of differential enthalpies for serum albumin–water and ribonuclease–water interactions obtained from adsorption and calorimetric data are found not to differ widely.[2,24]

In Fig. 9.6, we find that the curves for entropy and enthalpy possess inflections at the low water adsorption region. The adsorption and also desorption isotherms for most proteins below 80% relative humidity are found to differ from each other.[26] These results indicate the existence of hysteresis in protein gels in the low water adsorption region. In Fig. 9.7, the variation of $\Delta \bar{H}_1$ with n_1 for adsorption and desorption of water vapor are shown for egg albumin.[24,26] The thermodynamic parameters for adsorption and desorption

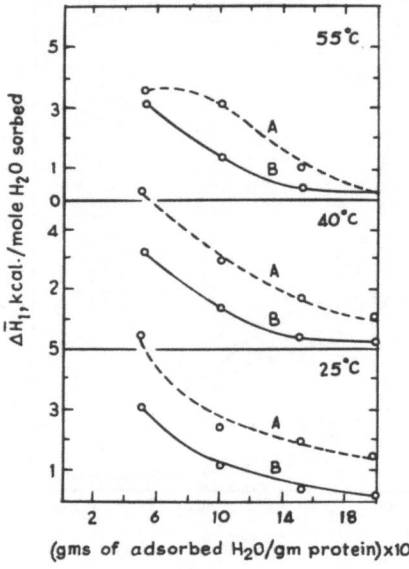

FIGURE 9.7. $\Delta \bar{H}_1$ negative of egg albumin at 25, 40, and 55°C. Curve A is for desorption while B represents adsorption. (Redrawn from Ref. 26.)

shown in this figure are different from each other. The existence of hysteresis suggests several possibilities. The protein system in the hysteresis region may be in nonequilibrium state; or there may exist more than one equilibrium state of equal free energy minimum. To elucidate these points, much more extensive work on water vapor adsorption and calorimetry is needed.[24]

9.9. WATER–PROTEIN INTERACTIONS IN THE PRESENCE OF ELECTROLYTE AND NEUTRAL SOLUTE

Frequently, the thermodynamic measurements relating to binding of water to a biopolymer by various physicochemical techniques have been carried out in the presence of neutral salts and denaturants. The concept of water-binding itself then becomes complex in these three-component systems,[27] since both water and salt now compete for the binding sites of biopolymers. A useful thermodynamic concept of preferential hydration of a biopolymer has been developed by Schachman and Lauffer.[28] Using this concept, preferential hydration of proteins and nucleic acids has been determined by various workers on the basis of extensive experiments on sedimentation and light-scattering.[29–31] The work has been critically reviewed by Kuntz and Kauzmann.[2]

Recently, preferential as well as absolute binding of water and solute to a biopolymer have been determined using isopiestic method of measurement.[32-36] The thermodynamic analysis of these data make uses of the concept of Gibbs surface excess defined originally for the derivation of the adsorption equation (vide Chapter 3).

9.10. EXCESS BINDING

In Section 2.8, experimental details for the water uptake by proteins in the presence of salts or denaturants have been discussed. At isopiestic equilibrium, one thus determines the molal concentration m_2' of the salt component 2 in the two-component reference system, whereby the mole ratio compositions X_1/X_2 or X_2/X_1 may be calculated. At state of equilibrium, direct analysis leads to the evaluation of the values of n_1' and n_2' for the sample containing protein, water, and salt. Here n_1' and n_2' stand for total moles of water and salt associated with one mole of dry protein. From analogy with equations (3.50) and (3.50a), we can write[32-36]

$$n_1' = n_1 + \Delta n_1 \tag{9.20}$$

$$n_2' = n_2 + \Delta n_2 \tag{9.21}$$

where Δn_1 and Δn_2 are, respectively, moles of water and solute remaining in the bound state with one mole of protein whereas n_1 and n_2 are moles of these components remaining in the free state in bulk solution in contact with one mole of biopolymer. If it is assumed that the hydrated protein in the mixture does not contribute to the relative vapor pressure p/p_0 of the system, then the solutions of the salt remaining free in the sample and that in the reference solution will have the same concentration at the same relative humidity p/p_0. The ratio n_1/n_2 (or X_1/X_2) becomes simply equal to $55.51/m_2'$, so that values of it or its reciprocal X_2/X_1 can be calculated from the known value of m_2'.

Based on the Gibbs concept of the surface excess, we now define[34-36] that a mole of protein is binding Γ_1^2 and Γ_2^1 moles of water and solute, respectively, as relative excesses per mole of protein such that

$$\Gamma_1^2 = n_1' - n_2' \frac{X_1}{X_2} \tag{9.22}$$

$$\Gamma_2^1 = n_2' - n_1' \frac{X_2}{X_1} \tag{9.23}$$

The superscript in Γ_1^2 indicates that binding of component 2 is arbitrarily put equal to zero. Similarly for Γ_2^1, binding of component 1 is arbitrarily put equal to zero.

Combining equations (9.22) (9.23), one obtains a relation similar in form to the Adam–Guggenheim equation:

$$\Gamma_1^2 X_2 + \Gamma_2^1 X_1 = 0 \tag{9.24}$$

This relation indicates that equations (9.22) and (9.23) are not independent but related to each other. If the value of Γ_1^2 is positive and definite at a given value of X_1/X_2, a negative value of Γ_2^1 becomes automatically fixed and vice versa. Values of Γ_1^2 and Γ_2^1 may be computed from the experimental data. In Figs. 9.8 and 9.9, Γ_2^1 and Γ_1^2 for egg albumin have been plotted against X_2/X_1 and X_1/X_2, for CsCl and urea, respectively. We note with interest that Γ_1^2 and Γ_2^1 are positive and negative excesses, respectively, for CsCl, whereas the reverse is the case with urea. In the latter case, urea is bound to the protein as positive excess and water as negative excess.

It appears that the plot of Γ_2^1 versus X_2/X_1 (and the plot of Γ_1^2 against X_1/X_2) for CsCl is linear in the entire range of the solution composition. The slope and intercept of this plot cannot be identified with n_1^t and n_2^t according to equation (9.23) since these quantities will always vary with X_2/X_1. If one inserts equations (9.20) and (9.21) in equations (9.22) and (9.23), then

$$\Gamma_1^2 = \Delta n_1 - \Delta n_2 \frac{X_1}{X_2} \tag{9.25}$$

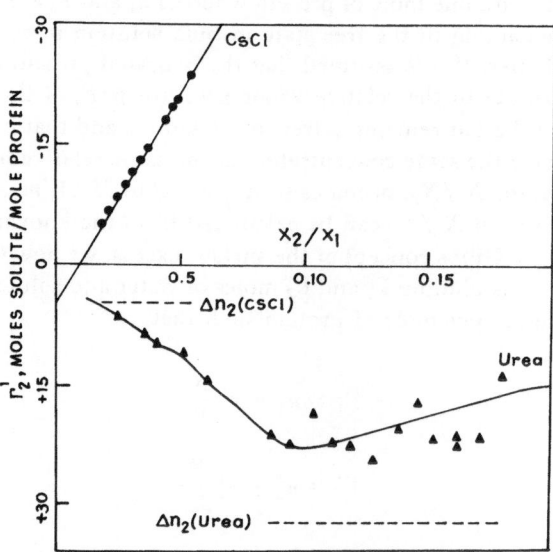

FIGURE 9.8. Plot of Γ_2^1 of egg albumin against X_2/X_1 for CsCl and urea.[34] (Reproduced with permission from *Indian Journal of Biochemistry and Biophysics*.)

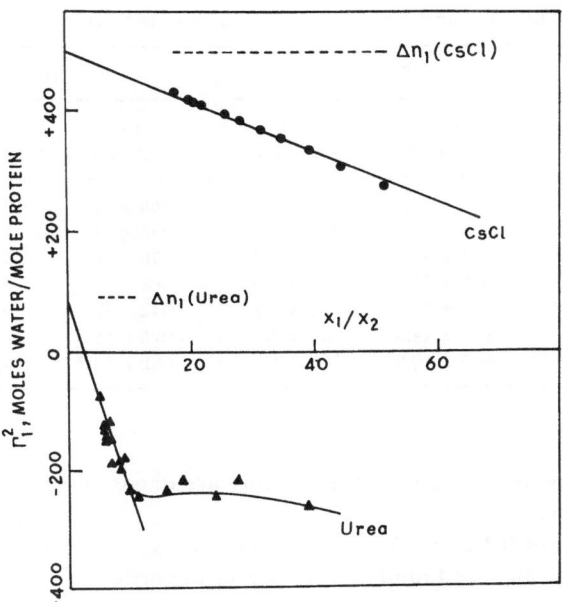

FIGURE 9.9. Plot of Γ_1^2 of egg albumin against X_1/X_2 for CsCl and urea.[34] (Reproduced with permission from *Indian Journal of Biochemistry and Biophysics.*)

$$\Gamma_2^1 = \Delta n_2 - \Delta n_1 \frac{X_2}{X_1} \qquad (9.26)$$

The linearity can easily be explained by these equations, if one assumes that with variation of X_2/X_1, Δn_1 and Δn_2 for the protein bound phase remain constant.[34–36] For urea, linearity exists in a limited range of concentrations, so that Δn_1 and Δn_2 can be estimated. Bull and Breese[32,33] for the first time evaluated Δn_1 and Δn_2 from the linear plot of the left-hand side of equation (9.27) against m_2':

$$m_2' n_1^i - 55.5 n_2^i = m_2' \Delta n_1 - 55.5 \Delta n_2 \qquad (9.27)$$

Equation (9.27) can be obtained from the combinations of equations (9.23) and (9.26) followed by replacement of the term X_2/X_1 by $m_2'/55.51$. Values of Δn_1 and Δn_2 for egg albumin in the presence of various salts are given in Table 9.4.

In Fig. 9.10, values of Γ_2^1 for bovine serum albumin[35] have been plotted against X_2/X_1. From the linear portions of the curves, Δn_1 and Δn_2 for several solutes have been evaluated in limited ranges of concentrations. In Table 9.5, values of Δn_1 and Δn_2 for serum albumin,[35] β-lactoglobulin,[36] and gelatin[36]

TABLE 9.4. Δn_1 and Δn_2 for Egg Albumin in the Presence of Salts[a]

	Corrected		Uncorrected	
Salt	Δn_1 (mol/mol)	Δn_2 (mol/mol)	Δn_1 (mol/mol)	Δn_2 (mol/mol)
LiCl	193 ± 36	5.6 ± 1.0	168 ± 31	4.2 ± 0.9
NaCl	211 ± 18	0.2 ± 0.7	216 ± 18	−1.4 ± 0.7
KCl	365 ± 36	2.4 ± 1.3	370 ± 32	1.3 ± 1.2
RbCl	409 ± 26	6.2 ± 1.0	402 ± 19	4.7 ± 0.8
CsCl	493 ± 21	5.2 ± 0.8	512 ± 37	4.6 ± 0.7
Urea	88.9 ± 14.6	32.5 ± 2.5	116.5 ± 15.5	32.4 ± 2.7
Guanidine HCl	40.7 ± 10.0	32.8 ± 1.3	44.0 ± 10.2	32.3 ± 1.4

[a] References 32, 33.

have been presented. However, these values are uncorrected for the effect of osmotic pressure.

From Tables 9.4 and 9.5, values of Δn_1 for egg albumin[32] and BSA[35] in the presence of salts are found to decrease in the order

$$LiCl < NaCl < KCl < RbCl < CsCl$$

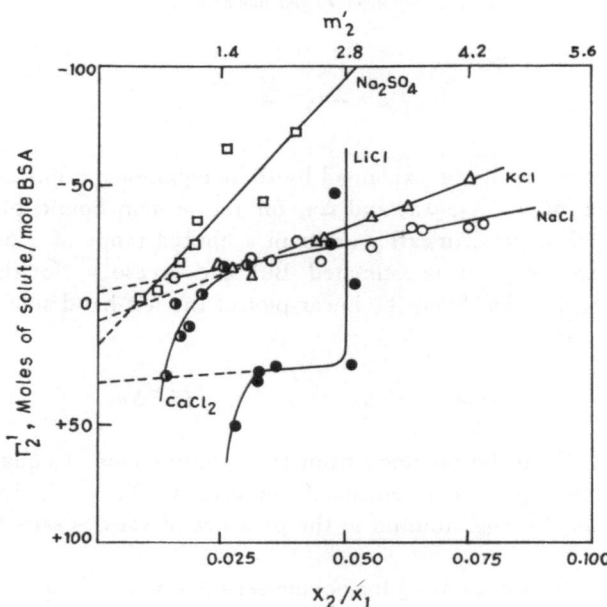

FIGURE 9.10. Plot of Γ_2^1 of BSA against X_2/X_1 and m_2' for NaCl, KCl, LiCl, Na_2SO_4, and $CaCl_2$[35]. (Reproduced with permission from *Indian Journal of Biochemistry and Biophysics*.)

TABLE 9.5. Δn_1 and Δn_2 for BSA[a], β-Lactoglobulin[b] and Gelatin[b] in the Presence of Salts

	BSA		β-lactoglobulin		Gelatin	
Salt	Δn_1 (mol/mol)	Δn_2 (mol/mol)	Δn_1 (mol/mol)	Δn_2 (mol/mol)	Δn_1 (mol/mol)	Δn_2 (mol/mol)
LiCl	195 ± 67	36.4 ± 14.4	—	—	—	—
NaCl	397 ± 121	−6.3 ± 5.7	712 ± 101	52.7 ± 9.2	2025 ± 911	6.0 ± 41.4
KCl	781 ± 70	6.7 ± 3.4	2181 ± 316	82.5 ± 15.1	7505 ± 1083	88.0 ± 35.4
CaCl$_2$	—	—	725 ± 108	53.7 ± 6.5	—	—
KSCN	1200 ± 11	137 ± 1	—	—	—	—
Na$_2$SO$_4$	2240 ± 418	17.4 ± 10.1	—	—	4742 ± 623	19.8 ± 9.9
KBr	1500 ± 221	70.4 ± 38.2	—	—	—	—
KI	2570 ± 336	288 ± 41	—	—	—	—
Gu-HCl	3660 ± 1060	615 ± 171	—	—	—	—
Urea	4030 ± 3400	915 ± 609	—	—	—	—

[a] Reference 35.
[b] Reference 36.

in accordance with the lyotropic series. For many electrolytes, the excess binding of salt is observed to be positive in a certain range of concentration, whereas in other concentration ranges, its value is negative (vide Fig. 9.10 for CaCl$_2$). Values of Δn_1 in the presence of Na$_2$SO$_4$ are relatively high. The water binding capacity of gelatin is unusually high, possibly because it assumes random-coil structure. For guanidine hydrochloride and urea, values of Δn_2 are considerably high; this may be related to their strong unfolding action on globular protein structure. The stabilizing and denaturing actions of electrolytes on proteins will be discussed in Section 9.15.

The hydration of casein in the presence and absence of milk fat has been recently studied with the addition of neutral salt in the aqueous medium (Sadhukhan and Chattoraj, unpublished). In Fig. 9.11, linear plots of Γ_2^1 for such system are found to be different. Values of Δn_1 and Δn_2 for casein for different systems are compared in Table 9.6. It appears that hydration of protein increases in the presence of fat, which itself is highly hydrophobic in character.

9.11. OSMOTIC CORRECTIONS

In equations (9.25) and (9.26), X_2/X_1 is put equal to $m_2'/55.51$ with the assumption that protein in the sample exerts negligible lowering of vapor pressure. In the sample bottle, protein in solution or gel state is highly concentrated. The osmotic pressure (or swelling pressure) π exerted by protein

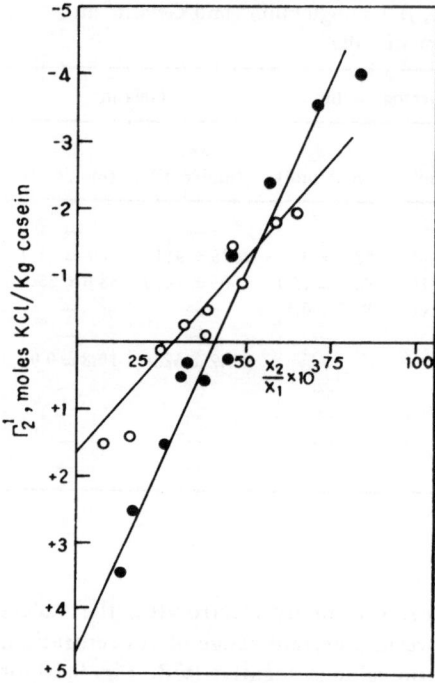

FIGURE 9.11. Plot of moles of KCl bound per kilogram of casein (Γ_2^1) against X_2/X_1. ●, Casein; ○, fat + casein. (B. Sadhukhan and D. K. Chattoraj, unpublished.)

TABLE 9.6. Values of Δn_1 and Δn_2 for Various Electrolytes in the Case of Casein without Fat and with Fat[a]

	Casein			
	without fat		with fat	
Electrolytes	Δn_1 $\left(\dfrac{\text{moles } H_2O}{\text{kg casein}}\right)$	Δn_2 $\left(\dfrac{\text{moles solute}}{\text{kg casein}}\right)$	Δn_1 $\left(\dfrac{\text{moles } H_2O}{\text{kg casein}}\right)$	Δn_2 $\left(\dfrac{\text{moles solute}}{\text{kg casein}}\right)$
NaCl	41 ± 7	1.5 ± 0.3	86 ± 8	5.2 ± 0.5
KCl	59 ± 7	1.6 ± 0.3	112 ± 8	4.6 ± 0.4
$CaCl_2$	52 ± 3	1.9 ± 0.1	146 ± 17	2.3 ± 0.4
Na_2SO_4	63 ± 8	1.4 ± 0.3	82 ± 7	1.7 ± 0.2
KSCN	105 ± 3	4.0 ± 0.1	119 ± 11	4.8 ± 0.7
LiCl	206 ± 21	11.0 ± 0.8	235 ± 13	9.2 ± 0.3
Urea	80 ± 4	5.3 ± 0.3	138 ± 5	5.7 ± 0.7

[a] B. Sadhukhan and D. K. Chattoraj (unpublished data).

is related to p'/p_0 by the following equation:

$$\ln \frac{p'}{p_0} = - \frac{\pi \bar{V}_1}{RT} \tag{9.28}$$

Here $\bar{V}_1$ is the partial specific volume of water and p'/p_0 is the fraction of vapor pressure lowered by a given quantity of protein alone in the sample mixture. Bull and Breese[32,33,37] have measured the osmotic and swelling pressures of protein solution or gel in the presence of concentrated solutions of inorganic salts, urea, and guanidine hydrochloride. From these experiments they have estimated p'/p_0 for a given composition of the biopolymer in the sample. The corresponding concentration (m_2^p) of the neutral salt of the same water activity p'/p_0 can be determined from the water-activity concentration curve of that salt drawn on the basis of known available data.[23,38] Obviously, the concentration m_2 of the free salt in the sample is equal to $m_2' - m_2^p$. Since m_2^p is small, m_2' and m_2 do not differ widely. Replacing m_2' by m_2 in equation (9.27), Bull and Breese[32,33] have given corrected values of Δn_1 and Δn_2 (vide Table 9.4). The osmotic correction for binding is found to be only of the second order in magnitude.

The measurement of osmotic pressure of highly concentrated protein solution is not an easy affair. Mitra and Chattoraj[35] have shown that π in such cases can be approximately estimated from the concentration of protein and its nonspherical size and shape parameters.[39]

9.12. FREE ENERGY OF EXCESS BINDING (TWO-PHASE MODEL)

Let us assume that solvated protein molecules mixed with free solution in the sample system form a two-phase colloidal suspension or gel at isopiestic equilibrium.[34] The Gibbs–Duhem equation for the free phase in the sample (with osmotic correction) may be written as

$$n_1 d\mu_1 + n_2 d\mu_2 = 0 \tag{9.29}$$

whereas for the protein bound inhomogeneous phase this equation assumes the form[34,35,40,41]

$$\Delta n_1 d\mu_1 + \Delta n_2 d\mu_2 + N A_p d\gamma = 0 \tag{9.30}$$

The chemical potentials μ_1 and μ_2 for the solvent and solute components are uniform in the bound and free phases, respectively. Eliminating $d\mu_1$ in equations (9.29) and (9.30), it can be shown that

$$-dG_{ws} = -NA_p d\gamma$$

$$= \left(\Delta n_2 - \Delta n_1 \frac{n_2}{n_1} \right) d\mu_2$$

$$= \Gamma_2^1 d\mu_2 \tag{9.31}$$

so that the total change of free energy $-\Delta G_{ws}$ can be calculated by integrating equation (9.31) between the limit of solute concentration zero to a definite value X_2. The value of Γ_2^1 varies in this situation from zero to Γ_2^1. Thus,

$$-\Delta G_{ws} = -\int_{\gamma_0}^{\gamma} NA_p d\gamma$$

$$= \int_0^{a_2} \Gamma_2^1 d\mu_2 \tag{9.32}$$

Here γ_0 refers to pure water whereas γ refers to a definite value of X_2. Equation (9.32) thus takes into account the free energy change due to the simultaneous interaction of solute and solvent to a mole of protein. γ here includes also other kinds of free energies of interactions such as conformational alterations, curvature change due to the change of vapor pressure, electrostatic interactions, etc. which may be separately included in equation (9.30) if it is so desired. Eliminating the $d\mu_2$ term in equations (9.29) and (9.30), it can similarly be shown that

$$-\Delta G_{ws} = \int_{a_1=1}^{a_1} \Gamma_1^2 d\mu_1 \tag{9.33}$$

For the nonionic solute, one can write

$$\mu_1 = \mu_1^0 + RT \ln f_1 X_1 \tag{9.34}$$

$$\mu_2 = \mu_2^0 + RT \ln f_2 X_2 \tag{9.35}$$

Here f_1 and f_2 are activity coefficients and μ_1 and μ_1^0 the standard chemical potentials of the solvent and solute in the Raoult's law or Henry's law rational scales of activities, respectively (vide Section 1.8).

Differentiating these equations at a constant temperature and then inserting the values in equations (9.32) and (9.33), we find[34]

$$-\Delta G_{ws} = RT \int_{f_2 X_2 = 0}^{f_2 X_2} \frac{\Gamma_2^1}{X_2} \, dX_2 \qquad (9.36)$$

or

$$-\Delta G_{ws} = RT \int_{f_1 X_1 = 1}^{f_1 X_1} \frac{\Gamma_1^2}{X_1} \, dX_1 \qquad (9.37)$$

For the evaluation of either of these integrals for nonelectrolytic solute such as urea, Γ_2^1/X_2 is plotted for various values of X_2 and the area under the curve is graphically evaluated. Difficulties in extrapolation of Γ_2^1/X_2 to zero value of $f_2 X_2$ have been discussed by Chattoraj and Mitra.[34] In such a situation, a Γ_2^1/X_2 versus X_2 curve in low concentration regions can be extrapolated assuming the validity of the Langmuir adsorption equation. Values of ΔG_{ws} for urea computed in this manner for egg albumin are shown in Fig. 9.13.

For an electrolyte $R_{\nu_+} Q_{\nu_-}$ which on complete dissociation in water gives ν_+ and ν_- number of cations and anions, respectively, $d\mu_2$ is given by equations (3.76) and (3.77). Inserting these values in equation (9.32), one obtains[41]

$$\Delta G_{ws} = (\nu_+ + \nu_-)RT \int_0^{f_\pm X_\pm} \Gamma_2^1 d \ln f_\pm X_\pm \qquad (9.38)$$

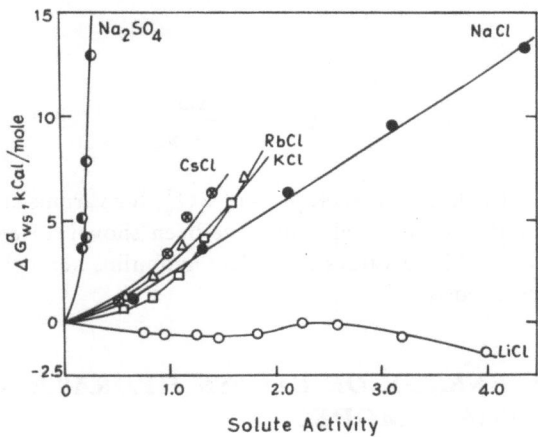

FIGURE 9.12. Plot of ΔG_{ws}^a of egg albumin against solute activity for various electrolytes.[34] (Reproduced with permission from *Indian Journal of Biochemistry and Biophysics*.)

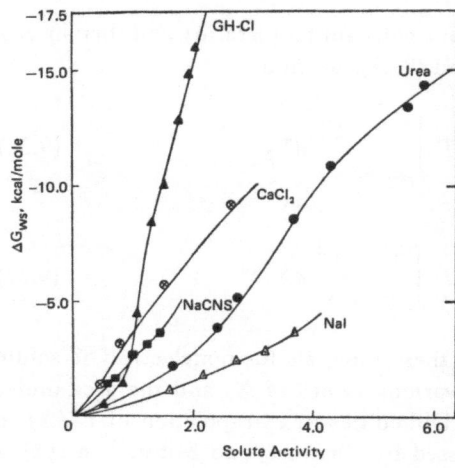

FIGURE 9.13. Plot of ΔG_{ws}^{a} (written as ΔG_{ws}) of egg albumin against solute activity for various destabilizers.[34] (Reproduced with permission from *Indian Journal of Biochemistry and Biophysics*.)

Calculations of the mean activity coefficient $f_{\pm}$ and mean mole fraction $X_{\pm}$ of a solute can be made from the experimental data using equations (3.81) and (3.80), respectively. Assuming $X_{\pm}$ to be roughly proportional to X_2, we find

$$\Delta G_{ws} = (\nu_{+} + \nu_{-})RT \int_{0}^{X_2 f_{\pm}} \frac{\Gamma_2^1}{f_{\pm} X_2} \, dX_2 f_{\pm}$$

$$= (\nu_{+} + \nu_{-})\Delta G_{ws}^{a} \qquad (9.39)$$

so that

$$\Delta G_{ws}^{a} = \frac{\Delta G}{\nu_{+} + \nu_{-}} \qquad (9.40)$$

In Figs. 9.12, 9.13, and 9.14, plots of ΔG_{ws}^{a} for various electrolytes against the solute activity in the molal scale have been shown for egg albumin[34] and serum albumin.[35] These values for β-lactoglobulin, hemoglobin, and gelatin have also been reported.[36]

9.13. FREE ENERGY OF EXCESS HYDRATION (ONE-PHASE MODEL)

In isopiestic experiments, proteins in the presence of a large number of inorganic salts form concentrated solutions, whereas in the presence of urea and guanidine hydrochloride etc., they form protein gels. In the one-phase

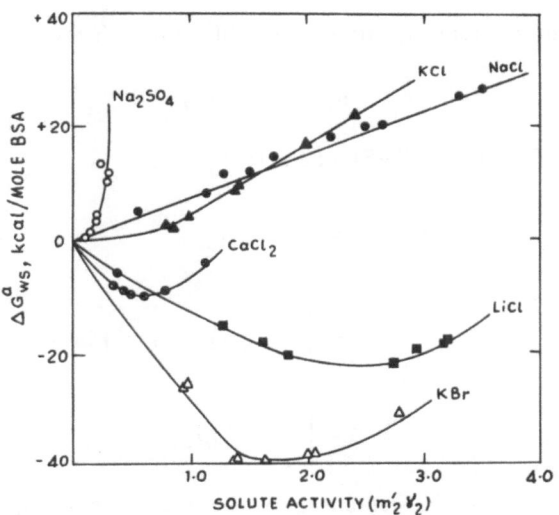

FIGURE 9.14. Plot of ΔG_{ws}^{a} of BSA against solute activity of NaCl, KCl, LiCl, Na$_2$SO$_4$, CaCl$_2$, and KBr.[35] (Reproduced with permission from *Indian Journal of Biochemistry and Biophysics*.)

model for these systems, the boundary between protein and water will not exist and the meaning of Δn_1 and Δn_2 will vanish. Equations (9.25) and (9.26) will then not remain valid. Excess binding of water and solute will still be defined by equations (9.22) and (9.23), respectively. One can define the Gibbs–Duhem equation for such a system containing one mole of protein in the following way[34,41]:

$$V\,dP + n_1'd\mu_1 + n_2'd\mu_2 + d\mu_p = 0 \qquad (9.41)$$

Here the excess pressure P for the sample arises due to the osmotic or swelling pressure π exerted by the protein. If V is the total volume of the sample, then

$$V = n_1'\bar{V}_1 + n_2'\bar{V}_2 + \bar{V}_p \qquad (9.42)$$

$\bar{V}_i$ refers to partial molal volume of component i in the mixture. Combining equations (9.41) and (9.42), and assuming $\bar{V}_i$ to be constant, it can be shown that

$$n_1'd(\mu_1 + P\bar{V}_1) + n_2'd(\mu_2 + P\bar{V}_2) + d(\mu_p + P\bar{V}_p) = 0 \qquad (9.43)$$

or

$$n_1'd\mu_1' + n_2'd\mu_2' + d\mu_p' = 0 \qquad (9.44)$$

where μ_i' is equal to $\mu_i + P\bar{V}_i$.

For the reference solution at equilibrium with the sample

$$n_1' d\mu_1' + n_2' d\mu_2' = 0 \tag{9.45}$$

Combining (9.44) and (9.45), one gets

$$-\Delta G_{ws}' = \int_{X_2=0}^{X_2} d\mu_p'$$

$$= \int_{X_2=0}^{X_2} \left(n_2' - n_1' \frac{n_2'}{n_1'} \right) d\mu_2'$$

$$= \int_{X_2=0}^{X_2} \Gamma_2' d\mu_2' \tag{9.46}$$

$\Delta G_{ws}'$ is the change in free energy without any osmotic pressure correction for Γ_2 and μ_2. This equation is exact and the free energy change includes both water and solute interactions with proteins as well as the osmotic work for bringing salt concentration X_2 from zero to X_2. This osmotic work for a fixed concentration of protein may be negligible, and under this condition, $\Gamma_2' \simeq \Gamma_2'$ and $\mu_2' \simeq \mu_2$, so that equations for "two-phase" and "one-phase" models become identical. In agreement with this, values of $\Delta G_{ws}'$ and ΔG_{ws} for egg albumin and serum albumin calculated from experimental data on the basis of the respective equations (9.38) and (9.46) are found to be quite close to each other.[34,35]

9.14. STABILITY OF PROTEIN IN CONTACT WITH WATER

For the theoretical treatment of hydrophobic forces in relation to the biopolymer structure and conformation, contributions of the surface energy and a few interfacial parameters have been considered earlier by Sinanoglu and Abdulnur[42] and Lewin.[43] From the discussion in Section 1.12, it becomes clear that surface tension or surface free energy is directly related to the intermolecular attraction within the condensed phase. In Table 9.7, the surface free energies of various air–liquid and liquid–liquid systems have been compared with one another.[10] Consider a droplet of an organic liquid (e.g., benzene or octane) surrounded either by air (or vapor) or water. From the values of γ given in Table 9.7, it is obvious that the surface free energy and hence the internal attraction within the droplet will increase significantly if the surrounding air (or vapor) is replaced by water. The formation and dispersion of

TABLE 9.7. Interfacial Energies[a] of Different Liquid Systems at 20°C

System	Interfacial energy (mN/m)	System	Interfacial energy (mN/m)
Benzene–air	28.8	n-Octyl alcohol–water	8.5
Toluene–vapor	28.5	Heptylic acid–water	7.0
m-Xylene–vapor	28.9	Ethyl ether–water	10.7
Phenol–vapor	40.9	Ethyl acetate–water	2.9
Aniline–vapor	42.9	n-Butanol–water	1.6
Benzaldehyde–vapor	40.0	Hexanol–water	6.8
n-Octane–vapor	21.8	Benzene–water	35.0
n-Octyl alcohol–air	27.5	Nitrobenzene–water	26.0
Hexane–water	51.1	Aniline–water	5.8

[a] Reference 10.

hydrophobic colloid and emulsion in aqueous media are for this reason energetically disfavored. On the other hand, when molecules of the surrounding liquid droplet are made polar by introduction of $-OH$, $-COOH$, $-NH_2$, $-NO_2$, and CH_3CO- groups in aliphatic and aromatic hydrocarbons, the surface free energy will decrease considerably due to the replacement of oil–air interface by oil–water interface. The dispersion of hydrophilic colloid will thus reduce the free energy of the system considerably due to the hydration of the hydrophilic groups at the interface of the oil drop.[10]

The molecules of globular protein such as serum albumin, hemoglobin, etc. are rigid and giant macromolecules whose boundary region contains a large amount of hydrophilic groups. By hydration of these groups, the free surface energy and hence the internal attraction within the folded molecules decrease considerably. There also exists 30% to 50% surface area of a globular protein molecule which is definitely made up of hydrophobic groups.[7,15] Contact with water in this hydrophobic interfacial region will eventually increase the surface free energy of the system. In fact, if one makes a balance sheet, then protein surrounded by a large amount of water can possess extremely low surface energy, the magnitude of which will be determined by the proportion of the hydrophobic and hydrophilic regions of the interface as well as interior of the biopolymer. An attempt has been very recently made to measure this low surface energy by Bull and Breese[44] using egg albumin. This will be discussed in Chapter 11. This residual interfacial energy is directly related to the free energy change due to the internal attraction within the macromolecule, which is largely responsible for the folded and compact structure of protein. It is expected that the free energy of denaturation of protein is closely related to this residual interfacial energy. Thus protein molecule in aqueous medium behaves as a rigid and solidlike particle. Energetically, its property is similar to that of a solid very close to its melting point. The rigid structure of the solidlike protein dispersed in aqueous medium can be destroyed by mild heat and other denaturing agents. By reducing the vapor pressure from unity, the hydration of protein decreases and the surface free energy increases and vice versa. Bull and Breese[15] have actually shown that hydration of protein is reversible up to water activity 0.8. The hysteresis regions below this activity value for proteins can also be explained suitably by the surface tension effect.[2,10]

Many proteins in aqueous medium exhibit distinct properties of solution forming a single phase. As a result of this mixing process in solution, the entropy of the system increases and corresponding free energy decreases.[45] The development of osmotic pressure though relatively low in a protein solution is another consequence. Proteins in the pure state have a fixed solubility which in a thermodynamic sense means that the protein–water system consists of a single phase. But if the protein solution is kept for a long time, even under sterile condition, aging takes place and protein molecules aggregate slowly with the formation of a precipitate. This aging is a colloidal property, as a result of which the surface free energy may decrease. It seems that the protein solution stands on the border line between the colloidal solution and true solution, thereby exhibiting dual properties.

9.15. STABILITY OF PROTEIN IN SALT SOLUTION

We have already assumed that protein in contact with large amounts of water is in the standard state of minimum (surface) free energy which can be arbitrarily put equal to zero. We will now examine the effect of addition of salt to the protein–water systems. In Figs. 9.12, 9.13, and 9.14, we note that ΔG_{ws} for egg albumin and serum albumin in the presence of NaCl, KCl, RbCl, CsCl, and Na_2SO_4 have positive values which increase with increase of solute activity. All these solutes introduce additional stability into the protein structure by increasing intersegmental attraction. These salts are, therefore, termed structure stabilizers.[34,35]

The degree of stability of the folded structure of a globular protein in solution can be qualitatively related to its melting temperature T_m which is determined by viscometric, pH-metric, calorimetric, and several other experiments.[46-50] With gradual increase of temperature, it is observed from all these experiments that a protein in aqueous medium maintains its folded structure up to 50°C or more; then at a narrow temperature range the compact globular shape of the biocolloid is disrupted by thermal forces. The biopolymer molecule is completely and sharply unfolded to a random-coil structure at this stage and a sharp change in physical properties of the solution is observed. The

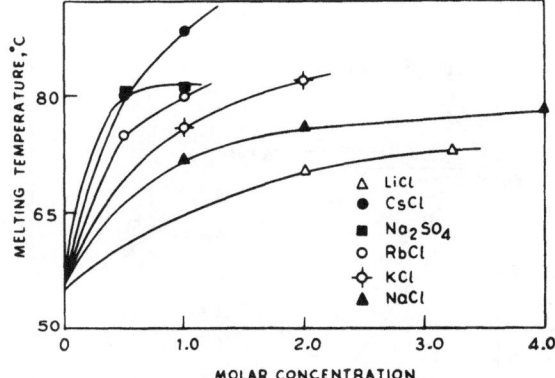

FIGURE 9.15. Plot of melting temperature (T_m) BSA against molarity of electrolytes.[50] (Reproduced with permission from *Indian Journal of Biochemistry and Biophysics*.)

melting temperature of serum albumin is observed to increase with increase of concentration of NaCl, KCl, CsCl, Na_2SO_4, and other stabilizing salts.[50] This is demonstrated in Fig. 9.15. The results are in agreement with the stabilizing action of these electrolytes as obtained from study of hydration.

The melting temperature of a protein decreases in presence of increasing concentration of denaturants such as urea, guanidine salts, KCNS, $CaCl_2$, etc. (vide Figs. 9.16 and 9.17). In Fig. 9.16, the melting temperature of ribonuclease[49] is shown to decrease with increase of guanidine hydrochloride in the medium. When bovine serum albumin is treated for a prolonged period with

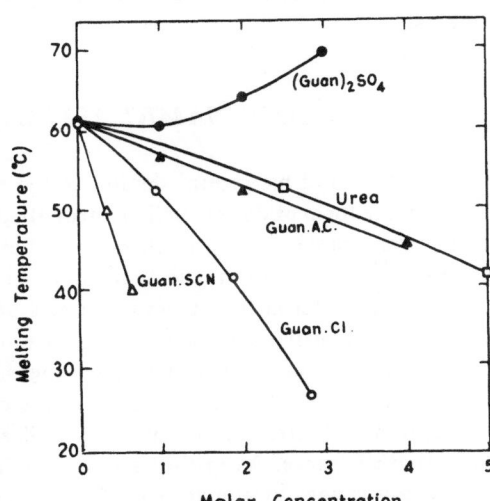

FIGURE 9.16. Melting temperatures of ribonuclease as a function of the concentration of urea and various guanidinium salts.[49] pH 7.0, 0.15 M KCl and 0.013 M sodium cacodylate. (Redrawn from Ref. 49.)

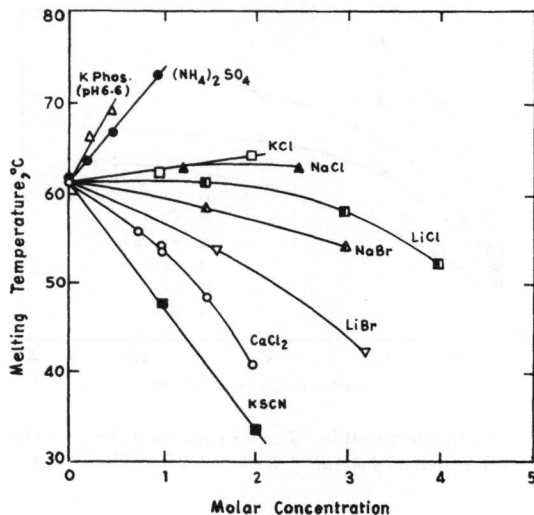

FIGURE 9.17. Melting temperatures of ribonuclease as a function of the concentration of various added salts. pH 7.0, 0.15 M KCl and 0.013 M sodium cacodylate. (Redrawn from Ref. 49.)

concentrated solutions of inorganic destabilizers such as KI, KCNS, $CaCl_2$, etc. and the solution is exhaustively dialyzed, considerable precipitation of denatured protein takes place.[50] ΔG_{ws} of BSA has been found to decrease substantially in the presence of these salts. The denaturation of biopolymers can thus be explained by the decrease of free energy of the protein–water system in the presence of salts, assuming the system to consist of one or two phases.

9.16. THERMODYNAMIC ASPECTS OF DNA HYDRATION

Wilkins and his group[4] demonstrated earlier that the X-ray structure of DNA in the form of double helix is strongly dependent on the relative humidity and water uptake by the nucleic acid fiber. From proton magnetic studies, Jacobson[51] observed that the Watson–Crick double helix structure fits nicely an ice I-like water lattice structure. The study of water–DNA interactions has been subsequently undertaken using various experimental techniques.[6,52–56]

Falk, Hartman, and Lord[52] have measured water uptake by sodium and lithium salts of DNA at different relative humidities. The corresponding adsorption isotherms shown in Fig. 9.18 are sigmoid shaped and resemble type II BET curve. As in the case of proteins, the BET equation is observed to fit at low range of water activity; Δn_m for monolayer formation [vide equation

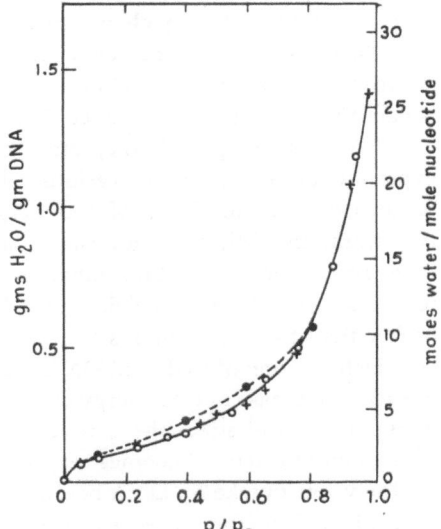

FIGURE 9.18. Adsorption (○) and desorption (●) of water vapor by a sample of calf-thymus NaDNA; adsorption of water by a sample of salmon-sperm NaDNA (+). (Redrawn from Ref. 52.)

(9.1)] is two moles of water per mole of nucleotide (mol. wt. 333 on the average). This amount of water, therefore, is strongly bound to one mole of DNA nucleotide. From the extrapolation of $n_\perp$ to p/p_0 equal to unity in the curve in Fig. 9.18, the maximum number of water molecules per DNA nucleotide is found to be approximately 40. This high value suggests that water uptake by DNA is related to the polyelectrolytic nature of DNA. An attempt has also been made by Falk *et al.*[52a] to examine various types of water bound to DNA forming the inhomogeneous region at the interface.

On the basis of the one-phase model for DNA gel, we can write the Gibbs–Duhem equation in the form

$$n_1^t d\mu_1 + n_N d\mu_N = 0 \tag{9.47}$$

Here n_N and n_1^t stand for moles of sodium salt of nucleotide and water, respectively, forming the DNA gel. If n_1^t is expressed as moles of water associated per mole of sodium salt of nucleotide then n_N becomes unity and the free energy of hydration ΔG_i can be calculated from the expression

$$\Delta G_i = \int_0^{n_1=\Delta n_1^0} d\mu_N$$

$$= -RT \int_{a_1=0}^{a_1=1} \frac{n_1^t}{a_1} da_1 \tag{9.48}$$

This integral free energy change includes the effect of the interaction of water with the phosphate site of DNA, counterion hydration, polyelectrolytic dissociation, conformational alteration of the polynucleotide, etc. It is difficult to isolate these effects and consider them separately.

From the graphical integration of equation (9.48) on the basis of the experimental data of n_1' for various values of a_1, it has been found that there occurs a decrease of 37.1 kJ of free energy per mole of sodium salt of nucleotide (Chattoraj and Mitra, unpublished data), when dry DNA is dissolved in a large amount of water. This enhanced value for the free energy of hydration (expressed per 100 g of nucleic acid in Table 9.3) is, as expected, considerably higher than those of proteins.

Dehydration of hydrated DNA is thus a nonspontaneous process, because it leads to increase in free energy of the system. This increase, with progressive dehydration, will affect the double helix structure of DNA; it is expected to be maximum when n_1 becomes very low. The hysteresis phenomenon observed for the water uptake by DNA below a_1 less than 0.75 (vide Fig. 9.18) can also be explained on this basis. Even in the presence of large amount of water, DNA assumes rodlike structure having more surface free energy than an equivalent sphere. This is due to the structural restrictions introduced by the extensive hydrogen bonding in the double helix structure. On denaturation, these bonds are broken to a large extent and the radius of gyration will tend to decrease considerably. It may also be possible that native DNA in water contains residual surface free energy as a result of which tertiary structure of DNA becomes wormlike, or circular, or supercoiled depending upon the environmental conditions. Various types of conformations of DNA occurring in nature have been discussed by Watson.[3]

9.17. BINDING OF WATER AND SOLUTE TO DNA

Isopiestic hydration of DNA in the presence of various inorganic salts, urea, and sucrose has been investigated[57] following the same experimental procedure as described in Section 2.8. At isopiestic equilibrium, the molality m_2' of the reference solution can be determined accurately from the chemical analysis. Also at this state, n_1' moles of water and n_2' moles of salts associated per mole of nucleotide in DNA-polyelectrolyte can be determined by gravimetric analysis. Since DNA itself is a polyelectrolyte, each phosphate site of its nucleotide is associated with n_p number of a metal cation. For sodium and magnesium salts of DNA, respective values of n_p are 1 and 0.5. Further, n_2' moles of the salt on complete dissociation may produce $\nu_+ n_2'$ moles of common metal cation and $\nu_- n_2'$ moles of anion. We thus find that one mole of nucleotide anion in the sample is associated with $(n_2')^\pm$ mean moles of an electrolyte, where

$$(n_2^t)^{\pm} = [(n_p + \nu_+ n_2^t)^{\nu_+}(\nu_- n_2^t)^{\nu_-}]^{1/\nu_+ + \nu_-} \qquad (9.49)$$

We can further assume that

$$(n_2^t)^{\pm} = n_2^{\pm} + \Delta n_2^{\pm} \qquad (9.50)$$

$$n_1^t = n_1 + \Delta n_1 \qquad (9.51)$$

where Δn_1 and $\Delta n_2^{\pm}$ are moles of components 1 and 2 bound, on an average, to a mole of nucleotide anion, whereas n_1 and $n_2^{\pm}$ are average moles of these components remaining free at isopiestic equilibrium. $n_2^{\pm}$ and $\Delta n_2^{\pm}$ can be defined as follows:

$$n_2^{\pm} = [(n_p + \nu_+ n_2)^{\nu_+}(\nu_- n_2)^{\nu_-}]^{1/\nu_+ + \nu_-} \qquad (9.52)$$

$$\Delta n_2^{\pm} = [(\Delta n_p + \nu_+ \Delta n_2)^{\nu_+}(\nu_- \Delta n_2)^{\nu_-}]^{1/(\nu_+ + \nu_-)} \qquad (9.53)$$

For such a complex system, we can operationally define Γ_1^2 and Γ_2^1 by the equations

$$\Gamma_1^2 = n_1^t - (n_2^t)^{\pm}(n_1/n_2^{\pm}) \qquad (9.54)$$

$$\Gamma_2^1 = (n_2^t)^{\pm} - n_1^t(n_2^{\pm}/n_1) \qquad (9.55)$$

From separate experiments on swelling pressure,[57] it has been observed that DNA contributes negligibly to vapor pressure so we can assume safely that at isopiestic equilibrium, the mean molalities of salt in the reference solution and in the free electrolyte solution within the gel are equal to each other, such that

$$\frac{n_2^{\pm}}{n_1} = \frac{\nu_p m_2'}{55.51} \qquad (9.56)$$

where

$$\nu_p = (\nu_+^{\nu_+} \cdot \nu_-^{\nu_-})^{1/(\nu_+ + \nu_-)} \qquad (9.57)$$

From the experimental data on n_1^t, n_2^t, n_p, and m_2', one can compute Γ_2^1 for various values of $n_2^{\pm}/n_1$ for DNA gel in the presence of an inorganic salt. Linear plots of Γ_2^1 against $n_2^{\pm}/n_1$ are shown in Figs. 9.19 and 9.20 for various electrolytes.

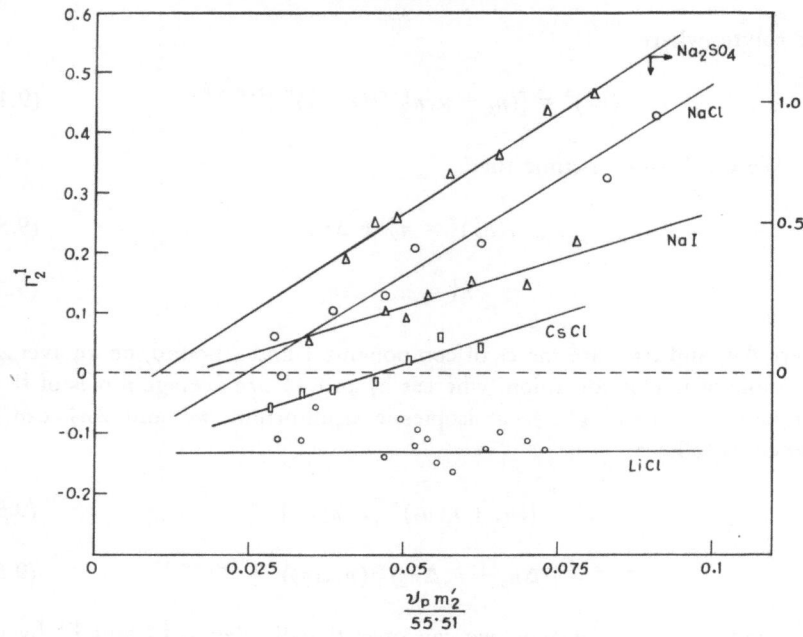

FIGURE 9.19. Linear plots of Γ_2^1 in moles solute per mole nucleotide against $\nu_p m_2'/55.51$ in presence of several salts[57] for calf thymus DNA. (Ordinate represents negative values of Γ_2^1.)

Inserting equations (9.50) and (9.51) in relations (9.54) and (9.55), it can be shown that

$$\Gamma_1^2 = \Delta n_1 - \Delta n_2^{\pm}(n_1/n_2^{\pm}) \qquad (9.58)$$

$$\Gamma_2^1 = \Delta n_2^{\pm} - \Delta n_1(n_2^{\pm}/n_1) \qquad (9.59)$$

The slopes and intercepts of the linear plots in Figs. 9.19 and 9.20 represent, therefore, Δn_1 and $\Delta n_2^{\pm}$, which are given in Table 9.8.

It is interesting to note that in the presence of LiCl, the binding of water to DNA is reduced practically to zero. The order of water bound by DNA in the presence of various salts does not follow the Hofmeister series. Δn_1 for all salts are always considerably less than Δn_1^0 in the complete absence of salt. The average binding of salt to DNA is small except that of $MgCl_2$ which may have some biological significance. A mole of nucleotide is able to bind nearly 8 and 5 moles of water molecules in the presence of urea and sucrose, respectively.

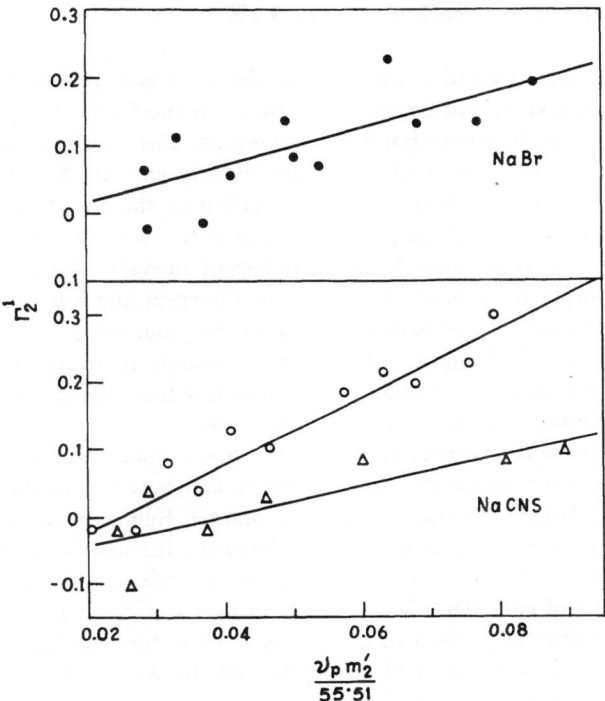

FIGURE 9.20. Linear plots of Γ_2^1 in moles solute per mole nucleotide against $\nu_p m_2'/55.51$ in presence of several salts[57] for calf thymus DNA. (Ordinate represents negative values of Γ_2^1.)

TABLE 9.8. Binding of Salt and Water to DNA Expressed in Moles per Mole of Nucleotide[a]

Type DNA	Salt	Δn_1	$\Delta n_2^{\pm}$
Li DNA	LiCl	0.16 ± 0.48	0.135 ± 0.024
Na DNA	NaCl	6.36 ± 0.44	0.154 ± 0.026
K DNA	KCl	5.12 ± 0.46	0.126 ± 0.023
Cs DNA	CsCl	3.34 ± 0.45	0.154 ± 0.021
Na DNA	NaBr	2.83 ± 1.04	0.041 ± 0.056
Na DNA	NaI	3.17 ± 0.50	0.049 ± 0.028
Na DNA	$NaClO_4$	3.64 ± 0.67	0.040 ± 0.041
Na DNA	NaCNS	2.21 ± 0.60	0.086 ± 0.033
Na DNA	Na_2SO_4	13.29 ± 0.72	0.131 ± 0.044
Mg DNA	$MgCl_2$ (High concentration)	1.19 ± 0.54	0.434 ± 0.028
Mg DNA	$MgCl_2$ (Low concentration)	5.16 ± 1.28	0.475 ± 0.046

[a] Reference 57.

9.18. SUMMARY AND COMMENTS

In living systems and in many other phenomena of practical importance (such as food, textile, and soils), water either in the form of vapor or liquid are in large interaction with polymeric materials. The results of adsorption of water vapor in the absence of solute are usually explained in terms of the kinetic approach involved in the BET equation or the statistical-mechanical approach presented by Guggenheim,[58] Anderson,[59] and de Boer.[60] All these treatments, however, assume that the water–protein system forms two separate phases. Attempts have been made also to interpret the adsorption data in terms of the statistical-mechanical theory for polymer solution orginally proposed by Flory and Huggins.[21] Any thermodynamic treatment for the study of water–protein interaction must therefore be free from the controversy about the "phase models" for the whole polymer system.

In the present chapter, it has been shown that the free energies of adsorption of water vapor by proteins can be computed from the integration of the Gibbs–Duhem equation in suitable manner. Such a procedure is shown to be independent of the phase models. It has been further shown that fixation of the reference state for the changes of free energies at infinite dilution of protein will be of considerable advantage in terms of the maintenance of the structural stability of protein in the presence of water. Further study of the role of entropy and enthalpy of adsorption on the biopolymer stability with reference to this new standard state may be of considerable biochemical importance. This aspect needs further investigation in future.

The study of water vapor adsorption by proteins in the presence of an electrolyte or nonelectrolyte indicates that both solvent and solute may compete significantly for binding sites of the biopolymer. The measured value of the extent of binding of a component represents the Gibbs excess per mole (or per gram) of biopolymer and the approach is independent of the phase models for the protein–water system. The excess binding of a component may be positive or negative and in many cases there exist an azeotropic point when excess binding of both solute and solvent components become zero (vide also Figs. 9.10 and 9.11). Assuming a two-phase model for the protein–solvent system, the extents of absolute binding of solute and water components for several proteins and protein–fat mixtures can be evaluated in a limited range of concentration. The Gibbs approach in slightly modified form appears to remain valid for excess binding of moles of solvent and mean moles of electrolyte bound to a mole of polynucleotide ion of deoxyribonucleic acid. All these observations indicate that the Gibbs definition for relative adsorption gives a clear concept about the complex binding phenomena observed in many biocolloid systems.

With reference to the infinitely dilute solution of biopolymer in water as the initial state, the free energy change ΔG_{ws} for water–protein interaction with increasing concentration of the electrolyte or nonelectrolyte may be calculated from the integration of the appropriate form of the Gibbs adsorption equation which can be derived on the basis of "two-phase" as well as "one-phase" models. Positive values of ΔG_{ws} calculated from the binding data increase with the increasing concentration (X_2) of many neutral inorganic salts. This may indicate the increase in the interfacial energy of the protein–water system so that the internal stability of protein should also increase with increase of X_2. This is in agreement with the increase of the melting point (T_m) and hence the thermal stability of protein with increase of X_2. Calculated negative values of ΔG_{ws} in the presence of denaturants such as urea, guanidine hydrochloride, KCNS, etc. increase with increase of X_2 in agreement with decreasing thermal stability of a protein with increase of X_2. It is highly desirable to mathematically relate T_M with ΔG_{ws} and its temperature coefficient for the interpretation of protein stability in aqueous media in a quantitative manner. Detailed physicochemical treatments of denaturation and stability of proteins in aqueous media have been developed only recently[61–66] based on calorimetric and other studies. It will be of interest to examine how evaluated values of ΔG_{ws} of various proteins in the presence of different salts at various temperatures, pH, ionic strengths, etc. fit this recent approach mainly based on calorimetric investigations. The agreement between the free energies of hydration of proteins from analytical and calorimetric studies are so far not satisfactory[2] possibly because the standard states in the two sets of measurements are not exactly the same.

CHAPTER 10

BINDING INTERACTIONS IN BIOLOGICAL SYSTEMS

10.1. INTRODUCTION

In the previous chapter we have shown that proteins, nucleic acids, and other biopolymers may adsorb water vapor both in the presence and absence of a third solute component in the system. The solute and water in the sample system are shown to compete with each other for binding interactions at vapor pressure equilibrium.

However, in living systems, binding interactions between biopolymers and solute frequently occur in the aqueous medium directly without involvement of water vapor. Thus, fatty acids and lipids may remain bound with proteins present in the blood serum, which are subsequently circulated inside the arterial system. The surface of the cell membrane formed by the complex association of proteins, lipids, and carbohydrates may bind sodium and potassium ions directly from the aqueous medium, and the process is basically responsible for the active or passive transport of these ions in the cellular system. Within the cell, proteins remain in the bound state with nucleic acids in chromosomal and ribosomal regions. Soluble ribonucleic acids in the cytoplasmic fluid bind various types of amino acids in the presence of an enzyme. The binding of highly complex biological components occurs at the ribosomes during protein synthesis.[1] Binding interactions between hormones and proteins, carbohydrates and proteins, enzymes and substrates, antibodies and antigens are known to be responsible for the occurrence of many processes of 'immense importance within the living system.[2,3]

The phenomenon of binding of a solute to a biopolymer in the aqueous medium has some remarkable similarities with the adsorption of a solute by solid particles suspended in water. Thus fatty acids initially dissolved in water may be either adsorbed on the surface of carbon particles or bound to a protein under suitable conditions. Like negative adsorption, negative binding is sometimes observed in experiments dealing with proteins and nucleic acid. There exist some operational differences between the two kinds of phenomena. In the case of adsorption, the adsorbent and the aqueous solution form two distinct phases. The surface area of the rigid solid particles during adsorption remains constant, so that adsorption can be expressed as amount of solute adsorbed per unit surface area. On the other hand, the biopolymer in the aqueous solution or in the gel phase is assumed for convenience to remain either as a single phase or in two phases. Further, the number of available binding sites (or interfacial area) of a biopolymer will alter with progressive binding of the solute, so that binding is expressed not in terms of surface area but per mole or per gram of the biopolymer.

Experimental and theoretical aspects of various types of biological binding have been extensively documented.[3-23] However, in most of these articles, the concept of the Gibbs surface excess has been neglected. A notable exception is the treatment of Bull,[24-26] who used this concept implicitly in analyzing the binding phenomenon. In the present chapter, we shall limit our discussion mostly to the interpretation of the binding phenomenon based on this concept.

10.2 GIBBS–DUHEM EQUATION AND EXCESS BINDING

Measurement of the extent of binding of organic solutes to proteins and other biopolymers may be carried out *in vitro* by using various physicochemical techniques.[3] The most convenient of them is the equilibrium dialysis technique described in detail in Section 2.9. According to this technique, the biopolymer dissolved in buffer medium is taken inside a dialysis bag which in turn is brought in contact with an outside solution of the same buffer solution. After attainment of equilibrium, the concentration of the solute in the dialysate is determined. This enables calculation of moles (Γ_R^1) of the ligand bound per mole (or per gram) of biopolymer present in the solution.

The organic solute bound to a biopolymer may be neutral, cationic, or anionic in character. The biopolymer itself may originally possess positive or negative charges depending upon the pH of the medium. At a specific pH termed the isoelectric point, the net charge of protein is zero. The study of binding has usually been carried out in the presence of an excess amount of a neutral inorganic salt (e.g., NaCl) or a buffer. Further, at dialysis equilibrium, the protein concentration within the dialysis bag is low so that the osmotic contribution to the chemical potentials is negligible.

Let us assume that n_p^t moles of a biopolymer of initial net charge zero at isoelectric pH are mixed with n_1^t, n_{NaCl}^t, and n_{RH}^t moles of water (component 1), salt, and a nonionic detergent (e.g., triton-X), respectively, at dialysis equilibrium. The superscript t refers to the total amount of a component present in the system. Neglecting the osmotic effect, one can write for such a system

$$n_p^t d\mu_p + n_1^t d\mu_1 + n_{\text{NaCl}}^t d\mu_{\text{NaCl}} + n_{\text{RH}}^t d\mu_{\text{RH}} = 0 \qquad (10.1)$$

Similarly for the outside dialysate solution containing, n_1, n_{NaCl}, and n_{RH} moles of H_2O, NaCl, and RH, respectively,

$$n_1 d\mu_1 + n_{NaCl} d\mu_{NaCl} + n_{RH} d\mu_{RH} = 0 \qquad (10.2)$$

Combining equations (10.1) and (10.2), we get

$$-d\mu_p = \frac{1}{n_p^t}\left(n_{RH}^t - n_1^t \frac{n_{RH}}{n_1}\right)d\mu_{RH}$$

$$+ \frac{1}{n_p^t}\left(n_{NaCl}^t - n_1^t \frac{n_{NaCl}}{n_1}\right)d\mu_{NaCl}$$

$$= \Gamma_{RH}^1 d\mu_{RH} + \Gamma_{NaCl}^1 d\mu_{NaCl} \qquad (10.3)$$

Γ_{RH}^1 and Γ_{NaCl}^1 by definition are, respectively, excess amounts of organic and inorganic solutes bound per mole of biopolymer, it being assumed that relative binding (Γ_1^1) of water component 1 is zero.

Since binding experiments are carried out at constant concentration (or constant activity) of the neutral salt, $d\mu_{NaCl}$ is zero. Further, molar concentration c_{RH} of the ligand is usually very small so that it may be taken equal to $X_{RH}/55.5$, where X_{RH} is the mole fraction concentration. Under this condition, one finds

$$-d\mu_p = RT\Gamma_{RH}^1 d \ln X_{RH}$$

$$= RT\Gamma_{RH}^1 d \ln c_{RH} \qquad (10.4)$$

If the organic solute is a cationic amphiphile RCl (e.g., cetylpyridinium chloride), then RH in equation (10.3) is to be replaced by RCl, at constant concentration of the netural salt, so that

$$-d\mu_p = \Gamma_{RCl}^1(d\mu_R + d\mu_{Cl^-})$$

$$+ \Gamma_{NaCl}^1(d\mu_{Na^+} + d\mu_{Cl^-}) \qquad (10.5)$$

Following the same arguments as discussed in Section 4.6, it can be shown that Γ_{RCl}^1 and Γ_{NaCl}^1 are, respectively, equal to Γ_R^1 and $\Gamma_{Na^+}^1$, and further $\Gamma_R^1 + \Gamma_{Na^+}^1$ is equal to $\Gamma_{Cl^-}^1$ [vide equation (4.21)]. At constant concentration of NaCl one can put $d\mu_{Na^+}$ equal to zero in equation (10.5), so that

$$-d\mu_p = \Gamma_R^1 RT d \ln c_R\left(1 + \frac{\Gamma_{Cl^-}^1}{\Gamma_R^1} \cdot \frac{d \ln c_{Cl^-}}{d \ln c_R}\right) \qquad (10.6)$$

Since the dialysate is also electroneutral, $(c_{Na^+} + c_R)$ becomes equal to c_{Cl^-} and at constant concentration of NaCl, $(c_R \, d \ln c_R)$ is equal $c_{Cl^-} \, d \ln c_{Cl^-}$. Inserting this in equation (10.6), we obtain

$$-d\mu_p = \Gamma_R^1 RT \, d \ln c_R \left(1 + \frac{\Gamma_{Cl^-}^1/\Gamma_R^1}{1 + c_{Na^+}/c_R} \right)$$

$$\simeq \Gamma_R^1 RT \, d \, (\ln c_R) \qquad (10.7)$$

since c_{Na^+}/c_R under the experimental condition is always very large. This equation will also remain valid irrespective of the cationic or anionic nature of the detergent in the presence of excess neutral salt.

In the case of nucleic acid or protein at the basic side of the isoelectric pH, the biopolymer itself is associated with counterions (Na^+ etc.). The binding of RCl in this case involves the exchange of metal cations for the ligand which in turn then interacts directly with the macroion. The term $d\mu_p$ in equation (10.7) takes account of all kinds of interactions between polyions and amphiphiles.

10.3. FREE ENERGY CHANGE DUE TO EXCESS BINDING INTERACTION

The similarities of form of the differential equations (10.4) and (10.7) are to be noticed. Accordingly we may use conventions for binding RCl and assume that relations derived for cation binding will be valid for the binding of neutral ligand. The total free energy change ΔG_{ws} for ligand–biopolymer binding interaction in bringing the bulk concentration of the ligand from zero to X_R can be given by the equation[24-28]

$$-\Delta G_{ws} = - \int_0^{\mu_R} d\mu_p$$

$$= RT \int_0^{X_R} \frac{\Gamma_R^1}{X_R} \, dX_R \qquad (10.8)$$

assuming the solution to be highly dilute in respect of RCl. $-\Delta G_{ws}$ can be estimated from the graphical integration using values of Γ_R^1 for various values of X_R. Replacing X_R by c_R equation (10.8) was first derived and applied to biological systems by Bull et al.[26]

For the calculation of the standard free energy of binding, we have to bring the bulk concentration of the ligand in the solution somehow to X_R

equal to unit mole fraction which is its reference state in the unitary scale of activity. Bull[24–26] has originally used the molar and molal scales for activities of the adsorbate in the solution. Following a similar method, this change has been brought about by the hypothetical dilution of the solute from X_R to unity in the unitary scale[27,28] of activity, keeping Γ_R^1 fixed. The free energy ΔG_d of dilution for this hypothetical process can be calculated from the relation

$$\Delta G_d = -\Gamma_R^1 RT \ln \frac{1}{X_R}$$

$$= \Gamma_R^1 RT \ln X_R \tag{10.9}$$

The total free energy of binding Γ_R^1 moles of ligand can thus be calculated from the relation[27]

$$\Delta G = -RT \int_0^{X_R} \left(\frac{\Gamma_R^1}{X_R}\right) dX_R + RT\Gamma_R^1 \ln X_R \tag{10.10}$$

The standard free energy change ΔG^0 for the transfer of one mole of ligand to the bound phase of the biopolymer is simply $\Delta G/\Gamma_R^1$, so that it can be computed from the experimental data.

In Fig. 10.1, the sigmoid shape of an isotherm for binding cationic and anionic detergents to a biopolymer has been shown. At high ligand concentrations, a state of saturation in binding has been reached for many systems, so that Γ_R^1 does not alter with further increase of ligand concentration. In the intermediate range of solute concentration, there exists a region of sharp inflection where binding takes place in a cooperative manner. For fully cooperative interaction,[29] the binding isotherm could be represented by the dashed curve; Γ_R^1/C_R versus C_R plot will generate an insignificantly small graphical area, so that the integral in equation (10.10) may be put equal to zero. The

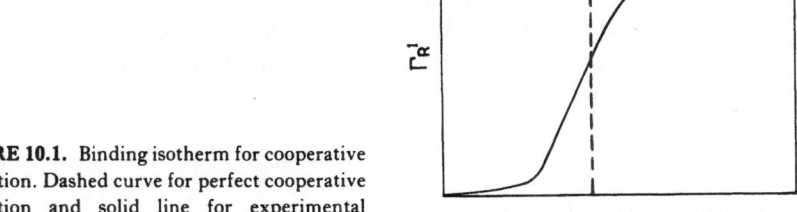

FIGURE 10.1. Binding isotherm for cooperative interaction. Dashed curve for perfect cooperative interaction and solid line for experimental observation.

Γ_R^1

Ligand concentration

standard free energy change ΔG_T^0 for the perfectly cooperative binding can be estimated by equation (10.11) put forward by Tanford[29]

$$\Delta G_T^0 = \frac{\Delta G}{(\Gamma_R^1)} = RT \ln X_R \qquad (10.11)$$

The integral in equation (10.10) thus takes care of the deviation of binding from the perfectly cooperative interaction as shown in Fig. 10.1. Equation (10.11) has been used frequently by Tanford and others[29,30] for the calculation of the standard free energy change whenever massive cooperative interaction takes place between protein and detergent ions.

In dealing with the thermodynamics of adsorption of proteins at the solid–liquid interfaces, Bull[24,25] has originally shown that the free energy change ΔG for binding Γ_R^1 moles of an adsorbent to the surface of a definite amount of solid powder can be calculated from the relation (vide also Section 8.14)

$$\Delta G = \Gamma_R^1 \Delta G_B^0$$

$$= -RT \int_0^{c_R} \left(\frac{\Gamma_R^1}{c_R}\right) dc_R + \Gamma_R^1 \ln c_R \qquad (10.12)$$

ΔG here stands for the energy change due to the transfer of one mole of an adsorbate when its molar (or molal) concentration is altered from zero to c_R and subsequently the bulk solution is hypothetically diluted from c_R to one molar (or one molal) concentration, keeping Γ_R constant. Equation (10.10), in fact, is identical in form with equation (10.12) except for the fact that the concentration in the former is expressed in the rational mole fraction or unitary scale, whereas in the latter, it is expressed in the molar or practical scale.

10.4. THE SCATCHARD EQUATION

In the last three decades, a considerable amount of work has been published on the binding of organic ligands to biopolymers; in most cases, an attempt has been made to calculate the free energies of binding on the basis of the Scatchard equation.[3,31] If a mole (or gram) of biopolymer P possesses Γ_R^S moles of independent binding sites, the equilibrium between ligand R and these sites may be expressed by the equation

$$P + nR \rightleftharpoons PR_n \qquad (10.13)$$

If Γ_R^1 represents moles of a ligand bound per mole of the macromolecule, then from detailed thermodynamic and statistical considerations, it can be shown[3] that

$$\theta = \frac{\Gamma_R^1}{\Gamma_R^S} = \frac{Kc_R}{1 + Kc_R} \qquad (10.14)$$

Here θ stands for the fraction of the total sites of macromolecules binding ligands when the molar concentration of the free ligand in solution is c_R. K stands for the macroscopic binding constant or the mass action equilibrium constant for the binding reaction (10.13). The Scatchard equation (10.14) is exactly similar in form to the Langmuir equation for the adsorption of solute by solid particles having fixed binding sites (vide Section 8.9). In the Langmuir equation, the adsorbate and adsorbent form separate phases, whereas in the derivation of the Scatchard equation, it is implied that the protein in the solution forms a single phase.

The Scatchard equation may be written in the linear form[3]

$$\left(\frac{\Gamma_R^1}{c_R}\right) = \Gamma_R^S K - \Gamma_R^1 K \qquad (10.15)$$

or

$$\frac{1}{\Gamma_R^1} = \frac{1}{\Gamma_R^S} + \frac{1}{\Gamma_R^S K} \cdot \frac{1}{c_R} \qquad (10.16)$$

These two linear equations are frequently used to calculate the number of statistical binding sites Γ_R^S and binding constant K. In Fig. 10.2, the Scatchard plot for binding myristyltrimethylammonium bromide to BSA is found to be linear[30] with negative slope in agreement with equation (10.15) only when Γ_R^1 is small. The same types of observations have also been noted by Sen *et al.*[28]

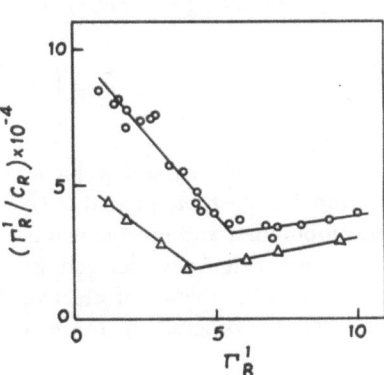

FIGURE 10.2. Plot of Γ_R^1/c_R against Γ_R^1 for the system BPA-MTAB. $\bigcirc$, 25°C; $\triangle$, 5°C. (Redrawn from Ref. 30.)

In many cases, more than one set of binding sites are found to exist on the surface of the macromolecule, such that set 1 has $^1\Gamma_R$ sites of intrinsic association constant K_1, set 2 has $^2\Gamma_R$ sites of an intrinsic association constant K_2, etc. Under such circumstances, the Scatchard equation will assume the generalized form[3]

$$\Gamma_R^1 = \frac{^1\Gamma_R K_1 c_R}{1 + K_1 c_R} + \frac{^2\Gamma_R K_2 c_R}{1 + K_2 c_R}$$

$$+ \cdots + \frac{^n\Gamma_R K_n c_R}{1 + K_n c_R} \tag{10.17}$$

Forms of the Scatchard equation for two ligands competing for the same site of the biopolymer have also been suggested.[3] Sometimes, the biopolymers and ligands are ionized so that binding alters the charge z and electrical potential around the macromolecule. The Scatchard plot in its original form has been justified by many workers on the basis of the assumption that the electrostatic effects due to the charge and potential of the macromolecule are reduced to a minimum if the binding takes place in the presence of excess neutral salt.[3]

The Scatchard equation for binding an electrolyte is actually shown to assume the form[3,4]

$$\Gamma_R^1 = \frac{\Gamma_R^S K c_R \exp(-2wz)}{1 + K c_R \exp(-2wz)} \tag{10.18}$$

This equation in the linear form can be written as

$$\ln \frac{\Gamma_R^1}{\Gamma_R^S - \Gamma_R^1} = \ln K - \ln c_R - 0.868wz \tag{10.19}$$

Here, z stands for the net charge of the macromolecule and

$$w = \left(\frac{\varepsilon^2}{2DRT}\right)\left(\frac{1}{b} - \frac{\kappa}{1 + \kappa a}\right) \tag{10.20}$$

Here, ε, D, R, T, snd κ have the same significance as those discussed in Section 4.9. Also, a is equal to $(b + b')$ where b and b' are the average radii of the biopolymer and inorganic ions, respectively. Equations (10.19) and (10.20) have been originally derived by Linderstrøm-Lang[32] on the basis of the Debye–Hückel theory of electrolytes. This has been further elaborated and modified by others.[4] If there are some conformational alterations in the macromolecule due to the binding of the ligand, the size of the biopolymer may be altered. For the case of hydrogen ion binding to protein, changes in the values of w and b have been studied extensively.[4] The equations for binding ionic solute to charged macromolecules have further been modified by Tanford *et al.*,[33] Hill,[34] and others.[4]

It appears that from the suitable fit of the binding data for various values of c_R in the case of many biopolymer systems, the values of the intrinsic binding constant K may be evaluated. The standard free energy change ΔG_S^0 will be equal to $(-RT \ln K)$, so that its value can be calculated from the Scatchard plot most conveniently using the experimental data.

Schwarz[35] has recently considered elaborately some pertinent aspects concerning the interpretation of the Scatchard plots in terms of basic model mechanisms. In his opinion,[16] deviations from linearity in the Scatchard plot are quite sensitive indications of cooperative behavior of different classes of binding sites and multiple contact binding.

10.5. ENTROPY AND ENTHALPY CHANGES DUE TO EXCESS BINDING

The standard enthalpy change ΔH^0 for the transfer of one mole of a ligand to one mole of free sites of a biopolymer may be calculated from the use of the thermodynamic relation[36]

$$\frac{d(\Delta G^0/T)}{d(1/T)} = \Delta H^0 \tag{10.21}$$

A reasonable way of determining ΔH^0 is to measure ΔG^0 at several temperatures and then plot $\Delta G^0/T$ against $1/T$. The slope of this curve at a given value of T then becomes equal to ΔH^0. Frequently, it is assumed that ΔH^0 remains constant in a small range of temperature,[27,28] so that one can integrate equation (10.21) to obtain the following relation:

$$\frac{\Delta G_2^0}{T_2} - \frac{\Delta G_1^0}{T_1} = \Delta H_{av}^0 \left(\frac{1}{T_2} - \frac{1}{T_1} \right) \tag{10.22}$$

Here T_1 and T_2 are a pair of temperatures which are close to each other. ΔG_1^0 and ΔG_2^0 are the standard free energies at these two temperatures whose values can be evaluated from the binding data. ΔH_{av}^0 stands for the change of enthalpy at an average temperature $\frac{1}{2}(T_1 + T_2)$ corresponding to an average free energy change ΔG_{av}^0 equal to $\frac{1}{2}(\Delta G_1^0 + \Delta G_2^0)$, respectively. The average values of the standard entropy of binding can be obtained from the relation

$$\Delta G_{av}^0 = \Delta H_{av}^0 - T\Delta S_{av}^0 \tag{10.23}$$

ΔS_{av}^0 can thus be evaluated from the experimental data.

It may be pointed out here that by replacing ΔG^0 in equation (10.21) by ΔG_T^0, ΔG_B^0, or ΔG_S^0 one can calculate standard changes of enthalpy and entropy due to the binding interaction which eventually will refer to different standard states of reference. The role of these standard states of reference will be further discussed in the next few sections.

10.6. DNA–SOLUTE BINDING INTERACTION

Sodium salt of deoxyribonucleic acid behaves as a polyelectrolyte, and ionizes into negatively charged polyphosphate ion and positively charged sodium ion. In the living cells, DNA remains in the bound state with histone, a basic protein, and thus exhibits certain important biological properties.[1,37–41] Lysine and arginine-rich histones usually contain many positively charged amino or imino groups in their polypeptide chains. The positively charged histone chains can bind themselves to the negatively charged DNA polyion by electrostatic attraction. The hydrophobic interactions also play a significant role in this binding process. The evaluation of this role, however, needs extensive thermodynamic analysis.

Binding of detergents such as long-chain amines, SDS, triton-X, fatty acid, etc. to DNA has also been investigated recently by Chatterjee et al.[27,42] using the equilibrium dialysis technique described in Section 2.9.

In Fig. 10.3, the isotherms for binding cetyltrimethylammonium bromide (CTAB), myristyltrimethylammonium bromide (MTAB), dodecyltrimethyl-ammonium bromide (DTAB), and cetylpyridinium chloride (CPCL) to deoxyribonucleic acid have been compared at constant temperature, pH, and ionic strength of the medium.[27] The number of CH_2 groups for CTAB, MTAB, and DTAB are 16, 14, and 12, respectively, and the head group in all cases is $-N^+(CH_3)_3$. Like CTAB, CPCL contains 16 CH_2 groups, but the head group for the latter is pyridinium cation. The binding isotherms in all cases are S-shaped, indicating lateral or cooperative interactions[29] of the adsorbed ligands with each other. At relatively high values of c_R, the extent of binding for each amine reaches a maximum value (Γ_R^m) due to the formation of saturated DNA–amine complexes. From Fig. 10.3 and also from Table 10.1, it is clear that Γ_R^1 and Γ_R^m depend on the chain length of the hydrocarbon tail of the cationic detergent. This is an example of the so-called hydrophobic effect. The binding ratios of CPCL and CTAB are different which means that

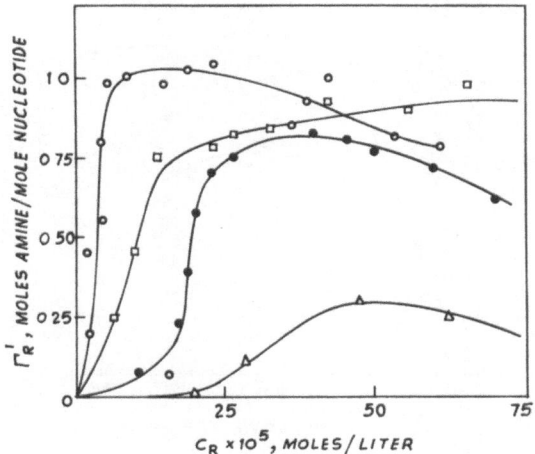

FIGURE 10.3. Γ_R^l vs. c_R plot of DNA. Phosphate buffer, ionic strength 0.0625 and 30°C for CTAB (O), CPCL (□), MTAB (●) and DTAB (△). (Redrawn from Ref. 27.)

the nature of the cationic head group of the amine has some influence on the binding interaction with DNA.

In Fig. 10.4, it is noted that MTAB–DNA binding shifts more and more to the right side of the figure as ionic strength of the medium increases. The initial slope of the binding isotherm is highest at 0.012 ionic strength but its

TABLE 10.1. Free Energy of Binding for DNA–Amine Complexes[a] at Ionic Strength 0.0625 and pH 6.0 (Phosphate Buffer)

Ligands	Temperature (°C)	Γ_R^m (Mole ligand per mole nucleotide)	$-\Delta G^0$ (kJ/mol nucleotide)	$-\Delta G_m^0$ (kJ/mol nucleotide)
CPCL	45	1.02	37.6	36.6
	30	0.95	31.2	30.3
	15	0.98	32.1	31.2
CTAB	30	1.03	37.0	37.0
MTAB	45	0.98	31.0	32.1
	30	0.83	27.0	24.9
	15	0.79	25.6	22.7
DTAB	30	0.30	9.97	11.1

[a] Reference 27.

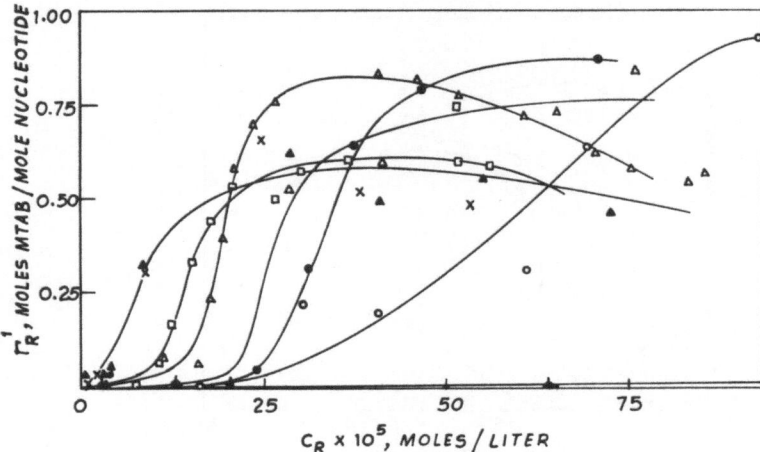

FIGURE 10.4. Γ_R^1 vs. c_R plots of DNA at pH 6.0 and temperature of 30°C. Ionic strengths (phosphate buffers): ▲, 0.0125; □, 0.025; △, 0.0625; ——, 0.1250; ●, 0.250. Ionic strengths (NaCl solution): ○, 0.5; ▲, 2.0; ×, 0.0125 (phosphate buffer + denatured DNA). (Redrawn from Ref. 27.)

value becomes nearly zero at 0.2 ionic strength, when the electrostatic effect is negligible. Initial slope in this case appears to be a qualitative measure of the electrostatic interaction. For treating this quantitatively, a detailed mathematical theory is needed. Even in the absence of the electrostatic effect at 0.2 ionic strength, Γ_R^m is quite high due to the hydrophobic or other types of cooperative interactions.

In Table 10.1, Γ_R^m for CTAB at 30°C is observed to be nearly unity so that each negatively charged phosphate group of DNA molecule can bind one positively charged amine cation to a state of saturation. Γ_R^m for MTAB and DTAB are 0.83 and 0.30, respectively. This means that the conformation of the biopolymer during the binding process is altered in such a manner that a portion of the phosphate binding sites remains unavailable for interaction with MTAB or DTAB. The binding ratio for DNA–MTAB complex increases with increase of temperature, the maximum being observed to depend on the pH and nature of the neutral salt present in the medium.

The isotherms for binding octanoic and hexanoic acids[42] to DNA are also sigmoid shaped, Γ_R^m in these two cases being, respectively, 1.30 and 3.20 moles per mole of nucleotide. The reversal of the chain-length effect suggests that the hydrophobic effect is insignificant. The neutral ligand triton-X binds appreciably[42] with DNA at 30°C but none at 45°C. This also suggests that in the binding process the hydrophobic effect is negligible.

10.7. HYDROPHOBIC EFFECT IN DNA–AMINE BINDING INTERACTION

The values of ΔG per mole of nucleotide due to the binding of CTAB, MTAB, and DTAB to DNA, respectively, can be calculated[27] from the graphical integration using equation (10.10). The mole fraction of the amine in the system is calculated from the assumption that X_R for a dilute solution is equal to $c_R/55.5$. In Fig. 10.5, ΔG is observed to increase with increasing ligand concentration until it becomes independent of X_R. ΔG at this stage reaches its maximum value ΔG_m^0. These values are presented in Table 10.1.

During the state of binding unsaturation, ΔG is found to vary linearly with Γ_R^1/Γ_R^m (vide Fig. 10.6.). From these plots, it is clear that

$$\Delta G^0 = \frac{\Delta G}{\Gamma_R^1/\Gamma_R^m} \tag{10.24a}$$

or

$$\Delta G = \Delta G^0 \frac{\Gamma_R^1}{\Gamma_R^m} \tag{10.24b}$$

The standard free energy change ΔG^0 for binding can be evaluated from the slope of the plot of ΔG against the fraction of the sites Γ_R^1/Γ_R^m bound to the ligand. At the state of saturation, Γ_R^1 is equal to Γ_R^m, so that according to

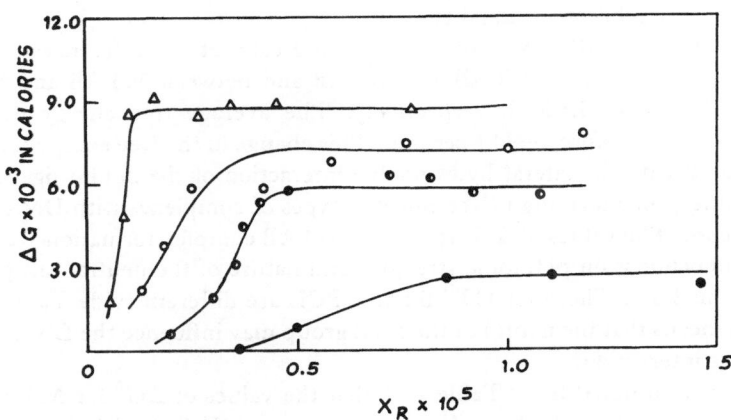

FIGURE 10.5. ΔG (negative) of binding of detergents with DNA vs. X_R plot. Phosphate buffer, pH 6.0, ionic strength 0.0625 and 30°C. $\triangle$, CTAB, $\bigcirc$, CPCL; $\textcircled{\tiny{\textbullet}}$, MTAB; and $\bullet$, DTAB. (Redrawn from Ref. 27.)

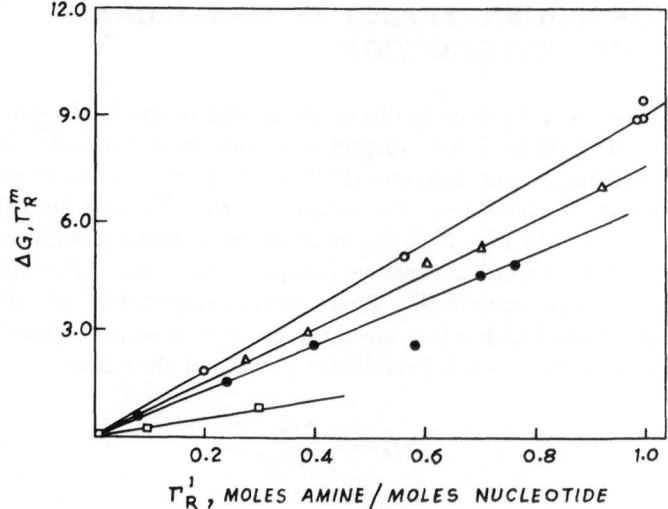

FIGURE 10.6. $\Delta G \ \Gamma_R^m$ negative vs. Γ_R^1 for CPCL ($\triangle$), CTAB ($\bigcirc$), MTAB ($\bullet$), and DTAB ($\square$) at 30°C; ionic strength 0.0625; pH = 6.0. (Redrawn from Ref. 27.)

equation (10.24), ΔG^0 is equal to ΔG_m^0. In agreement with this, ΔG^0 and ΔG_m^0 for different systems are found to be equal (vide Table 10.1). ΔG^0 or ΔG_m^0 may thus be regarded as the free energy change due to the transfer of one mole of a ligand from the solution to one mole of binding sites of DNA nucleotide remaining free at (hypothetical or real) states of binding saturation so that Γ_R^1/Γ_R^m becomes equal to unity.

From Table 10.1, we also note with interest that the differences between the values of ΔG_m^0 for CTAB and MTAB and between MTAB and DTAB are 12,100 and 13,800 J, respectively. The average free energy decrease $-\Delta G_{CH_2}^0$ is, therefore, 6490 J per mol. This change in the free energy is mostly associated with the lateral hydrophobic interaction of the bound ligand with each other, thus forming mixed micellar types of complexes with DNA as the inner core. The values of ΔG^0 for DNA–MTAB complex formation are found to be dependent on pH, ionic strength, and nature of the neutral salt present in the medium. Those for CTAB and CPCL are different (vide Table 10.1), which means that the nature of the head group may influence the DNA–amine binding interaction.

It is also noted from Table 10.1 that the values of ΔG^0 for MTAB and CPCL are significantly dependent on temperature. Values of ΔH_{av}^0 and ΔS_{av}^0 obtained with the help of equations (10.22) and (10.23) are presented in Table 10.2. It is found from the data that the major contribution to ΔG^0 is made by $T\Delta S_{av}^0$ but the contribution of ΔH_{av}^0 is small. This quantitatively proves

TABLE 10.2. Entropy and Enthalpy of Binding for DNA–Ligand Complexes[a]

Ligand	T_{av} (K)	$-\Delta H^0_{av}$ (kJ/mol)	$-\Delta G^0_{av}$ (kJ/mol)	$-\Delta S^0_{av}$ (kJ/mol/deg)	$-T_{av}\Delta S^0_{av}$ (kJ/mol)
MTAB	310.5	−11.6	+28.4	−0.129	−40.0
	295.5	−1.96	+23.8	−0.087	−25.7
CPCL	310.5	−9.69	+33.5	−0.139	−43.2
	295.5	+4.58	+30.7	−0.089	−26.1

[a] Reference 27.

that the binding of these amines to DNA is largely guided by the entropy-controlled hydrophobic interaction. However, on the basis of similar thermodynamic analysis for triton-X and fatty acid binding with DNA, the contribution of $T\Delta S^0_{av}$ is found to be negligible, whereas that of ΔH^0_{av} is appreciable, proving thereby that the hydrophobic effect is insignificant in such binding interactions.[42]

10.8. HISTONE–DNA BINDING INTERACTION

In chromosomes of higher forms of living systems, five or more types of basic proteins called histones are bound with deoxyribonucleic acid so that nucleohistones are formed.[1] The bead structure of the nucleohistones has been established from the electron-micrograph.[37,38] As a result of this, the length of DNA becomes considerably shortened. The whole structure of the nucleohistone under this condition can be accommodated within a small region of the nucleus of the cell in a relatively compact form.

For a long time, the study of binding of various histones to DNA *in vitro* has been of interest to many workers.[43–45] A simple technique for the study of such macromolecular binding[46] has been described in Section 2.9. Binding isotherms of lysine-rich F2B histone and arginine-rich F3 histone at different ionic strengths have been presented in Figs. 10.7 and 10.8, respectively. The histones differ from one another in amino acid composition and amino acid sequence. They also contain significant amounts of lysine or arginine residues having NH_3^+ or imino cationic groups which can electrostatically bind negatively charged phosphate groups of DNA. Like DNA–amine binding, histone–DNA binding is also strongly dependent on the ionic strength of the medium (vide Figs. 10.7 and 10.8). The binding isotherms for F2b and F3 histones under identical conditions are widely different.[46] This is in agreement with the observation made in the previous section that the nature of the head group of an amine has considerable influence on the nature of binding with DNA. The side-chains of the different histones contain different numbers of CH_2

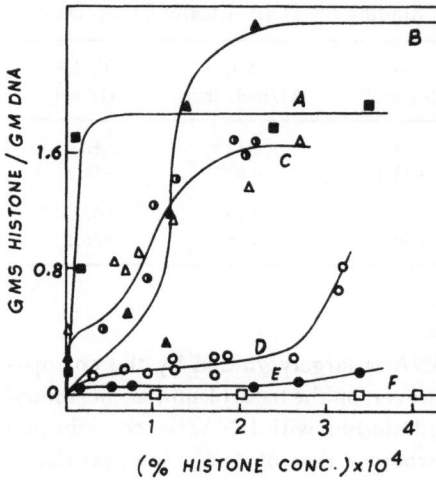

FIGURE 10.7 F2b histone–DNA binding isotherm. Variation of binding ratio gm histone/gm DNA as function of equilibrium histone concentration. Curve A, 0.01 M NaCl; curve B, 0.10 M NaCl; curve C, 0.25 M NaCl; ◑, normal experiment, △, reversible experiment; curve D, 0.5 M NaCl; curve E, 1.0 M NaCl; curve F, 2.0 M NaCl. (Redrawn from Ref. 46.)

groups which are sequenced differently in a polypeptide chain. The hydrophobic interactions of these groups will then be different for different histones. DNA complexed with a histone can undergo conformational alterations to various degrees depending upon the solution environment and the nature of the protein.

The standard free energy changes ΔG^0 of binding (D. K. Chattoraj and R. Chatterjee, unpublished) calculated similarly with the help of equations (10.10) and (10.24b) have been presented in Table 10.3 Values of ΔG^0 for the histone–DNA system are considerably lower than those for the amine–DNA system. The low values can be accounted for, since only a very small fraction

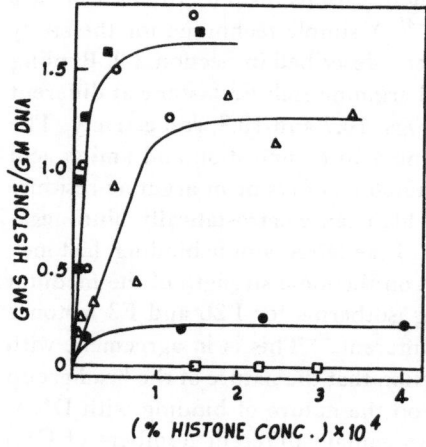

FIGURE 10.8. F3 histone–DNA binding isotherm. Variation of binding of binding ratio gm histone/gm DNA as a function of the equilibrium histone concentration. ○, 0.01 M NaCl; ■, 0.10 M NaCl; △, 0.50 M NaCl; ●, 1.00 M NaCl; □, 2.00 M NaCl. (Redrawn from Ref. 46.)

TABLE 10.3. Standard Free Energies of Binding for Histone–DNA Complexes[a]

Histone	Ionic strength	Γ_R^m (moles histone per mole nucleotide)	$-\Delta G^0$ (kJ/mol nucleotide)
F2B	0.50	0.006	0.313
	0.25	0.0396	2.03
	0.10	0.0593	2.96
	0.01	0.0426	2.49
F3	1.00	0.0065	0.37
	0.50	0.0350	1.88
	0.10	0.0470	2.66
	0.01	0.0470	2.66
F1	0.50	0.0090	0.042
	0.25	0.0150	0.76

[a] D. K. Chattoraj and R. Chatterjee (unpublished).

of a mole of histone is actually bound to one mole of nucleotide; further, many DNA binding sites remain buried within the structural matrix of the macromolecular complex and are unavailable for binding interactions. The significance of this observation will be further discussed in the later sections of this chapter.

10.9. PROTEIN–LIGAND BINDING INTERACTION

Using the equilibrium dialysis method, Chattoraj and co-workers[28,47] have studied in detail binding interaction of long-chain amines and sodium dodecyl sulfate (SDS) with bovine serum albumin and gelatin under varying conditions of pH, ionic strength, and solute concentrations of the aqueous medium. In Figs. 10.9, 10.10 and 10.11, the binding isotherms of quaternary amines with BSA and gelatin have been presented at fixed temperature, pH, and ionic strength of the medium. At temperatures below 50°C, values of Γ_R^m for BSA–amine interactions[28] vary from 1 to 40 depending upon the nature of the ligand, pH, and temperature of the medium (vide Table 10.4). Such variations do not depend regularly on the chain length of the ligand,[28,48] pH, ionic strength, and temperature of the system. The behavior of protein in this respect is distinctly different from that of DNA. These differences in binding interaction are in all probability related to the heterogeneous structure of the boundary as well as bulk regions of a protein. The proteins are made up of 20 or more different amino acid residues having different types of hydrophobic and hydrophilic groups. These groups are unevenly distributed in the boundary and bulk regions of a protein as a result of which the surface and the interior

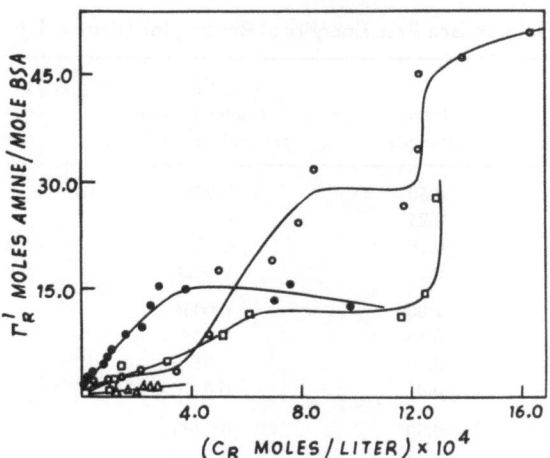

FIGURE 10.9. Γ_R^1 vs. c_R plot for binding of CTAB, MTAB, and DTAB with BSA. Phosphate buffer pH 6.0, ionic strength, 0.125. ●, CTAB (30°C); ○, MTAB (30°C); □, DTAB (30°C); △, CTAB (15°C). (Redrawn from Ref. 28.)

of a protein molecule become highly heterogeneous. This kind of heterogeneity in the double helix structure is, however, comparatively less pronounced.

At 65°C, the globular structure of BSA is completely destroyed, and as a result of this, as high as 840 CTAB molecules become bound per BSA molecule,[28] which in its turn contains only 590 amino acid residues (vide Fig. 10.10). Most probably each peptide group binds one CTAB molecule; besides, many side-chain groups are able to bind amines. The values of Γ_R^m for MTAB and CTAB are high at this temperature, but, curiously, binding even at this

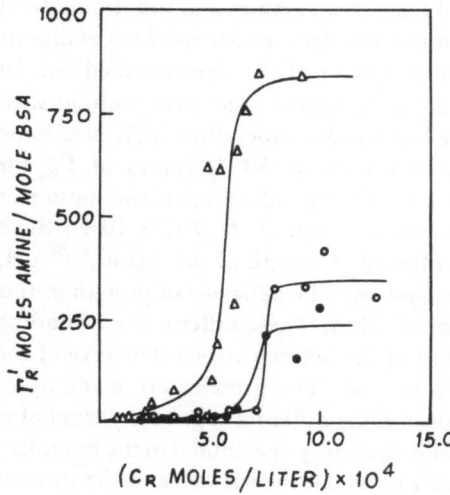

FIGURE 10.10. Γ_R^1 vs. c_R plot for binding of CTAB, MTAB , and DTAB with BSA at temperature 65°C, phosphate buffer pH 6.0, ionic strength = 0.125, △, CTAB; ●, MTAB; ○, DTAB. (Redrawn from Ref. 28.)

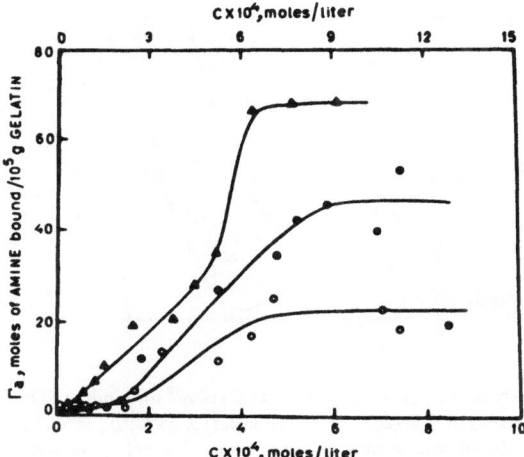

FIGURE 10.11. Γ_R^i (written as Γ_a) vs. c_R (written as C) plot[48b] for binding amines to gelatin at 30°C. Phosphate buffer (pH 6.0), ionic strength = 0.125. Left lower scale: (▲), CTAB; (○), MTAB. Left upper scale: (●), DTAB. (Reproduced with permission from *Indian Journal of Biochemistry and Biophysics*.)

denatured state does not follow the sequence of amine chain lengths. Gelatin is a denatured protein, but at ordinary or high temperatures it binds only a small amount of amines.[48a] From this, one can conclude that the large number of prolyl groups occurring in gelatin peptide link hinder the binding of amines by steric or other types of effects. The complete lack of binding of amines to polyvinyl pyridine (Sadhukhan and Chattoraj, unpublished) strongly supports this proposition.

TABLE 10.4. Standard Free Energies of Binding of BSA–Amine Complexes[a] at pH 6.0 and Ionic Strength 0.125

Ligands (amines)	Temperature (°C)	Γ_R^m (moles amine per mole BSA)	$-\Delta G_S$ (kJ/mol amines)	$-\Delta G_B^0$ (kJ/mol amines)	$-\Delta G^0$ (kJ/mol BSA)	$-\Delta G_m^0$ (kJ/mol BSA)
CTAB	65	836	—	34	27 870	28 290
MTAB	65	235	—	31	7 496	7 076
DTAB	65	339	—	34	12 060	11 650
CTAB	45	15.1	22	33	541	415
MTAB	45	40.2	26	33	1 373	1 289
DTAB	45	9.0	26	31	291	311
CTAB	30	14.8	24	33	541	511
CPCL	30	12.0	20	31	412	403
CTAB	15	1.0	20	22	33.2	33.2

[a] Reference 28.

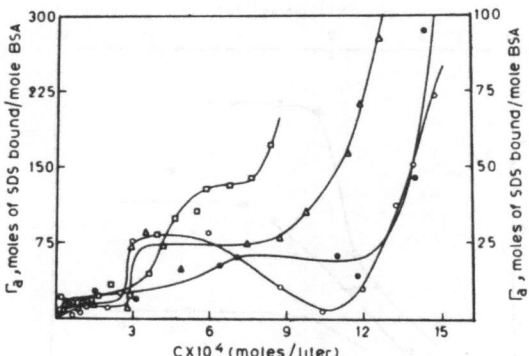

FIGURE 10.12. Γ_R^1 (written as Γ_a) vs. c_R (written as C) plot[47] for binding SDS to BSA. Phosphate buffer (pH 6.0), ionic strength = 0.125. Left lower scale: O, 65°C; $\triangle$, 45°C; $\square$, 30°C. Right lower scale: ●, 15°C. (Reproduced with permission from *Indian Journal of Biochemistry and Biophysics*.)

The binding of SDS to serum albumin has been studied in much detail in the last three decades.[3,7,8,29] In Fig. 10.12, several typical isotherms[47] for binding SDS to BSA at different temperatures have been presented. Values of Γ_R^m for SDS–BSA binding isotherms have been included in Table 10.5. The maximum ratios for binding SDS are usually high. For many situations, Γ_R^1 even increases without limit when c_R is significantly high (Fig. 10.12). All these characteristics suggest that the globular protein molecule unfolds in steps with progressive binding of ligands, as a result of which more groups will be available for binding with SDS.

In Figs. 10.13 and 10.14, the isotherms for binding SDS to 12 different types of globular proteins have been compared.[7,8] Binding in these figures is

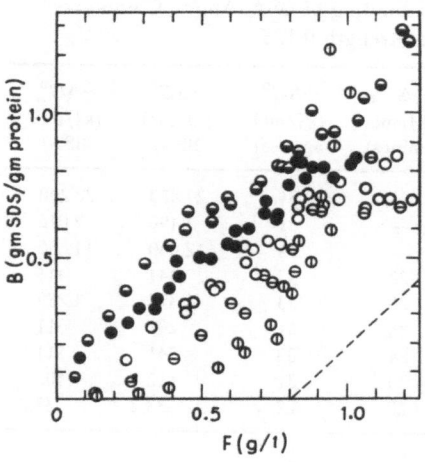

FIGURE 10.13. The amount of SDS bound (B) in grams per gram at 20°C by five proteins as a function of the free SDS concentration (F) at equilibrium.[7] The diagonal line at the lower right represents the dye solubilized when no protein is present. Symbols: Human serum albumin (O); apoferritin (⊕; IgG ⊖); metmyoglobin (◐); aldolase (●). (Reproduced with permission from the American Chemical Society.)

Table 10.5. Standard Free Energies of Binding[a] of BSA–SDS Complexes at Different Temperatures, Ionic Strengths, and pH

Ionic strength	pH	Temperature (°C)	Γ_R^m (moles SDS bound/mole BSA)	$-\Delta G_S^0$ (kJ/mol SDS)	$-\Delta G_B^0$ (kJ/mol SDS)	$-\Delta G_B^0$ (kJ/mol BSA)	$-\Delta G_m^0$ (kJ/mol BSA)	$-\Delta G_T^0$ (kJ/mol SDS)
0.625	6.0	30	212	26	31	707	695	31.4
0.125	6.0	30	133	26	36	461	407	29.7
0.0625	6.0	30	175	24	34	581	524	31.4
0.0125	6.0	30	32	27	34	110	112	29.7
0.625	4.0	30	89	20	34	295	305	
0.125	6.0	65	81	—	33	286	283	
0.125	6.0	45	77	20	34	297	279	
0.125	6.0	15	18	20	33	615	569	

[a] Reference 47.

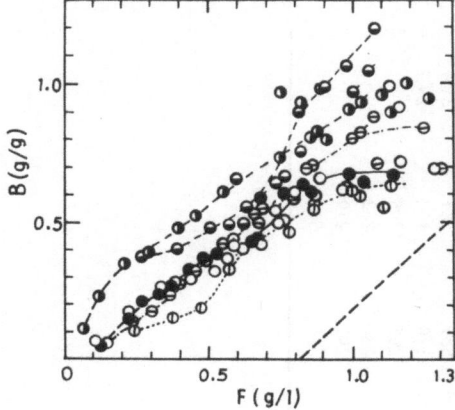

FIGURE 10.14. The amount of SDS bound (B) in grams at 20°C; data obtained with proteins.[7] Symbols: conalbumin (○); ovalbumin (●); β-lactoglobulin (⊖); transferrin (◌); lysozyme (◓); methenoglobin (◑). (Reproduced with permission from the American Chemical Society.)

expressed in grams (B) of SDS bound per gram of a protein. The equilibrium concentrations of SDS are similarly expressed as grams (F) of SDS per liter of dialysate. From the knowledge of molecular weights of different proteins, Γ_R^1 and c_R may be easily calculated from the data. In each case, the binding is observed to reach a state of saturation near cmc. These saturation values Γ_R^m for the proteins are different from each other. In Fig. 10.15, Γ_R^m have been plotted against molecular weights of different proteins. It seems that the bigger the molecular size of the protein, the greater the number of sites appearing on the protein surface (Birdi, unpublished).

Γ_R^m for a protein is strongly dependent on the nature of the inorganic salt in the medium, even when the ionic strength is maintained constant.[28,47,48] This means that the hydrated layer covering the protein surface contributes to the binding of ligands to a biopolymer. The dependence of Γ_R^m for various proteins and ligands on temperature, pH, and ionic strength of the medium has been discussed by Chattoraj *et al.*[28,47,48]

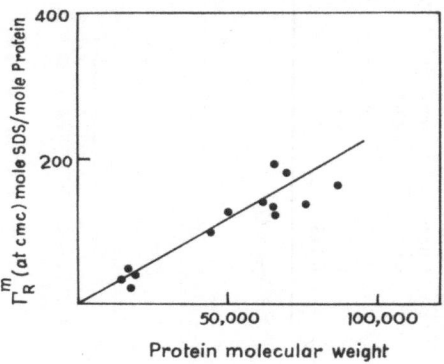

FIGURE 10.15. Plot of Γ_R^m (as mole SDS/mole protein) vs. molecular weight of protein (Birdi *et al.*, unpublished).

10.10. STANDARD FREE ENERGY CHANGE FOR PROTEIN–LIGAND BINDING INTERACTION

In Fig. 10.16, ΔG for BSA–amine binding obtained with the help of equation (10.10) has been plotted against X_R. Maximum values of the free energy termed as ΔG_m^0 thus evaluated for various types of binding interactions,[28,47,48] have been incorporated in Tables 10.4 and 10.5. In a state of unsaturation, ΔG can be plotted as a function of Γ_R^1/Γ_R^m (vide Fig. 10.17) and the slopes of the straight lines thus obtained will represent ΔG^0. Values of ΔG^0 and ΔG_m^0 in these tables are close to each other as expected.

We know from thermodynamics that the affinity or extent of conversion of a chemical reaction is closely related to the standard free energy change ΔG^0. This may be expressed by the equation

$$\Delta G^0 = -RT \ln K_b \tag{10.25}$$

Here K_b is the equilibrium constant for the overall binding reaction considered in previous sections. The higher the value of the maximum binding ratio Γ_R^m, the higher the values of the K_b and $-\Delta G^0$. In other words, the order of Γ_R^m and that of $-\Delta G^0$ for several systems undergoing binding interactions should be the same under similar standard conditions. This is found to be true for

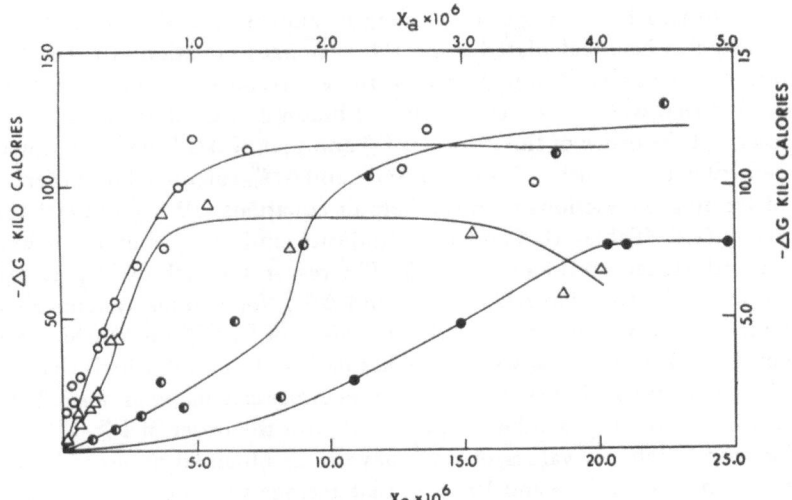

FIGURE 10.16. Plot[28] of free energy of binding (ΔG) of amines to BSA vs. X_R (written as X_a), Phosphate buffer pH 6.0, ionic strength = 0.125. Left lower scale: O, 30°C (CTAB); △, 30°C (CPCL). Left upper scale: ◑, 45°C (CTAB). Right upper scale: ●, 15°C (CTAB). (Reproduced with permission from Elsevier.)

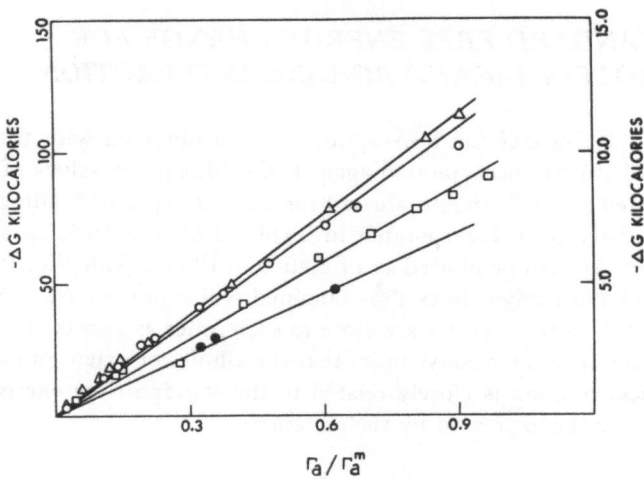

FIGURE 10.17. Free energy[28] of binding (ΔG) vs. Γ_R^1/Γ_R^m (written as Γ_a/Γ_a^m), phosphate buffer pH 6.0, ionic strength = 0.125. Left lower scale: $\bigcirc$, 30°C; $\triangle$, 45°C (CTAB); $\square$, 30°C (CPCL). Right upper scale: $\bullet$, 15°C (CTAB). (Reproduced with permission from Elsevier.)

different types of binding systems considered in Tables 10.4 and 10.5. This correlation is valid so long as ΔG^0 is expressed with reference to Γ_R^1/Γ_R^m (or fraction θ) equal to unity.

The standard free energies of binding of amines and SDS to BSA and gelatin have also been calculated using the Scatchard equation (10.15). The Scatchard plot is observed to be linear for these systems only when c_R is very low. We find that ΔG_S^0 for different types of binding interactions are close to each other but the order of the values of Γ_R^m and that of ΔG_S^0 are in complete disagreement with each other. Values of ΔG_T^0 and ΔG_B^0 calculated for different types of binding interactions with the help of equations (10.11) and (10.12), respectively (vide Tables 10.4 and 10.5), indicate similar types of anomalous behavior with respect to the order of Γ_R^m. The reason for such discrepancy in all cases lies in the fact that ΔG_S^0, ΔG_B^0, and ΔG_T^0 for various systems have been compared with each other at different values of Γ_R^1/Γ_R^m (or θ) which are less than unity. It can be easily shown (Chattoraj *et al.*, unpublished) that the order of magnitudes of all these standard free energies calculated or normalized at Γ_R^1/Γ_R^m equal to unity will be in agreement with the order of Γ_R^m.

For many systems, values of ΔG^0 have been obtained at several temperatures (vide Tables 10.4 and 10.5), so that average values of ΔH^0 and ΔS^0 have been evaluated using equations (10.22) and (10.23). It is obvious from Table 10.6 that both ΔH_{av}^0 and $T\Delta S_{av}^0$ make significant contributions to the standard free energies of binding interactions. This means that entropy-controlled hydrophobic effect as well as enthalpy-controlled interactions of

TABLE 10.6. Standard Entropy and Enthalpy Changes for BSA–Amine Binding Interaction at pH 6.0 and Ionic Strength 0.125[a]

Ligand (amines)	Average temperature (°K)	$-\Delta G^0$ (average) (kJ/mol)	ΔH^0 (average) (kJ/mol)	$T\Delta S^0$ (average) (kJ/mol)	ΔS^0 (average) (kJ/deg mol)
CTAB	328	14 150	416 000	428 500	1373
MTAB	328	4158	91 520	95 680	291
DTAB	328	6238	187 200	191 400	581
CTAB	310	541	−83.1	457	1.39
MTAB	310	1290	707	1992	6.29
DTAB	310	332	−1705	−1373	−4.19

[a] Reference 28.

various types are involved in the ligand–protein binding interactions. For this reason also, Γ_R^m and ΔG^0 for the series of amines (vide Table 10.4) are not changing regularly with the gradual increase of chain length of the cationic ligand. These irregularities are possibly the consequences of heterogeneity of the boundary and bulk structures of the protein. As against these, the amine–DNA binding interactions have been shown to be controlled largely by the so-called hydrophobic effect.

10.11. BINDING OF LIGAND TO PROTEIN–PROTEIN AND FAT–PROTEIN MIXTURES

In biological systems, a protein remains mixed with various other proteins in the aqueous medium. Thus blood serum contains five or more different varieties of proteins. To be of any practical relevance, the study of binding by a mixture of proteins is likely to be biologically relevant. Any serious study with a mixed system has not so far been made possible because of the difficulty in the physicochemical and thermodynamic interpretations of the experimental data.

Recently, studies on binding of SDS and amines by a mixture of BSA and gelatin in 1 : 1 weight ratio has been carried out by the equilibrium dialysis method.[48] In Fig. 10.18, the binding ratios Γ_R^1 for CTAB per hundred kilogram of mixed proteins have been compared[48b] with those obtained per hundred kilogram of the individual proteins. From Table 10.7, it appears that the values of Γ_R^m for the mixed proteins are always considerably less than those to be expected on the basis of the additivity rule. Similar deviations for various weight ratios of the binary mixtures of casein and BSA are shown in Fig. 10.19 (B. Sadhukhan and D. K. Chattoraj, unpublished).

Values of ΔG^0 for such mixed systems per kilogram of total protein can be easily calculated using equations (10.10) and (10.24). These values are

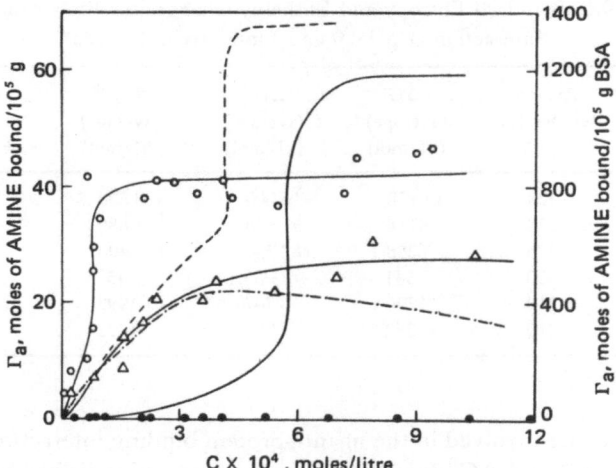

FIGURE 10.18. Γ_R^i (written as Γ_a) vs. C (equal to c_R) plot[48b] for CTAB. Phosphate buffer (pH 6.0), ionic strength = 0.125. Left lower scale: (----), gelatin at 30°C; ●, gelatin at 65°C; -----, BSA at 30°C; ○, BSA + gelatin (65°C); △, BSA + gelatin (30°C). Right lower scale: (——), BSA (65°C). (Reproduced with permission from *Indian Journal of Biochemistry and Biophysics*.)

always less than those calculated from the additivity rule (vide Table 10.7). The difference between the two values may represent the free energy change due to the protein–protein interaction in the mixed system. These are presented in Fig. 10.19. From the data on binding by the mixed systems at different temperatures, ΔS^0 and ΔH^0 for protein–protein interaction can be evaluated in the same manner as done with the single-protein system.

In milk, fat is in combination with various proteins. In Fig. 10.20, isotherms of CTAB binding by milk casein and by milk casein–fat mixture have been compared. Fat itself binds a negligible amount of CTAB but Γ_R^m for casein is considerably increased in the presence of fat. ΔG^0 for the protein–fat system is always higher than that to be expected from the additivity rule. Thus in contrast to protein–protein interaction, protein–fat interaction can increase the standard free energy due to their mutual interaction (B. Sadhukhan and D. K. Chattoraj, unpublished).

10.12. NEGATIVE BINDING OF INORGANIC ELECTROLYTES TO A PROTEIN

Recently, Bull and Breese[49] have determined the absolute binding of water and electrolytes to egg albumin, serum albumin, and hemoglobin using the gravimetric method of equilibrium dialysis technique (vide Section 2.9).

TABLE 10.7. ΔG^0 for Binding Ligands to Protein Mixtures[a] at pH 6.0 and Ionic Strength 0.125

Protein mixture	Proportion by wt.	Ligand	Temp. (°C)	Γ_R^m moles ligand bond per 100 kg of protein		ΔG^0 (kJ/kg protein)	
				Experimental	Theoretical (additivity)	Experimental	Theoretical (additivity)
Gelatin + BSA	1:1	CTAB	30	27.0	45.0	8.80	14.6
Gelatin + BSA	1:1	CTAB	65	40.2	615.0	15.5	206.0
Gelatin + BSA	1:1	SDS	30	52.0	258.0	18.8	85.0
Casein + BSA	1:1	CTAB	30	26.0	33.2	9.08	12.2
Casein + BSA	1:2	CTAB	30	17.3	29.1	5.98	10.6
Casein + BSA	1:3	CTAB	30	10.5	27.2	3.72	9.88
Casein + BSA	1:1	SDS	30	56.0	149.0	24.1	52.5
Casein + BSA + Gelatin	1:1:1	CTAB	30	18.0	27.2	6.11	15.3
Casein + BSA + Gelatin	1:2:1	CTAB	30	30.2	38.8	10.9	13.4

[a] B. Sadhukhan and D. K. Chattoraj; Data presented at International Symposium on Surfactants in Solution, University of Lund, Sweden 1982.

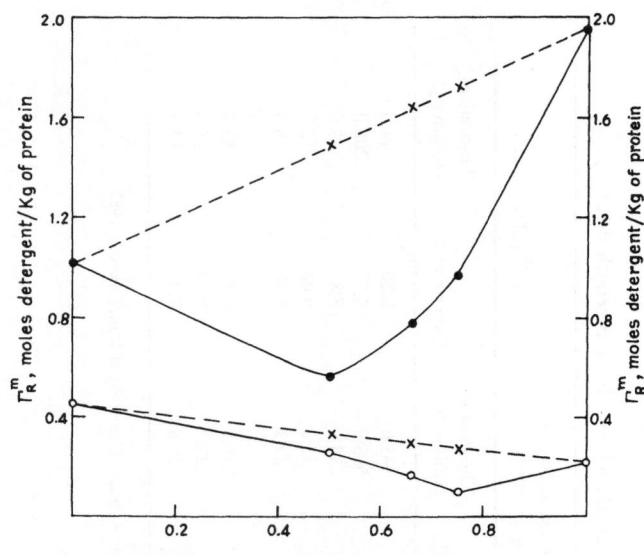

FIGURE 10.19. Plot of Γ_R^m vs. weight fraction of BSA in the (casein + BSA) mixture at pH 6.0, ionic strength 0.125 and temperature 30°C. (- × - × -), Ideal system (CTAB or SDS). ○, CTAB (experimental). ●, SDS (experimental). (Redrawn from Ref. 48a.)

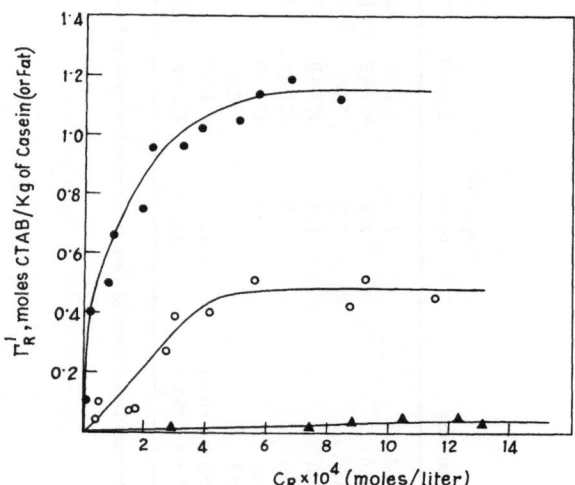

FIGURE 10.20. Plot of Γ_R^l vs. c_R for CTAB binding at pH 6.0, temperature 30°C and ionic strength 0.125. Defatted casein: ○; casein with fat: ●; fat: ▲. (Sadhukhan and Chattoraj, unpublished.)

The molalities m_2' and m_2 of the electrolyte (component 2) in the dialysate solution before and after binding have been determined which are related[50] to Γ_2^1 by the following equation:

$$\Gamma_2^1 = \left(\frac{n_1' M_1}{1000}\right)(m_2' - m_2) \tag{10.26}$$

Here M_1 is the molecular weight of the water component 1. n_1' and n_2' are the total number of moles of component 1 and solute component 2, respectively, present in the system per mole of protein. For the case of inorganic electrolyte, m_2' is always found to be less than m_2 so that Γ_2^1 is negative. This indicates that water is bound to the protein in relatively larger amounts than the electrolytes. Replacing m_2' and m_2 with $1000\, n_2'/M_1 n_1'$ and $1000\, n_2/M_1 n_1$, respectively, one can easily write[50]

$$\Gamma_2^1 = n_2' - n_1'\left(\frac{X_2}{X_1}\right) \tag{10.27}$$

Here X_1 and X_2 are the mole fractions of water and solute components in the dialysate at dialysis equilibrium. Equation (10.27) can be rearranged to give an identical relation:

$$n_1' - n_2'\frac{X_1}{X_2} = -\Gamma_2^1\left(\frac{X_1}{X_2}\right)$$

$$= \Gamma_1^2 \tag{10.28}$$

Here Γ_1^2 is termed as the excess binding of component 1 per mole of protein. Values of Γ_1^2 may be obtained by multiplying Γ_2^1 by X_1/X_2. From the experimental values of m_2' and m_2, those of Γ_1^2 and Γ_2^1 can thus be computed.

At dialysis equilibrium,[50] the chemical potential μ_2^i of the protein solution inside the dialysis bag is related to the chemical potential μ_2 of the dialysate by the equation

$$\mu_2 = \mu_2^i + \bar{V}_2 \pi \tag{10.29}$$

Here $\bar{V}_2$ is the partial molal volume of the solute component in the protein solution. Since μ_2^0 for the inside and outside solutions are the same,

$$RT \ln f_2^0 m_2 = RT \ln f_2^i m_2^i + \bar{V}_2 \pi \tag{10.30}$$

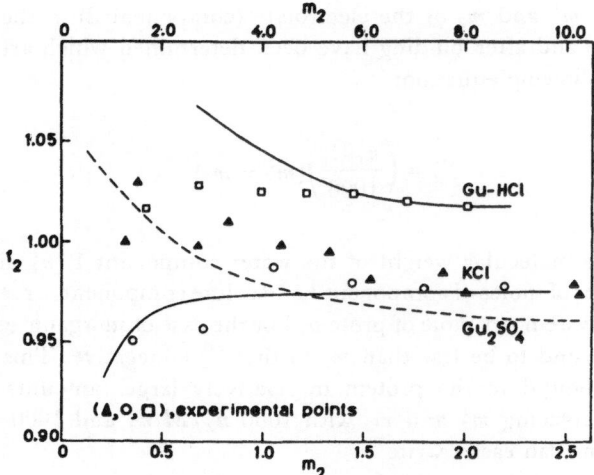

FIGURE 10.21. Plot[50] of f_2 against molality m_2 (Lower abscissa for KCl and Gu$_2$SO$_4$ and upper abscissa for Gu-HCl). $\square$, $\blacktriangle$, $\bigcirc$: experimental values obtained from m_2^i and m_2. Broken line calculated from evaluated values of Δn_1 and Δn_2, for Gu$_2$SO$_4$ and the solid lines for KCl and Gu-HCl, respectively. (Reproduced with permission from *Indian Journal of Biochemistry and Biophysics*.)

Here f_2^i and f_2^0 are the respective activity coefficients of the solute inside and outside the dialysis bag. Writing ($RT \ln f_\pi$) for $\bar{V}_2 \pi$ in equation (10.30), one finds

$$m_2^i = f_2 m_2 \qquad (10.31)$$

Here f_2 stands for the activity coefficient, which multiplied by the concentration of the dialysate gives the concentration of the same component inside the dialysis bag. The value of f_2 is equal to $f_2^0/f_2^i f_\pi$. If the osmotic effect on the activity coefficient is neglected ($f_\pi = 1$), then f_2^0/f_2^i becomes equal to f_2. Bull and Breese[49] have measured m_2^i and m_2 directly in their binding experiments, so that f_2 can be estimated.[50] Using their data, plots of f_2 for KCl, guanidine hydrochloride (Gu-HCl), and guanidine sulfate (Gu$_2$SO$_4$) as functions of m_2 are shown in Fig. 10.21. In the case of KCl, f_2 is less than unity whereas for Gu-HCl, it is greater than unity. For Gu$_2$SO$_4$, f_2 is greater than unity when m_2 is large, but at relatively lower values of m_2, its value becomes greater than unity. From the theoretical standpoint, the value of f_2 depends on the interionic attraction, ion-association, ionic hydration, ionic size, competitive interactions of water and electrolyte with the biopolymer etc.

10.13. FREE ENERGY OF EXCESS HYDRATION

The respective Gibbs–Duhem equations for the inside protein solution and the outside dialysate solution can be written in the form[50]

$$n_1^i d\mu_1^i + n_2^i d\mu_2^i + n_p^i d\mu_p^i = 0 \tag{10.32}$$

and

$$n_1 d\mu_1 + n_2 d\mu_2 = 0 \tag{10.33}$$

Here, n_1^i, n_2^i and n_p^i stand, respectively, for moles of water, solute, and protein components in the dialysis bag and μ_1^i, μ_2^i, and μ_p^i are the respective chemical potentials. Similarly, n_1 and n_2 are the respective moles of water and solute in the dialysate and μ_1 and μ_2 are their respective chemical potentials. Differentiating (10.29) at constant $\bar{V}_2$, we obtain

$$d\mu_2^i = d\mu_2 - \bar{V}_2 d\pi \tag{10.34}$$

Similarly, for component 1,

$$d\mu_1^i = d\mu_1 - \bar{V}_1 d\pi \tag{10.35}$$

where $\bar{V}_1$ is the partial molar volume of water, which is taken as a constant.

Combining equations (10.32), (10.33), (10.34), and (10.35), it can be shown[50] that the free energy change ΔG for changing the mole fraction of the solvent from unity to X_1 (or changing the solute mole fraction from zero to X_2) can be calculated from the following equation:

$$
\begin{aligned}
-\Delta G &= -\int_1^2 d\mu_p^i \\
&= \int_1^2 \Gamma_2^1 d\mu_2 - (V_t^i - \bar{V}_p)(\pi_s - \pi_0) \\
&= \int_1^2 \Gamma_1^2 d\mu_1 - (V_t^i - \bar{V}_p)(\pi_s - \pi_0) \tag{10.36}
\end{aligned}
$$

Here V_t^i is the total volume of the solution per mole of protein inside the bag and $\bar{V}_p$ the partial molal volume of the protein in the solution. π_s and π_0 stand for the osmotic pressures of protein solution in the presence and absence of the neutral salt, respectively. If one neglects the protein–protein attraction,[51] in the solution in the absence of the neutral salt of the medium, then π_s will be very close to π_0 and $\pi_s - \pi_0$ becomes zero. Replacing μ_2 in equation (10.36) by the chemical potentials of the ions of the solute (vide Section 9.12), it can be shown[50] that

$$-\Delta G = (\nu_+ + \nu_-)RT \int_0^{f_\pm X_\pm} \Gamma_2^1 \, d \ln f_\pm X_\pm \qquad (10.37)$$

10.14. EVALUATION OF ΔG, Δn_1, AND Δn_2

Bull and Breese[49] have carried out experiments on the binding of water and electrolytes to a protein at electrolyte concentrations higher than 0.2 molar. They have also assumed that n_1^i and n_2^i for such a system can be written as $n_1 + \Delta n_1$ and $n_2 + \Delta n_2$, where Δn_1 and Δn_2 are the absolute amounts of water and electrolytes bound per mole of protein, respectively. Here n_1 and n_2 are moles of components in the free phase inside the bag, so that n_1/n_2 is equal to X_1/X_2, the mole ratio composition of the dialysate solution. As discussed in Chapter 9, such consideration remains valid only when the protein–water mixture behaves as a two-phase system. Considering the gelatin–water system on this model (vide Chapter 8), one can relate Γ_1^2 and Γ_2^1 to Δn_1 and Δn_2 by equations (8.9) and (8.10). These equations can be converted into the following form, originally derived by Bull et al.[49]

$$n_1^i m_2 - 55.51 n_2^i = \Delta n_1 m_2 - 55.51 \Delta n_2 \qquad (10.38)$$

so that from a suitable linear plot of the experimental data, Δn_1 and Δn_2 can be evaluated. Values of Δn_1 and Δn_2 are presented in Table 10.8 for bovine serum albumin, egg albumin, and hemoglobin in the presence of several inorganic and organic salts.

Using the weight method of equilibrium dialysis technique, Bull and Breese[52] have extensively studied hydration of egg albumin in the presence of increasing concentrations of monohydric alcohols (e.g., CH_3OH, C_2H_5OH, C_3H_7OH, C_4H_9OH) and polyhydric alcohols such as sucrose, glycerol, mannitol, ethylene glycols, etc. Depending upon the chain length, the monohydric alcohols are bound to the protein in the aqueous media as positive excesses as a result of which the protein structure is destabilized. On the other hand, polyhydric alcohols are bound to the globular protein as negative excesses as a result of which the structural stability of the protein has been increased considerably. This explanation is consistent with that given for the stability of protein in aqueous media in terms of ΔG_{ws} (vide Section 9.15).

Chothia[53] has recently analyzed the fraction (f_a) of each kind of amino-acid residue buried in the interior of a globular protein molecule so that this fraction

TABLE 10.8. Free Energy Change for the Hydration of Proteins in the Presence of Electrolytes[a]

Salt	Δn_1 (moles H_2O/ mole protein)	Δn_2 (moles salt/ mole protein)	ΔG_M^0 (kJ/mole protein)	Γ_1^m (moles H_2O/ kg protein)	Γg_M^0 (kJ/kg protein)
		Egg albumin			
Na_2SO_4	1320	4.30	2090	29.3	46.5
Gu_2SO_4	850	8.56	1600	18.9	35.6
Li_2SO_4	775	0.12	1338	17.2	29.7
CsCl	450	−1.19	753	9.99	16.7
KCl	400	−3.22	669	8.88	14.9
NaCl	300	−1.47	498	6.68	11.0
LiCl	125	−0.41	209	2.77	4.65
NaBr	75	−2.42	129	1.66	2.88
Gu-HCl	0	41.0	−527		
		Hemoglobin			
Na_2SO_4	1550	0.98	1600	22.8	23.6
		Bovine serum albumin			
Na_2SO_4	2310	7.17	4019	33.9	59.1

[a] Reference 50.

subtracted from unity is obviously representing the fraction of each residue exposed to the solvent. Using the high-resolution NMR spectra of the polypeptide–water system frozen at temperatures between −20 and 60°C, Kuntz[54] has estimated the hydration of different amino-acid residues. Bull[55] has theoretically estimated the hydration of the exposed amino acids of globular protein combining the observations of Chotia and Kuntz. These are given in Table 10.9. From the equilibrium dialysis techniques, Bull has also determined values of Δn_1 and Δn_2 for different proteins in the presence of varying concentrations of sucrose using basically equation (10.38) in linear form. Δn_2 in all cases are observed to be zero. Grams of water thus bound per gram of a protein (estimated from Δn_1) are presented in Table 10.9. These values are not too far off from the theoretically estimated values of hydration. This observation establishes the fact that hydration of the globular protein takes place mostly at the protein–water interface.

In the "one-phase" model for the protein solution, Δn_1 and Δn_2 have no physical significance but these may be regarded only as empirical activity parameters. Using these values of constants, one can calculate Γ_1^2 or Γ_2^1 now defined empirically by equations (8.9) and (8.10), respectively.[50]

Bull and Breese[49] have carried out binding experiments at electrolyte concentrations (m_2) equal to 0.2 molal or higher so that ΔG can be calculated

TABLE 10.9. Experimental and Calculated Values of Protein Hydration[a]

Protein	pH of protein solutions	Grams H₂O bound per gram protein	
		Experiment	Theoretical
Conalbumin	6.65	0.426	0.33
Myoglobin	6.89	0.342	0.35
Lysozyme	8.74	0.545	0.28
Carbon monoxide Hemoglobin	6.96	0.337	0.31
β-lactoglobulin	5.72	0.450	0.28
Bovine serum albumin	5.00	0.356	0.36
Ovomucoid	4.10	0.542	0.39
Ribonuclease	9.20	0.577	0.32
Egg albumin	4.80	0.257	0.31

[a] Reference 55.

using equation (10.37) from the plot $\Gamma_2^1/f_\pm X_\pm$ against $f_\pm X_\pm$ and estimation of the graphical area between the limits zero and $f_\pm X_\pm$. From m_2 less than 0.2, integration may be carried out on the assumption that Γ_2^1 in dilute solution decreases linearly with decrease of $f_\pm X_\pm$. Values of ΔG thus calculated are shown in Fig. 10.22 for egg albumin[50] in the presence of several electrolytes.

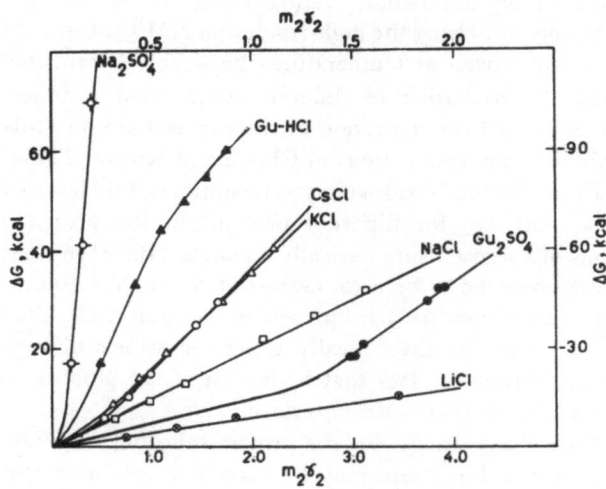

FIGURE 10.22. Plot of ΔG (kcal/mole egg albumin)[50] against solute activity (shown as $m_2\gamma_2$). Left ordinate and lower abscissa for LiCl, NaCl, KCl, CsCl and Na₂SO₄. Left ordinate and upper abscissa for Gu₂SO₄. Right ordinate and lower abscissa for Gu-HCl. (Reproduced with permission from *Indian Journal of Biochemistry and Biophysics*.)

ΔG in the case of all the salts except Gu-HCl is found to be positive, its magnitude increasing with increase of the salt concentration of the medium. As discussed in Chapter 9, these results indicate that the structural stability of a globular protein increases with increase of the concentration of the structure-stabilizing salts in the medium. For destabilizing reagents such as guanidine hydrochloride, ΔG is negative; moreover, its magnitude increases with increase of the solute concentration due to the gradual decrease of internal attraction of the protein in the presence of the destabilizer (vide Chapter 9).

10.15. STANDARD FREE ENERGY CHANGE FOR EXCESS HYDRATION

In Section 10.12, we have shown that inorganic electrolytes usually remain bound as negative excesses due to the (positive) excess hydration of the protein. From isopiestic experiments also, discussed elaborately in Chapter 9, Γ_2^1 for several proteins in the presence of inorganic salts are found to be negative for the same reason. Γ_2^1 for gelatin in the presence of aqueous solution of alcohols and glycols is also negative (vide Chapter 8).

From the thermodynamic treatment based either on "one-phase" or "two-phase" models for the protein–water system, Chattoraj *et al.*[50,56] have shown that the free energy change ΔG_h for excess hydration due to the change of the solvent activity a_1 from unity to a_1^{St} can be conveniently calculated with the help of the following equation:

$$\Delta G_h = -RT \int_1^{a_1} \left(\frac{\Gamma_1^2}{a_1}\right) da_1 - RT\Gamma_1^2 \ln\left(\frac{a_1^{St}}{a_1}\right) \tag{10.39}$$

The integral part here represents ΔG [vide equations (10.36) and (10.37)], which can be computed from the experimental data.

The second part of the right-hand side of equation (10.39) represents the free energy change ΔG_d for the hypothetical dilution of the solution at constant Γ_1^2 so that the solvent activity changes from a_1 to a_1^{St}. Chattoraj *et al.*[50,56] have arbitrarily assumed that the activity a_1^{St} of the solvent in the final state of reference is 0.5 mole fraction (or unit mole ratio composition of the solvent). This may be regarded as a practical scale of activity for solvent component in solution; ΔG_d can also be calculated from the experimental data.

In all cases of excess hydration of proteins in the presence of a salt, ΔG_h changes with increase of X_2 until it reaches a fixed value ΔG_M^0 (vide Table 10.8) when Γ_1^2 also attains a constant value Γ_1^M.

Values of ΔG_M^0 for BSA and egg albumin in Table 10.8 cannot be compared with each other because the proteins differ in molecular weights (denoted by M_p). After multiplying ΔG_M^0 by $1000/M_p$, the specific free energy change Δg_M^0 per kilogram of protein can be calculated. These values also given in Table 10.8 for different proteins are comparable to each other. In the same table are given the values Γ_1^m equal to $1000 \; \Gamma_1^M/M_p$ for several proteins. This parameter represents the maximum change in excess binding of water per kilogram of protein and its magnitude may be a relative measure for the affinity of the hydration process in the presence of the salt. The order of the Γ_1^m and Δg_M^0 in Table 10.8 are same so that the reactivity scale in terms of Δg_M^0 is established. Δg_M^0 in the presence of pure water is zero whereas in the presence of salt, their values are positive. The thermodynamic scale for Δg_M^0 thus proposed (Chattoraj *et al.*, unpublished) is independent of the type of the protein. It is not dependent also on the physical state of protein–water mixture (e.g., gel state or solution state). The comparison of Δg_M^0 in the scale is valid only for one kind of solvent used, such as water. Δg_M^0 for gelatin calculated for aqueous and nonaqueous solutions are not comparable with each other because of the difference of initial and final standard states of reference for the solvent compound present (Chattoraj *et al.*, unpublished).

For excess binding of solute like CTAB and SDS to a protein, the standard free energy change ΔG^0 has been calculated using equations (10.10) and (10.24) in which the hypothetical standard state of the solute in the bulk after binding is a unit mole fraction. Thus ΔG^0 and ΔG_M^0 are expressed in different scales and they are not comparable to each other. However Δg_M^0 for these systems can be calculated by computing the free energy change ΔG_S per mole for excess solute binding[50] from the use of the following equation:

$$\Delta G_{\text{solute}} = -RT \int_0^{a_2} \left(\frac{\Gamma_2^1}{a_2}\right) da_2 - RT\Gamma_2^1\left(\ln \frac{a_2^{\text{st}}}{a_2}\right) \qquad (10.40)$$

From the plot of ΔG_{solute} against a_2 (solute activity), maximum or standard value ΔG_M^0 for the free energy of binding may be calculated which on multiplication by factor $1000/M_p$ gives the free energy change Δg_M^0 per kilogram of protein. Here a_2^{st} is again taken to be equal to 0.5 mole fraction of solute.[50] At this standard state of the solute, mole fraction of solvent is also 0.5 so that these states of reference for the excess hydration and excess salt binding for a protein become nearly identical. Values of Δg_M^0 for binding SDS, CTAB, etc. to BSA given in Table 10.10 are all negative. Δg_M^0 given in Tables 10.8 and 10.10 are expressed in the same thermodynamic scale and the affinities of the solute for proteins are comparable to each other in terms of Δg_M^0. Further, since the scale is independent of the molecular weight, Δg_M^0 for proteins and nucleic acids for binding ligands are also comparable to each other (Chattoraj *et al.*, unpublished). Δg_M^0 represents the change of free energy during binding interaction when Γ_2^1/Γ_2^m or Γ_1^2/Γ_1^m becomes unity. Any conformational alteration of the biopolymer during binding interaction has been

TABLE 10.10. Δg_M^0 for Binding Amines and SDS to BSA and Gelatin; pH 6.0 and Ionic Strength 0.125

Protein	Ligand	Temperature (°C)	Γ_R^m (moles ligand/ kg protein)	$-\Delta g_M^0$ (kJ/kg protein)
BSA	CTAB	65	12.3	412
BSA	DTAB	65	4.98	178
BSA	MTAB	65	3.45	110
BSA	MTAB	45	0.591	20.3
BSA	CTAB	45	0.222	7.99
BSA	CPCL	30	0.176	6.09
BSA	CTAB	15	0.015	0.49
BSA	SDS	30	1.95	68.3
Gelatin	CTAB	30	0.68	22
Gelatin	MTAB	30	0.21	6.3
Gelatin	CTAB	15	0.12	3.3

considered in its computation so that the whole binding scale on this basis is independent of such conformational alteration. The scale appears to be universal since it is valid for positive values of Γ_1^2 as well as for positive values of Γ_2^1. The scale is also found to be useful for comparison of Δg_M^0 for binary or ternary protein mixtures (B. Sadhukhan and D. K. Chattoraj, unpublished).

10.16. FREE ENERGY OF COOPERATIVE BINDING AND TRANSDUCTION

For massive cooperative interaction, Tanford[29] has proposed that protein may act as a nucleus for the formation of a bound or micellar phase of ligand due to the excess transfer of free amphiphile from the bulk solution. For such a process, ΔG_T^0 in equation (10.11) can be written as

$$\Delta G_T^0 = (\mu_R^0)_b - \mu_R^0 \qquad (10.41)$$

Here $(\mu_R^0)_b$ and μ_R^0 are the standard chemical potentials of the ligand in the bound micellar phase and the aqueous phase, respectively. In Fig. 10.23, values of ΔG_T^0 calculated from binding experiments for anionic amphiphiles to serum albumin have been plotted[29] against the number (n) of carbon atoms present in the ligand molecule. Such a plot is expected to be linear if the hydrophobic effect contributes mainly to the binding interaction. The deviation from linearity for higher values of n indicates that many other contributions are included in the binding interactions besides the hydrophobic effect. These are extensively discussed in Sections 10.9 and 10.10.

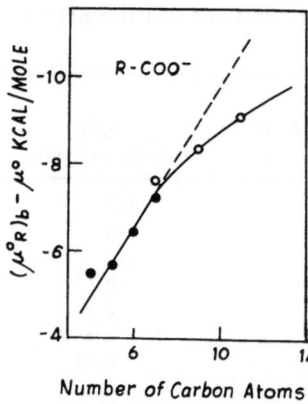

FIGURE 10.23. Free energy of association $(\mu_R^0)_b - \mu^0$ between anionic amphiphiles and native BSA. For $RCOO^-$: $\bigcirc$, data 1°C, ionic strength 0.2; $\bullet$, 2°C ionic strength 0.03. (Redrawn with changes from Ref. 29.)

During the transfer of one mole of ligand to protein, its conformation may alter significantly. ΔG_{con}^0 is expressed in joules per mole of protein whereas ΔG_T^0 is expressed in joules per mole of ligand (vide Table 10.5). According to Tanford,[29]

$$\Delta G_T^0 = (\Delta G_T^0)_{int} + \left(\frac{\Delta G_{con}^0}{\Gamma_R^m}\right) \tag{10.42}$$

at the completion of the massive binding interaction. Here $(\Delta G_T^0)_{int}$ is the intrinsic free energy of association per mole of a ligand to Γ_R^m binding sites. Equation (10.42) can be written in the form

$$\Gamma_R^m \Delta G_T^0 = \Gamma_R^m (\Delta G_T^0)_{int} + \Delta G_{con}^0 \tag{10.42a}$$

$\Gamma_R^m \Delta G_T^0$ expressed in joules per mole of protein is equal to ΔG in equation (10.11) when the binding interaction is perfectly cooperative. ΔG thus contains all other effects besides the effect due to the intrinsic interaction.

Recently Tanford[57] has considered further the physicochemical interpretation of $(\mu_R^0)_b$ in equation (10.41) from elaborate thermodynamic and statistical considerations. The chemical potential $(\mu_R)_b$ of the bound ligand is found to be related to $(\mu_R^0)_b$ by

$$(\mu_R)_b = (\mu_R^0)_b + RT \ln \frac{\theta}{1 - \theta} \tag{10.43}$$

where θ is the fraction of the total mole of sites remaining in the bound state. Tanford[57] has further extended this treatment for binding simultaneously two different ligands to their protein binding sites which may be situated in the biological membrane. The thermodynamic treatment is shown to be useful in the elucidation of the mechanism of biological free energy transduction. Free energy transduction involves the transfer of free energy from one molecule to another. The actual transfer may often occur while both molecules are bound to the transducer enzyme, which means the chemical potential of one bound ligand increases at the expense of that of the other.

10.17. PROTEIN-DETERGENT COMPLEXES

One important point which is still ambiguous in the literature is the description of the systems of protein–detergent complexes on the basis of various thermodynamic parameters. The arguments as to how ligand is associated has been described in literature.[3,29] In these and some other studies,[58] the possibility of "protein–micelle" complex formation has invariably been mentioned. Studies of Steinhardt *et al.*[7,8] have on the contrary clearly shown on the basis of experimental data and thermodynamic analysis that no protein-micelle complex can be formed. It has been shown that water-insoluble organic compounds do not solubilize in micelles and in detergent–protein complexes with the same standard free energy of solubilization. Further, the effect of electrolyte on the micellar systems is different from that found in detergent-protein systems.

For the formation of micelle R_n from an association of n monomers (R), the standard free energy ΔG_D^0 for the transfer of one mole of monomer from aqueous to micellar phase can be given by

$$\Delta G_D^0 = -RT \ln \left[\frac{R_n}{(R)^n} \right] \qquad (10.44)$$

Let us consider further the equilibrium for binding detergent R to protein P thus:

$$P + qR \rightleftharpoons PR_q \qquad (10.45)$$

Here q represents the number of moles of ligand bound to a mole of protein (q is actually equal to Γ_R^1). The free energy change for the transfer of one mole of R from aqueous to the boundary region of the protein–detergent complex may be given as

$$\Delta G_{PD}^0 = -RT \ln \frac{PR_q}{(PR)^q} \qquad (10.46)$$

One can further split up ΔG_D^0 and ΔG_{PD}^0 into the free energy changes due to intrinsic (int), electrical (el), and hydration (hy) interactions and write (Birdi and Chattoraj, unpublished)

$$\Delta G_D^0 = (\Delta G_D^0)_{int} + (\Delta G_D^0)_{el} + (\Delta G_D^0)_{hy} \qquad (10.47)$$

and

$$\Delta G_{PD}^0 = (\Delta G_{PD}^0)_{int} + (\Delta G_{PD}^0)_{el} + (\Delta G_{PD}^0)_{hy} \qquad (10.48)$$

$(\Delta G_{PD}^0)_{int}$ will differ from $(\Delta G_D^0)_{int}$ since the possibility would exist that the binding of an amphiphile could lead to interaction between hydrophobic alkyl chain of detergent and protein (to some nonpolar side chains) whereas in the formation of pure detergent micelles, ΔG_D^0 arises only from the interaction between two or more detergent molecules of some kind. The assertion made by several workers for the existence of protein–micelle complexes is possibly not valid on strict thermodynamic grounds. The finding that $(\Delta G_D^0)_{int} \neq (\Delta G_{PD}^0)_{int}$, as found from solubilization energy (next section) supports this conclusion.

10.18. HYDROPHOBIC CHARACTER OF PROTEIN–DETERGENT COMPLEXES

Even though a vast number of studies on the detergent–protein binding equilibria has been reported in the literature, there have been only a few studies reported on the hydrophobicity of the protein–detergent complexes in comparison to that of protein and detergent micelles.[6–8] The amphiphiles with simple structure are well characterized as regards the relation between the alkyl part of the molecule and any physical property which allows one to determine its effect on the hydrophobicity (vide Chapter 11). The standard free energy of solubilization (i.e., transfer of probe from water phase to micellar phase) ΔG_D^0 varies linearly with change in the number of carbon atoms in linear alkyl chain of detergents (vide Chapter 11).

The solubility behavior of a dye in a protein solution with varying SDS concentration is given in Fig. 10.24. The binding of SDS to protein proceeds as increasing amount of SDS is added to the aqueous solution (vide Figs. 10.13 and 10.14). At the limit of binding as noticed by a plateau in the plot of (B) versus (F), the free energies of binding and of micelle formation are going to be the determining factors. These data show that micelles are present after or near the plateau region. This is obviously not going to be a sharp transition, as is also found experimentally.

In region 1 for the protein–SDS curve in Fig. 10.24 there is no solubilization increase other than the solubility of the probe in buffer without SDS.

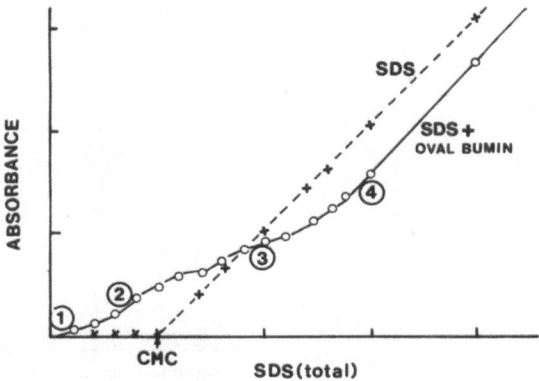

FIGURE 10.24. The solubilization (increase in absorbance) of dimethylaminoazobenzene (probe) by ovalbumin–SDS complexes in pH 7.4 (0.033 ionic strength phosphate buffers) at 25°C. (Redrawn from ref. 7.)

In region 2, enhanced solubilization of probe takes place by protein–detergent complex (in the absence of the micelles since SDS concentration is far below CMC). In region 3, the curve is leveling off and crosses the pure SDS curve at region 4. At concentrations higher than that present at region 4, the SDS curve and protein–SDS curves are parallel. This indicates that in this region, only micelles are being formed for additional amount of SDS.

Different SDS–protein complexes show varying solubilization characteristics or hydrophobicity. The proteins studied do not show any solubilization of probes when no detergent is present. Below CMC, therefore, solubilization takes place in the protein–SDS complex (Figs. 10.25 and 10.26).

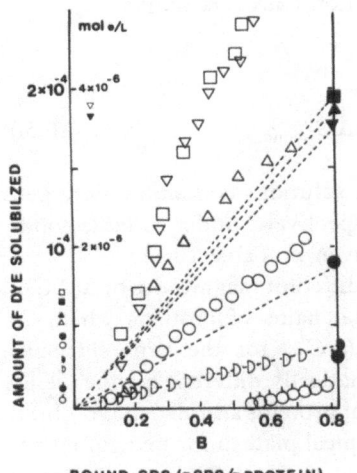

FIGURE 10.25. The dependence of probe solubilization (in mol/liter) on amounts of SDS (B) by 0.1% solutions of bovine serum albumin. ○, ●, Orange OT; ○, ●, dimethylaminoazobenzene; ◗, ◗, Sudan II; △, ▲, azobenzene; □, ■, naphthalene; ▽, ▼, anthracene. Open symbols: bovine serum albumin + SDS, closed symbols: SDS. The abscissa in the case of SDS micellar solutions (i.e., the closed symbols) refers to concentrations above the critical micelle concentration (i.e., B equivalent to SDS concentration in micellar form). (Redrawn from Ref. 8.)

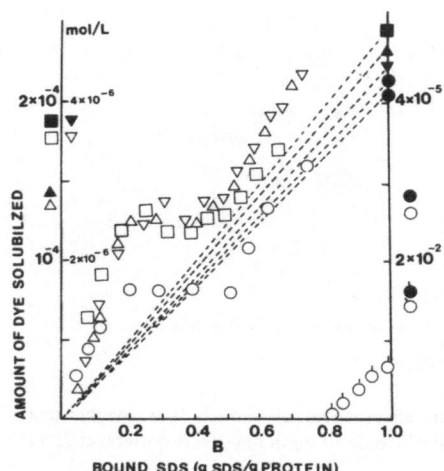

FIGURE 10.26. The dependence of probe solubilization (in mol/liter) on amounts of SDS bound (B) by 0.1% solutions of myoglobin. Open symbols: myoglobin + SDS, closed symbols: SDS. (Redrawn from Ref. 8.) (Symbols as in Fig. 10.25.)

At F less than cmc, the standard free energy ΔG_{P+D}^0 of solubilization in the protein detergent system can be written as

$$\Delta G_{P+D}^0 = -RT \ln\left(\frac{C_{P+D}^{probe}}{C_{aqueous}^{probe}}\right) \tag{10.49}$$

Analogous expressions for the free energy of solubilization ΔG_D^0 in the detergent micelles in the absence of protein are given in Chapter 11.

One can define then $\Delta(\Delta G_P^0)$ by the relation (Birdi, unpublished):

$$\Delta(\Delta G_P^0) = \Delta G_D^0 - \Delta G_{P+D}^0$$

$$= \Delta G_{micelle}^0 - \Delta G_{bound}^0 \tag{10.50}$$

From the measurement of absorbance of the solutions containing detergent micelles and detergent–protein complexes, respectively, under suitable conditions, $\Delta(\Delta G_P^0)$ can be estimated. These are given in Table 10.11.

The different probes are solubilized with different magnitudes of $\Delta(\Delta G_P^0)$. This may indicate that the hydrophobic side-chains of proteins which are taking part are specific for each protein. $\Delta(\Delta G_P^0)$ for the heme proteins, hemoglobin and myoglobin, behave in a dramatically different manner. Here $\Delta(\Delta G_P^0)$ is different between these two classes of proteins and also the solubility of probe versus SDS concentration shows a typical plateau for myoglobin and hemoglobin.

TABLE 10.11. Differences in Standard Free Energy of Solubilization, $-\Delta(G_P^0)$ (in kJ/mol) for Different Proteins for Various Probes[a]

Protein	Dmab	Sudan II	Azobenzene	Naphthalene	Anthracene
BSA (or HSA)	236	573	512	1550	1692
Ovalbumin	236	−522	−781	1207	1415
β-Lactoglobumin	168	−522	−584	798	936
Hemoglobin	−1264	—	—	210	1163
Myoglobin	−2465	—	—	302	452

[a] Data from Ref. 8.

Referring to the fact that the free energy of transfer of a water-insoluble probe from water† to micelle change by 220 calories per mole CH_2 in the detergent molecule, $\Delta(\Delta G_P^0)$ values can be given quantitiative dimensions. If $\Delta(\Delta G_P^0)$ were zero, then we would conclude that detergent solubilizes with the same free energy in micellar form as in detergent–protein complex. It is thus seen that the different protein–SDS complexes behave as if these were equivalent to SDS micelles or some equivalents of CH_2 groups different from SDS micelles when values of $\Delta(\Delta G_P^0)$ are positive or negative (Birdi, unpublished).

10.19. SUMMARY AND COMMENTS

From the discussions presented in this chapter, it will be clear that the binding of solute and solvent to biopolymers occurring in the living system can be quantitatively represented by the Gibbs excesses. The theoretical analysis is mainly based on the binding data obtained from the equilibrium dialysis experiments. The approach in all probability will remain valid for the binding data obtained from other types of experiments also. The solution of the biopolymer present in the dialysis bag is assumed to form a single phase, and this is in equilibrium with the outside dialysate solution. With the help of the Gibbs–Duhem equations, the excess binding of a component to a mole (or gram) of a biopolymer can be related to the free energy (ΔG_{ws}) of interaction of these components with the macromolecule. The similarity of form of these equations with those derived for adsorption at the liquid interface of a multi-component solution may be noted with considerable interest.

Based on the Gibbs approach, a very suitable equation for calculating the standard free energy change (ΔG^0) for binding interaction due to the change of ligand concentration from zero to unit mole fraction can be derived. It has been further shown that for quantitative comparison of ΔG^0 for various systems, these values should refer to saturation value Γ_R^m which can be determined from the binding experiments. The importance of the saturation values of adsorption of a solute at the liquid interface in computing many thermodynamic quantities has been discussed by Lucassen-Reynders (vide Section 5.15). The salient features of various equations derived by Scatchard, Bull, and Tanford for the calculation of the free energy of binding interaction have been discussed and their limitations and range of applicabilities have been mentioned. ΔG^0 calculated with the new reference state is free from many of these limitations.

ΔG^0 for binding cationic ligands to DNA varies regularly with change of temperature, ionic strength, hydrocarbon chain length, etc. From quantitative calculations, the binding interactions in these systems are shown to be mostly controlled by the entropy-driven "hydrophobic effect." In the case of binding cationic and anionic ligands to proteins, regularities in the change of the magnitude of ΔG^0 with temperature, ionic strength, and chain length are not observed, possibly due to the gradual exposition of the heterogeneous binding sites of the biopolymer with the progress of ligand interaction when the ligand concentration in the bulk solution is increased. From the calculation of ΔG^0 for mixed proteins as well as fat–protein systems, standard free energy change for protein–protein and protein–fat mixtures may be evaluated. Based on this approach, the thermodynamic features of protein–protein, protein–nucleic acid, and protein–polysaccharide interactions in the presence or absence of the ligand may be critically examined in future investigations. Such complex interactions actually occurs in cell membranes, chromosomes, ribosomes, blood serum, and many other multicomponent biological systems.

In the presence of many inorganic salts, proteins bind water as positive excess so that excess electrolyte binding becomes negative. The standard free energy change (Δg_M^0) for excess hydration can be calculated with the help of a thermodynamically derived equation so that these are comparable to each other irrespective of the nature of the proteins. This free energy change for excess hydration which are all positive refers to a practical scale of 0.5 mole fraction or unit mole ratio composition of solvent in the mixture. For positive excess binding of ligand to a protein, values of Δg_M^0 (reference state being 0.5 mole fraction of solute) in this universal scale are negative. The order of the maximum extent of positive excess binding of water or ligand, respectively, to a kilogram of protein is in complete agreement with that of Δg_M^0. The applicability of this reactivity scale for various types of binding interactions occurring in the biological system should be examined elaborately in the future.

Tanford has recently expressed the opinion that the standard chemical potential of the ligand bound to a biopolymer is different from that of the free ligand in solution. This gives a satisfactory explanation of the transport of ligand across the biological membrane of heterogeneous structure. This concept has close similarity with that proposed originally by Butler for the interfacial phase attached with a multicomponent solution of the liquid (vide

Section 3.17). The thermodynamic behavior of the ligand bound to protein has been examined beyond cmc. It appears from this study that ligand–ligand and ligand–protein interactions are both operative in this bound phase besides water interaction. Further study of such interactions in future may be important in understanding the interaction phenomena occurring in cell membranes.

MISCELLANEOUS SYSTEMS

11.1. INTRODUCTION

In several chapters of this book, we have elaborately discussed the concept of the Gibbs surface excess to well-defined specific systems in order to understand the feature of the competitive interactions of the solute and the solvent components with the interface. In the present chapter, we shall discuss the applicability of the same concept for micelle formation, hydration of powdered detergents, adsorption of inorganic electrolytes and gases, kinetic phenomena related to the adsorption process, etc. The adsorption equation derived from the statistical approach will be compared with that obtained by Gibbs from the application of classical thermodynamics. A summary of the existing information and the future directions for the application of the Gibbs excess concept will be discussed at the end.

11.2. COLLOIDAL MICELLES

It has been mentioned in Section 1.3 that the long-chain ionic and nonionic detergent molecules associate to form colloidal aggregates or micelles when the critical micellar concentration in the solution is exceeded. An ionic detergent RNa can associate at cmc to form the micelle M. This can be represented by the equation[1-3]

$$pR^- + qNa^+ = M^{-(p-q)} \qquad (11.1)$$

Here p stands for the average number of the monomer present in micellar unit. Frequently it is referred to as the aggregation number. The charge density near the micelle under this condition will be considerably high so that q number of Na^+ counterions will remain in the bound state with the micellar unit as a result of the intense electrostatic effect. At a given temperature T, the association constant K_M for the micellar reaction can be written as

$$K_M = \frac{a_M}{(a_{R^-})^p (a_{Na^+})^q} \qquad (11.2)$$

Here a_M, a_{R^-}, and a_{Na^+} represent the active concentrations of the micelle, organic ion as free monomer, and free inorganic counterion, respectively. The law of mass action thus applied will remain valid if the reaction takes place in a single phase and further micelle is monodispersed. Critical analysis, however, in recent years has revealed that the micellar system is mostly polydispersed in nature.

If ΔG^0 is the total standard free energy change for the micellar reaction, then the standard free energy change ΔG_m^0 for the incorporation of one mole of monomer within the micelles will be equal to $\Delta G^0/p$ so that from equation (1.37)

$$\Delta G_m^0 = -\frac{RT}{p} \ln K_M$$

$$= RT \ln a_{R^-} - \frac{1}{p}(RT \ln a_M - q \ln a_{Na^+}) \qquad (11.3)$$

When p is large and a_M and q are low in magnitude, the last term on the right-hand side of equation (11.3) is small compared to $RT \ln a_{R^-}$, so that[1,4]

$$\Delta G_m^0 = (\mu_2^0)_m - \mu_2^0$$

$$\simeq RT \ln a_{R^-}$$

$$\simeq RT \ln [\text{cmc}] \qquad (11.4)$$

Here (μ_2^0) is the chemical potential of the detergent in the monomeric form remaining free in the solution whereas $(\mu_2^0)_m$ is that present in the micelle formed in the aqueous phase (vide Section 1.10). This equation remains valid at the critical micelle concentration so that a_{R^-} may be replaced by cmc. The value of cmc can be determined accurately from various types of physico-chemical experiments so that ΔG_m^0 can be evaluated. If the value of cmc is expressed in the mole fraction unit, ΔG_m^0 is referred to the unitary scale. The use of this scale has been strongly advocated by Tanford[4] for the micellar system but criticized by others[5] on various theoretical grounds. It is also to be noted that for more precise calculation of ΔG_m^0 with the help of equation (11.3), values of p, q, and a_{Na^+} are to be determined from separate experiments. The effect of ion-atmosphere formation, counterion binding,[1,4] hydration of the ionic species,[4] micelle–micelle interaction,[6] etc. are implicitly included in ΔG_m^0 through the activity coefficient terms associated with the ionic species present in aqueous phase.

11.2.1. SURFACE TENSION AND CMC

The critical micelle concentration of ionic or nonionic detergents can be conveniently determined[1,7] from the measurement of the surface tension γ of the solvent as a function of the total concentration ($c_{R^-}^t$) of the organic solute in the medium. Below cmc, γ decreases sharply with increase of c_{R^-} whereas above this concentration, γ practically remains constant (vide Figs. 4.20 and 11.1). The Gibbs adsorption equation in the generalized form at cmc can be written as[8-10]

$$-d\gamma = RT[\Gamma_{R^-}d \ln a_{R^-} + \Gamma_M d \ln a_M + \Gamma_{Na^+}d \ln a_{Na^+}] \qquad (11.5)$$

Here Γ_{R^-}, Γ_M, and Γ_{Na^+} are respective values of the surface excesses of R^-, M, and Na^+. Following the theoretical arguments presented in Section 4.10, it has been shown that equation (11.5) can assume the form[8-10]

$$d\gamma = mRT\Gamma_{R^-}^t d \ln c_{R^-} \qquad (11.6)$$

Here $\Gamma_{R^-}^t$ is the total amount of the surface excesses equal to $\Gamma_{R^-} + p\Gamma_M$. For a large value of p of the micelle, m can be shown to attain a negligibly small value. Also above cmc, γ practically remains unchanged. It can therefore be concluded from equation (11.6) that above cmc, c_{R^-} remains constant so that only c_M increases with increase of $c_{R^-}^t$. Also the micelle formed in the bulk phase is surface inactive.

In Fig. 11.1, the γ–$c_{R^-}^t$ curve for highly pure SDS sample has been compared with that for an impure sample.[11] The expected impurity in this latter sample is dodecanol which is a product of the hydrolysis of SDS. γ for the impure sample decreases rapidly with the increase of c_R^t, and further, close to cmc, the curve exhibits a deep minimum. Beyond this minimum, γ increases with increase of c_R^t until it reaches a nearly constant value. At the concentration where a minimum occurs, $d\gamma/dc_{R^-}^t$ becomes zero so that Γ_2^1 should tend to zero also on the basis of the Gibbs adsorption equation. In this apparent azeotropic state, the concentration of SDS in the bulk and the interfaces would become identical. Sweeping the interface with a movable barrier, γ is found, however, to increase due to the surface-cleaning.[12] This indicates that the azeotropic state is not attained in the system and the reason for the minimum is different.

It is therefore concluded by Crisp[13] and others[14,15] that the presence of lauryl alcohol (as component 3) in the impure sample is the cause of the minimum observed in Fig. 11.1. For such a ternary system, the Gibbs equation reads

$$\frac{d\gamma}{dc_{R^-}^t} = -\Gamma_2^1 \frac{d\mu_{R^-}}{dc_{R^-}^t} - \Gamma_M^1 \frac{d\mu_M}{dc_{R^-}^t} - \Gamma_3^1 \frac{d\mu_3}{dc_{R^-}^t} \qquad (11.7)$$

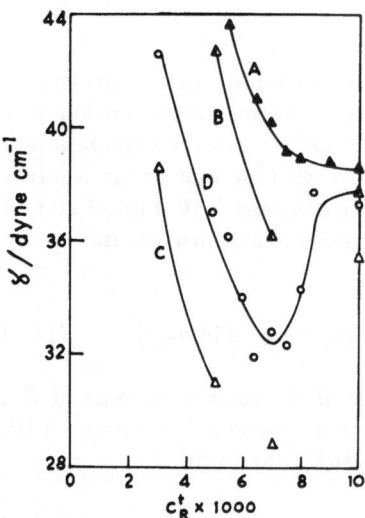

FIGURE 11.1. Effect of the addition of dodecanol on the surface tension of SDS. A, pure SDS; B, SDS with 0.1% dodecanol; C, SDS with 0.5% dodecanol; D, impure salt. (Redrawn from Ref. 11.)

Above cmc, c_{R^-} and μ_{R^-} almost remain constant and also $d\mu_M$ and Γ_M^1 are small. However, component 3 produced from the hydrolysis dissolves partly in the SDS micelles formed beyond cmc. The free concentration of the alcohol thus formed decreases with increase of $c_{R^-}^t$ as a result of which $d\gamma/dc_{R^-}^t$ becomes positive and γ beyond cmc for the impure sample increases with increase of $c_{R^-}^t$. The surface tension experiments with mixture of pure SDS and lauryl alcohol also (vide Fig. 11.1, curves B and C) support this explanation. Thus purity of a detergent sample can be tested in terms of the presence or absence of the minimum in the γ–$c_{R^-}^t$ curve.

11.2.2. TWO-PHASE MODEL AND MICELLAR EQUILIBRIUM

Many workers[1,4,16–20] have assumed that a micelle of colloidal dimension formed in the aqueous medium itself behaves as a separate phase like an oil drop. For an ionic system this separate phase is charged. The activity of the detergent placed in this pure phase is taken as unity by convention. At cmc, the aqueous phase in contact with this micellar phase is assumed to remain saturated with the monomeric detergent molecules and the concentration at the state of saturation is [cmc]. The distribution constant K_d for the nonionic or ionic detergents for the partition of the surfactant in the micellar and the water phases is 1/[cmc]. Using equation (1.43), one can then derive relation (11.4) in a straightforward manner.

The two-phase model is very convenient for the straightforward calculation of ΔG_m^0. However, the model is not satisfactory for various reasons.[1,20–22] On the basis of this model, a physical property of the system containing a detergent solution should change sharply at cmc due to the phase change. But actually such a change is observed to be continuous and less sharp.

Tanford[4] has shown from a detailed thermodynamic consideration that the difference between the "two-phase" and "one-phase" models will disappear when aggregation number for the colloidal micelle is high. Very recently Rusanov[23] has presented a general equation based on the thermodynamics of micellar systems in which he has demonstrated that "one-phase" and "two-phase" models are in complete agreement with each other. Relations between the size distribution of the micelles, the surface tension, and cmc have been established by Rusanov. In dealing with this theory, it appears that the thermodynamic relations similar to that used in the Gibbs adsorption equations have been used.

The hydrophilic groups of the micelle oriented towards the aqueous phase is extensively hydrated. A monomer is to compete with water for its placement in the micellar region, and from the intuitive point of view, the concept of the Gibbs excess is expected to be useful in dealing with the thermodynamics of micelle formation. In Chapters 9 and 10, the Gibbs excess concepts for the "one-phase" and "two-phase" models for bipolymer systems are found to be in agreement with each other.

From the knowledge of the net micellar charge and potential of the electrical double layer formed around spherical micelles, the contribution $(\Delta G_{el}^0)_m$ for the electrical free energy change can be computed.[1,4] There will also be free energy change due to the counterion binding and hydration of the micelles. All these free energy terms are to be added to ΔG_m^0 for obtaining total free energy change due to the formation of the micelles in the aqueous environment at and above cmc. However, the estimation of the thermodynamic parameters becomes complicated due to the recent observations that the micelles are polydispersed and the shape of a micelle is nonspherical.[4]

From equation (11.3) written in a simple form, it can be shown[24] that

$$\ln(\text{cmc}) = K' - \frac{q}{p}\ln(\text{cmc} + c_{Na^+}) \tag{11.8}$$

Here K' is a constant. The experimental data show that for all systems of ionic micelles without exception, a plot of $\ln(\text{cmc})$ versus $\ln(\text{cmc} + c_{Na^+})$ is linear. The slope q/p is equal to the degree of neutralization α of the micellar charge by the counterions. The fraction of the Na^+ counterions remaining in the bound state with the micellar phase is found to vary between 70% and 85% for various systems of ionic detergents.

11.2.3. HYDROPHOBIC INTERACTION

From the measurement of the values of cmc at a given temperature, values of ΔG_m^0 have been calculated[4] with the help of equation (11.4) for

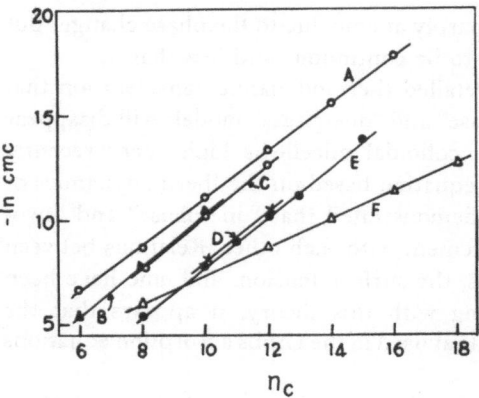

FIGURE 11.2. Plots of $-\ln$ cmc versus n_c. A, Alkyl hexaoxyethylene glycol monoethers at 25°C; B, alkyl sulfinyl alcohols at 25°C; C, alkyl glucosides at 25°C; D, alkyl trimethylammonium bromides in 0.5 M NaBr at 25°C; E, N-Alkyl betaines at 25°C; F, alkyl sulfates in the absence of added salt at 40°C. (Redrawn from Ref. 4.)

different homologous series of ionic, nonionic, and zwitter-ionic surfactants dissolved in the aqueous medium. From Fig. 11.2, one finds that ln [cmc] proportional to ΔG_m^0 increases linearly with increase in the number of carbon atoms (n_c) in the hydrocarbon chain of the detergents belonging to a particular homologous series.[4,25-29] All these linear curves fit the general equation

$$\Delta G_m^0 = RT \ln[\text{cmc}]$$

$$= (\Delta G_m^0)_{\text{hyl}} + (\Delta G_m^0)_{\text{CH}_2} \cdot n_c \qquad (11.9)$$

$(\Delta G_m^0)_{\text{CH}_2}$ is the standard free energy change due to the transfer of one mole of CH_2 groups of the detergent from the aqueous to the micellar environment. The net charge of the homologous series of the compounds called betaines, $RN(CH_3)_2^+CH_2COO^-$, is zero. The average value of $(\Delta G_m^0)_{\text{CH}_2}$ for such a system has been found to be -3060 J/mol obtained from the slope of the linear plot[4] in Fig. 11.2. This value is slightly lower than the free energy change -3300 J/mol for the transfer of one mole of CH_2 groups of an organic acid[30] from the bulk to the oil–water interface (vide Section 5.14). The value in the latter case is computed from Traube's rule using Langmuir's equation for adsorption. $(\Delta G_m^0)_{\text{hyl}}$ is essentially representing the change in the free energy due to the placement of one mole of the hydrophilic groups of the detergent from the aqueous to the micellar phases. It may also contain the end effect for the hydrocarbon group and other correction terms. Its average value for betaine is $-12\,250$ J/mol.

$(\Delta G_m^0)_{\text{CH}_2}$ for nonionic micelle as calculated by Mukerjee[1] from cmc data of a homologous series of straight-chain alcohols is -2860 J/mol. Tanford[4] has pointed out that there exist difficulties in estimating this quantity for the nonionic detergents. $(\Delta G_m^0)_{\text{CH}_2}$ for alkyl sulfate and other ionic detergents in the absence of the neutral salt are all close to -1750 J/mol. The free energy of transfer of a mole of CH_2 groups of sodium salt of a fatty acid at the oil–water interface is -2300 J/mol.[30] The lower values of $(\Delta G_m^0)_{CH_2}$ for ionic and nonionic surfactants possibly indicate that the chains inside the micelles are more constrained than in liquid hydrocarbon.[1]

For a homologous series of detergents, the results in Fig. 11.2 indicate that ΔG_m^0 depends on the regular increase in the length of the hydrocarbon chain. This behavior is an important characteristic of the hydrophobic effect, sometimes also termed the hydrophobic interaction. This force is mainly responsible for the association of the surfactant molecules into the micelles.

From the evaluation of ΔG_m^0 of a detergent at several temperatures, the enthalpy of micellization ΔH_m^0 can be calculated using equation (10.21). For ionic and zwitterionic micelles, ΔH_m^0 is practically zero.[4] The values of ΔH_m^0 for nonionic micelles are slightly positive.[4] ΔG_m^0 is therefore close to $-T\Delta S_m^0$ so that micelle formation from the monomer is an entropy controlled process.

Various statistical and thermodynamic theories have been proposed for the hydrophobic interaction in terms of the ordered structure of water near the micellar surface.[5,31] Thermodynamics of the micelle formation on the basis of the hydrophobic interactions have been elaborated in recent years.[32-35]

In order to determine the hydrophobicity of the interior alkyl part of the micelle by solubilization experiments, the excess of a probe (water-insoluble organic compound) is brought in equilibrium with a detergent solution. One can write

$$\mu_{probe}^{solid} = \mu_{probe}^{aqueous} = \mu_{probe}^{micelle} \tag{11.10}$$

The standard free energy of transfer ΔG_D^0 is given as

$$\Delta G_D^0 = -RT \ln \left[\frac{c_{pr}^m}{c_{pr}^w} \right] \tag{11.11}$$

Here C_{pr}^m and C_{pr}^w are the concentrations of the probe in the micelle and water phases, respectively. It has been shown that ΔG_D^0 varies linearly with number of carbon atoms in the alkyl chain of the detergent molecule.[33] The slope of this plot representing the free energy of transfer per CH_2 group is -837 J/mol in all cases regardless of the detergent of the different homologous series or the probes. The values of ΔG_D^0 for various probes in SDS are given in Table 11.1. ΔG_D^0 is the main quantity which determines the hydrophobicity of the alkyl part of the amphiphile.

The dialkylphosphatidyl cholines $(Cn)_2PL$ dispersed in the aqueous phase are known to form vesicles (vide Chapter 6). The alkyl chain length in these lipids gives rise to a variety of physical characteristics which are attributed to the hydrophobicity. The phase transition temperature and the enthalpy of melting of vesicles ranging from $(C_{22})_2PL$ to $(C_{14})_2PL$ are found to vary

TABLE 11.1. Standard Free Energy of Solubilization ΔG_D^0 of Various Probes in Aqueous Micellar Solutions of SDS[a]

Probe	$-\Delta G_D^0$ (kJ/mol)	Mole SDS/Mole probe $(1/C_{pr}^m)$
Dimethylaminozobenzene	10.83	80
Sudan II	8.48	31
Azobenzene	6.52	14
Naphthalene	6.52	14
Anthracene	16.45	780
Phenanthrene	9.53	47

[a] At 25°C, ionic strength 0.033, pH = 7.4.

linearly with n_c. The lower homologues of lecithins $(C_7)_2PL$, $(C_8)_2PL$, $(C_{10})_2PL$, and $(C_{16})_2PL$ are reported to form micelles having characteristic cmc.[36] It is seen that the change in $(\Delta G_v^0)_{CH_2}$ for the vesicle formation is approximately of the same value as $(\Delta G_m^0)_{CH_2}$ as observed for nonionic micelles. When the hydrocarbons of the amphiphile molecules are nonlinear, modified equations are to be used for the calculation of $(\Delta G_v^0)_{CH_2}$ and $(\Delta G_m^0)_{CH_2}$ which will have term containing area of contact with alkyl group and water. Such corrections are also necessary for the determination of the hydrophobicity of straight-chain compounds.[33,37,38]

Tanford[39] has recently presented a simple approach for the interpretation of the hydrophobic effect on the basis of the interfacial energies associated with water and a liquid hydrocarbon with and without contact with each other. Suppose two immiscible liquids A and B with an interface between them is present in a tube of 1 cm^2 cross-sectional area. The work required to separate these two layers (thus producing 2 cm^2 of liquid–air surface) is called the work of adhesion W_{AB}. From the Dupre equation [vide equation (7.1)]

$$W_{AB} = \gamma_A + \gamma_B - \gamma_{AB} \qquad (11.12)$$

The surface tension γ_A at the air–water interface at 25°C is 72 mN/m whereas the tensions (γ_B) of hexane–air and octane–air interfaces at 25°C are 17.9 and 21.2 mN/m, respectively. The interfacial tensions γ_{AB} for hexane–water and octane–water systems are 39.5 and 42.0 mN/m, respectively. The average value of W_{AB} for hexane–water and octane–water interfaces calculated from equation (11.12) is +51 mN/m. If the pure hydrocarbon and pure water from air is brought in contact with each other in the tube, then the free energy will change by an amount −51 erg/cm^2 (−0.051 J/m^2). This process of contact (based on forces of attraction) is therefore a spontaneous process.

From the measurement of the solubilities of hexane and octane in water at 25°C, the incremental free energy change per square angstrom of surface per mole of hydrocarbon transferred are found to be -84 and -128 J/mol, which correspond to -14 and -23 erg/cm^2 (-0.014 and -0.023 J/m^2) of the actual contact area.[40–42] These quantities thus represent the change in free energy due to the contact of hydrocarbon molecules with water in the bulk phase. The surface curvature in the contact region in the molecular level is maximum in this case. Because of the curvature effect, these values of free energy are considerably less than 51 erg/cm^2 (0.051 J/m^2), the value obtained with negligible curvature effect when the macroscopic contact area is 1 m^2 in the tube. The curvature effect on the interfacial energy has been discussed by several authors.[43,44]

11.3. EXCESS HYDRATION OF POWDERED DETERGENTS

The solid detergent mixed with a relatively small amount of water forms wet gel system. The ionic head groups of the detergent have strong affinities for water molecules. It has been shown by Luzzati[45,46] that the lyotropic liquid crystal is formed under this condition. The structure of the liquid crystal depends on the proportion and nature of the compounds present in the system. With increase of the water content, different types of phase changes may occur within the system, indicating alternations in the liquid crystalline structure.[45] At certain stages, the hydrocarbon parts of the detergent molecules remain associated due to the hydrophobic forces thus forming lamellar bilayer structure. Water present in the channels between two such bilayers is in interaction with the hydrophilic groups of the associated detergent molecules.[45,47b] With increase of the proportion of water, rodlike micelles of detergent formed are converted to globular spherical and nonspherical micelles.[4]

Using the isopiestic vapor pressure technique (vide Section 2.8), adsorption of the water vapor by the cationic and anionic detergents has been studied by Sadhukhan, Chattoraj, and Birdi.[47a] An examination of the isotherms shown in Fig. 11.3 will indicate that the curves for CTAB and MTAB resemble type III BET isotherm whereas for SDS and DTAB, the curves resemble type II BET isotherm. At low values of p/p_0, two molecules of water are found to be tightly bound with a detergent molecule in the primary surface layer of the detergent powder. At higher values of p/p_0, the extent of hydration is considerably high and further its magnitude increases with increase in the length of the hydrocarbon chain associated with detergent molecule.

The binding interaction of water with the solid detergent has also been investigated by Sadhukhan et al.[47a] in the presence of inorganic electrolytes, sugar, and urea. The isopiestic experiments used for such a system have also

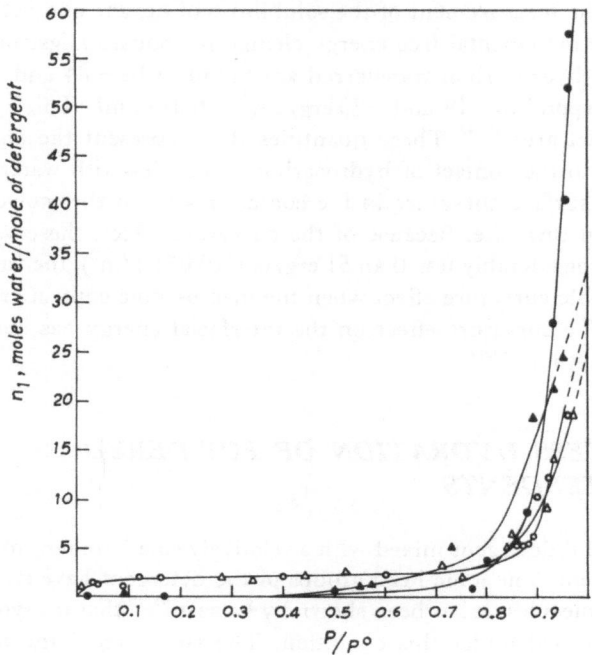

FIGURE 11.3. Plot of n_1 vs. p/p_0 at 25°C. ●, CTAB; ▲, MTAB; ○, DTAB; △, SDS. (Redrawn from Ref. 47a.)

been discussed in Section 2.8. In most cases, the excess hydration Γ_1^2 per mole of detergent present in the powder varies linearly with X_1/X_2 (vide Figs. 11.4 and 11.5) in agreement with the equation

$$\Gamma_1^2 = \Delta n_1 - \Delta n_2 \frac{X_1}{X_2} \qquad (11.13)$$

Values of Δn_1 and Δn_2 (moles of solvent and solute bound per mole of detergent, respectively) obtained from the intercept and slope of these linear plots are shown in Table 11.2. X_1/X_2 are the mole ratio composition of the solvent and the solute components in the reference solution at isopiestic equilibrium. It is assumed that the wet detergent powder in the gel state does not contribute to the vapor pressure to any significant extent. The counterions of the detergent in the gel are observed to be in the bound state with the detergent ion. Magnitudes of Δn_1 and Δn_2 depend upon the nature of the electrolyte and the length of the hydrocarbon chain of the detergent. Urea dehydrates the detergent whereas sugar does not affect the detergent hydration to a significant extent.

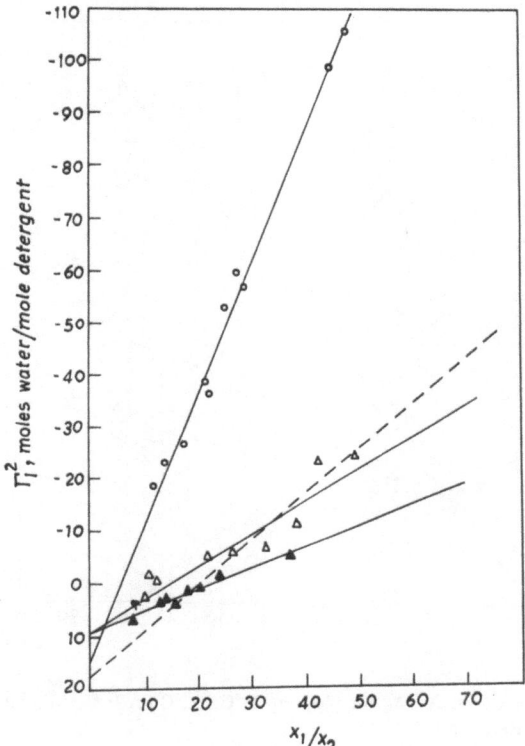

FIGURE 11.4. Plot Γ_1^2 vs. X_1/X_2 for SDS at 25°C. $\bigcirc$, LiCl; (- - -), NaCl; $\triangle$, KCl; $\blacktriangle$, CsCl. (Redrawn from Ref. 47a.)

TABLE 11.2. Values of Δn_1 (Moles Water/Mole Detergent) and Δn_2 (Moles Solute/Mole Detergent) in the Presence of Various Electrolytes Sucrose and Urea[a]

Electrolytes, sucrose or urea	SDS		CTAB	
	Δn_1	Δn_2	Δn_1	Δn_2
LiCl	13.9	2.50	16.2	2.30
NaCl	17.6	0.46	13.4	0.40
KCl	9.3	0.63	15.0	0.44
CsCl	9.2	0.40	—	—
Na_2SO_4	10.9	0.12	10.3	0.07
NaBr	7.3	0.41	8.6	1.05
NaI	5.2	0.08	1.5	0.02
Urea	8.5	0.21	—	—
Sucrose	20.2	0.19	—	—

[a] Reference 47a.

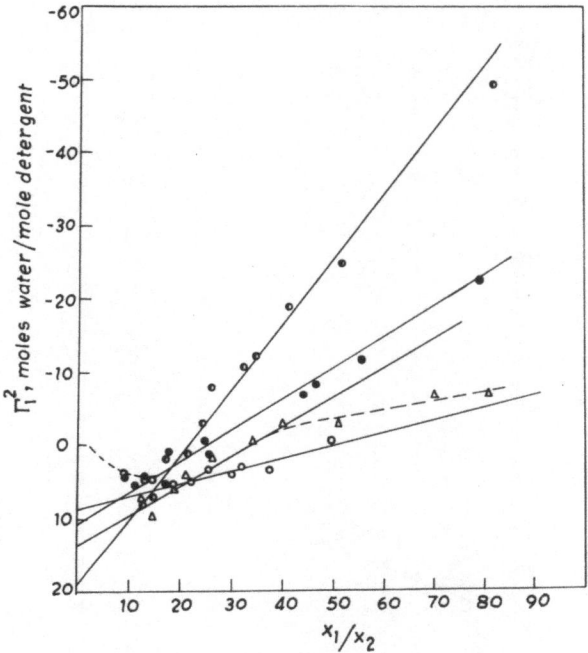

FIGURE 11.5. Plot of Γ_1^2 vs. X_1/X_2 for NaCl at 25°C. $\triangle$, CTAB; ●, MTAB; ○, DTAB; ◖, SDS. (Redrawn from Ref. 47a.)

11.4. ADSORPTION OF INORGANIC ELECTROLYTES AT SOLID–WATER INTERFACES

In Chapter 8, we have discussed some features of the adsorption of long-chain organic electrolytes, ionic dyes, and organic polyelectrolytes at the solid–water interfaces. The affinities of these electrolytes for the solid surfaces are relatively high compared to that of water. Determination of the extent of the adsorption of these substances by analytical method becomes easier for this reason. The relative affinities of the inorganic electrolytes for the surface of the powdered solid suspended in the aqueous medium have, on the other hand, not been tested so far very seriously. The experimental difficulties for such a study arise due to the relatively lower affinity of these electrolytes for the solid surface in general.

Nag, Sadhukhan, and Chattoraj (unpublished) have recently studied the excess accumulation of inorganic electrolytes by powdered barium sulfate sample using the isopiestic vapor pressure technique described in Section 2.8. The results shown in Fig. 11.6 indicate that the electrolytes in many cases are accumulated as negative surface excesses due to the preferential adsorption

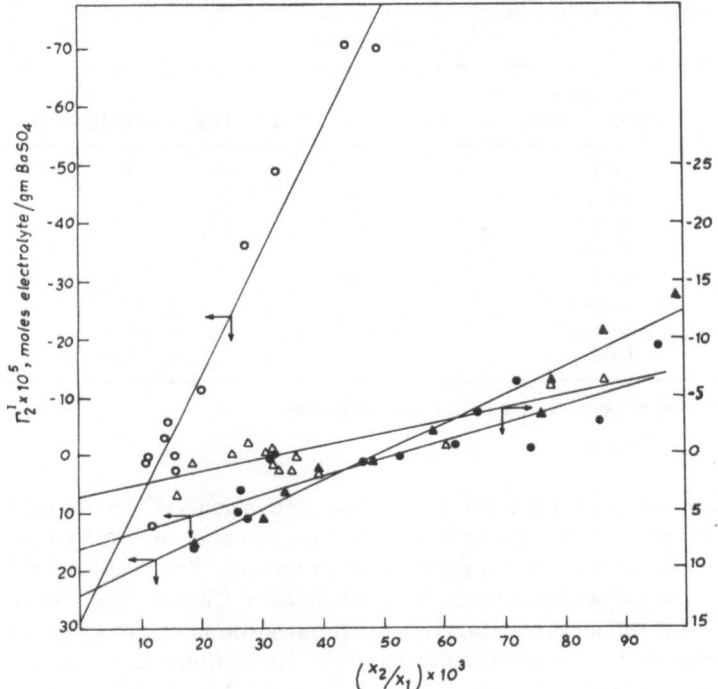

FIGURE 11.6. Plot of Γ_2^1 vs. X_2/X_1 for $BaSO_4$ at 25°C. ●, NaCl; △, KCl; ▲, RbCl; ○, Na_2SO_4.

of water at the solid–water interface. In a few cases, the azeotropic point of zero Γ_2^1 and Γ_1^2 are found to exist. This is the characteristic feature of the type IV isotherm formulated by Schay (vide Section 8.5). Γ_2^1 in all cases are found to vary linearly with X_2/X_1. From the slopes and intercepts of these curves, absolute amounts of water and electrolyte components Δn_1 and Δn_2 bound per gram of powdered barium sulfate have been calculated (vide Table 11.3). These values are found to be dependent on the nature of the electrolytes. The results appear to be consistent with those obtained for the adsorption of different inorganic electrolytes at the air–liquid interface (vide Section 3.14).

11.5. GAS ADSORPTION AT SOLID AND LIQUID INTERFACES

In Chapter 9 and also in Sections 11.3 and 11.4, the thermodynamic aspects of the adsorption of the water vapor by powdered biopolymer, detergents, and solid powder have been extensively discussed. Equation (9.6)

TABLE 11.3. Hydration of $BaSO_4$ in the Presence of Electrolytes, Urea, and Sucrose[a]

Electrolytes, urea or sucrose	Δn_1 (mg H_2O/g $BaSO_4$)	Δn_2 (mg of solute/g $BaSO_4$)
LiCl	189	24
NaCl	55	9.2
KCl	20	2.9
RbCl	90	29
CsCl	94	3.9
Na_2SO_4	381	40
Sucrose	324	68
Urea	31	2.8

[a] A. Nag, B. Sadhukhan, and D. K. Chattoraj (unpublished).

derived on thermodynamic grounds can be used in general[48] for the calculation of the change in the free energy due to the adsorption of any gas by a powdered solid material at a given temperature and pressure. The problems related to the standard states have already been discussed in Chapter 9. It has also been shown[48] that the ideal equation of state for adsorption of gases by solids may be represented by relation (5.5) if the Gibbs adsorption equation is combined with the form of the Langmuir equation at the low-pressure region. Such combination at the high-pressure region will lead to the equation of state of the form (5.54). When the adsorbed gas molecules exhibit interchain cohesional forces the equation of state in modified form is obtained. Combining this modified equation with the Gibbs equation and further assuming adsorption to be monomolecular, it can be shown[48] that

$$\ln Kp = \frac{\theta}{1 - \theta} + \ln \frac{\theta}{1 - \theta} - C\theta \qquad (11.13)$$

Here θ is the fraction of the surface covered by the gas and K is a constant. This complex equation supports the S-shaped nature of the adsorption isotherm sometimes obtained on the basis of the experimental data. The value of C is related to the intermolecular attraction term. Ross and Winkler[49] found that the adsorption of nitrogen on carbon black at 778 K obeyed equation (10.14). Several other types of isotherms obtained by the combination of various equations of states of the adsorbed film of gas with the Gibbs adsorption equation have been subsequently proposed.[48] Harkins and Jura[50] deduced an equation representing an isotherm for the adsorption of gases based mainly on the Gibbs adsorption equation. At higher values of p/p_0 this equation fits the experimental data better than the BET equation.

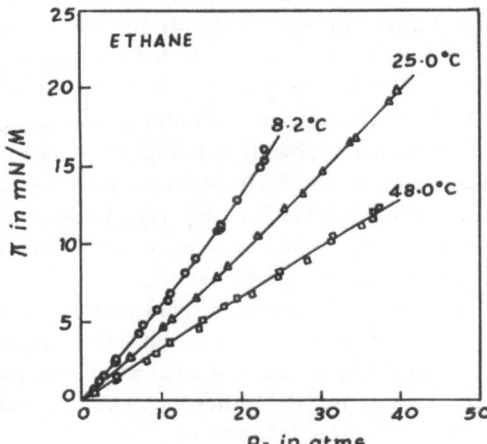

FIGURE 11.7. Surface pressures of ethane on water. (Redrawn from Ref. 51.)

Recently an attempt has also been made to measure the surface tension of water as a function of the pressure of a compressed gas with which the liquid is in contact.[51] In Figs. 11.7 and 11.8, the results of such measurements for ethane and isobutane are shown. Neglecting the presence of water vapor in the system at high pressure region, one can write the Gibbs equation in the form

$$\left(\frac{\partial \pi}{\partial p}\right)_T = \Gamma_2^1 \frac{\bar{Z}RT}{p} \qquad (11.14a)$$

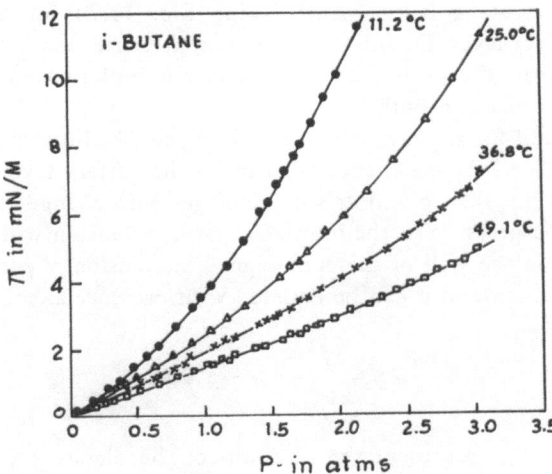

FIGURE 11.8. Surface pressures of isobutane on water. (Redrawn from Ref. 51.)

Here Γ_2^1 is the excess amount of the gas adsorbed per square centimeter of the liquid surface. $\bar{Z}$ stands for the compressibility factor of the gas at a pressure p which accounts for the fugacity of the gas. The effect of water vapor on p is negligibly small. Plots of Γ_2^1 against p for compressed methane, ethylene, ethane, normal butane, isobutane, and carbon dioxide are shown as a function of pressure ranging up to saturation atmosphere for condensable gases and up to 68 atmospheres with methane. The shapes of the isotherms are in most cases either BET type I or type III. With the exception of methane, the low-temperature adsorption isotherms for each gas all pass into the multi-layer regime at high pressures.

The thermodynamic parameters characterizing the adsorption of these gases can be evaluated from either the surface pressure or adsorption isotherms. The equilibrium heats of adsorption calculated from the data vary from -7.9 to -19.3 kJ/mol for different gases. The entropies of adsorption for various gases range from -25 to 63 J/°Cmol. The standard free energies of adsorption at low coverages has also been calculated using the Kemball–Rideal conversion[52] for the standard states when p is equal to one atmosphere and π equal to 0.608 mN/m for the surface.

11.6. STATISTICAL MODELS AND THE SURFACE TENSION

From the molecular point of view based upon the concepts of statistical mechanics, between two fluid bulk phases such as a liquid in contact with its vapor, there exists a layer AA'BB' (vide Fig. 11.9b) of finite thickness called the surface layer. Desnity, composition, pressure, etc. in this thin region vary rapidly from the value characteristic of one bulk phase to that valid for the other fluid phase in bulk.

Let p stand for the uniform pressure of the two bulk phases in contact when the surface between them is planar. In the surface layer, however, the pressure p^x is less than p and its value changes with change of x. Due to the decrease of the pressure in the interfacial layer, a tension will be exerted by the interface on the wall of the container. This tension γ per unit length is the surface tension and it can be equated with pressure as follows:[12,43,53,54]

$$\gamma = \int_{x_{AA'}}^{x_{BB'}} (p - p^x)\, dx \qquad (11.15)$$

Here $x_{AA'}$ and $x_{BB'}$ represent the positions of the planes AA' and BB' with respect to the position of the Gibbs dividing plane GG' (Fig 11.9a) placed at x equal to zero. For many mathematical operations, the positions of these two

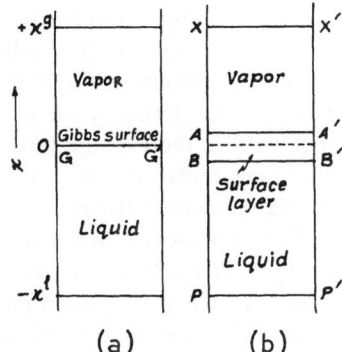

FIGURE 11.9. Surface of pure liquid in contact with vapor, (a) Gibbs dividing surface, (b) surface layer.

planes are set at $-\infty$ and $+\infty$. The pressure characteristic in the inhomogeneous surface region has been elaborately discussed by Goodrich.[55]

Using the radial distribution method based on the distribution functions of the bulk liquid in contact with its vapor, Kirkwood and Buff[12,53,54] have shown that equation (11.15) can be written in the following relatively simple form:

$$\gamma = \frac{\pi}{8} \int_0^\infty \left(\frac{du}{dR}\right) R^4 \rho^{(2),1}(R)\, dR \tag{11.16}$$

Here R is the distance between a pair of molecules in the system and u, the potential energy of interaction between them. The value of u for liquid argon has been calculated[12,53] from the Lennard-Jones equation for the potential energy. Also $\rho^{(2),1}(R)$ is the pair distribution function for the homogeneous liquid phase far from the dividing surface. Its value, depending on R, can be calculated for liquid argon at 90 K from the X-ray diffraction measurements. γ estimated from equation (11.16) for argon at 90 K is 14.9 nM/m, the experimental value being 11.9 nM/m.

11.6.1. CELL MODEL AND SURFACE TENSION

In the cell model[12] for a liquid having structure of the short-range order, a molecule in the fluid remains in a cell or a cage formed by Z number of neighboring molecules. The coordination number Z for a molecule in the bulk liquid is higher than that when the same molecule is present at the interface. From the partition functions of a number of such molecules forming quasicrystalline structures in the bulk and at the interface, it can be shown that

$$\gamma = \frac{1}{A}\left[kT \ln\frac{\nu'}{\nu} - \frac{1}{2}m\omega(0)\right] \tag{11.17}$$

Here A is the area occupied by one molecule at the interface. Also ν and ν' represent the frequencies of vibration of the molecule in the cages (or cells) present in the liquid in the bulk and at the interfaces respectively. The term m related to the lattice plane has been explained in Section 6.4. The cohesive energy $\omega(0)$ is the average minimum potential of the molecule at the center of the cage remaining in the bulk phase. From the known values of the quantities present in the right-hand side of equation (11.17), γ of liquid argon at 85 K has been calculated to be 9 mN/m. The experimental value of this quantity is 13.2 mN/m. Assuming the existence of the occupied and vacant sites of the cell in the incomplete monolayer model proposed by Prigogine and Saraga[12], calculated value of γ has also been found to be 13 mN/m in agreement with that obtained from the experiment.

11.6.2. CELL MODEL AND THE GIBBS ADSORPTION EQUATION

Applying the statistical approach involved in the cell model to the system of a regular binary solution in bulk in contact with an interfacial phase of monomolecular thickness, one can derive[12] the following relations:

$$\gamma = \gamma_1^{0} + \frac{RT}{A_1} \ln \frac{X_1^m}{X_1} + \frac{\alpha l}{A_1}[(X_2^m)^2 - X_2^2] - \frac{\alpha_m}{A_1}(X_2)^2$$

$$= \gamma_2^0 + \frac{RT}{A_2} \ln \frac{X_2^m}{X_2} + \frac{\alpha l}{A_2}[(X_1^m)^2 - X_1^2] - \frac{\alpha_m}{A_2}(X_1)^2 \qquad (11.18)$$

Here X_1 and X_2 are respective mole fractions of components 1 and 2 in the bulk and X_1^m and X_2^m those present in the monolayer phase. γ_1^0 and γ_2^0 are the surface tensions of pure liquid components 1 and 2, respectively. Also m and l are the parameters associated with the planes of the quasicrystalline lattice (vide Section 6.4); α has been defined previously by equation (6.22). Here A_1 and A_2 stand for the partial molal surface areas which remain unchanged. Differentiating equation (11.18) with respect to X_1 and X_2 and assuming $A_1 = A_2$, one finds

$$\left(\frac{\partial \gamma}{\partial X_2}\right)_{T,p} = \frac{1}{A_2}\left[RT\left(\frac{X_1^m}{X_1} - \frac{X_2^m}{X_2}\right) - 2\alpha(1-m)(X_1^m X_2 - X_1 X_2^m)\right]$$

$$(11.19)$$

Since the surface area A_m of the monolayer is $A_2(n_1^m + n_2^m)$ according to equation (6.24),

$$\frac{d\gamma}{dX_2} = \frac{RT}{A_m}\left[\frac{n_1^m}{X_1} - \frac{n_2^m}{X_2}\right] - \frac{2\alpha(1-m)}{A_m}(X_2 n_1^m - X_1 n_2^m) \qquad (11.20)$$

If the Gibbs dividing surface GG' in the liquid column in Fig. 11.9 is placed between monolayer and the bulk liquid, then n_1^m/A_m and n_2^m/A_m can be replaced by Γ_1^x and Γ_2^x so that[12]

$$-\frac{d\gamma}{dX_2} = \frac{RT}{X_2}\left(\Gamma_2^x - \Gamma_1^x\frac{X_2}{X_1}\right) - 2\alpha X_1(1-m)\left(\Gamma_2^x - \Gamma_1^x\frac{X_2}{X_1}\right) \qquad (11.21)$$

Since the relative surface excess Γ_2^1 is given by relation (3.37), equation (11.21) can be converted into the form[12]

$$-d\gamma = \Gamma_2^1 d\ln X_2 RT\left[1 - \frac{2\alpha(1-m)X_1 X_2}{RT}\right] \qquad (11.22)$$

The Gibbs adsorption equation derived on the basis of the classical thermodynamics can be written in the form

$$-d\gamma = RT\Gamma_2^1 d\ln f_2 X_2$$

$$= \Gamma_2^1 d\ln X_2 RT\left[1 + X_2\frac{d\ln f_2}{dX_2}\right] \qquad (11.23)$$

For a regular solution, it can be shown that

$$RT\ln f_2 = \alpha(1 - X_2)^2 \qquad (11.24)$$

Combining equations (11.23) and (11.24), the Gibbs adsorption equation (11.25) for the regular solution is obtained:

$$-d\gamma = \Gamma_2^1 d\ln X_2 RT\left[1 - \frac{2\alpha X_1 X_2}{RT}\right] \qquad (11.25)$$

When α becomes zero, the solution becomes ideal and perfect so that equation (11.22) based on the cell model becomes the same as the Gibbs adsorption equation (11.25). However, for a regular solution, equation (11.24) based on the monolayer model is not the same as the Gibbs adsorption equation. It is therefore concluded that the monolayer model assumed in the statistical derivation is very crude for the case of the regular solution. Defay *et al.*[12]

have suggested that two layers instead of one forming the interfacial phase should be considered for the statistical consideration. Ono *et al.*[43,56,57] have assumed multilayer model for the interface in the derivation of the adsorption equation based or statistical mechanics. They have also shown that as the number of the transition layers at the interfacial region approaches infinity, the form of these derived equations becomes closer to the Gibbs adsorption equation.

11.7. THE ELECTROCAPILLARY SYSTEM

In Chapters 4 and 5, we have discussed various forms of the Gibbs adsorption equations for the calculation of the adsorption of the organic ions at the air–water or the oil–water interfaces. Ions in such systems can be frequently transferred from the bulk to the surface phases or vice versa. In the case of the mercury–water interface, the transfer of electrons will occur under the applied electrical field. The electrons should be regarded as a component involved in the system under consideration.[12]

An alternative and useful approach in dealing with such systems is the replacement of the chemical potential μ_i terms in the surface or bulk phases by the electrochemical potential $\bar{\mu}_i$ so that the Gibbs equation at a constant temperature can now be written in the form[12]

$$d\gamma = -\sum_i \Gamma_i^x d\bar{\mu}_i \qquad (11.26)$$

Here Γ_i^x stands as usual for the surface excess of the ionic species as defined by Gibbs [vide equation (3.28)]. The electrochemical potential $\bar{\mu}_i$ is related to the chemical potential μ_i by the equation

$$\bar{\mu}_i = \mu_i + z_i \bar{F} \phi \qquad (11.27)$$

Here ϕ is the inner electrical potential at a point in the boundary phase adjacent to the bulk. $\bar{F}$ stands for faraday and z_i is the valency of the ion which remains adsorbed at the interface. Miller and Frommer[58] have argued that equation (11.26) with mole fraction scale for the chemical potential should be used for the calculation of the adsorption of polyelectrolyte at the liquid interface. They have also shown that respective values of the coefficient m of the Gibbs equation for polyelectrolyte adsorption become 1 and $(1 + z_i)$ on the basis of this equation in the presence and absence of excess neutral salts in the bulk medium in agreement with the conclusions drawn earlier (vide Chapter 4). Further, they have derived a general equation for m on the basis of equation (11.26) which differs from equation (4.40) derived by Chattoraj.[59]

Defay *et al.*[12] and also Chattoraj[59] have shown that for the air–water and oil–water interfaces, if one replaces the $\bar{\mu}_i$ term in equation (11.26) by equation (11.27) and applies the electroneutrality condition as expressed by relation (4.21), relation (11.26) will be converted to the form

$$-d\gamma = \sum_{i=1}^{i} \Gamma_i^x d\mu_i \qquad (11.28)$$

Further if the Gibbs dividing plane GG′ (vide Section 3.6) is now placed in a position when Γ_1^x becomes zero, then

$$-d\gamma = \sum_{i=2}^{i} \Gamma_i^1 d\mu_i \qquad (11.29)$$

Here i stands for the different ionic species of the solute and Γ_i^1 is the relative surface excess of the solute component i as defined by equation (3.75). The general form of the coefficient m of the Gibbs equation given by equation (4.40) and supported from experiments thus remains valid.

Mercury in contact with aqueous solution forms the polarized electrode. The system can be represented by the following scheme[12]:

	A		B		C	
Bulk phase I	¦	Surface phase		Surface phase	¦	Bulk phase II
(Hg)	¦	(Hg)		(solution)	¦	(solution)
	A′		B′		C′	

For polarized electrode, one can imagine that concentration of Hg_2^{++} in the aqueous solution is practically nil so that the electrode potential does not depend on its concentration. The potential difference $\phi^I - \phi^{II}$ can be varied without changing the composition of the solution. For such a polarized electrode, one can imagine that the surface phase can be divided into two parts, i.e., AA′BB′ and BB′CC′. These two parts may be imagined to be separated by a membrane which is not permeable to all components of the system. Applying the Gibbs adsorption equation (11.26) and then combining with equation (11.27), it can be shown that at constant temperature

$$-d\gamma = \sum_i \Gamma_i d\mu_i^I + \sum_j \Gamma_j d\mu_j^{II} + qd(\phi^I - \phi^{II}) \qquad (11.30)$$

Here Γ_i and Γ_j are the surface excesses of the ith and jth components in phase I and phase II, respectively, whereas μ_i^I and μ_i^{II} are the values of their chemical potentials. Also $(\phi^I - \phi^{II})$ are the potential drop in the system and q is the charge per unit area of the metal surface. At constant composition of the system, one can then write[12]

$$d\gamma = -qd(\phi^{I} - \phi^{II}) \tag{11.31}$$

which is the Lippmann equation.

In the experimental arrangement for the variation of $\phi^{I} - \phi^{II}$, mercury is connected by a platinum wire to a metal terminal A (made of copper) whereas the solution is connected to another terminal B so that the potential drop at A and B can be measured since

$$E = \phi^{A} - \phi^{B} \tag{11.32}$$

By changing E at constant bulk composition, one finds

$$dE = d(\phi^{A} - \phi^{B})$$

$$= d(\phi^{I} - \phi^{II}) \tag{11.33}$$

Combining equations (11.31) and (11.33), we get

$$\left(\frac{\partial\gamma}{\partial E}\right)_{T,\mu} = -q \tag{11.34}$$

The electrocapillary curve relates values of γ for different values of E so that from the slope of the curve at a value of E, q can be evaluated.

The surface of metallic mercury in contact with an electrolyte solution is positively charged. In the experiment on the electrocapillarity, this charge at the mercury surface is decreased by the applied potential whereby the interfacial tension γ between mercury solution system will increase. The electrocapillary curve obtained from the plot of γ against applied potential forms parabola. The Gibbs adsorption equation (11.26) is very useful in dealing with the electrocapillary phenomena observed for the polarized mercury electrode in contact with the aqueous solution. The microscopic picture of the electrical double layer at the metal solution interface (vide Fig. 5.22) emerges in details from the analysis of the data based on electrocapillary experiments.[60]

11.8. SIZE AND STABILITY OF MICROEMULSION

Oil can be dispersed in water in the presence of an ionic surfactant whereby turbid oil–water emulsion will be formed. The free energy of the system, however, increases due to the enormous increase of the surface area during the emulsification process so that both oil in water and water in oil emulsions are unstable from the thermodynamic standpoint.

However, Hoar and Schulman[61] earlier showed that with the addition of a nonionic cosurfactant hexanol to an oil–water emulsion stabilized by a surfactant, the turbid system becomes clear due to the formation of the microemulsion. For microemulsion, the γ at the oil–water interface in the presence of the surfactant and cosurfactant may be as low as 0.01 mN/m or it may be even slightly negative so that the formation of the microemulsion becomes spontaneous.[61-65] The microemulsion system at equilibrium is thermodynamically stable. The size of the globule in the microemulsion may vary between 8 and 80 nm.

An elegant theory on the stability of the microemulsion based on the Gibbs adsorption equation has recently been presented by Ruckenstein and Krishnan[66] for the presence of the surfactant and cosurfactant components 2 and 3 in the oil–water system. For the nonionic components or for the ionic components in the presence of the excess neutral salt, one can write the Gibbs adsorption equation in the form

$$-d\gamma = \Gamma_2^1 d\mu_2 + \Gamma_3^1 d\mu_3 \tag{11.35}$$

This on integration gives

$$\gamma = \gamma_0 - kT \int_1^2 \Gamma_2^1 \, d \ln c_2 - kT \int_1^2 \Gamma_3^1 \, d \ln c_3 \tag{11.36}$$

Here Γ_2^1 and Γ_3^1 are relative surface excesses and c_2 and c_3 the concentrations of the components in the bulk liquid phases, respectively. γ_0 and γ are the interfacial tensions of the oil–water interface in the absence and presence of solute components 2 and 3.

From the consideration of the adsorption–desorption equilibrium, Ruckenstein *et al.*[66] have also obtained alternative relations between Γ_2^1 and Γ_3^1 and c_2 and c_3, which on combination with equation (11.36), will finally read

$$\gamma = \gamma_0 - N_R kT \ln \left[1 + (K_2 c_2)^{1/p_2} + (K_3 c_3)^{1/p_3} \right] \tag{11.37}$$

K_2 and K_3 are distribution constants for adsorption whose values can be estimated from experiments. Also p_2 and p_3 are the number of surface sites occupied by one molecule of surfactant and cosurfactant, respectively. N_R is the inverse of the average area occupied by a site.

The Helmholtz free energy change ΔF of the microemulsion formation per unit volume for a radius R of the globule can be divided into two parts:

$$\Delta F = \Delta F_d + \Delta F_e \tag{11.38}$$

The initial reference state of the free energy is selected as the hypothetical nondispersed state in which the surfactant and cosurfactant are distributed between oil and water.[66] ΔF_d is the Helmholtz free energy change for the formation of the interface followed by the dilution of the components to the mole fractions in the final state of equilibrium. ΔF_e is defined as the free energy change due to the globule dispersion. Expressions relating ΔF_d and ΔF_e have been obtained[66] so that these quantities can be estimated as function of R when equations (11.37) and (11.38) are combined together.

In Fig. 11.10, ΔF and γ, respectively, computed from equations (11.37) and (11.32) have been plotted as a function of the radius of the globule. When the interfacial tension γ is close to zero, ΔF vs. R plot exhibits minimum and at this equilibrium state, the value of R (represented by R_e) becomes 20 nm. This is in agreement with experimental observations of the average size of the microemulsion. Further beyond this value of R, γ becomes negative and the system also existing beyond free energy minimum is in the nonequilibrium state. R in this region must decrease and more globules must be formed to reach the state of equilibrium again.

Extending this approach further[67] for the system forming microemulsion containing a cationic surfactant R^+A^- in the presence of a neutral salt X^+A^- and a nonionic cosurfactant (s),

$$\gamma = \gamma_0 - N_R kT \ln{(1 + K_2 c_R\, e^{-\varepsilon\psi/kT})^{1/p_2}}$$

$$+ (K_2 c_s)^{1/p_3} - \int \sigma\, d\psi \tag{11.38a}$$

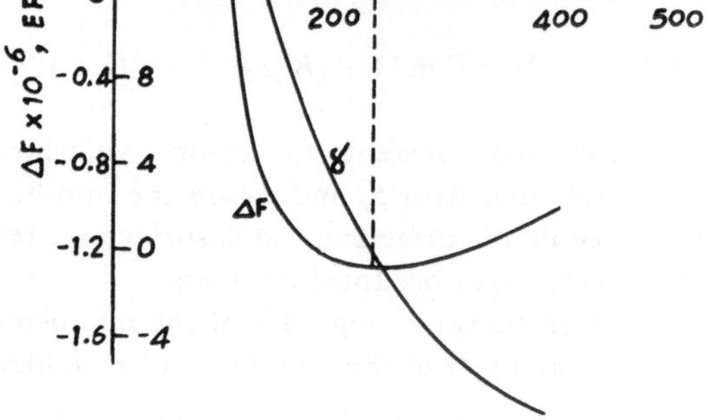

FIGURE 11.10. Free energy of formation vs. the radius R. Dependence of interfacial tension on the radius R. (Redrawn from Ref. 66.)

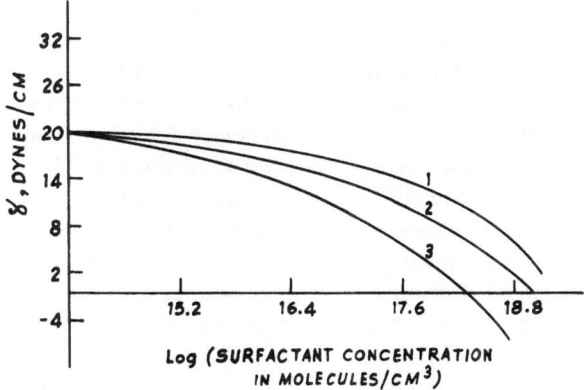

FIGURE 11.11. Interfacial tensions as function of ionic surfactants in the presence of a co-surfactant for different salt concentrations. Salt concentrations 5×10^{18}, 5×10^{19}, and 1×10^{21} molecules/cm^3 for curves 1, 2, and 3, respectively. (Redrawn from Ref. 67.)

Here σ is the charge density at the surface of the globule and ψ is the surface potential in the electrical double layer. In Fig. 11.11, values of γ thus calculated have been plotted against the log of the concentration of the surfactant when the cosurfactant is also present. γ is found to be very low and even negative for the microemulsion as expected.

11.9. ADSORPTION AND NONEQUILIBRIUM STATES

In many naturally occurring processes, the extents of adsorption at the solid and liquid interfaces do not actually attain the state of equilibrium within a reasonable time period. Thus, *in vitro* study indicates that adsorption of hydrochloric acid by different cereals require about 15 h time for the attainment of equilibrium.[68] However, after taking a cereal meal, the food remaining in the stomach region of a human being for 2 to 4 h adsorbs secret hydrochloric acid when the system is not in a state of adsorption equilibrium. Many proteins adsorbed on the biosurfaces alter conformation by surface denaturation and this process depends significantly on time. The study of the adsorption in nonequilibrium state is of considerable biological importance.

The surface-active amphiphiles and proteins are spontaneously transferred from the bulk to the interface of a clean liquid at a certain rate which depends on diffusional property of the adsorbate and other factors. From the application of the Ficks law of diffusion, it can be shown[69,70] that

$$\Gamma_p = 2c_p \left(\frac{Dt}{\pi}\right)^{1/2} \frac{N}{1000} \tag{11.39}$$

Here c_p is the concentration of the adsorbate in the bulk and Γ_p the number of the adsorbed molecules per square centimeter of the surface at time t seconds. D here stands for the diffusion coefficient of the adsorbate molecule. The equation is derived on the assumption that the adsorbed molecules cannot desorb from the surface and the activation energy for adsorption is zero. Hence it is valid only at the initial stage of the adsorption process. The above equation has further been modified by Ward and Tordai,[71] which includes the effect of desorption. Using equation (11.39) with plausible value of D, it can easily be shown that the biological polymers may take two hours time or more to reach the state of adsorption equilibrium whereas for small molecules like alcohols, this time may be of the order of one second or even considerably less than that.

For a protein, Γ_p at a given value of c_p and t can be directly measured using radiotracer experiments so that D can be estimated using equation (11.39). These values are compared with literature values in Table 11.4. In recent experiments of Graham and Phillips,[74] D obtained for several proteins from equation (11.39) are found to be higher in order than those found in literature because of the stirring effect during adsorption studies.

In Fig. 11.12, $d\gamma/dt^{1/2}$ for hemoglobin and myoglobin are plotted against c_p. For constant Γ_p, the slope of this line is proportional to D. The data also fit the virial equation (11.40) at a given value of c_p thus[73]:

$$\gamma(t) = a_1 t^{1/2} + a_2 t^4 + a_3 t^3 + a_4 t^2 + a_5 t + a_6 \qquad (11.40)$$

Values of these coefficients for hemoglobin and myoglobin are given by Birdi.[73]

The rate of reduction of surface pressure with time at the liquid interface possibly involves following three steps[75-77]: (i) diffusion of protein to the interface, (ii) surface denaturation of protein, and (iii) reconformation of

TABLE 11.4. Diffusion Constants as Determined from Adsorption Kinetics

Protein	D_{exp} (m²/s)	$D_{literature}$ (m²/s)
β-Casein[a]	3.3×10^{-10}	1.0×10^{-10}
K-Casein[a]	1.5×10^{-10}	1.0×10^{-10}
BSA[b]	0.6×10^{-10}	0.1×10^{-10}
Hemoglobin[c]		0.7×10^{-10}
Myoglobin[c]		1.0×10^{-10}

[a] Reference 72.
[b] Reference 69.
[c] Reference 73.

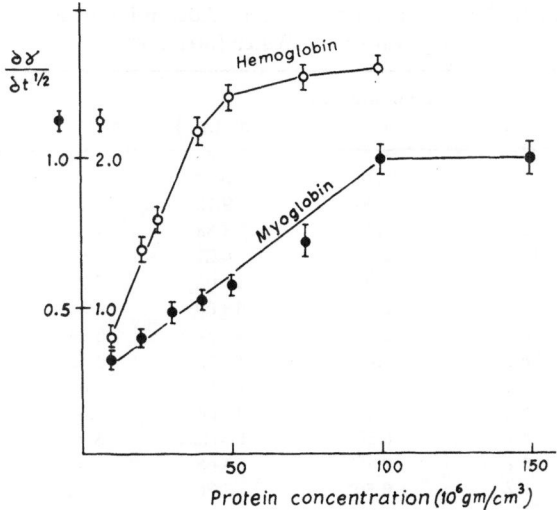

FIGURE 11.12. Plot of $d\gamma/dt^{1/2}$ against concentration of proteins.[73]

adsorbed protein. If this rate follows the first-order equation,[78] then it should obey the following relation:

$$\ln \frac{\pi_{ss} - \pi_t}{\pi_{ss} - \pi_0} = -Kt \qquad (11.41)$$

Here π_0 and π_t are interfacial pressures at time equal to zero and t, respectively. π_{ss} is the value of π at a time when steady state is reached.[74] Its value has been arbitrarily taken as one hour.[79] In Fig. 11.13, are shown the plots of the left-hand side of equation (11.41) against t for BSA adsorbed at the oil–water

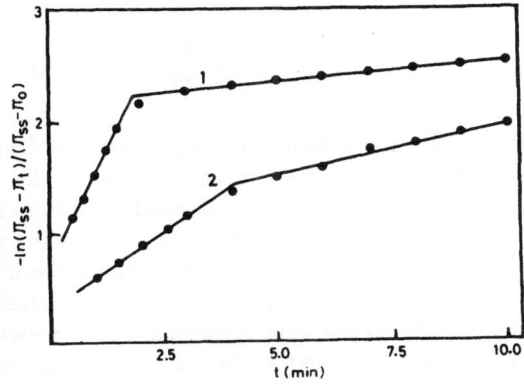

FIGURE 11.13. Plot[79] of $-\ln(\pi_{ss} - \pi_t)/(\pi_{ss} - \pi_0)$ vs. time for BSA solutions at varying concentrations at the peanut oil–water interface (1, 0.0025%; 2, 0.00125%). (Reproduced with permission from *Indian Journal of Biochemistry and Biophysics.*)

TABLE 11.5. Kinetic Parameters of BSA Adsorption from Solution to the Peanut Oil–Water Interface[a]

Conc. (%)	pH	Ionic strength (NaCl)	K_1 (h^{-1})	K_2 (h^{-1})	ΔA (Å^2)
0.00125	1.0	0.10	NA[b]	—	—
0.00125	2.2	0.00	9.18	5.10	275
0.00125	8.3	0.00	8.88	4.26	250
0.00125	12.7	0.00	6.21	5.07	305
0.00125	6.2	0.00	16.02	5.28	205
0.00125	5.0	0.00	18.00	4.80	185
0.00250	6.2	0.02	41.34	3.68	190
0.00500	6.2	0.02	51.72	3.72	—
0.03000	6.2	0.02	Too fast	3.64	—
0.00125	6.2	0.10	18.06	3.54	225
0.00125	6.2	0.50	Too fast	3.96	—
0.00125	6.2	0.50[c]	4.68	—	830
0.00125	6.2	0.50[d]	7.20	4.12	520

[a] Reference 79.
[b] NA = No appreciable time variation.
[c] KSCN = 0.5 M.
[d] KSCN = 0.1 M + NaCl = 0.40 M.

interface.[79] The slopes of two straight lines thus obtained represent the rate constants K_1 and K_2 for diffusion followed by denaturation process and reconformation step, respectively. These values are given in Table 11.5 for BSA at various pH and ionic strength of the medium.

Ward and Tordai[71] related the increase in Γ_R for simple surfactants to the kinetics of adsorption and desorption where there exists an energy barrier for adsorption. The relation can be used in modified form[80] by the following equation for the rate of adsorption:

$$\ln \frac{d\pi}{dt} = \ln (K_w c_b) - \frac{\pi \Delta A}{kT} \tag{11.42}$$

Here K_w is a constant and ΔA represents the area of the liquid surface swept out by residues of a protein molecule. The plot of $\ln (d\pi/dt)$ against π is linear so that ΔA can be evaluated.[79] Values of ΔA for BSA at polar oil–water interfaces included in Table 11.5 are in the range 1.80–8.30 nm^2/molecule. This value is considerably lower than 100 nm^2/molecule, the area covered by fully extended protein molecule. The surface denaturation is thus not that extensive and the bulk dimension of the molecule at the interface is considerably lower than that of the fully extended chain.[79]

TABLE 11.6. Comparison of Γ_p^1 and
$(\Gamma_p^1)_{\text{Gibbs}}$ for β-Casein Films Adsorbed at
A/W Interface[a]

$c_p \times 10^6$ (wt.%)	$(\Gamma_p)_{\text{exp}}$ (mg/m^2)	$(\Gamma_p^1)_{\text{Gibbs}}$ (mg/m^2)
3	0.25	0.1
6	0.5	0.5
8	0.75	1.0
10	1.00	2.0
20	1.50	5.0
50	2.50	10.0

[a] References 73, 80–82.

Recently Graham and Phillips[74,81–83] have made extensive studies of proteins adsorbed at liquid interfaces using radiotracer and ellipsometric methods. There are indications for multilayer adsorption from such studies. They have also measured π for the adsorbed protein layer as function of c_p after allowing sufficient time for adsorption to reach a state close to equilibrium. The activity coefficients of the β-casein, a random-coiled protein, in salt solution have been estimated by them from the second virial coefficients, so that the surface excess $(\Gamma_p^1)_{\text{Gibbs}}$ has been calculated using the Gibbs equation

$$(\Gamma_p^1)_{\text{Gibbs}} = \frac{1}{kT} \frac{d\pi}{d \ln a_p} \tag{11.43}$$

where a_p stands for the activity of the protein. The order of $(\Gamma_p^1)_{\text{Gibbs}}$ is not far off from Γ_p obtained directly from the experiments (vide Table 11.6). The difference in the magnitudes is possibly due to the lack of true equilibrium between surface and bulk phases.

Adsorption of BSA and gelatin at polar oil–water interfaces has been studied analytically using the emulsification technique by Das and Chattoraj[84] as a function of biopolymer concentration at a given temperature, pH, and ionic strength of the medium (vide Fig. 11.14). The saturation value (Γ_p^m) for the adsorption is observed to be affected by the nature of the neutral salts in the media. The standard free energies of adsorption have been calculated using equation (8.55) based on the validity of the Gibbs adsorption equation (vide Table 11.7). However, results in Fig. 11.15 will indicate that even when Γ_p reaches the constant value Γ_p^m, the value of π never becomes independent of time. This means that the adsorbed protein in the saturated state alters conformation and the system is not in the equilibrium state. The effect of this for the calculation of ΔG for a system in equilibrium is, however, small.

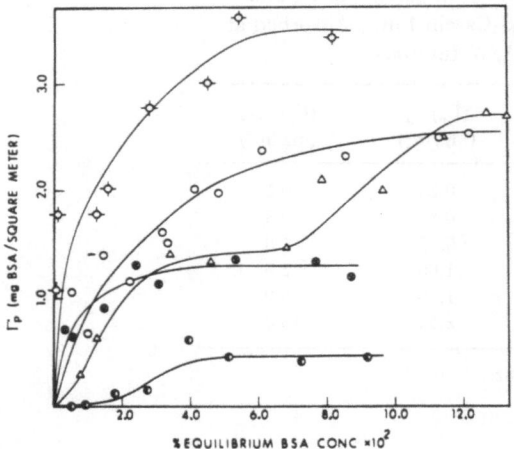

FIGURE 11.14. Plot[84] of Γ_p vs. percent concentration of BSA in the presence of various neutral salts at constant ionic strength of 0.10 and temperature 30°C. —○— $CaCl_2$; △ LiCl; ○ NaCl; ⊗ Na_2SO_4; ◑ KSCN. (Reproduced with permission from Academic Press, New York.)

Bull[85] has demonstrated that only a fraction of a globular protein in the adsorbed monolayer undergoes surface denaturation in a certain period of time so that these cannot be desorbed back in the bulk solution. Earlier Ghosh and Bull[86] also observed that enzyme chymotrypsin adsorbed on the paraffin–water interface becomes inactive by surface denaturation.

The phenomenon of adsorption at the liquid interface is studied when the system is not in equilibrium. It has been observed that the values of γ for solutions containing amphiphiles measured by the dynamic method do not always agree with those for the static method. In the dynamic method of

TABLE 11.7. Standard Free-Energy Change ΔG_B^0 due to Adsorption of BSA at the Peanut Oil–Water Interface[a]

Salt	pH	Ionic strength	Temp. (°C)	Γ_p^m (mg/m²)	ΔG_B^0 (kJ/mol)
NaCl	4.0	0.10	30	0.70	36.78 ± 2.96
NaCl	5.0	0.10	30	2.54	35.24 ± 1.21
NaCl	6.0	0.10	30	1.52	36.78 ± 4.26
NaCl	5.0	0.01	30	2.38	37.41 ± 1.96
NaCl	5.0	0.10	10	4.00	27.63 ± 2.63
NaCl	5.0	0.10	45	0.70	43.05 ± 3.17
KSCN	5.0	0.10	30	0.46	41.56 ± 5.06
Na_2SO_4	5.0	0.10	30	1.30	36.36 ± 3.34
LiCl	5.0	0.10	30	2.68	31.35 ± 1.17
$CaCl_2$	5.0	0.10	30	3.48	38.03 ± 4.18

[a] Reference 84.

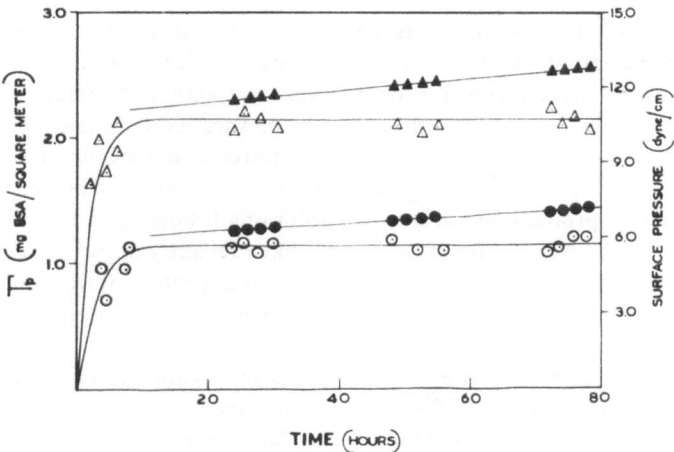

FIGURE 11.15. Plot[84] of Γ_p (left scale) and interfacial pressure (right scale) against time. $\odot$ Initial BSA concentration 0.02% ($\triangle$) initial BSA concentration 0.1%. Open symbols for Γ_p and solid symbols for π. For all cases pH = 5.0. (Reproduced with permission from Academic Press, New York.)

measurement, the system does not exist strictly in the state of thermodynamic equilibrium. From the application of thermodynamics, a general form of the Gibbs adsorption equation has been derived which is claimed to remain valid both for equilibrium and nonequilibrium states. This at a constant temperature reads[12]

$$-d\gamma = \sum_i \Gamma_i^x d\mu_i^x - c \sum_i \varepsilon_i dx_i - c' \sum_i \varepsilon_i' dx_i' \qquad (11.44)$$

Here μ_i^x is equal to $df^x/d\Gamma_i^x$, where f^x is the excess Helmholtz free energy per unit surface area. Also c and c' are total molar concentrations of all components in the α and β phases, respectively, (vide Chapter 3) and c_i and c_i' are the molar concentrations of the ith component in the two phases, respectively. The terms x_i and x_i' are, respectively, equal to c_i/c and c_i/c'. The terms ε_i and ε_i' are, respectively, df^x/dc_i and df^x/dc_i'. Equation (11.44) is very useful in interpreting the deviation of the values of γ obtained by the dynamic method from the corresponding value obtained by the static method.[12]

11.10. PROTEIN–WATER INTERFACIAL TENSION

In Section 9.15, for the interpretation of the stability and solubility properties of proteins in the presence of stabilizers and denaturants, it has been assumed that there exists some residual surface free energy at the boundary region of the biopolymers in contact with water in bulk. Melander and Horvath[87] have recently proposed that the extent of increase of surface tension of water per mole of the addition of various neutral salts represents the relative capacity of increase in energy required to make holes in the aqueous solution to accommodate the protein molecules. This can be related to the solubilities of proteins in various salt solutions following the order assigned in the lyotropic series. Bull and Breese[88] have recently argued that the interfacial tensions between protein molecules and protein–water interface are crucial factors in controlling the order for the extents of solubility of a protein in various types of salt solutions.

In Table 11.8, the interfacial tensions of decane–water systems in the presence and absence of 0.025% egg albumin have been presented. The solvent in all cases is 0.1 M NaCl to which other electrolytes are added to form $1(M)$ concentration of the salt. From this, it is obvious that order of magnitude of interfacial tensions of the various salt solutions with or without protein agrees with each other with the exception of LiCl. According to the concept of Bull *et al.*,[88] the placement of the protein in the salt solution makes holes in water. The solubility of protein depends on the process of separation of the protein molecules from hydrated protein mass by the translational kinetic motion and the reverse process controlled by the interfacial tension of protein molecules in solution.

For the work W_α of adhesion of protein molecules during the solution process, Bull *et al.*[88] have written

$$W_\alpha = \bar{A}(\gamma_a + \gamma_b - \gamma_{ab}) \tag{11.45}$$

Here γ_a is the interfacial tension between dissolved protein molecule and water. γ_b is the corresponding tension between water and the site of the undissolved protein from which the dissolved protein molecule has come. γ_b is very close to γ_a. γ_{ab} is the interfacial tension between the hydrated but undissolved protein molecules and water. Its value, mainly depending on water–water contact, is close to zero. $\bar{A}$ is the area of contact between the undissolved molecules and the dissolving molecule. This is equal to the new area exposed during the dissolution of the molecule. Thus W_a is very close to $2\bar{A}\gamma_a$, the work of cohesion. W_α can be obtained from the estimation of the solubility data and approximate calculation of the value of $\bar{A}$ from the molecular dimension of protein so that γ_a for egg albumin is found to be 0.09 erg/cm^2. This value is considerably lower than γ for decane–water interface containing a highly viscous monolayer of protein (vide Table 11.8). Native and folded protein molecule in aqueous medium have low interfacial tension due to the curvature, hydration, and other effects. It is of interest to note that the order of γ for protein–water system and that of a droplet of oil

TABLE 11.8. Interfacial Tension (γ) between Decane and
$1(M)$ Salt Solution at 30°C by Pendant Drop Method[a]

Salt	γ in mN/m (without EA)	γ in mN/m (with 0.25% EA)
0.1 M NaCl	51.85	27.97
LiCl	48.28	28.64
Na_2SO_4	52.60	28.58
NaBr	51.96	28.27
CsCl	52.33	28.10
NaCl	52.66	27.74
$BaCl_2$	50.96	25.57
KCl	52.89	27.10
0.5 M $KClO_3$	50.81	26.56
NaCNS	48.44	21.07
$K_3Fe(CN)_6$	49.87	20.98
Gu-HCl	49.59	19.17
KCNS	49.14	16.34

[a] Reference 88.

forming the microemulsion are similar. A small change in the interfacial tension of protein by addition of salts may lead to the change in the solubility of protein possibly following the order given in the lyotropic series.[88]

11.11. PRESSURE COEFFICIENT OF SURFACE TENSION

In Fig. 5.11, it has been shown that γ for a binary solution at constant temperature changes with pressure p. The pressure coefficient $(\partial\gamma/\partial p)_T$ has been related by Motumura *et al.*[89] to the volume of the interface formation. Hansen[90] has assumed that the surface phase of total area $\bar{\mathscr{A}}$ possesses volume V^x so that the ratio $V^x/\bar{\mathscr{A}}$ may be regarded as the excess thickness τ^x of the interface. The pressure coefficient possesses the dimension of length and its magnitude has been estimated by Defay *et al.*[12] from experimental data. The value is found to be less than 10^{-6} cm.

Because of the neglect of the volume of the interface in the derivation of the Gibbs equation (vide Section 3.6), the pressure coefficient term does not occur in equation (3.23). However, for a system of pure liquid, Gibbs[91] introduced a distance parameter λ_1' which can be defined as

$$\lambda_1' = \frac{\Gamma_1'}{c_1^\alpha - c_1^\beta} \tag{11.46}$$

Here Γ_1' is the interfacial excess of component 1, with respect to the surface of tension t and c_1^α and c_1^β are the bulk concentrations of the components in the α and β phases, respectively. For a one-component system, λ_1^i is the distance of the Gibbs dividing plane from the surface of tension.

The application of this approach for a two-phase binary system has been recently made by Good.[92] According to Good, let x_1 and x_2 be the locations of the Gibbs dividing plane GG' for $\Gamma_1^x = 0$ and $\Gamma_2^x = 0$, respectively (vide Figs. 3.6 and 3.7). For a binary solution, one then has

$$\lambda_2^1 = x_2 - x_1 \tag{11.47}$$

$$\lambda_1^2 = x_1 - x_2 \tag{11.48}$$

Good has further shown that

$$\lambda_2^1 = \Gamma_2^1/\Delta c_2$$

$$= -\Gamma_1^2/\Delta c_1 \tag{11.49}$$

where Δc_1 and Δc_2 are equal to $c_1^\alpha - c_1^\beta$ and $c_2^\alpha - c_2^\beta$, respectively,

$$\lambda_2^1 = \left(\frac{\partial \gamma}{\partial p} \right)_{T,\text{sat}} \tag{11.50}$$

The physical meaning of the subscript "sat" is that for both components, $\mu_1^\alpha = \mu_i^\beta$. It has been shown by Good[93] that the parameter λ_2^1 is equivalent to τ^x, the thickness parameter introduced by Hansen.[90]

11.12. CONCLUDING REMARKS

In different chapters of this book, we have emphasized the applications of the concept of the Gibbs surface excess to interpret the phenomena of adsorption at the interfaces of liquid, solid, biopolymer, etc. The Gibbs excess quantities are essentially extensive properties of the system but these become intensive when expressed per unit surface area, per gram of the solid powder, per mole of biopolymer, etc. All these intensive quantities are uniformly expressed in terms of Γ_2^1 or Γ_1^2 etc. This makes derived equations quite simple in appearance, and under this circumstance, the usefulness of the concept of the surface excess appears to be prominent. Many workers prefer to use different symbols for Γ_2^1, Γ_1^2 etc. to indicate differences in the magnitudes of the surface excesses with reference to the different unit quantities of the system. Under such circumstance, the derived equations based on the Gibbs excesses appear to be complicated in form.

The Gibbs surface excesses can be evaluated easily for various types of multicomponent systems on the basis of suitable experiments. The surface excess is composed of two quantities related to the properties of the interfacial region whose values can be evaluated either with the further assumption that the boundary phase is a monolayer or from the observation that the excess quantity varies linearly with mole ratio of composition of a binary solution. Fortunately for many systems, either or both of these conditions appear to be valid so that the concept of the surface excess becomes physically meaningful. For many other systems these two conditions are not valid. Further, for the biopolymers dissolving in aqueous solution, the physical concept for the boundary region itself vanishes. Even under these situations, the operational interpretation of the excess concept remains valid and equivalent.

From the Gibbs adsorption equation, one finds that the surface excess of a component is proportional to the rate of change of the interfacial free energy with change of the bulk chemical potential of a component in the system. The use of the Gibbs adsorption equation in integrated form can lead to the evaluation of the change of the free energy in the boundary region of the various systems under consideration which occurs due to the change in the bulk composition of the medium. The free energy change thus evaluated includes the effect of interaction of all the components in the interfacial region which is of interest. For the hydration and binding phenomena occurring in the biological systems, the standard free energy change for such an interaction can be evaluated based on a universal scale so that comparisons of the nature of interactions can be made quantitatively. However, for the adsorption at liquid and solid interfaces, the evaluation and comparison of the free energy changes due to the interaction have not yet been made on the basis of a uniform scale.

In recent years, the changes in enthalpy or entropy of the system due to the adsorption or binding interaction at solid or biopolymer interfaces have been measured directly from the calorimetric method. From the measurement of the surface excesses of the system at various temperatures, values of the standard change in enthalpy or entropy can also be calculated under nearly identical conditions. The magnitudes of various thermodynamic parameters obtained by the calorimetric methods and those obtained from the measurement of the surface excesses differ widely from each other, possibly because the reference states for comparison in the two methods are not same. This point needs further clarification.

Use of the concept of the Gibbs excess in the biological system is not very extensive up till now. Since biopolymers and biosurfaces are extensively hydrated, competitive effect of water must be included in the binding, hydration, and other related phenomena occurring in the biological system. The use of the concept of the Gibbs excess seems to be essential for the energetics involved in the study of the model biological systems.

The status of the concept of the Gibbs surface excess for systems not in equilibrium has not been examined in detail so far and it is hoped that more studies will be made in this direction in the future. Such investigation will be immensely useful in understanding interaction phenomena occurring in the living system existing in the nonequilibrium state.

REFERENCES

CHAPTER 1

1. A. W. Adamson, *Physical Chemistry of Surfaces*, Third Edition (John Wiley & Sons, New York, 1976).
2. N. D. Parkyns and K. S. W. Sing, *Colloid Science*, Vol. 2 edited by D. H. Everett (The Chemical Society, London, 1975), p. 1.
3. J. W. Gibbs, *The Collected Works of J. W. Gibbs*, Vol. 1 (Longmans, Green, New York, 1931).
4. C. Tanford, *The Hydrophobic Effect*, Second Edition (John Wiley & Sons, New York, 1980).
5. K. L. Mittal and P. Mukherjee, in *Micellization, Solubilization and Microemulsions*, Volume 1, edited by K. L. Mittal (Plenum Press, New York, 1977), p. 1.
6. R. H. Haschemeyer and A. E. V. Haschemeyer, *Proteins* (John Wiley and Sons, New York, 1973).
7. J. W. Watson, *Molecular Biology of Genes*, Third Edition (W. A. Benjamin, Inc., Menlo Park, California, 1977).
8. L. Pauling and R. B. Corey, *Proc. Natl. Acad. Sci. U.S.A.*, **37**, 251, 729 (1951).
9. L. Pauling, R. B. Corey, and H. R. Branson, *Proc. Natl. Acad. Sci. U.S.A.*, **37**, 205 (1951).
10. G. N. Ramachandran and V. Sasisekharan, *Adv. Protein Chem.*, **21**, 283 (1968).
11. B. Pullman and A. Pullman, *Adv. Protein Chem.*, **28** (1974).
12. C. Tanford, *Adv. Protein Chem.*, **23**, 122 (1970).
13. H. B. Bull, *An Introduction to Physical Biochemistry*, Second Edition (F. A. Davis, Philadelphia, 1971), p. 237.
14. E. A. Guggenheim, *Thermodynamics*, Third Edition (North-Holland, Amsterdam, 1957).
15. K. Denbigh, *The Principles of Chemical Equilibrium*, Third Edition (Cambridge University Press, Cambridge, 1971).
16. I. M. Klotz, *Chemical Thermodynamics* (Prentice-Hall Inc., Englewood Cliffs, New Jersey, 1950).
17. R. Defay, I. Prigogine, and A. Bellemans, *Surface Tension and Adsorption*, D. H. Everett trans. (Longmans Green, London, 1966).
18. E. A. Guggenheim, *Trans. Faraday Soc.*, **36**, 398 (1940).
19. G. S. Schay, in *Physical Chemistry: Enriching Topics from Colloid and Surface Science*, edited by H. von Holphen and K. J. Mysels (Therorex, La Jolla, California, 1975).

CHAPTER 2

1. A. Chatterjee and D. K. Chattoraj, *J. Colloid Interface Sci.*, **26**, 1 (1968).
2. S. J. Rehfeld, *J. Phys. Chem.*, **71**, 738 (1967).
3. A. W. Adamson, *Physical Chemistry of Surfaces*, Third Edition (John Wiley & Sons, New York, 1976).
4. Lord Rayleigh (J. W. Streett), *Proc. R. Soc. London*, **A92**, 184 (1915).

5. S. Sugden, *J. Chem. Soc.*, 1483 (1921).
6. J. F. Padday and A. Pitt, *J. Colloid Interface Sci.*, **38**, 323 (1972).
7. J. E. Lane, *J. Colloid Interface Sci.*, **42**, 145 (1973).
8. G. Jones and W. A. Ray, *J. Amer. Chem. Soc.*, **59**, 187 (1937).
9. H. B. Bull and K. Breese, *Arch. Biochem. Biophys.*, **202**, 116 (1980).
10. H. B. Bull and K. Breese, *Arch. Biochem. Biophys.*, **161**, 665 (1974).
11. W. D. Harkins and F. E. Brown, *J. Am. Chem. Soc.*, **41**, 499 (1919).
12. L. Wilhelmy, *Ann. Phys. (Leipzig)*, **119**, 177 (1863).
13. R. Ryussen, *Rec. Trav. Chim.*, **65**, 580 (1946).
14. J. F. Padday, *Proc. Second Int. Congr. Surface Activity London*, **1**, 1 (1957).
15. H. B. Bull, *J. Colloid Interface Sci.*, **41**, 305 (1972).
16. K. S. Birdi and A. Nikolov, *J. Phys. Chem.*, **83**, 365 (1979).
17. K. S. Birdi, unpublished results.
18. K. Johnson and J. C. Eriksson, *J. Colloid Interface Sci.*, **40**, 398 (1972).
19. D. N. Furlong and S. Hartland, *J. Colloid Interface Sci.*, **71**, 301 (1979).
20. H. W. Fox and C. H. Chrisman, *J. Phys. Chem.*, **56**, 284 (1952).
21. B. B. Freud and H. Z. Freud, *J. Am. Chem. Soc.*, **52**, 1772 (1930).
22. M. Bartmuss and C. P. Kurzeudorfer, *Colloid Polymer Sci.*, **258**, 99 (1980).
23. J. M. Andreas, E. A. Hauser, and W. B. Tucker, *J. Phys. Chem.*, **42**, 1001 (1938).
24. C. E. Stauffer, *J. Phys. Chem.*, **69**, 1933 (1965).
25. S. Ghosh and H. B. Bull, *Biochim. Biophys. Acta.*, **66**, 150 (1963).
26. G. L. Gaines, *Insoluble Monolayers at Liquid–Gas Interfaces* (Interscience Publishers Inc., New York, 1966).
27. I. Langmuir, *J. Am. Chem. Soc.*, **39**, 1848 (1917).
28. I. M. Jalal, G. Zografi, A. K. Rakshit, and F. D. Gunstone, *J. Colloid Interface Sci.*, **76**, 146 (1980).
29. A. W. Neumann and D. Renzow, *Z. Phys. Chem.*, *Neue Folge*, **68**, 11 (1969).
30. S. Ghosh, K. Breese, and H. B. Bull, *J. Colloid Sci.*, **19**, 457 (1964).
31. J. J. Kipling, *Adsorption from Solutions of Nonelectrolytes* (Academic Press, London, 1965).
32. R. Chatterjee and D. K. Chattoraj, *Indian J. Biochem. Biophys.*, **16**, 158 (1979).
33. C. Orr and J. M. Dallaville, *Fine Particle Measurement, Size, Surface and Pore Volume* (The MacMillan Company, New York, 1959).
34. J. W. McBain and R. C. Swain, *Proc. R. Soc. London Ser. A*, **154**, 608 (1936).
35. J. W. McBain and C. W. Humphreys, *J. Phys. Chem.*, **36**, 300 (1932).
36. K. P. Das and D. K. Chattoraj, *J. Colloid Interface Sci.*, **78**, 422 (1980).
37. K. P. Das, "Physico-chemical studies of polar oil/water emulsions stabilised by proteins and other emulsifiers", Ph.D. thesis, Jadavpur University (1982).
38. D. Salley, A. J. Wieth Jr., A. A. Argyle, and J. K. Dixon, *Proc. R. Soc. London Ser. A*, **203**, 42 (1950).
39. K. Tajima, M. Muramatsu, and T. Sassaki, *Bull. Chem. Soc., Jpn*, **43**, 1991 (1970).
40. M. Muramatsu, in *Surface and Colloid Science*, edited by E. Matijević (Wiley-Interscience, New York, 1973), Vol. 6, p. 101.
41. H. B. Bull and K. Breese, *Arch. Biochem. Biophys.*, **128**, 488 (1968).
42. H. B. Bull and K. Breese, *Arch. Biochem. Biophys.*, **137**, 299 (1970).
43. D. K. Chattoraj and H. B. Bull, *Arch. Biochem. Biophys.*, **142**, 363 (1972).
44. S. P. Mitra, D. K. Chattoraj, and M. N. Das, *Indian J. Biochem. Biophys.*, **14**, 101 (1977).
45. J. Steinhardt and J. A. Reynolds, *Multiple Equilibria in Proteins* (Academic Press, New York, 1970).
46. R. Chatterjee and D. K. Chattoraj, *Biopolymers*, **18**, 147 (1979).
47. M. Sen, S. P. Mitra, and D. K. Chattoraj, *Colloids and Surfaces*, **2**, 259 (1981).
48. F. O. Akinrimisi, J. Bonner, and P. O. Ts'o, *J. Mol. Biol.*, **11**, 128 (1965).

49. H. B. Bull and K. Breese, *Biopolymers*, **15**, 1573 (1976).
50. D. K. Chattoraj and S. P. Mitra, *Indian J. Biochem. Biophys.*, **16**, 406 (1979).
51. D. K. Chattoraj, H. B. Bull, and R. Chalkley, *Arch. Biochem. Biophys.*, **152**, 778 (1972).

CHAPTER 3

1. A. W. Adamson, *Physical Chemistry of Surfaces*, Third Edition (John Wiley & Sons, New York, 1976).
2. J. W. Gibbs, *The Collected Works of J. W. Gibbs*, Vol. 1 (Longmans, Green, New York, 1931).
3. G. Schay, in *Surface and Colloid Science*, edited by E. Matijevic (Wiley-Interscience, New York, 1969), Vol. 2, p. 150.
4. J. D. van der Waals, "Thermodynamische Theorie der Capillarität," *Verh. K. Akad. V. Wet., Amsterdam* (1893).
5. G. Bakker, "Kapillarität und Oberflächenspannung," *Wien-Harms Handbuch der Experimentalphysik, Leipzig*, Vol. 6, 1928.
6. J. E. Verschaffelt, *Acad. R. Belgique Bull. Classe Sci.*, **22**, 373, 390, 402 (1936).
7. E. A. Guggenheim, *Trans. Faraday Soc.*, **36**, 398 (1940).
8. R. C. Tolman, *J. Chem. Phys.*, **16**, 758 (1948).
9. F. M. Fowkes, *J. Phys. Chem.*, **65** 355 (1961).
10. F. M. Fowkes, *J. Phys. Chem.*, **66**, 385 (1962).
11. R. C. Hansen, *J. Phys. Chem.*, **66**, 410 (1962).
12. A. I. Rusanov, *Kolloid Zhur.*, **24**, 309 (1962).
13. J. C. Eriksson, *Arkiv. Kemi*, **25**, 331 (1966).
14. F. C. Goodrich, *Trans. Faraday Soc.*, **64**, 3403 (1968).
15. K. Motomura, *J. Colloid Interface Sci.*, **64**, 348 (1978).
16. R. Defay, I. Prigogine, and A. Bellemans, in *Surface Tension and Adsorption*, D. H. Everett, trans (Longmans Green, London, 1966).
17. S. Ono and S. Kondo, *Handbuch der Physik*, Vol. X (Springer-Verlag, Berlin, 1960).
18. N. K. Adam, *The Physics and Chemistry of Surfaces*, Third Edition (Oxford University Press, London, 1941).
19. E. A. Guggenheim and N. K. Adam, *Proc. R. Soc. London Ser. A*, **139**, 218 (1933).
20. D. K. Chattoraj and S. P. Moulik, *Indian J. Chem.*, **15A**, 73 (1977).
21. D. K. Chattoraj, *Indian J. Chem.*, **20A**, 941 (1981).
22. J. W. McBain and R. C. Swain, *Proc. R. Soc. London Ser. A*, **154**, 608 (1936).
23. J. W. McBain and C. W. Humphreys, *J. Phys. Chem.*, **36**, 300 (1932).
24. K. P. Das and D. K. Chattoraj, *Colloids Surf.* **7**, 53 (1983).
25. D. J. Salley, A. J. Weith, A. A. Argyle, and J. K. Dixon, *Proc. R. Soc. London Ser. A*, **203**, 42 (1950).
26. K. Tajima, *Bull. Chem. Soc. Jpn*, **43**, 3063 (1970).
27. J. E. B. Randles, *Advances in Electrochemistry and Electrochemical Engineering*, Vol. 3, edited by P. Delahay, p. 1 (Interscience, New York, 1963).
28. L. Onsager and N. N. T. Samaras, *J. Chem. Phys.*, **2**, 528 (1934).
29. H. S. Harned and B. B. Owen, *The Physical Chemistry of Electrolyte Solutions*, Third Edition (Reinhold, New York, 1958).
30. R. A. Robinson and R. H. Stokes, *Electrolyte Solutions*, Second Edition (Butterworths, London, 1959).
31. D. K. Chattoraj and S. P. Moulik, in *Adsorption at Interfaces*, edited by K. L. Mittal, ACS Symposium Series 8 (American Chemical Society, Washington D.C., 1975).
32. R. Defay, *Bull. Acad. R. Belg.*, **16**, 1249 (1930).
33. L. Dufour, *Med. Kon. Vl. Acad. Belg.*, **13**, 14 (1951).

34. F. O. Koenig, *J. Chem. Phys.*, **18**, 449 (1950).
35. J. A. V. Butler, *Proc. R. Soc. London Ser. A*, **135**, 348 (1932).
36. G. Schay, *Surface and Colloid Science*, Vol. 2, edited by E. Matijevic (Wiley-Interscience, New York, 1969), p. 194.
37. L. G. Nagy and G. Schay, *Acta Chim. Acad. Sci. Hung.*, **39**, 365 (1963).
38. J. J. Kipling, in *Adsorption from Solutions of Nonelectrolytes* (Academic Press, London, 1965), p. 244.
39. A. Schuchowitzky, *Acta Physicochemika, U.R.S.S.*, **19**, 176 (1944).
40. T. P. Hoar and B. A. Melford, *Trans. Faraday Soc.*, **53**, 315 (1957).
41. E. A. Guggenheim, *Trans. Faraday Soc.*, **45**, 150 (1945).
42. I. W. Belton and M. G. Evans, *Trans. Faraday Soc.*, **41**, 1 (1945).
43. J. C. Eriksson, *Arkiv. Kemi*, **26**, 49 (1966).
44. F. B. Sprow and J. M. Prausnitz, *Trans. Faraday Soc.*, **53**, 1105 (1966).
45. M. Patel and V. Ramakrishna, *J. Colloid Interface Sci.*, **76**, 166 (1980).
46. K. Motomura, N. Matubayasi, M. Aratono, and R. Matuura, *J. Colloid Interface Sci.*, **64**, 356 (1978).
47. P. M. Nassonov, *Russ. J. Phys. Chem.*, **45**, 1593 (1971).
48. P. M. Nassonov, *Russ. J. Phys. Chem.*, **45**, 1595 (1971).
49. D. H. Everett, *Colloid Science*, Vol. 1, edited by D. H. Everett (The Chemical Society, London, 1973), p. 49.

CHAPTER 4

1. D. K. Chattoraj and R. P. Pal, *Indian J. Chem.*, **10**, 410 (1972).
2. K. S. Birdi, K. P. Das, and D. K. Chattoraj, ready for communication.
3. H. Helmholtz, *Wied Ann.*, **7**, 337 (1879).
4. H. A. Abramson, "Electrokinetic Phenomena and their Application to Biology and Medicine." ACS Monograph Series (The Chemical Catalog Company Inc., New York, 1934).
5. G. Gouy, *J. Phys. (Paris)*, **9**, 457 (1910).
6. G. Gouy, *Ann. Phys. (Leipzig)*, **7**, 129 (1917).
7. D. T. Chapmann, *Phil. Mag.*, **25**, 475 (1913).
8. O. Stern, *Z. Electrochem.*, **30**, 508 (1924).
9. D. C. Grahame, *Chem. Rev.*, **41**, 441 (1947).
10. R. Parsons and M. A. V. Devanathan, *Trans. Faraday Soc.*, **49**, 673 (1953).
11. J. OM. Bockris and A. K. N. Reddy, *Modern Electrochemistry* (Plenum Press, New York, 1970).
12. D. K. Chattoraj, *J. Phys. Chem.*, **70**, 2687 (1966).
13. D. K. Chattoraj, *J. Colloid Interface Sci.*, **29**, 399 (1969).
14. D. K. Chattoraj and R. P. Pal, *Indian J. Chem.*, **10**, 417 (1972).
15. S. Voyutsky, *Colloid Chemistry* (Mir Publishers, Moscow, 1978).
16. A. W. Adamson, *Physical Chemistry of Surfaces*, Third Edition (John Wiley & Sons, New York, 1976).
17. H. R. Kruyt, *Colloid Science*, Vol. 1 (Elsevier, New York, 1952).
18. P. Debye and E. Hückel, *Phys. Z.*, **24**, 185 (1923).
19. K. P. Das and D. K. Chattoraj, *Colloids and Surfaces*, **7**, 53 (1983).
20. R. H. Bijsterbosch and H. van den Hul, *J. Phys. Chem.*, **71**, 1169 (1967).
21. I. R. Miller and M. A. Frommer, *J. Phys. Chem.*, **72**, 1834 (1968).
22. D. K. Chattoraj, *J. Phys. Chem.*, **71**, 3709 (1967).
23. R. A. Robinson and R. H. Stokes, *Electrolyte Solutions*, Second Edition (Butterworths, London, 1959).
24. D. K. Chattoraj and L. Ghosh, *Indian J. Chem.*, **16A**, 383 (1978).
25. D. K. Chattoraj, *J. Colloid Interface Sci.*, **26**, 379 (1968).

26. E. A. Guggenheim, *Thermodynamics*, Third Edition (North-Holland Publishing Co., Amsterdam, 1958).
27. A. P. Brady, *J. Phys. Chem.*, **53**, 56 (1949).
28. B. A. Pethica, *Trans. Faraday Soc.*, **50**, 413 (1954).
29. A. J. Salley, A. A. Weith, A. A. Argyle, and J. A. Dixon, *Proc. R. Soc.* (*London*), **A203**, 42 (1950).
30. C. P. Roe and P. C. Brass, *J. Am. Chem. Soc.*, **76**, 4703 (1954).
31. J. T. Davies, *Trans. Faraday Soc.*, **48**, 1052 (1952).
32. G. Nielson, *J. Phys. Chem.*, **61**, 1135 (1957).
33. H. L. Rosano and G. Karg, *J. Phys. Chem.*, **63**, 1692 (1959).
34. A. M. Mankowich, *J. Am. Oil Chemists' Soc.*, **43**, 615 (1966).
35. K. Tajima, M. Muramatsu, and T. Sasaki, *Bull. Chem. Soc., Jpn*, **43**, 1991 (1970).
36. K. Tajima, *Bull. Chem. Soc., Jpn*, **43**, 3063 (1970).
37. D. K. Chattoraj, *Indian J. Chem.*, **6**, 309 (1968).
38. W. Kling and H. Lange, *Proc. 2nd Intern. Congr. Surface Activity* (Butterworths, London, 1957), Vol. 1, p. 295.
39. D. A. Haydon and F. H. Taylor, *Phil. Trans. R. Soc. London*, **A252**, 225 (1960).
40. A. K. Chatterjee and D. K. Chattoraj, *J. Colloid Interface Sci.*, **26**, 140 (1968).
41. K. Tajima, *Bull. Chem. Soc., Jpn*, **44**, 1767 (1971).
42. J. Sasaki, K. Tajima, and T. Sasaki, *Bull. Chem. Soc., Jpn*, **45**, 348 (1972).
43. T. Okumura, A. Nakamura, K. Tajima, and T. Sasaki, *Bull. Chem. Soc., Jpn*, **47**, 2986 (1974).
44. W. R. West and W. R. Todd, *Textbook of Biochemistry* (McMillan Publishing Company, Inc., New York, Indian Edition, 1974).
45. A. Nakamura, K. Tajima, and T. Sasaki, *Bull Chem. Soc., Jpn*, **48**, 214 (1975).
46. K. Tajima, T. Sassaki, and M. Iwahashi, *Bull Chem. Soc., Jpn*, **44**, 3251 (1971).
47. K. Tajima and M. Iwahashi, *Bull Chem. Soc., Jpn*, **48**, 214 (1975).
48. K. Tajima, *Bull. Chem. Soc., Jpn*, **49**, 3403 (1976).
49. S. Ikeda, *Bull. Chem. Soc., Jpn*, **50**, 1403 (1977).
50. S. Ikeda, M. Tsunoda, and H. Maeda, *J. Colloid Interface Sci.*, **67**, 336 (1978).
51. S. Ozaki, M. Tsunoda, and S. Ikeda, *J. Colloid Interface Sci.*, **64**, 28 (1978).
52. J. Rodakiewicz-Nowak, *Z. Nauk. Univ. Jagiellonskiego*, **25**, 114 (1980).
53. J. Roadkiewicz-Nowak, *Z. Nauk. Univ. Jagiellonskiego*, **25**, 126 (1980).

CHAPTER 5

1. I. Langmuir, *J. Am. Chem. Soc.*, **38**, 2221 (1916).
2. I. Langmuir, *J. Am. Chem. Soc.*, **39**, 1848 (1917).
3. J. T. Davies and E. Rideal, *Interfacial Phenomena* (Academic Press, Inc., New York, 1961).
4. A. W. Adamson, *Physical Chemistry of Surfaces*, Third Edition (John Wiley & Sons, New York, 1976).
5. B. A. Pethica, *Trans. Faraday Soc.*, **50**, 413 (1954).
6. T. Yamashita and H. B. Bull, *J. Colloid Interface Sci.*, **27**, 19 (1968).
7. H. B. Bull, *J. Colloid Interface Sci.*, **41**, 305 (1972).
8. A. Chatterjee and D. K. Chattoraj, *J. Colloid Interface Sci.*, **26**, 1 (1968).
9. I. Traube, *Ann.* **265**, 27 (1891).
10. W. D. Harkins and H. S. King, *J. Am. Chem. Soc.*, **41**, 970 (1919).
11. E. Hutchinson, *J. Colloid Sci.*, **3**, 219 (1948).
12. D. A. Haydon and F. H. Taylor, *Phil. Trans. R. Soc.* (*London*), **A252**, 225 (1960).
13. J. J. Jasper and R. D. Vandell, *J. Phys. Chem.*, **69**, 481 (1965).
14. R. P. Pal, A. Chatterjee, and D. K. Chattoraj, *J. Colloid Interface Sci.*, **52**, 46 (1975).

15. N. K. Adam, J. F. Danielli, and J. B. Harding, *Proc. R. Soc. London*, **A147**, 491 (1934).
16. G. Weitzel, A. M. Fretzdorff, S. Heller, and H. Sewlars, *Z. Physiol. Chem.*, **303**, 14 (1956).
17. H. J. Trurnit, *Z. Naturforsch.*, **26**, 258 (1947).
18. J. L. Shereshepsky, H. T. Carter, E. Nichols, and P. L. Robinson, *J. Phys. Chem.*, **66**, 1846 (1962).
19. P. M. Jeffers and J. Dean, *J. Phys. Chem.*, **66**, 1846 (1962).
20. H. B. Bull, *An Introduction to Physical Biochemistry* (F. A. Davies, Philadelphia, 1971).
21. L. Ter Minassian-Saraga and I. Prigogine, *Mem. Serv. Chim. Etat*, **38**, 109 (1953).
22. F. M. Fowkes, *J. Phys. Chem.*, **68**, 3515 (1964).
23. E. H. Lucassen-Reynders and M. van den Tempel, Proc. IVth Int. Cong. Surface Active Substances, Brussels, 1964, Vol. 2, p. 779.
24. K. Motomura, N. Matubayasi, M. Aratono, and R. Matuura, *J. Colloid Interface Sci.*, **64**, 356 (1978).
25. K. Motomura, *J. Colloid Interface Sci.*, **64**, 348 (1978).
26. R. A. Hansen, *J. Phys. Chem.*, **66**, 410 (1962).
27. K. Motomura, M. Aratono, N. Matubayasi, and R. Matuura, *J. Colloid Interface Sci.*, **67**, 247 (1978).
28. M. Aratono, M. Yamanaka, N. Matubayasi, K. Motomura, and R. Matuura, *J. Colloid Interface Sci.*, **74**, 489 (1980).
29. A. K. Chatterjee and D. K. Chattoraj, *J. Colloid Interface Sci.*, **26**, 140 (1968).
30. J. Guastala, *J. Chim. Phys.*, **43**, 184 (1946).
31. J. T. Davies, *J. Colloid Sci.*, **11**, 377 (1956).
32. J. T. Davies, *Proc. R. Soc. London*, **A245**, 429 (1958).
33. J. T. Davies, *Proc. R. Soc. London*, **A245**, 417 (1958).
34. J. T. Davies, *Proc. R. Soc. London*, **A208**, 224 (1951).
35. J. M. Phillips and E. K. Rideal, *Proc. R. Soc. London*, **A232**, 159 (1954).
36. D. A. Haydon and J. N. Phillips, *Trans. Faraday Soc.*, **54**, 698 (1958).
37. D. A. Haydon, *J. Colloid Sci.*, **13**, 159 (1958).
38. I. D. Robb and A. E. Alexander, *J. Colloid Interface Sci.*, **28**, 1 (1968).
39. D. K. Chattoraj, *J. Colloid Interface Sci.*, **29**, 407 (1969).
40. S. Hachisu, *J. Colloid Interface Sci.*, **33**, 445 (1970).
41. B. Deryaguin and L. Landeau, *Acta Phys. Chem.*, *USSR*, **14**, 633 (1941).
42. E. J. W. Verwey and J. Th. G. Oberbeck, *Theory of the Stability of the Lyophobic Colloids* (Elseviers, New York, 1948).
43. G. S. Hartley and J. W. Roe, *Trans. Faraday Soc.*, **36**, 101 (1940).
44. D. K. Chattoraj and H. B. Bull, *J. Phys. Chem.*, **63**, 1809 (1959).
45. A. K. Chatterjee and D. K. Chattoraj, *Koll. Z.*, *Z. Polymere*, **233**, 966 (1969).
46. R. P. Pal and D. K. Chattoraj, *J. Colloid Interface Sci.*, **52**, 56 (1975).
47. G. M. Bell, S. Levine, and B. A. Pethica, *Trans. Faraday Soc.*, **58**, 904 (1962).
48. S. Levine, J. Mingins, and G. M. Bell, *J. Phys. Chem.*, **67**, 2095 (1963).
49. S. Levine, G. M. Bell, and B. A. Pethica, *J. Phys. Chem.*, **40**, 2304 (1964).
50. S. Levine, J. Mingins, and G. M. Bell, *Canadian J. Chem.*, **43**, 2834 (1965).
51. G. R. Feat and S. Levine, in *Monolayers* edited by E. D. Goddard, Advances in Chemistry Series, Vol. 144 (Am. Chem. Soc., Washington D.C., 1975).
52. E. Matijevic and B. A. Pethica, *Croat. Chem. Acta*, **29**, 431 (1957).
53. Th. A. J. Payens, *Phillips Res. Rep.*, **10**, 425 (1955).
54. M. Plaisance and L. Ter-Minassian Saraga, *J. Colloid Interface Sci.*, **59**, 113 (1977).
55. E. D. Goddard, *Adv. Colloid Interface Sci.*, **4**, 45 (1974).
56. J. Mingins, N. F. Owens, J. A. G. Taylor, J. H. Brooks, and B. A. Pethica, in *Monolayers* edited by E. D. Goddard, Advances in Chemistry Series, Vol. 144 (Am. Chem. Soc., Washington, D.C., 1975), pp. 14–27.

57. C. Tanaka and H. Fukutome, *J. Colloid Interface Sci.*, **66**, 492 (1978).
58. R. O. James, *Colloids Surf.*, **2**, 201 (1981).
59. D. C. Grahame, *Chem. Rev.*, **41**, 441 (1947).
60. J. O'M. Bockris and A. K. N. Reddy, *Modern Electrochemistry* (Plenum Press, New York, 1970).
61. J. O'M. Bockris, M. A. V. Devanathan, and K. Muller, *Proc. R. Soc. London*, **274**, 55 (1963).
62. N. K. Adam, *The Physics and Chemistry of Surfaces*, Third Edition (Oxford University Press, London, 1941).
63. A. K. Chatterjee and D. K. Chattoraj, *Kolloid Z. Z. Polymere*, **234**, 1053 (1969).
64. D. K. Chattoraj and A. K. Chatterjee, *J. Colloid Interface Sci.*, **21**, 159 (1966).
65. N. L. Gershfeld, *J. Colloid Interface Sci.*, **28**, 240 (1968).
66. R. Defay, I. Prigogine, A. Bellemans, and D. H. Everett, *Surface Tension and Adsorption* (Longmans, Green and Co., London, 1966).
67. K. Denbigh, *The Principles of Chemical Equilibrium*, Third Edition (Cambridge University Press, Cambridge, 1971), p. 119.
68. I. Langmuir, *J. Chem. Phys.*, **1**, 756 (1933).
69. K. Tajima, M. Muramatsu, and T. Sasaki, *Bull. Chem. Soc., Jpn*, **43**, 1991 (1970).
70. K. Tajima, *Bull Chem. Soc., Jpn*, **43**, 3063 (1970).
71. B. N. Ghosh, *J. Indian Chem. Soc.*, **45**, 1120 (1968).
72. B. N. Ghosh, *J. Indian Chem. Soc.*, **47**, 557 (1970).
73. A. F. H. Ward, *Trans. Faraday Soc.*, **42**, 399 (1946).
74. B. V. Szyszkowski, *Z. Phys. Chem.*, **64**, 385 (1908).
75. M. Temkin, *J. Phys. Chem., Moscow*, **15**, 296 (1941); *Chem. Abs.*, **36**, 6392 (1942).
76. J. H. de Boer, *The Dynamic Character of Adsorption* (Oxford University Press, Oxford, 1953).
77. D. A. Haydon and F. H. Taylor, *Phil. Trans. R. Soc. London*, **A253**, 255 (1960).
78. S. Levine, G. M. Bell, and B. A. Pethica, *J. Chem. Phys.*, **40**, 2304 (1964).
79. H. B. Bull, *Biochim. Biophys. Acta*, **19**, 464 (1956).
80. H. B. Bull, *Arch. Biochem. Biophys.*, **68**, 102 (1957).
81. H. B. Bull and K. Breese, *Arch. Biochem. Biophys.*, **161**, 665 (1974).
82. Y. Nozaki and C. Tanford, *J. Biol. Chem.*, **246**, 2211 (1971).
83. W. Kauzmann, *Adv. Prot. Chem.*, **14**, 1 (1959).
84. J. J. Betts and B. A. Pethica, *Trans. Faraday Soc.*, **56**, 1515 (1960).
85. A. M. Posner, J. R. Anderson, and A. E. Alexander, *J. Colloid Sci.*, **7**, 623 (1952).
86. W. R. Gillap, N. D. Weiner, and M. Gibaldi, *J. Phys. Chem.*, **72**, 2218 (1968).
87. J. Stauff and J. Rasper, *Kolloid Z.*, **151**, 148 (1957).
88. F. Van Voorst Vander, T. F. Eikens, and M. van den Tempel, *Trans. Faraday Soc.*, **60**, 1170 (1964).
89. J. L. Zatz, *J. Colloid Interface Sci.*, **56**, 179 (1976).
90. M. L. Rosen and S. Aronson, *Colloids Surf.*, **3**, 201 (1981).
91. E. H. Lucassen-Reynders, *J. Phys. Chem.*, **70**, 1777 (1966).

CHAPTER 6

1. B. Franklin, in *The Ingenious Dr. Franklin*, edited by Goodman (1931).
2. Lord Rayleigh, *Proc. R. Soc.*, **47**, 364 (1890).
3. A. Pockels, *Nature*, **43**, 437 (1891).
4. I. Langmuir, *J. Am. Chem. Soc.*, **39**, 1848 (1917).
5. K. S. Birdi, to be published.
6. N. K. Adam, *The Physics and Chemistry of Surfaces*, Third Edition (Oxford University Press, London, 1941).

7. W. D. Harkins, *The Physical Chemistry of Surfaces Films* (Reinhold, New York, 1952).
8. A. W. Adamson, *Physical Chemistry of Surfaces*, Third Edition (John Wiley, New York, 1976).
9. J. T. Davies and E. K. Rideal, *Interfacial Phenomena* (Academic Press, New York, 1961).
10. T. Smith, *J. Colloid Interface Sci.*, **23**, 27 (1967).
11. I. M. Jalal, thesis, University of Wisconsin, U.S.A. (1978).
12. K. S. Birdi, to be published.
13. N. L. Gershfeld, *J. Colloid Interface Sci.*, **32**, 167 (1970); N. L. Gershfeld and R. E. Pagano, *J. Phys. Chem.*, **76**, 1238 (1972).
14. J. W. King, A. Chatterjee, and B. L. Karger, *J. Phys. Chem.*, **76**, 972 (1972).
15. D. K. Chattoraj and A. Chatterjee, *J. Colloid Interface Sci.*, **21**, 1959 (1966).
16. N. L. Gershfeld, *J. Colloid Interface Sci.*, **28**, 240 (1968). (a) A. K. Rakshit, G. Zografi, I. M. Jalal, and F. D. Gunstone, *J. Colloid Interface Sci.*, **80**, 466 (1981).
17. L. Salem, *J. Chem. Phys.*, **37**, 2100 (1962).
18. (a) K. S. Birdi, *J. Colloid Interface Sci.*, **57**, 2288 (1976). (b) K. S. Birdi and K. Sorensen, *Colloid Polymer Sci.*, **257**, 942 (1979).
19. E. Shapiro and S. Ohki, *J. Colloid Interface Sci.*, **4**, 38 (1974).
20. K. S. Birdi, to be published.
21. M. Lundquist, *Chem. Scripta*, **1**, 197 (1971); *Prog. Chem. Fats and other Lipids*, Vol. 16, pp. 101–124 (Pergamon Press, London, 1978).
22. R. Defay, I. Prigogine, A. Bellemans, and D. H. Everett, *Surface Tension and Adsorption* (Longmans, Green & Co., London, 1966).
23. F. C. Goodrich, *Surface and Colloid Science*, edited by E. Matijevic, Vol. I and II (John Wiley & Sons, New York, 1974).
24. J. C. Eriksson, *J. Colloid Interface Sci.*, **37**, 659 (1971).
25. K. Motomura, *Adv. Colloid Interface Sci.*, **12**, 1 (1980).
26. T. Hill, *Introduction to Statistical Thermodynamics* (Addison Wesley, New York, 1962).
27. F. H. Ree and W. G. Hoover, *J. Chem. Phys.*, **40**, 939 (1964).
28. N. L. Gershfeld and C. S. Patlak, *J. Phys. Chem.*, **70**, 286 (1966).
29. E. H. Lucassen-Reynders and M. van den Tempel, Proceedings of the IVth International Congress on Surface-Active Substances, Brussels, 1964.
30. F. M. Fowkes, *J. Phys. Chem.*, **66**, 385 (1962).
31. H. B. Bull, *An Introduction to Physical Biochemistry*, Second Edition (F. A. Davis, Philadelphia, 1971).
32. H. B. Bull, *J. Biol. Chem.*, **185**, 127 (1950).
33. D. J. Crisp, in *Surface Phenomena in Chemistry and Biology*, edited by J. F. Danielli, K. G. A. Pankhurst, and A. Riddiford (Pergamon Press, New York, 1958).
34. S. J. Singer, *J. Chem. Phys.*, **16**, 872 (1948).
35. M. L. Huggins, *J. Phys. Chem.*, **46**, 151 (1942).
36. P. J. Flory, *J. Chem. Phys.*, **10**, 51 (1942).
37. J. T. Davies, *J. Colloid Sci. Suppl.*, **1**, 9 (1954).
38. J. T. Davies and J. Lopis, *Proc. R. Soc. (London)*, **A227**, 537 (1955).
39. K. Motomura and R. Matuura, *J. Colloid Sci.*, **18**, 52 (1963).
40. H. L. Frisch and R. Simha, *J. Chem. Phys.*, **24**, 652 (1956); *J. Chem. Phys.*, **27**, 702 (1957).
41. M. L. Huggins, *Makromol. Chem.*, **87**, 119 (1965).
42. G. Gabrielli, M. Puggelli, M. Puggelli, and R. Faccioli, *J. Colloid Interface Sci.*, **37**, 213 (1971).
43. G. Gabrielli and M. Puggelli, *J. Appl. Poly. Sci.*, **16**, 2427 (1972).
44. K. S. Birdi, G. Gabrielli, and M. Puggelli, *Kolloid Z. Z. Polym.*, **250**, 591 (1972).
45. K. S. Birdi and G. D. Fasman, *J. Polym. Sci.*, **10**, 2483 (1972); *J. Polym. Sci.*, **42**, 1099 (1973).
46. Th. A. J. Payens, *Phillips Res. Rep.*, **10**, 425 (1955).
47. T. M. Saraga and C. Thomas, *J. Colloid Interface Sci.*, **48**, 42 (1974).
48. J. Mingins, J. A. G. Taylor, N. F. Owens, and J. H. Brooks, *Advances Chem. Series*, Vol. 144, *Monolayers*, edited by E. D. Goddard (American Chemical Society, 1975).

49. R. O. Jones, *Colloid Surface*, **2**, 201 (1981).
50. T. W. Healy and E. Wjite, *Adv. Colloid Interface Sci.*, **9**, 303 (1978).
51. G. L. Gaines, *J. Chem. Phys.*, **69**, 2627 (1978).
52. G. M. Bell, S. Levine, and D. Stephens, *J. Colloid Interface Sci.*, **33**, 482 (1970).
53. A. Vrij, *Proc. Kon. Vlam. Akad. Vetensh.*, p. 13 (1965).
54. K. S. Birdi, *Kolloid Z. Z. Polym.*, **250**, 222 (1972).
55. K. S. Birdi and A. Nikolov, *J. Phys. Chem.*, **83**, 365 (1979).
56. S. Chen, Ph.D. thesis, City University of New York, New York (1976).
57. J. Kubicki, H. D. Ohlenbush, E. Schroeder, and A. Wollmer, *Biochemistry*, **15**, 5698 (1976).
58. K. S. Birdi, in Procedd. Membrane Transport, Biophysics School, Wisla, Poland, 1977.
59. C. J. Epstein, R. F. Goldberger, and C. B. Anfinsen, Cold Spring Harb. Symp., *Quant. Biol.*, **28**, 439 (1963).
60. B. Robson and R. H. Pain, *J. Mol. Biol.*, **58**, 237 (1971).
61. A. W. Finkelstein and D. B. Pititsyn, *J. Mol. Biol.*, **62**, 613 (1971).
62. M. Charton, *J. Theor. Biol.*, **91**, 115 (1981).
63. A. W. Burgess, P. K. Ponnuswamy, and H. A. Scheraga, *Israel J. Chem.*, **12**, 239 (1974).
64. G. L. Amidon, S. H. Yalkowski, and S. Leung, *J. Phar. Sci.*, **63**, 3225 (1974).
65. K. S. Birdi, A. C. S. Symposium Series *Micellization, Solubilization and Microemulsions*, edited by K. L. Mittal (Plenum Press, New York, 1976).
66. B. Chatterjee and A. K. Ghosh, *J. Indian Chem. Soc.*, **49**, 751 (1972).
67. G. M. Bell, J. Mingins, and A. G. Taylor, *J. Chem. Soc., Faraday Trans. II*, **74**(2), 223 (1978).
68. J. Mingins, N. F. Owens, J. A. G. Taylor, J. H. Brooks, and B. A. Pethica, Advances in Chemistry Series, Vol. 144, *Monolayers* (American Chem. Soc., Washington, D.C., 1975), p. 14.
69. B. Y. Yue, C. M. Jackson, J. A. G. Taylor, J. Mingins, and B. A. Pethica, *J. Chem. Soc., Faraday Trans I*, **72**, 2685 (1976).
70. C. M. Jackson and B. Y. J. Yue, Advances in Chemistry Series, Vol. 144, *Monolayers* (American Chem. Soc., Washington, D.C., 1975), p. 202.
71. D. J. Crisp, *Surface Chemistry* (Butterworths, London, 1949).
72. F. C. Goodrich, Proc. Intern. Congr. Surface Activity II, London (1957), p. 85.
73. F. M. Fowkes, *J. Phys. Chem.*, **66**, 1863 (1962).
74. G. L. Gaines, *J. Colloid Interface Sci.*, **21**, 315 (1966).
75. P. Joos, *Bull. Soc. Chem. Beleges*, **78**, 207 (1969).
76. M. C. Phillips and P. Joos, *Kolloid Z. Z. Polym.*, **238**, 499 (1970).
77. D. Chapman, in *Biological Membranes*, edited by D. Chapman and D. F. Wallach, Vol. 2 (Academic Press, New York, 1972).
78. S. Ohki and C. B. Ohki, *J. Theor. Biol.*, **62**, 389 (1976).
79. K. R. Srinivasan and J. F. Nagle, *Biochemistry*, **13**, 3494 (1974).
80. J. F. Nagle, *Ann. Rev. Phys. Chem.*, **31**, 157 (1980).
81. L. Lin and R. L. Kay, *Biochemistry*, **16**, 3484 (1977).
82. D. Chapman, N. F. Owens, and D. A. Walker, *Biochim. Biophys. Acta*, **120**, 148 (1966).
83. M. C. Phillips and D. Chapman, *Biochim. Biophys. Acta*, **163**, 301 (1968).
84. F. Villalonga, *Biochim. Biophys. Acta*, **163**, 290 (1968).
85. R. A. Demel, L. M. G. van Kessel, and L. L. M. van Deenen, *Biochim. Biophys. Acta*, **266**, 26 (1972).
86. M. Hayashi, T. Muaramatsu, and I. Hana, *Biochim. Biophys. Acta*, **291**, 335 (1973).
87. O. Albrecht, H. Gruler, and E. Sackmann, *J. Phys. (Paris)*, **39**, 301 (1978).
88. D. G. Dervichian, *J. Chem. Phys.*, **7**, 931 (1939).
89. A. G. Bois, J. L. Firpo, J. J. Dupin, G. Albinet, and J. F. Baret, *Colloid Polymer Sci.*, **257**, 512 (1979).
90. J. J. Dupin, J. L. Firpo, G. Alvinet, A. G. Bois, L. Casalta, and J. F. Baret, *J. Chem. Phys.*, **70**, 2357 (1979).

91. N. Albon and J. M. Sturtevant, *Proc. Natl. Acad. Sci., U.S.A.*, **75**, 2258 (1978).
92. G. A. Hawkins and G. B. Benedek, *Phys. Rev. Lett.*, **32**, 524 (1974).
93. H. W. Kim and D. S. Connell, *Phys. Rev. Lett.*, **35**, 889 (1975).
94. K. S. Birdi and B. Sander, in *Surface Chemistry*, edited by K. S. Birdi (Holte, Denmark, 1981).
95. N. L. Gershfeld, *Ann. Rev. Phys. Chem.*, **27**, 349 (1976).
96. A. Cary and E. Rideal, *Proc. R. Soc., London*, **A109**, 318 (1925).
97. G. E. Boyd, *J. Phys. Chem.*, **62**, 537 (1958).
98. J. Brooks and A. Alexander, *J. Phys. Chem.*, **66**, 1851 (1962).
99. A. Alexander and F. Goodrich, *J. Colloid Interface Sci.*, **19**, 468 (1964).
100. I. M. Jalal, G. Zografi, A. K. Rakshit, and F. D. Gunstone, *J. Colloid Interface Sci.*, **76**, 146 (1980).
101. P. Joos, *Bull. Soc. Belges*, **79**, 645 (1979).
102. N. L. Gershfeld and K. Tajima, *Nature*, **279**, 708 (1979).
103. K. S. Birdi and B. Sander, to be published.
104. H. L. Scott and W. H. Cheng, *J. Colloid Interface Sci.*, **62**, 125 (1977).
105. T. L. Hill, *Thermodynamics of Small Systems*, I (W. A. Benjamin, New York, 1963).
106. G. T. Barnes, *Colloid Science*, Vol. 2, *Specialist Periodical Reports*, edited by D. H. Everett (The Chemiscal Society, London, 1975), p. 173.
107. K. Motomura, *J. Colloid Interface Sci.*, **48**, 307 (1974).
108. K. Motomura, T. Yano, M. Ikematsu, H. Matao, and R. Matuura, *J. Colloid Interface Sci.*, **69**, 209 (1979).
109. G. L. Gaines, *Insoluble Monolayers at Liquid–Gas Interface* (Interscience, New York, 1966).
110. A. K. Rakshit and G. Zographi, *J. Colloid Interface Sci.*, **80**, 474 (1981).
111. Y. Hendrix, *J. Colloid Interface Sci.*, **69**, 493 (1979).
112. C. Thomas and L. Ter-Minassian-Saraga, *J. Colloid Interface Sci.*, **51**, 122 (1975).
113. I. S. Costin and G. T. Barnes, *J. Colloid Interface Sci.*, **51**, 122 (1975).
114. K. J. Bacon and G. T. Barnes, *J. Colloid Interface Sci.*, **67**, 70 (1978).
115. W. D. Harkins and R. T. Florence, *J. Chem. Phys.*, **6**, 847 (1938).
116. R. T. Florence and W. D. Harkins, *J. Chem. Phys.*, **6**, 856 (1938).
117. K. Tajima and N. L. Gershfeld, in *Monolayers* edited by E. Goddard, Advances in Chemistry Series, No. 144 (Amer. Chem. Soc., Washington D.C., 1975).
118. G. L. Gaines, Jr., *J. Colloid Interface Sci.*, **21**, 315 (1966).
119. D. A. Cadenhead and R. J. Demchak, *J. Colloid Interface Sci.*, **30**, 76 (1969).
120. E. H. Lucassen-Reynders, *J. Colloid Interface Sci.*, **42**, 554 (1973).
121. K. Tajima and N. Gershfeld, *Biophys. J.*, **22**, 489 (1978). (a) E. Gorter and F. Grendel, *J. Exp. Med.*, **41**, 439 (1925). (b) A. G. Loewy and P. Siekevitz, *Cell Structure and Function*, Second Edition (Holt, Reinhart & Winston, New York, 1971).
122. C. F. Fox, *Sci. Amer.*, Feb. 31 (1972).
123. S. J. Singer and G. L. Nicholson, *Science*, **175**, 720 (1972).
124. R. E. Dickerson and I. Geis, *Structure and Action of Proteins* (Harper and Row, New York, 1969).
125. C. Tanford, *Hydrophobic Effect*, Second Edition (John Wiley & Sons, New York, 1980).
126. R. M. C. Dawsan, *Biological Membranes* edited by D. Chapman (Academic Press, New York, 1968).
127. J. H. Schulman and E. K. Rideal, *Proc. R. Soc. London Ser. B*, **122**, 46 (1937).
128. D. D. Eley and D. G. Hedge, *J. Colloid Sci.*, **11**, 445 (1956); **12**, 419 (1957).
129. P. J. Quinn and R. M. C. Dawson, *Biochem. J.*, **116**, 671 (1970).
130. D. O. Shah, *Biochim. Biophys. Acta*, **193**, 217 (1979).
131. H. B. Bull, *J. Am. Chem. Soc.*, **67**, 10 (1945).
132. J. T. Pearson and A. E. Alexander, *J. Colloid Interface Sci.*, **27**, 53 (1968).
133. J. T. Pearson, *J. Colloid Interface Sci.*, **27**, 64 (1968).

134. M. N. Ydenfreund and P. Becher, *Monolayers*, edited by E. D. Goddard, Advances in Chemistry Series, No. 144 (Amer. Chem. Soc., Washington, D.C., 1975), p. 192.
135. G. Kemp and C. E. Wenner, *Biochim. Biophys. Acta*, **282**, 1 (1972).
136. G. Kemp and C. Wenner, *Biochim. Biophys. Acta*, **323**, 161 (1973).
137. K. Motomura, *J. Colloid Interface Sci.*, **48**, 307 (1974).
138. K. Motomura, K. Sekita, and R. Matuura, *J. Colloid Interface Sci.*, **48**, 319 (1974).
139. K. Sekita, M. Nakamura, K. Motomura, and R. Matuura, *J. Colloid Interface Sci.*, **57**, 52 (1976).
140. K. Motomura, T. Yano, M. Ikematsu, H. Matuo, and R. Matuura, *J. Colloid Interface Sci.*, **69**, 209 (1979).
141. F. M. Fowkes, *J. Phys. Chem.*, **65**, 355 (1961).
142. F. M. Fowkes, *J. Phys. Chem.*, **66**, 1863 (1962).
143. F. M. Fowkes, *J. Phys. Chem.*, **68**, 3515 (1964).
144. E. H. Lucassen-Reynders, *J. Colloid Interface Sci.*, **42**, 554 (1973).
145. E. H. Lucassen-Reynders, *J. Colloid Interface Sci.*, **41**, 156 (1972).

CHAPTER 7

1. A. Dupre, *Théorie mécanique de la chaleur* (Paris, 1869), p. 207.
2. T. Young, *Phil. Trans. R. Soc. London*, **95**, 65 (1805).
3. N. K. Adam, *The Physics and Chemistry of Surfaces* (Oxford University Press, London, 1941).
4. J. E. McNutt and G. M. Andes, *J. Chem. Phys.*, **30**, 1300 (1959).
5. R. Defay, I. Prigogine, A. Bellemans, and D. H. Everett, *Surface Tension and Adsorption* (Longmans, Green & Co., London, 1966).
6. J. J. Bikerman, Proc. 2nd Intern. Congr. Surface Activity, London, 1957, Vol. III, p. 125.
7. J. J. Bikerman, *Surface Chemistry* (Academic Press Inc., New York, 1958).
8. R. H. Dettre and R. E. Johnson, Symp. Contact Angle, Bristol, U.K., 1966).
9. R. E. Johnson and R. H. Dettre, *Surface and Colloid Science*, Vol. II, edited by E. Matijevic (Wiley Interscience, New York, 1969).
10. F. C. Goodrich, *Surface and Colloid Science*, Vol. I, edited by E. Matijevic (Wiley Interscience, New York, 1969).
11. T. Kihara, *Intermolecular Forces* (John Wiley, New York, 1978).
12. R. J. Good, *Aspects of Adhesion*, edited by D. J. Alner and K. W. Allen, Vol. 7, p. 182 (Transcripta Press, London, 1973).
13. J. N. Israelachvili, *J. Chem. Soc.*, *II*, **69**, 1729 (1973).
14. J. Stefan, *J. Ann. Phys.*, **29**, 655 (1886).
15. F. M. Fowkes (editor), *Hydrophobic Surfaces, Proceed. 155th Amer. Chem. Soc. Meeting*, 1968 (Academic Press, New York, 1969).
16. F. M. Fowkes, *Chemistry and Physics of Interfaces*, II (American Chemical Society Publications, Washington, 1971).
17. F. M. Fowkes, *Treatise on Adhesion and Adhesives*, Vol. 1, edited by R. L. Patrick (Marcel Dekker, New York, 1967).
18. R. J. Good, *Chemistry and Physics of Interfaces*, II (American Chemical Society Publications, Washington, 1971).
19. R. J. Good, *Treatise on Adhesion and Adhesives*, Vol. 1, edited by R. L. Patrick (Marcel Dekker, New York, 1967).
20. F. London, *Trans. Faraday Soc.*, **33**, 8 (1936).
21. A. C. Zettlemoyer, *J. Colloid Interface Sci.*, **28**, 343 (1968); *Hydrophobic Surfaces, Proceed. 155th American Chemical Society Meeting*, 1968, Washington (Academic Press, New York, 1969).

22. J. H. Hildebrand and R. L. Scott, *The Solubility of Nonelectrolytes* (Dover Publications, New York, 1964).
23. G. Antonow, *J. Chem. Phys.*, **5**, 372 (1907).
24. R. E. Johnson and R. H. Dettre, *J. Colloid Interface Sci.*, **21**, 610 (1966).
25. W. R. Gillap, N. D. Weiner, and M. Gibaldi, *J. Amer. Oil Chem. Soc.*, **44**, 71 (1967).
26. R. Aveyard, *J. Colloid Interface Sci.*, **52**, 621 (1975).
27. J. F. Padday and N. D. Uffindell, *J. Phys. Chem.*, **72**, 1407 (1968).
28. G. L. Amidon, S. H. Yalkowski, and S. Leung, *J. Phar. Sci.*, **63**, 3225 (1974).
29. K. S. Birdi, A. C. S. Symposium Series, *Micellization, Solubilization and Microemulsions*, edited by K. L. Mittal (Plenum Press, New York, 1976).
30. W. A. Zisman, *Ind. Eng. Chem.*, **55**, 19 (1963).
31. K. S. Birdi and J. Jeppesen, *Colloid Polymer Sci.*, **256**, 261 (1978).
32. K. S. Birdi, *J. Dentistry, U.K.*, **7**, 230 (1979).
33. K. S. Birdi, *J. Theor. Biol.* (1981).
34. K. S. Birdi, *Tooth Surface Interactions* (edited by S. Leach (IRL, U.K., 1981)).
35. D. H. Bangham and R. I. Razouk, *Trans. Faraday Soc.*, **33**, 1459 (1937).
36. B. V. Derjaguin and Z. M. Zorin, Proceed. 2nd Intern. Congress Surface Activity, London, Vol. II, p. 145, 1957.
37. K. S. Birdi, *J. Colloid Interface Sci.*, **43**, 545 (1973).
38. S. Wu, *J. Polymer Sci., C*, **34**, 19 (1971).
39. K. S. Birdi, *J. Colloid Interface Sci.*, **88**, 290 (1982).
40. R. D. Baguall, J. A. D. Annis, and S. J. Sheolika, *J. Biomed. Mat. Res.*, **14**, 1 (1980).
41. T. Nguyen, "Polar Dispersion Force Contributions of Wood," *Wood Sci. Technol.*, **12**, 63 (1978).
42. H. D. von Wehle, *Holztechnologie*, **20**, 4 (1979).
43. M. E. Ginn, C. M. Noyes, and E. Jungermann, *J. Colloid Interface Sci.*, **26**, 146 (1968).
44. H. Schott, *J. Pharm. Sci.*, **60**, 1893 (1971).
45. A. Rosenberg, R. Williams, and G. Cohen, *J. Pharm. Sci.*, **62**, 920 (1973).
46. A. El-Shimi and E. D. Goddard, *J. Colloid Interface Sci.*, **48**, 242 (1974).
47. J. L. Zatz, *J. Pharm. Sci.*, **64**, 1080 (1975).
48. Y. Tamai, K. Makundi, and M. Suzuki, *J. Phys. Chem.*, **71**, 4176 (1967).
49. W. C. Hamilton, *J. Colloid Interface Sci.*, **40**, 219 (1972).
50. K. S. Birdi, to be published.
51. J. O. Melrose, *Adv. Chem. Ser.*, **43**, 158 (1964).
52. W. Adamson, *Physical Chemistry of Surfaces*, Third Edition (Wiley-Interscience, New York, 1976).
53. A. W. Neuman and R. J. Good, *J. Colloid Interface Sci.*, **38**, 341 (1972); J. D. Eick, R. J. Good, and A. W. Neumann, *J. Colloid Interface Sci.*, **53**, 235 (1975).
54. L. S. Penn and B. Miller, *J. Colloid Interface Sci.*, **78**, 238 (1980).
55. N. G. Maroudas, *J. Theor. Biol.*, **29**, 101 (1979).

CHAPTER 8

1. J. J. Kipling, *Adsorption from Solutions of Nonelectrolytes* (Academic Press, London, 1965).
2. D. H. Everett, *Colloid Science*, Vol. 1, edited by D. H. Everett (The Chemical Society, London, 1973), p. 49.
3. C. E. Brown and D. H. Everett, *Colloid Science*, Vol. 2, edited by D. H. Everett (The Chemical Society, London, 1975), p. 52.

4. A. W. Adamson, *Physical Chemistry of Surfaces*, Third Edition (John Wiley & Sons, New York, 1976), p. 385.
5. K. L. Mittal, *Adsorption at Interfaces*, ACS Symposium Series, No. 8 (American Chemical Society, Washington D.C., 1975).
6. R. Chatterjee and D. K. Chattoraj, *Indian J. Biochem. Biophys.*, **16**, 158 (1979).
7. G. Schay, *Surface and Colloid Science*, Vol. 2, edited by E. Matijevic (Wiley-Interscience, New York, 1969), p. 155.
8. G. D. Parfitt and P. C. Thompson, *Trans. Faraday Soc.*, **67**, 3371 (1971).
9. L. G. Nagy and G. Schay, *Magyr. Kem Folyoirat*, **66**, 31 (1960).
10. G. Schay and L. G. Nagy, *Peridica Polytechn. Budapest*, **4**, 45 (1960).
11. G. Schay and L. G. Nagy, *J. Chim. Phys.*, 149 (1961).
12. I. Dckany, F. Szanto, L. G. Nagy, and G. Foti, *J. Colloid Interface Sci.*, **50**, 265 (1975).
13. I. Dekany, L. G. Nagy, and G. Schay, *J. Colloid Interface Sci.*, **66**, 197 (1978).
14. R. Defay and I. Prigogine, *Trans. Faraday Soc.*, **46**, 199 (1950).
15. S. Ono and S. Kondo, in *Handbuch der Physik*, edited by S. Flugge (Springer, Berlin, 1960), Vol. 1, p. 134.
16. A. M. Williams, *Medd K. Vetenskapsalead Nobelinst*, **2**, No. 27 (1913).
17. J. J. Kipling and D. A. Tester, *J. Chem. Soc.*, **4123**, 1952.
18. A. Blackburn and J. J. Kipling, *J. Chem. Soc.*, **1493**, 1955.
19. F. G. Tryhorn and W. F. Wyatt, *Trans. Faraday Soc.*, **21**, 1925 (1925).
20. D. C. Jones and L. Outridge, *J. Chem. Soc.*, **1574**, 1930.
21. M. R. A. Rao, *J. Indian Chem. Soc.*, **12**, 345 (1933).
22. D. H. Everett, *Trans. Faraday Soc.*, **60**, 1803 (1964).
23. D. H. Everett, *Trans. Faraday Soc.*, **61**, 2478 (1965).
24. S. G. Ash, D. H. Everett, and G. H. Findenegg, *Trans. Faraday Soc.*, **66**, 708, 1970.
25. C. Orr and J. M. Dallaville, *Fine Particle Measurement, Size, Surface and Pore Volume*, pp. 207–215 (The MacMillan Company, New York, 1959).
26. Somasundaran and Fuerstenau, *J. Phys. Chem.*, **68**, 3562 (1964).
27. S. P. Mitra and D. K. Chattoraj, *Indian J. Biochem. Biophys.*, **15**, 147 (1978).
28. J. J. Kipling and E. H. M. Wright, *J. Chem. Soc.*, 855 (1962).
29. P. V. Cornford, J. J. Kipling, and E. H. M. Wright, *Trans. Faraday Soc.*, **58**, 74 (1962).
30. E. L. Cook and N. Hackerman, *J. Phys. Chem.*, **55**, 549 (1951).
31. G. Venturello and A. M. Ghee, *Gazzetta*, **89**, 1181 (1959).
32. R. N. Smith, C. F. Geiger, and C. Pierce, *J. Phys. Chem.*, **57**, 382, 1953.
33. J. J. Kipling and E. H. M. Wright, *J. Chem. Soc.*, 855 (1962).
34. C. H. Giles, T. H. MacEwan, S. N. Nakhwa, and D. Smith, *J. Chem. Soc.*, 3973 (1960).
35. W. B. Weiser, *Colloid Chemistry* (Wiley, New York, 1950).
36. S. Voyutsky, *Colloid Chemistry* (Mir Publishers, Moscow, 1978).
37. E. J. W. Verwey and J. Th. G. Overbeek, *Theory of the Stability of Lyophobic Colloids* (Elsevier, New York, 1948).
38. B. N. Ghosh, S. C. Rakshit, and D. K. Chattoraj, *Trans. Faraday Soc.*, **50**, 729 (1954).
39. B. N. Ghosh and D. K. Chattoraj, *Kolloid Z.*, **154**, 48 (1957).
40. I. D. Rattee and M. M. Breuer, *The Physical Chemistry of Dye Adsorption* (Academic Press, New York, 1974).
41. S. Iyer, S. R. Srinivasan, and N. T. Baddi, *Textile Res. J.*, **38**, 693 (1968).
42. C. H. Giles, *Text. Research J.*, **31**, 141 (1961).
43. S. R. S. Iyer, A. S. Chanekar, and G. Srinivasan, in *Adsorption at Interfaces*, edited by K. L. Mittal, ACS Symposium Series (American Chemical Society, Washington, D.C., 1975).
44. G. R. F. Rose, A. S. Weatherburn, and C. H. Bayley, *Textile Res. J.*, **21**, 427 (1951).
45. M. L. Corrin, E. L. Lind, A. Roginsky, and W. D. Harkins, *J. Colloid Sci.*, **4**, 485 (1949).
46. A. Fava and H. Eyring, *J. Phys. Chem.*, **60**, 890 (1956).

47. R. D. Vold and N. H. Sivaramakrishnan, *J. Phys. Chem.*, **62**, 984 (1958).
48. B. Tamamushi and K. Tamaki, *Proceedings of the Second International Congress of Surface Activity* (Butterworths, London, 1957), Vol. III, p. 449.
49. B. Tamamushi and K. Tamaki, *Trans. Faraday Soc.*, **55**, 1007 (1959).
50. P. Connor and R. H. Ottewill, *J. Colloid Interface Sci.*, **37**, 595 (1964).
51. P. Mukerjee and A. Anavil, in *Adsorption at Interfaces*, edited by K. L. Mittal, ACS Sympsoim Series No. 8 (Amer. Chem. Soc., Washington D.C., 1975), p. 107.
52. P. Somasundaran and D. H. Fuerstenau, *J. Phys. Chem.*, **70**, 90 (1966).
53. P. Somasundaran, T. W. Healy, and D. W. Fuerstenau, *J. Phys. Chem.*, **68**, 3562 (1964).
54. W. R. Foster, *J. Petrol. Technol.*, February, 205 (1973).
55. V. K. Bansal and D. O. Shah, in *Micellization, Solubilization and Micro-emulsions*, Vol. 1, edited by K. L. Mittal (Plenum Press, New York, 1977).
56. M. S. French, G. W. Keys, G. L. Stegemeier, R. C. Uber, A. Abrams, and H. J. Hill, *J. Petrol. Technol.*, February, 195 (1973).
57. F. J. Trogus, R. S. Schechter, and W. H. Wade, *J. Colloid Interface Sci.*, **70**, 293 (1979).
58. R. Ulman, J. Koral, and F. R. Eirich, Proc. Intern. Cong. Surface Activity, 2nd, London, 1957, Vol. III, p. 485.
59. F. Rowland, R. Bulas, E. Rothstein, and F. R. Eirich, Ind. Eng. Chem., September 1965, p. 46.
60. S. G. Ash, in *Colloid Science*, Vol. 1, edited by D. H. Everett (The Chemical Society, London, 1973), p. 103.
61. V. K. Dunn and R. D. Vold, *Adsorption at Interfaces*, edited by K. L. Mittal, ACS Symposium Series No. 8 (Amer. Chem. Soc., Washington, D.C., 1975), p. 96.
62. F. S. Chan, P. S. Minhas, and A. A. Robertson, *J. Colloid Interface Sci.*, **33**, 586 (1970).
63. F. R. Eirich, R. Bulas, E. Rothstein, and F. Rowland, in *Chemistry and Physics of Interfaces*, edited by D. E. Guestve, p. 109 (American Chemical Society, Washington, D.C., 1965).
64. R. Ulman, J. Koral, and F. R. Eirich, Proc. International Congress of Surface Activity, 2nd, London, Vol. III, p. 485 (1957).
65. W. Heller, *Pure Appl. Chem.*, **12**, 249 (1966).
66. B. J. Fontana and J. R. Thomas, *J. Phys. Chem.*, **65**, 480 (1961).
67. R. Simha, H. L. Frisch, and F. R. Eirich, *J. Phys. Chem.*, **57**, 584 (1953).
68. H. L. Frisch and R. Simha, *J. Phys. Chem.*, **58**, 507 (1954).
69. H. L. Frisch, *J. Phys. Chem.*, **59**, 633 (1955).
70. H. L. Frisch, M. Y. Hellman, and J. L. Lundberg, *J. Polymer Sci.*, **38**, 441 (1959).
71. A. Silberberg, *J. Phys. Chem.*, **66**, 1872 (1962).
72. A. Silberberg, *J. Chem. Phys.*, **48**, 2835 (1968).
73. A. Silberberg, *J. Colloid Interface Sci.*, **38**, 217 (1972).
74. C. A. Hoeve, E. A. Dimarzio, and P. Peyser, *J. Chem. Phys.*, **42**, 2558 (1965).
75. C. A. Hoeve, *J. Polym. Sci.*, **34**, 1 (1971).
76. S. G. Ash, D. Everett, and G. H. Findenegg, *Trans. Faraday Soc.*, **66**, 708 (1970).
77. J. M. Anderson, *Nature Lond.*, **253**, 536 (1975).
78. B. Pesosac and V. Defendi, *Science*, **175**, 898 (1972).
79. H. P. Schneble and M. M. Burger, *Proc. Nat. Acad. Sci.*, *USA*, **69**, 3825 (1972).
80. G. Stoner and S. Srinivasan, *J. Phys. Chem.*, **74**, 1088 (1970).
81. J. L. Brash and D. J. Lyman, *The Chemistry of Biosurfaces*, edited by M. L. Hair, p. 1977 (Marcel Dekker, New York, 1971).
82. J. M. Singer, *Bull. Rheum. Dis.*, **24**, 762 (1974).
83. Y. Kobayashi and S. Aomine, *Soil Sci. Pl. Nutr.*, **13**, 189 (1967).
84. G. R. Stark, edited, *Biochemical Aspects of Reactions on Solid Supports* (Academic Press, New York, 1971).
85. H. B. Bull, *Biochim. Biophys. Acta*, **19**, 464 (1956).

86. H. B. Bull, Arch. Biochem. Biophys., **68**, 102 (1957).
87. D. K. Chattoraj and H. B. Bull, *J. Amer. Chem. Soc.*, **81**, 5128 (1959).
88. I. R. Miller and D. Bach, *Surface and Colloid Science*, Vol. 6, edited by E. Matijevic (Wiley Interscience, New York, 1973).
89. W. J. Dilman and I. R. Miller, *J. Colloid Interface Sci.*, **44**, 221 (1973).
90. B. W. Morrissey and R. R. Stromberg, *J. Colloid Interface Sci.*, **44**, 221 (1974).
91. S. P. Mitra and D. K. Chattoraj, *Indian J. Biochem. Biophys.*, **15**, 147 (1978).
92. V. Hlady and H. Furedi-Milhofer, *J. Colloid Interface Sci.*, **69**, 460 (1979).
93. W. Norde and J. Lyklema, *J. Colloid Interface Sci.*, **71**, 350 (1979).
94. W. Norde and J. Lyklema, *J. Colloid Interface Sci.*, **66**, 257, 266, 277, 285, 295 (1978).
95. W. Norde, "Proteins at Interface," doctoral thesis (1976), Agricultural University, Wageningen, The Netherlands.
96. S. P. Mitra and D. K. Chattoraj, *Indian J. Biochem. Biophys.*, **15**, 147 (1978).
97. C. L. Riddiford and B. R. Jennings, *Biochim. Biophys. Acta*, **126**, 71 (1966).
98. D. Lang and M. Mitani, *Biopolymers*, **9**, 373 (1970).
99. A. K. Kleinschmidt, *Biochim. Biophys. Acta*, **61**, 857 (1962).
100. J. D. Watson, *Molecular Biology of the Gene*, Third Edition (W. A. Watson, Benzamin Inc., Menlo Park, California, 1977), p. 246.
101. I. R. Miller, *J. Mol. Biol.*, **3**, 229, 357 (1961).
102. I. R. Miller, *Biochim Biophys. Acta*, **103**, 219 (1965).
103. M. A. Frommer and I. R. Miller, *J. Phys. Chem.*, **72**, 2862 (1968).
104. D. K. Chattoraj, P. Chowrashi, and K. Chakravarti, *Biopolymers*, **5**, 173 (1967).
105. P. Chowrashi, D. K. Chattoraj, and K. Chakravarti, *Biopolymers*, **6**, 97 (1968).
106. M. A. Frommer and I. R. Miller, *J. Colloid Interface Sci.*, **21**, 245 (1966).
107. S. N. Upadhyay and D. K. Chattoraj, *Biochim. Biophys. Acta*, **161**, 561 (1968).
108. D. J. Crisp, *J. Colloid Sci.*, **11**, 356 (1956).
109. S. G. Daniel, *Trans. Faraday Soc.*, **47**, 1345 (1951).
110. H. B. Bull, *An Introduction to Physical Biochemistry* (F. A. Davis Co., Philadelphia).
111. H. Freundlich, *Colloid and Capillary Chemistry* (Methuen, London, 1926).
112. D. C. Henry, *Phil Mag.*, **44**(6), 689 (1922).
113. J. Oscik, A. Dabrowski, M. Jaronec, and W. Rudzinski, *J. Colloid Interface Sci.*, **56**, 403 (1976).
114. J. Oscik, A. Dabrowski, and W. Rudzinski, *Colloid Polymer Sci.*, **255**, 50 (1977).
115. B. Chopra and V. Ramakrishna, *Indian J. Chem.*, **10**, 187 (1972).
116. M. T. Coltrap and N. Hackerman, *J. Colloid Interface Sci.*, **43**, 185 (1976).
117. B. R. Puri, O. P. Mahajan, and B. C. Kaistha, *Indian J. Chem.*, **12**, 161 (1974).
118. S. Sircar and A. L. Myers, *AIChE J.*, **19**, 159 (1973).
119. A. Dabrowski, J. Oscik, W. Rudzinski, and M. Jaroniec, *J. Colloid Interface Sci.*, **69**, 287 (1979).
120. M. Borowko, M. Jaroniec, J. Oscik, and R. Kusak, *J. Colloid Interface Sci.*, **69**, 311 (1979).

CHAPTER 9

1. J. L. Finney, *Phil. Trans. R. Soc., London*, **B778**, 3 (1977).
2. I. D. Kuntz and W. Kauzmann, *Adv. Protein Chem.*, **28**, 239 (1974).
3. J. D. Watson, *Molecular Biology of the Gene*, Third Edition (W. A. Benjamin Inc., Menlo Park, California, 1977).
4. D. R. Davies, *Ann. Rev. Biochem.*, **36**, 321 (1967).
5. J. D. Watson and F. H. C. Crick, *Nature*, **177**, 964 (1953).

6. D. Eagland in *Water, a Comprehensive Treatise*, Vol. 5, edited by F. Franks (Plenum Press, New York, 1975).

6a. C. van den Berg, "Vapor Sorption Equilibria and Other Water–Starch Interactions; A Physico-chemical Approach," Doctoral thesis, Agricultural University, Wagenigen, the Netherlands, 1981.

7. C. Tanford, *The Hydrophobic Effect*, Second Edition (John Wiley & Sons, New York, 1980).

8. A. Ben Naim, *The Hydrophobic Interactions* (Plenum Press, New York, 1980).

9. H. B. Bull, *J. Am. Chem. Soc.*, **66**, 1499 (1944).

10. D. K. Chattoraj and S. P. Mitra, *Indian J. Biochem. Biophys.*, **14**, 1 (1977).

11. A. W. Adamson, *Physical Chemistry of Surfaces*, Third Edition (Academic Press, New York, 1976).

12. L. Pauling, *J. Am. Chem. Soc.*, **67**, 555 (1945).

13. E. F. Mellon, A. H. Korn, and S. R. Hoover, *J. Am. Chem. Soc.*, **70**, 3040 (1948).

14. I. M. Klotz, *Science*, **128**, 815 (1958).

15. H. B. Bull and K. Breese, *Arch. Biochem. Biophys.*, **128**, 488 (1968).

16. Y. I. Frenkel, *Kinetic Theory of Liquids* (The Clarendon Press, Oxford; reprinted by Dover Publication, New York, 1955).

17. G. D. Halsey, Jr., *J. Chem. Phys.*, **16**, 931 (1948).

18. T. L. Hill, *Advan. Catalysis*, **4**, 211 (1952).

19. S. W. Benson, D. A. Ellis, and R. W. Zanzig, *J. Am. Chem. Soc.*, **72**, 2102 (1950).

20. J. A. Cutler and A. D. McLaren, *J. Polymer Sci.*, **3**, 792 (1948).

21. P. J. Flory, *Principles of Polymer Chemistry* (Cornell Univ. Press, Ithaca, New York, 1953).

22. R. L. D'Arcy and I. C. Watt, *Trans. Faraday Soc.*, **66**, 1236 (1970).

23. H. B. Bull, *Introduction to Physical Biochemistry* (F. A. Davis Company, Philadelphia, 1971).

24. W. B. Bryan, *J. Theoret. Biol.*, **87**, 639 (1980).

25. T. L. Hill, *J. Am. Phys.*, **17**, 520 (1949).

26. R. L. Altman and S. W. Benson, *J. Phys. Chem.*, **64**, 851 (1960).

27. C. Tanford, *J. Mol. Biol.*, **39**, 539 (1969).

28. H. K. Schachman and M. A. Lauffer, *J. Am. Chem. Soc.*, **71**, 536 (1949).

29. W. Wales and J. W. Williams, *J. Polymer Sci.*, **8**, 449 (1952).

30. J. E. Hearst and J. Vinograd, *Proc. Nat. Acad. Sci., U.S.*, **47**, 999 (1961).

31. S. N. Timasheff, *Accounts Chem. Research*, **3**, 62 (1970).

32. H. B. Bull and K. Breese, *Arch. Biochem. Biophys.*, **137**, 299 (1970).

33. H. B. Bull and K. Breese, *Arch. Biochem. Biophys.*, **139**, 93 (1970).

34. D. K. Chattoraj and S. P. Mitra, *Indian J. Biochem. Biophys.*, **14**, 7 (1977).

35. S. P. Mitra, D. K. Chattoraj, and M. N. Das, *Indian J. Biochem. Biophys.*, **14**, 101 (1977).

36. S. P. Mitra, D. K. Chattoraj, M. N. Das, and A. Sen, *Indian J. Biochem. Biophys.*, **15**, 153 (1978).

37. H. B. Bull and K. Breese, *Arch. Biochem. Biophys.*, **149**, 164 (1972).

38. R. A. Robinson and R. H. Stokes, *Electrolyte Solutions* (Butterworths Publishers, London, 1959).

39. D. K. Chattoraj and R. Chatterjee, *J. Colloid Interface Sci.*, **54**, 364 (1976).

40. R. Chatterjee and D. K. Chattoraj, *Indian J. Biochem. Biophys.*, **16**, 158 (1979).

41. D. K. Chattoraj and S. P. Mitra, *Indian J. Biochem. Biophys.*, **16**, 406 (1979).

42. O. Sinanoglu and S. Abdulnur, *Photochem. Photobiol.*, **3**, 333 (1964).

43. S. Lewin, *Nature (New Biol.), London*, **80**, 231 (1971).

44. H. B. Bull and K. Breese, *Arch. Biochem. Biophys.*, **202**, 116 (1980).

45. C. Tanford, *The Physical Chemistry of Macromolecules* (John Wiley, New York, 1961).

46. P. H. Von Hippel and K. Y. Wong, *Biochemistry*, **2**, 1399 (1963).

47. H. B. Bull and K. Breese, *Arch. Biochem. Biophys.*, **156**, 604 (1973).

48. J. M. Steim, *Arch. Biochem. Biophys.*, **156**, 604 (1973).

49. J. H. Von Hippel and T. Schleich, *Structure and Stability of Biological Macromolecules*, edited by M. Timasheff and G. D. Fasman (Marcel Dekker Inc., New York, 1969), p. 213.

50. S. P. Mitra and D. K. Chattoraj, *Indian J. Biochem. Biophys.*, **15**, 239 (1978).
51. B. Jacobson, *Nature, London*, **172**, 666 (1953).
52. M. Falk, K. A. Hartman, and R. G. Lord, *J. Am. Chem. Soc.*, **84**, 3843 (1962).
52a. M. Falk, K. A. Hartman, and R. C. Lord, *J. Am. Chem. Soc.*, **85**, 387, 391 (1963).
53. J. Depireux and D. Williams, *Nature, London*, **195**, 699 (1962).
54. H. B. Gray, V. A. Bloomfield, and J. E. Hearst, *J. Chem. Phys.*, **46**, 1493 (1967).
55. B. Tunis and J. E. Hearst, *Biopolymers*, **6**, 1325, 1345 (1968).
56. D. K. Chattoraj and H. B. Bull, *J. Colloid Interface Sci.*, **35**, 220 (1971).
57. D. K. Chattoraj and H. B. Bull, *Arch. Biochem. Biophys.*, **142**, 363 (1971).
58. E. A. Guggenheim, *An Application of Statistical Mechanics* (Clarendon Press, Oxford, 1966).
59. R. B. Anderson, *J. Am. Chem. Soc.*, **68**, 686 (1946).
60. J. H. de Boer, *The Dynamical Character of Adsorption* (Clarendon Press, Oxford, 1968).
61. W. Pfeil and P. L. Privalov, *Biophysical Chem.*, **4**, 23 (1976).
62. W. Pfeil and P. L. Privalov, *Biophysical Chem.*, **4**, 33 (1976).
63. W. Pfeil and P. L. Privalov, *Biophysical Chem.*, **4**, 41 (1978).
64. C. Tanford, *Adv. Protein Chem.*, **23**, 121 (1968).
65. C. Tanford, *Adv. Protein Chem.*, **23**, 1 (1970).
66. H. Edelhoch and J. C. Osborne, *Adv. Protein Chem.*, **30**, 183 (1976).

CHAPTER 10

1. J. W. Watson, *Molecular Biology of Genes*, Third Edition (W. A. Benjamin, Inc., Menlo Park, California, 1977).
2. E. S. West, W. R. Todd, H. S. Mason, and J. T. van Bruggen, *A Textbook of Biochemistry* (McMillan Publishing Comp., Indian Edition, 1974).
3. J. Steinhardt and J. A. Reynolds, *Multiple Equilibria in Proteins* (Academic Press, New York, 1969).
4. C. Tanford, *Physical Chemistry of Macromolecules* (John Wiley and Sons, New York, 1961).
5. J. Steinhardt, in *Protein–Ligand Interactions*, edited by H. Sund and G. Blauer (W. de Gruyter, Berlin–New York, 1974).
6. J. Steinherdt, N. Stocker, D. Caroll, and K. S. Birdi, *Biochemistry*, **13**, 4461 (1974).
7. J. Steinherdt, J. R. Scott, and K. S. Birdi, *Biochemistry*, **16**, 718 (1977).
8. K. S. Birdi and J. Steinhardt, *Biochem. Biophys. Acta*, **534**, 219 (1978); and unpublished.
9. P. Coassolo, M. Sarrazin, and J. C. Sari, *Ann. Biochem.*, **104**, 37 (1980).
10. A. A. Spector and E. C. Santos, *Ann. N.Y. Acad. Sci.*, **226**, 247 (1973).
11. P. Coassolo, M. Briand, Bourdeaux, and J. C. Sari, *Biochim. Biophys. Acta*, **538**, 512 (1978).
12. Y. Inone, S. Shigeru, C. Riichiro, S. Nagaoka, and M. Sogami, *Biopolymers*, **18**, 373 (1979).
13. S. J. Gill, H. T. Gaud, J. Wyman, and B. G. Barias, *Biophys. Chem.*, **8**, 53 (1978).
14. R. Epstein, *Biophys. Chem.*, **8**, 327 (1978).
15. J. D. McGhee and P. H. von Hippel, *J. Mol. Biol.*, **86**, 469 (1974).
16. G. Schwarz, *Biophys. Chem.*, **6**, 65 (1977).
17. G. Merino, M. Menendez, and J. Laynez, *Biophys. Chem.*, **9**, 263 (1979).
18. S. Stankowski, J. Engel, and R. A. Berg, *Biophys. Chem.*, **9**, 383 (1979).
19. G. Barisas and G. Gill, *Biophys. Chem.*, **9**, 235 (1979).
20. M. D. Reboiras, H. Pfister, and H. Pauly, *Biophys. Chem.*, **9**, 37 (1978).
21. G. S. Manning, *Biophys. Chem.*, **9**, 65 (1978).
22. J. Ramstein, M. Leng, and N. R. Kallenbach, *Biophys. Chem.*, **5**, 319 (1976).
23. T. Gilanyi and E. Wolfram, *Colloids Surfaces*, **3**, 181 (1981).
24. H. B. Bull, *Introduction to Physical Biochemistry*, Second Edition (F. A. Davis, Philadelphia, 1971).

25. H. B. Bull, *Biochim. Biophys. Acta*, **19**, 464 (1956).
26. H. B. Bull and K. Breese, *Biopolymers*, **15**, 1573 (1976).
27. R. Chatterjee and D. K. Chattoraj, *Biopolymers*, **18**, 147 (1979).
28. M. Sen, S. P. Mitra, and D. K. Chattoraj, *Colloids Surfaces*, **2**, 259 (1981).
29. C. Tanford, *The Hydrophobic Effect*, Second Edition (John Wiley and Sons, New York, 1980).
30. K. Hiramatsu, C. Ueda, K. Iwata, K. Arikawa, and K. Aoki, *Bull. Chem. Soc., Jpn*, **50**, 368 (1977).
31. G. Scatchard, *Ann. N.Y. Acad. Sci.*, **51**, 660 (1949).
32. K. Linderstrom-Lang, *Compt. Rend. Trav. Lab. Carlsberg* **15**, No. 7 (1924).
33. C. Tanford and J. G. Kirkwood, *J. Am. Chem. Soc.*, **79**, 5333 (1957).
34. T. L. Hill, *Arch. Biochem. Biophys.*, **57**, 299 (1955).
35. G. Schwarz, *Biophys. Struct. Mechanism*, **2** (1976).
36. K. Denbigh, *The Principles of Chemical Equilibrium* (Cambridge University Press, 1971).
37. A. L. Olins and D. E. Olins, *Science*, **183**, 330 (1974).
38. F. H. Crick, *Nature*, **234**, 25 (1971).
39. C. G. Sahasrabuddhe and K. E. Van Holde, *J. Biol. Chem.*, **249**, 152 (1974).
40. R. J. Clark and G. Felsenfeld, *Nature (New Biol.)*, **229**, 101 (1971).
41. B. Sollner-Webb and G. Felsenfeld, in *Chromosomal Proteins and Their Role in the Regulation of Gene Expression*, edited by G. S. Stein and L. J. Kleinsmith (Academic Press, New York, 1975), p. 213.
42. R. Chatterjee, S. P. Mitra, and D. K. Chattoraj, *Indian J. Biochem. Biophys.*, **16**, 22 (1979).
43. R. H. Stellwagen and R. D. Cole, *Ann. Rev. Biochem.*, **38**, 951 (1969).
44. F. O. Akinrimisi, J. Bonner, and P. O. Ts'o, *J. Mol. Biol.*, **11**, 128 (1965).
45. T. Y. Shih and G. D. Fasman, *Biochemistry*, **10**, 1675 (1971).
46. D. K. Chattoraj, H. B. Bull, and R. Chalkley, *Arch. Biochem. Biophys.*, **152**, 778 (1972).
47. M. Sen, S. P. Mitra, and D. K. Chattoraj, *Indian J. Biochem. Biophys.*, **17**, 370 (1980).
48. (a) B. K. Sadhukhan and D. K. Chattoraj, "Surfactants in Solution," in *Surfactants in Solution*, edited by K. L. Mittal and B. Lindman, Plenum Press, New York, Forthcoming; (b) M. Sen, S. P. Mitra, and D. K. Chattoraj, *Indian J. Biochem. Biophys.*, **17**, 406 (1980).
49. H. B. Bull and K. Breese, *Biopolymers*, **15**, 1573 (1976).
50. D. K. Chattoraj and S. P. Mitra, *Indian J. Biochem. Biophys.*, **16**, 406 (1979).
51. D. K. Chattoraj and R. Chatterjee, *J. Colloid Interface Sci.*, **54**, 364 (1976).
52. H. B. Bull and K. Breese, *Biopolymers*, **17**, 2121 (1978).
53. C. Chothia, *Nature (London)*, **254**, 304 (1975).
54. I. D. Kuntz, *J. Am. Chem. Soc.*, **93**, 514 (1971).
55. H. B. Bull, *Arch. Biochem. Biophys.*, **208**, 229 (1981).
56. R. Chatterjee and D. K. Chattoraj, *Indian J. Biochem. Biophys.*, **16**, 158 (1979).
57. C. Tanford, *Proc. Natl. Acad. Sci., USA*, **78**, 270 (1981).
58. J. Reynolds, *Methods Enzymol.*, **61**, 58 (1979).

CHAPTER 11

1. P. Mukerjee, *Adv. Colloid Interface Sci.*, **1**, 241 (1967).
2. E. R. Jones and C. R. Bury, *Phil. Mag.*, **4**, 841 (1927).
3. G. S. Hartley, *Aqueous Solution of Paraffin Chain Salts* (Hermann et Cie, Paris, 1936).
4. C. Tanford, *The Hydrophobic Effect* (John Wiley and Sons, New York, 1980).
5. A. Ben Naim, *J. Phys. Chem.*, **82**, 792 (1978).
6. D. K. Chattoraj, K. S. Birdi, and S. Dalsagar, *Solution Behavior of Sulfactants Vol. 1*, edited by K. L. Mittal (Plenum Press, New York, 1982), p. 505.

7. P. Mukerjee and K. J. Mysels, *Critical Micellar Concentration of Aqueous Surfactant Systems* (National Bureau of Standards, Washington D.C., 1971).
8. D. K. Chattoraj, *J. Phys. Chem.*, **71**, 455 (1967).
9. D. K. Chattoraj, *J. Colloid Interface Sci.*, **29**, 399 (1969).
10. D. K. Chattoraj and R. P. Pal, *Indian J. Chem.*, **10**, 417 (1972).
11. G. D. Miles and L. Shedlovsky, *J. Phys. Chem.*, **48**, 57 (1944).
12. R. Defay, I. Prigogine, and A. Bellemans, *Surface Tension and Adsorption* (John Wiley and Sons, Inc., New York, 1966).
13. D. J. Crisp, *Trans. Faraday Soc.*, **43**, 815 (1947).
14. A. E. Alexander, *Trans. Faraday Soc.*, **38**, 248 (1942).
15. D. Reichenberg, *Trans. Faraday Soc.*, **43**, 467 (1947).
16. K. J. Mysels and L. H. Princen, *J. Colloid. Sci.*, **12**, 594 (1957).
17. G. Stainsby and A. E. Alexander, *Trans. Faraday Soc.*, **46**, 587 (1950).
18. E. D. Goddard and G. C. Benson, *Can. J. Chem.*, **35**, 986 (1957).
19. E. Matijevic and B. A. Pethica, *Trans. Faraday Soc.*, **54**, 587 (1958).
20. P. White and G. C. Benson, *Trans. Faraday Soc.*, **55**, 1025 (1959).
21. E. D. Goddard, C. A. Hoeve, and G. C. Benson, *J. Phys. Chem.*, **61**, 593 (1957).
22. A. Ben-Naim, *Hydrophobic Interactions* (Plenum Press, New York, 1979).
23. A. I. Rusanov, *J. Colloid Interface Sci.*, **85**, 157 (1982).
24. P. Mukerjee, *J. Phys. Chem.*, **71**, 4166 (1967).
25. P. Becher, *Nonionic Surfactants*, edited by M. J. Schick (Marcel Dekker, Inc., New York).
26. M. F. Emerson and A. Holtzer, *J. Phys. Chem.*, **71**, 1898 (1967).
27. R. D. Geer, E. H. Eylar, and E. W. Anacker, *J. Phys. Chem.*, **75**, 369 (1971).
28. J. Swarbrick and J. Daruwala, *J. Phys. Chem.*, **73**, 2627 (1969).
29. H. C. Evans, *J. Chem. Soc.*, 579 (1956).
30. A. K. Chatterjee and D. K. Chattoraj, *J. Colloid Interface Sci.*, **26**, 1 (1968).
31. H. S. Frank and M. W. Evans, *J. Chem. Phys.*, **13**, 507 (1945).
32. C. Tanford, in *Micellization, Solubilization and Microemulsions*, Vol. 1, edited by K. L. Mittal (Plenum Press, New York, 1977), p. 159.
33. K. S. Birdi, in *Micellization, Solubilization and Microemulsions*, Vol. 1, edited by K. L. Mittal (Plenum Press, New York, 1977), p. 151.
34. E. Ruckenstein and R. Nagarajan, in *Micellization, Solubilization and Microemulsions*, Vol. 1, edited by K. L. Mittal (Plenum Press, New York, 1977), p. 133.
35. J. N. Israelachvili, D. J. Mitchell, and B. W. Ninham, *J. Chem. Soc., Faraday Trans.*, **29**, 1523 (1976).
36. R. J. M. Tausk, J. Karmiggelt, C. Oudshoorn, and J. Th. C. Overbeek, *Biophys. Chem.*, **1**, 175 (1974).
37. G. L. Amidon, S. H. Yalkowsky, S. T. Anik, and S. C. Valvan, *J. Phys. Chem.*, **79**, 2239 (1975).
38. G. L. Amidon, S. H. Yalkowsky, and S. Tenng, *J. Pharm. Sci.*, **64**, 1858 (1974).
39. C. Tanford, *Proc. Natl. Acad. Sci., USA*, **76**, 4175 (1979).
40. R. B. Hermann, *J. Phys. Chem.*, **71**, 2754 (1972).
41. R. B. Hermann, *Proc. Natl. Acad. Sci., USA*, **74**, 4144 (1977).
42. J. A. Reynolds, D. B. Gilbert, and C. Tanford, *Proc. Natl. Acad. Sci., USA*, **71**, 2925 (1974).
43. S. Ono and S. Kondo, in *Handbuch der Physik*, Vol. 10, edited by S. Fluegge (Springer, Berlin), pp. 130–280.
44. D. S. Choi, M. S. John, and N. Eyring, *J. Chem. Phys.*, **53**, 2608 (1970).
45. V. Luzzati, in *Biological Membranes*, edited by D. Chapman (Academic Press, New York, 1968), Chap. 3.
46. F. Reiss-Husson and V. Luzzati, *J. Colloid Interface Sci.*, **21**, 534 (1966).
47. (a) B. Sadhukhan, D. K. Chattoraj, and K. S. Birdi, *Indian J. Chemistry*, **22A**, 741 (1983); (b) K. Larsson and I. Lundstrom, in *Lyotropic Liquid Crystals*, edited by Stig Friberg, Advances in Chemistry Series 152 (American Chem. Soc., Washington, D.C., 1976), p. 43.

48. A. W. Adamson, *Physical Chemistry of Adsorption*, Third Edition (Wiley-Interscience, New York, 1976), pp. 571–575.
49. S. Ross and W. Winkler, *J. Colloid Sci.*, **10**, 319 (1955).
50. W. D. Harkins and G. Jura, *J. Am. Chem. Soc.*, **66**, 1366 (1944).
51. C. Jho, D. Nealon, S. Shogbola, and A. D. King, Jr., *J. Colloid Interface Sci.*, **65**, 141 (1978).
52. C. Kemball and E. K. Rideal, *Proc. R. Soc., London, Ser.*, **A187**, 53 (1946).
53. J. G. Kirkwood and F. P. Buff, *J. Chem. Phys.*, **17**, 338 (1949).
54. R. C. Tolman, *J. Chem. Phys.*, **16**, 758 (1948).
55. F. Goodrich, in *Surface and Colloid Science*, edited by E. Matijevic, Vol.   , (Wiley-Interscience, New York), p. 1.
56. S. Ono, *Mem. Fac. Eng. Kyushu Univ.*, **10**, 195 (1947).
57. T. Murakami, S. Ono, M. Tamura, and M. Kurata, *J. Phys. Soc., Jpn*, **6**, 309 (1951).
58. I. R. Miller and M. A. Frommer, *J. Phys. Chem.*, **72**, 1834 (1968).
59. D. K. Chattoraj, *J. Phys. Chem.*, **72**, 1835 (1967).
60. J. O'M. Bockris and A. K. N. Reddy, *Modern Electrochemistry* (Plenum Press, New York, 1970).
61. T. P. Hoar and J. H. Schulman, *Nature (London)*, **152**, 102 (1943).
62. K. Shinoda and S. Friberg, *Adv. Colloid Interface Sci.*, **4**, 281 (1975).
63. D. O. Shah, *J. Colloid Interface Sci.*, **37**, 744 (1971).
64. M. L. Rabins, in *Micellization, Solubilization and Microemulsions*, Vol. 2, edited by K. L. Mittal (Plenum Press, New York, 1977).
65. S. G. Frank and G. Zografi, *J. Colloid Interface Sci.*, **29**, 27 (1967).
66. E. Ruckenstein and R. Krishnan, *J. Colloid Interface Sci.*, **76**, 188 (1980).
67. E. Ruckenstein and R. Krishnan, *J. Colloid Interface Sci.*, **76**, 201 (1980).
68. K. N. Jalan, T. K. Maitra, D. Mahalanobis, S. K. Agarwal, M. L. Chakravarti, D. K. Chattoraj, and S. P. Moulik, *J. Food Sci.*, **42**, 1675 (1977).
69. F. Mac Ritchie and A. E. Alexander, *J. Colloid Sci.*, **18**, 453 (1963).
70. J. G. Petrow and R. Miller, *Colloid Polymer Sci.*, **255**, 669 (1977).
71. A. F. H. Ward and L. Tordai, *J. Chem. Phys.*, **14**, 453 (1946).
72. J. Benjamins, J. A. De Feijter, M. T. A. Evans, D. E. Graham, and M. C. Phillips, *Discuss. Faraday Soc.*, **59**, 218 (1975).
73. K. S. Birdi, to be published.
74. D. E. Graham and M. C. Phillips, *J. Colloid Interface Sci.*, **70**, 403 (1979).
75. L. K. James and B. R. Ray, *J. Colloid Interface Sci.*, **38**, 477 (1972).
76. C. Cumper and A. E. Alexander, *Trans. Faraday Soc.*, **46**, 235 (1950).
77. J. T. C. Bohm, "Adsorption of Polyelectrolytes at Liquid–Liquid Interface and its Effect on Emulsification," Ph.D. thesis, p. 19, Agricultural University, Wageningen, The Netherlands.
78. H. L. Frisch and S. Al Madfai, *J. Am. Chem. Soc.*, **80**, 3561 (1958).
79. K. P. Das, J. Chowdhury, and D. K. Chattoraj, *Indian J. Biochem. Biophys.*, **18**, 387 (1981).
80. F. McRitchie and A. E. Alexander, *J. Colloid Sci.*, **18**, 458 (1963).
81. D. E. Graham and M. C. Phillips, *J. Colloid Interface Sci.*, **70**, 415 (1979).
82. D. E. Graham and M. C. Phillips, *J. Colloid Interface Sci.*, **70**, 427 (1979).
83. M. C. Phillips, M. N. Jones, C. P. Patrick, N. B. Jones, and M. Rogers, *J. Colloid Interface Sci.*, **72**, 98 (1979).
84. K. P. Das and D. K. Chattoraj, *J. Colloid Interface Sci.*, **78**, 422 (1980).
85. H. B. Bull, *J. Colloid Interface Sci.*, **41**, 305 (1972).
86. S. B. Ghosh and H. B. Bull, *Biochim. Biophys. Acta*, **66**, 150 (1963).
87. W. Melander and C. Horvath, *Arch. Biochem. Biophys.*, **183**, 200 (1977).
88. H. B. Bull and K. Breese, *Arch. Biochem. Biophys.*, **202**, 116 (1980).
89. K. Motomura, M. Matubayasi, M. Aratono, and R. Matuura, *J. Colloid Interface Sci.*, **64**, 356 (1978).

90. R. A. Hansen, *J. Phys. Chem.*, **66**, 410 (1962).

91. J. W. Gibbs, *The Collected Works of J. W. Gibbs*, Vol. 1 (Longmans, Green, New York, 1931).

92. R. J. Good, *J. Colloid Interface Sci.*, **85**, 128 (1982).

93. R. J. Good, *J. Colloid Interface Sci.*, **85**, 141 (1982).

94. R. Defay, *Sortir de l'Equilibre Thermodynamique et Capillarité* (Université de Bruxelles, Belgique, 1981).

AUTHOR INDEX

SUBJECT INDEX